Advances in Carbon Management Technologies

Volume 2

Biomass Utilization, Manufacturing, and Electricity Management

Editors

Subhas K Sikdar
Retired, Cincinnati, OH, USA
formerly Associate Director for Science
National Risk Management Research Laboratory
US Environment Protection Agency, Cincinnati, Ohio, USA

Frank Princiotta
Retired, Chapel Hill, North Carolina, USA
formerly Director, Air Pollution Prevention and Control Division
National Risk Management Research Laboratory
US Environment Protection Agency, Research Triangle Park, NC, USA

CRC Press is an imprint of the
Taylor & Francis Group, an **informa** business
A SCIENCE PUBLISHERS BOOK

CRC Press
Taylor & Francis Group
6000 Broken Sound Parkway NW, Suite 300
Boca Raton, FL 33487-2742

Version Date: 20200617

International Standard Book Number-13: 978-0-367-52049-6 (Hardback)

Visit the Taylor & Francis Web site at
http://www.taylorandfrancis.com

and the CRC Press Web site at
http://www.routledge.com

Dedication

To the fond memories of my mother, Biva; to my elder sister, Ratna; and my daughter, Manjorie; all of whom have deeply affected my professional and social attitude.

—Subhas Sikdar

To my children Thomas, Elizabeth and John. They, their generation and generations to follow, will reap from seeds we sow. If humanity doesn't get its act together in a hurry, climate change will drastically degrade the habitability of the home planet for them and their fellow species.

—Frank Princiotta

Preface

Average global temperatures, computed from measurements on land, ocean, and via satellites, have steadily increased since the Industrial Revolution, the time we started extracting and burning fossil fuels. Keeping pace with this, greenhouse gases in the atmosphere, principally CO_2, have been accumulating monotonically. Experts have developed climate models which suggest that in the absence of near-term dramatic emission reductions, warming over time could yield impacts jeopardizing life on Earth as we know it. To avoid such potential catastrophic impacts, experts, including the U.S. National Academy of Sciences, have prescribed replacement of fossil fuels, such as coal, petroleum, and natural gas with renewable and non-carbon fuels, principally solar, wind, and nuclear. This aspiration has been called the principal grand challenge of this century. This book, Advances in Carbon Management Technologies, was designed to take a stock of the state-of-the-science development along this journey.

Advances in carbon Management Technologies comprises 43 chapters, in 2 volumes, contributed by experts from all over the world. Volume 1 of the book, containing 22 chapters, discusses the status of technologies capable of yielding substantial reduction of carbon dioxide emissions from major combustion sources. Such technologies include renewable energy sources that can replace fossil fuels, and technologies to capture CO_2 after fossil fuel combustion or directly from the atmosphere, with subsequent permanent long-term storage. The introductory chapter emphasizes the gravity of the issues related to greenhouse gas emission-global temperature correlation, the state of the art of key technologies and the necessary emission reductions needed to meet international warming targets. Section 1 deals with global challenges associated with key fossil fuel mitigation technologies, including removing CO_2 from the atmosphere, and emission measurements. Section 2 presents technological choices for coal, petroleum, and natural gas for the purpose of reducing carbon footprints associated with the utilization of such fuels. Section 3 deals with promising contributions of alternatives to fossil fuels, such as hydropower, nuclear, solar photovoltaics, and wind.

Volume 2 of Advances in Carbon Management Technologies has 21 chapters. It presents the introductory chapter again, for framing the challenges that confront the proposed solutions discussed in this volume. Section 1 presents various ways biomass and biomass wastes can be manipulated to provide a low-carbon footprint of the generation of power, heat and co-products, and of recovery and reuse of biomass wastes for beneficial purposes. Section 2 provides potential carbon management solutions in urban and manufacturing environments. This section also provides state-of the-art of battery technologies for the transportation sector. The chapters in section 3 deals with electricity and the grid, and how decarbonization can be practiced in the electricity sector.

The overall topic of advances in carbon management is too broad to be covered in a book of this size. It was not intended to cover every possible aspect that is relevant to the topic. Attempts were made, however, to highlight the most important issues of decarbonization from technological viewpoints. Over the years carbon intensity of products and processes has decreased, but the proportion of energy derived from fossil fuels has been stubbornly stuck at about 80%. This has occurred despite very rapid development of renewable fuels, because at the same time the use of fossil fuels has also increased. Thus, the challenges are truly daunting. It is hoped that the technology choices provided here will show the myriad ways that solutions will evolve. While policy decisions are the driving forces for technology development, the book was not designed to cover policy solutions.

As editors, we are thankful to the contributing authors for their great efforts in delivering the chapters in a timely fashion. We commend the chapter reviewers for great engagement with the topics and for providing constructive comments in each case. Without the utmost cooperation of the authors and reviewers we would not be able to meet the deadline to produce this timely book. Throughout the formative stages of this development, publisher's representative, Mr. Vijay Primlani, has provided encouragement and assisted us in every way possible to complete the project. We owe our gratitude to him.

Subhas K Sikdar
Cincinnati, Ohio

Frank Princiotta
Chapel Hill, North Carolina

Contents

Section 2. Manufacturing and Construction (Batteries, Built Environment, Automotive, and other Industries)

Section 3. Electricity and the Grid

INTRODUCTION

What Key Low-Carbon Technologies are Needed to Meet Serious Climate Mitigation Targets and What is their Status?

Frank Princiotta

1. Introduction

Since the industrial revolution, humanity has emitted Gigaton quantities of carbon dioxide (CO_2) and other greenhouse gases. Figure 1 (NASA GISS, 2019) shows that the warming that has occurred since 1880 has been in the order of 1.1 degrees centigrade higher than pre-industrial levels. As a result of manmade emissions, carbon dioxide concentrations have dramatically spiked to unprecedented levels when viewed from an 800,000-year perspective. Current concentrations of carbon dioxide are now approximately 410 PPM, relative to the 280-ppm level just before the industrial revolution. Note that in the absence of a serious global emission reduction program, CO_2 concentrations are projected to rise as high as 1000 ppm later this century.

Figure 2 compares actual warming (NASA GISS, 2019) to model projections. The model used was the Model for the Assessment of Greenhouse Gas Induced Climate Change (MAGICC), using middle of the road model assumptions and assuming a fossil fuel intensive emission trajectory (A1FI). It is important to note the close correlation of the actual warming relative to the model projections. As can be seen, if we continue on this fossil fuel intensive emission path and the model continues to accurately predict warming, the planet would be 1.5 ºC warmer by 2035 and 2 ºC warmer by 2045. These warming levels are particularly relevant since the international community has set a warming target of no greater than 2 ºC and optimally below 1.5 ºC by the end of this century. If we were to continue on our current fossil fuel intensive trajectory, 2100 warming is projected to be greater than 4 ºC and rising.

To put the significance of such warming in perspective, Figure 3 was generated based on data from a reconstruction of global temperature for the last 11,300 years (Marcot, 2013), with more recent warming data and model projections included. When current and projected warming is viewed from this long-term perspective, it becomes clear that humankind has, in just 240 years, fundamentally changed the heat transfer characteristics of the planet, with even more dramatic change projected. Note that, as of 2017, warming was about 0.2 ºC warmer than any time in the last 11,300 years. If we continue on our fossil fuel intensive emission trajectory, warming is projected to be in the order of 3.5 ºC greater than any time over this period.

Retired Research Director, USEPA, 100 Longwood Drive, Chapel Hill, NC 27514.
Email: fprinciotta@msn.com

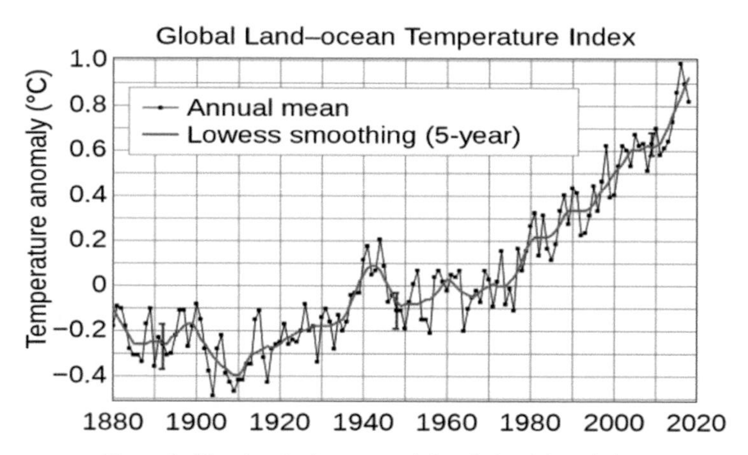

Figure 1. Warming that has occurred since industrial revolution.

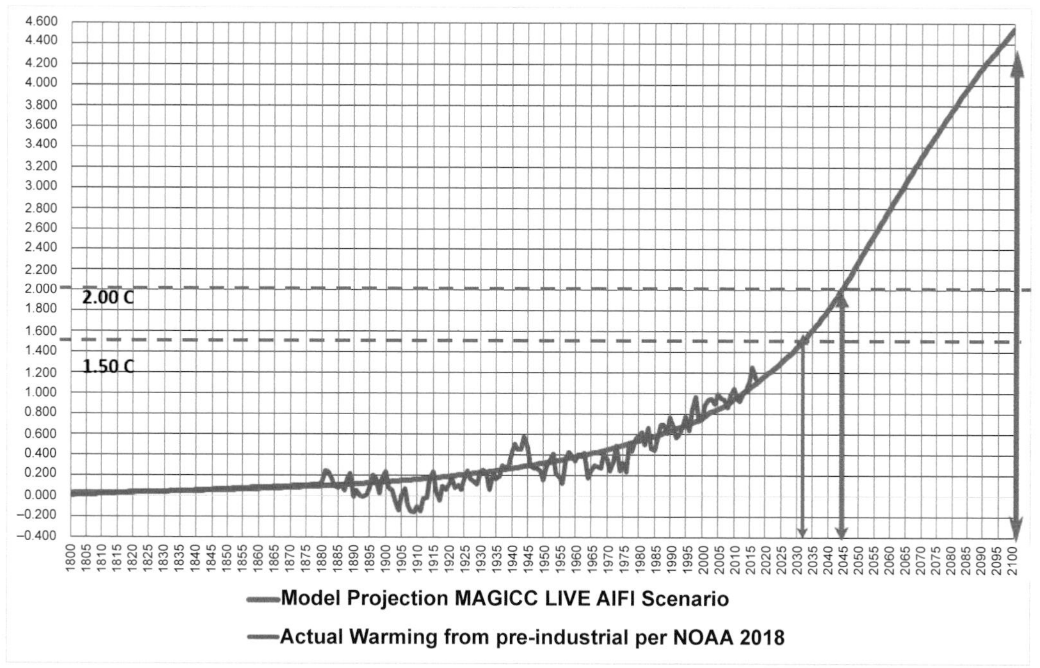

Figure 2. Actual global warming compared to a model projection.

2. Only Emissions of Greenhouse Gases (GHGs) Can Account for the Warming Experienced Since the Industrial Revolution

As discussed, it is clear that the planet has warmed considerably since the industrial revolution. A legitimate question Is whether such warming is a result of human emissions of greenhouse gases or the result of natural factors, such as solar variations and volcanic eruptions. Such eruptions can cool the planet after the reflective particles are driven into the stratosphere, while the planet could warm back up after the particles have settled out of the atmosphere. Figure 4 (USEPA, 2017) illustrates that when

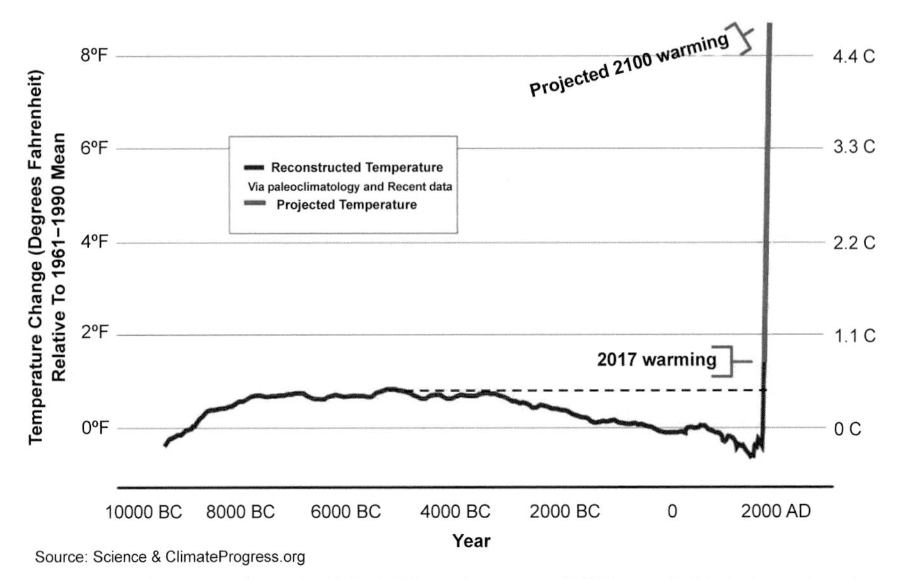

Figure 3. Temperature change (relative to the 1961–1990 mean) over past 11,300 years (in blue) plus projected warming this century on humanity's current emissions path (in red).

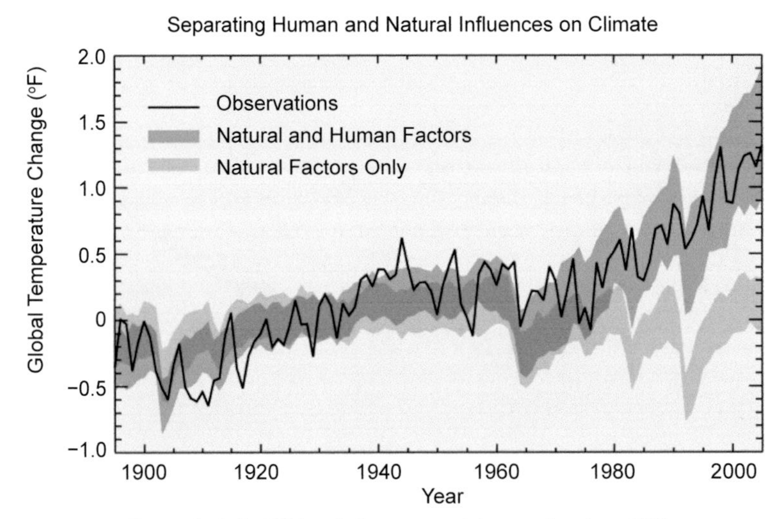

Figure 4. Only GHG emissions can explain warming since 1900.

comparing actual warming to warming predicted by models when only natural factors are considered versus when one accounts for the greenhouse gas impacts, it is clear that human GHG emissions have provided the driving force for the observed warming. Note that Figure 2 reinforces this conclusion, since model warming projections, assuming only GHG emission impacts, yield results consistent with the actual warming.

3. The Heat Added by Anthropogenic Emissions of GHGs is Already Yielding Major Impacts, with More to Come

It is not possible to change the heat balance of the earth so substantially, as humanity has done by adding large quantities of GHGs, without major impacts. It has been calculated (Skeptical Science, 2019) that

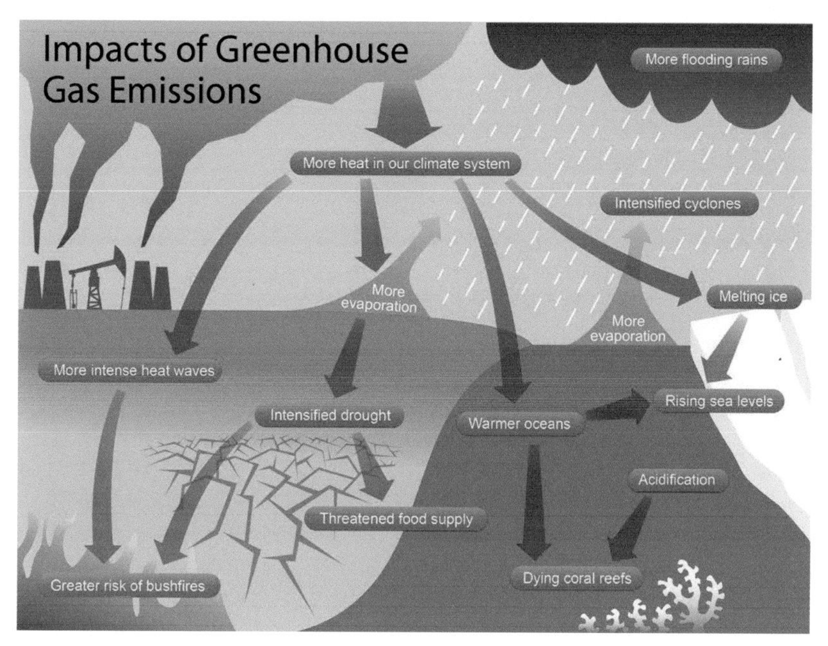

Figure 5. The impacts of greenhouse gas emissions.

the added heat associated with elevated levels of GHGs in the atmosphere since 1998 is equivalent to the detonation of 2.74 billion Hiroshima size atomic bombs. Currently, heat equivalent to four such bomb detonations is being added each second. Ten bombs per second of heat is projected to be added later this century, in the absence of serious global mitigation efforts. Figure 5 (Cook, 2017) illustrates the impacts of all this heat being added to the atmosphere. More heat means both higher temperatures in the atmosphere as well as greater rates of evaporation, yielding more flooding rains. The impacts of higher temperatures and greater evaporation rates are depicted in Figure 5. They include: More intensive heat waves and drought, potentially threatened food supplies, greater risk of wild fires, melting ice yielding seawater rise and more intense weather events, such as more dangerous cyclones. The ocean's ecosystems are also at risk due to a combination of ocean warming and acidification, since about 90% of all the heat and 25% of the CO_2 ends up in the oceans. CO_2 absorbed in the ocean generates carbonic acid which increases the ocean's acidity.

As Figure 2 illustrated, we face the prospects of warming at the 4 °C level later this century. Warren (2010) summarized the implications of a 4 °C warmer world as follows: "Enormous adaptation challenges in the agricultural sector, with large areas of cropland becoming unsuitable for cultivation. ... large losses in biodiversity, forests, coastal wetlands... supported by an acidified and potentially dysfunctional marine ecosystem. Drought and desertification would be widespread, with large numbers of people experiencing increased water stress. ... Human and natural systems would be subject to increasing levels of agricultural pests and diseases, and increases in the frequency and intensity of extreme weather events."

4. There is the Danger of a Runaway Situation if Warming occurs Too Rapidly and Activates Tipping Points Associated with Amplifying Feedbacks

It is important to note, that current projection models do not account for the possibility that there could be accelerating warming due to "tipping points" associated with driving forces that could yield a point in time when the global climate changes from one stable state to another, a threshold which reaches a point of "no return" that can change the planet irreversibly. Such points could cascade, yielding a "hothouse Earth". Figure 6 (an updated/upgraded figure from Climate Change Knowledge (2014)) illustrates the

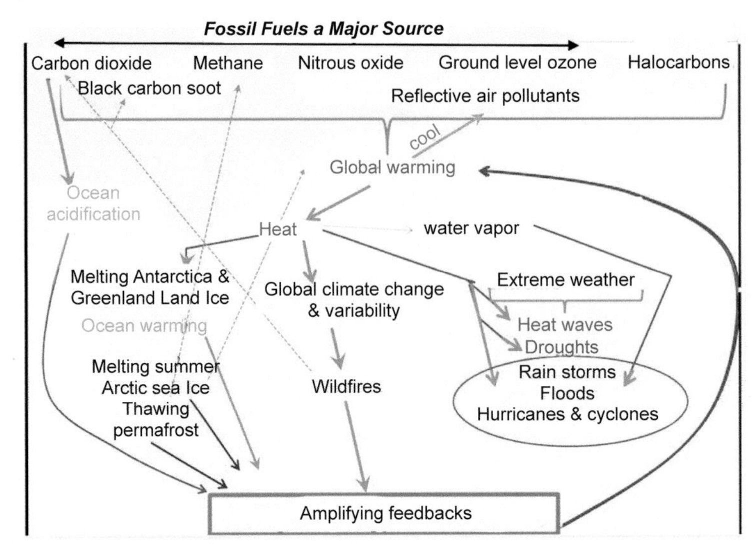

Figure 6. Amplifying feedbacks could yield "runaway" warming.

relationship between GHG emissions and potential impacts with a focus on Amplifying Feedbacks. Examples of such feedbacks, that if cascaded could contribute to such a runaway state include:

- Accelerated Melting of Arctic sea ice and Antarctica/Greenland land ice (such melting would decrease Earth's reflectivity, allowing more heat to be absorbed by the atmosphere).
- Melting of permafrost in Siberia, Canada and Alaska (large additional source of CO_2 & CH_4 not accounted for in models).
- Ocean warming and acidification along with increasing number and intensity of wildfires (these would have the effect of weakening CO_2 sinks, absorption on land and in the ocean, yielding an acceleration of growth of GHG concentrations in the atmosphere).

A recent study (Steffen, 2018) examined this issue and concluded that potential planetary thresholds yielding accelerating and potentially irreversible warming could occur at a temperature rise as low as 2.0 °C above preindustrial levels. They concluded that limiting warming to a maximum of 1.5 °C would dramatically lower the risk of this potentially catastrophic instability.

Although not discussed in the study, it follows that warming in the vicinity of 3–4 °C would substantially raise the probability of such tipping points, yielding a "hothouse earth".

5. Growing Global Emissions, the Result of a Growing Population Demanding an Expanding Array of Resource Intensive Goods and Services

The dramatic growth in GHG emissions since the industrial revolution are driven by two key drivers. First, world population has been growing relentlessly. World population is now at 7.5 billion, has tripled since 1950, and is expected to grow to over 9 billion by 2050. Second, in developed nations, people have expanded their list of "needs" to include personal transportation, residences with energy-intensive heating, cooling, and lighting, a diet heavily oriented toward meat consumption, and an ever-growing array of consumer goods and services. Developing countries are moving in the same direction, albeit at earlier stages. World population has been growing annually at about 1.2% and CO_2 emissions at 2.3% over the last 17 years.

Figure 7 illustrates the factors responsible for the challenges to long term sustainability with a focus on climate change, the most serious sustainability threat. The middle of the figure indicates that these human needs are met by means of a large array of industrial, agricultural, and energy technologies and

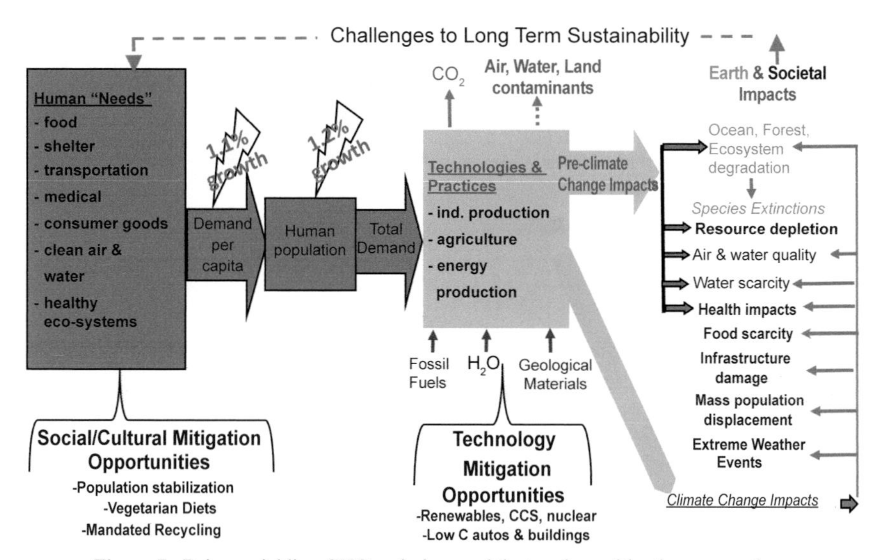

Figure 7. Drivers yielding GHG emissions and the two key mitigation approaches.

practices. Although, there are a multitude of sustainability impacts associated with these "technologies and practices", independent of climate change. The major threats are shown color coded in two categories: Earth and Societal impacts. These include, degradation of air and water quality, depletion of minerals and fresh water supplies and ecosystem damage. Unique climate change impacts are listed on the right side of the figure and include: Potential food scarcity, infrastructure damage, mass population displacement and extreme and damaging weather events. As indicated by the red return arrows, in addition to such unique impacts, climate change has the potential to exacerbate impacts associated with other human activities, such as ocean and forest degradation. The bottom of the figure indicates that there are two classes of mitigation opportunities. The most commonly considered approach is replacing/upgrading current technologies and practices. Another, less discussed, but potentially important if technology modifications alone are insufficient to avoid serious climate impacts, would be to modify social and cultural behavior toward energy-efficient and resource-intensive lifestyles.

6. Each Country Has a Unique GHG Emission Trajectory and Mitigation Challenge

When one examines GHG emissions on a country by country basis, fundamental differences in emission characteristics and mitigation challenges are observed. Table 1 (generated based on databases from Global Carbon Atlas (2018)) summarizes CO_2 emission data for the 14 largest emitters in 2017. They are positioned in the order of the magnitude of their emissions. China, the EU and the U.S. are by far the largest emitters. The developed countries are identified by the normal font, while those in various stages of economic transition are in the bold font. Let us briefly discuss the situation in key developed countries and then in developing countries. At this point it should be noted that the IPCC (2013) has concluded that, in order to have a chance of limiting warming to no greater than 2 °C, global per capita emissions should be between 1.1 and 2.2 t/person in 2050 and zero in 2100.

China is by far the largest emitter, passing the U.S. in 2006. It is considered somewhere between a developing and a developed country. Their 17-year emission growth rate at 6.6% is unmatched by any other country. They have rapidly transformed from a low-end developing country to a country with unprecedented economic growth via rapid urbanization, industrialization (supported by an unprecedented power generation expansion, primarily based on coal), and major growth of their on-road transportation fleet. Population growth at 0.6% has been only a minor factor in influencing their rapid emission growth. Their per capita emission has grown dramatically to 7 t/p and still growing.

Table 1. CO_2 emission data for countries with the greatest emissions in 2017.

Country	2017 Emissions GT CO_2	2017 per Capita Emissions tonnes/person	2017 Population millions	2000 to 2017 Annual Emission Growth Rate	2000 to 2017 Annual Population Growth Rate
China	*9863*	*7.0*	*1409*	*6.6%*	*0.6%*
USA	5184	16.0	324	–0.9%	0.8%
EU	3544	7.0	507	–1.0%	0.2%
India	***2446***	***1.8***	***1359***	***5.2%***	***1.5%***
Russia	1716	12.0	143	0.8%	–0.1%
Japan	1207	9.5	127	–0.3%	0.0%
Iran	***672***	***8.3***	***81***	***3.6%***	***1.2%***
Saudi Arabia	627	19.0	33	4.5%	2.7%
S. Korea	616	12.0	51	1.9%	0.5%
Canada	592	16.0	37	0.2%	1.0%
Brazil	***481***	***2.3***	***209***	***2.3%***	***1.0%***
Indonesia	***475***	***1.8***	***264***	***3.5%***	***1.3%***
S. Africa	456	8.0	57	1.1%	1.3%
Australia	408	17.0	24	0.9%	1.4%
Rest of World	***7866***	***2.6***	***2995***	***3.7%***	***2.2%***
Total	36153	4.7	7620	2.3%	1.3%

IPCC: Per Capita target for 2 °C maximum warming = 1.1 to 2.2 in 2050 and near Zero in 2100.

The U.S. and the EU are both highly developed with similar standards of living, yet the EU's per capita emissions are less than half those of the U.S. The EU has been the most conscientious regarding minimizing GHG emissions, has had a culture of treating energy conservation seriously, and uses less electricity per capita, in part because of very high electric rates. They have also has been leaders in utilizing wind and solar power. As a more population dense area with less dependence on large low-efficiency cars and trucks, their transportation emissions are much lower as well. Finally, the very low population growth has also been a factor. However, for both the EU and especially the US, reducing per capita emissions to the 1.1 to 2.2 level by 2050 will be a monumental challenge.

The developing world is in a fundamentally different situation. Per Table 1, countries in this category would be include India, Iran, Brazil, Indonesia and many African, South American and Asian countries in the "Rest of the world" designation. It is estimated that over 4 billion people fall in this category. They generally have similar characteristics: Low standards of living, modest per capita incomes, high birth rates and relatively low per capita emissions. The challenge for developing countries is to improve their economies while at the same time lowering or at least not raising their per capita emissions. India is a particularly important case since its population is close to that of China, with a fast-growing economy and a major increase in emissions (and per capita emissions) in recent years. This is not the direction this sector of the world's economy should be heading if we are serious about avoiding unacceptable climate change.

7. Greenhouse Gas Emissions are Associated with All Energy, Industrial and Agricultural/Land Use Sectors

In order to appreciate the scope of the mitigation challenge, it is important to understand the relationships between: The energy, industrial and agricultural/land use sectors, the related activities that are needed for the desired societal end uses ("needs") and the resulting GHG emissions. Figure 8 (WRI, 2007) quantifies these relationships for the world as of 2005, the last year such a chart was published. An example of these relationships follows: If the end use/activity is heating, cooling and lighting residential buildings, the relevant sectors are Electricity and Other Fuel Combustion. This is the case since some residences are heated with electricity (heat pumps or resistance heating) and others via direct combustion of a fuel, such as natural gas. So, both electricity and fuel combustion sectors contribute to CO_2 emissions associated with residential buildings, as indicated in Figure 8.

As can be seen, key end uses that drive energy-related emissions include road travel, residential and commercial building cooling, heating and lighting and the production of chemicals, cement, and iron & steel needed for production of goods. The net result are huge emissions of CO_2, most of which are associated with the combustion of coal, oil and natural gas. Also, the agricultural sector is responsible for the majority of emitted methane, the second most important greenhouse gas. Land use change was another important contributor to raising CO_2 concentrations, however less active deforestation in recent years has decreased the importance of this sector. Note that in 2005, 77% of the anthropogenic warming was associated with CO_2, with methane and N_2O contributing 15% and 7%, respectively. Also note, that the term CO_2 equivalent [CO_2(e)] emissions, is the amount of CO_2 which would have the equivalent global warming impact when accounting for CO_2 plus the other GHG gases. For 2005 that number globally was 44 Gt(e).

8. Major Emission Mitigation from All Sectors and GHGs is Required Immediately in Order to Have a Chance of Meeting International Targets

In order to have a chance of limiting warming to between 1.5 and 2.0 °C per the international community's stated target, it will be necessary to drastically reduce emissions as soon as possible. Figures 9 and 10 have been generated in order to help quantify this monumental challenge. The previously mentioned MAGICC model was used. Warming projected should be considered as "best guess" values, considering that there are uncertainties in such values, especially for long term projections. Both figures assess the

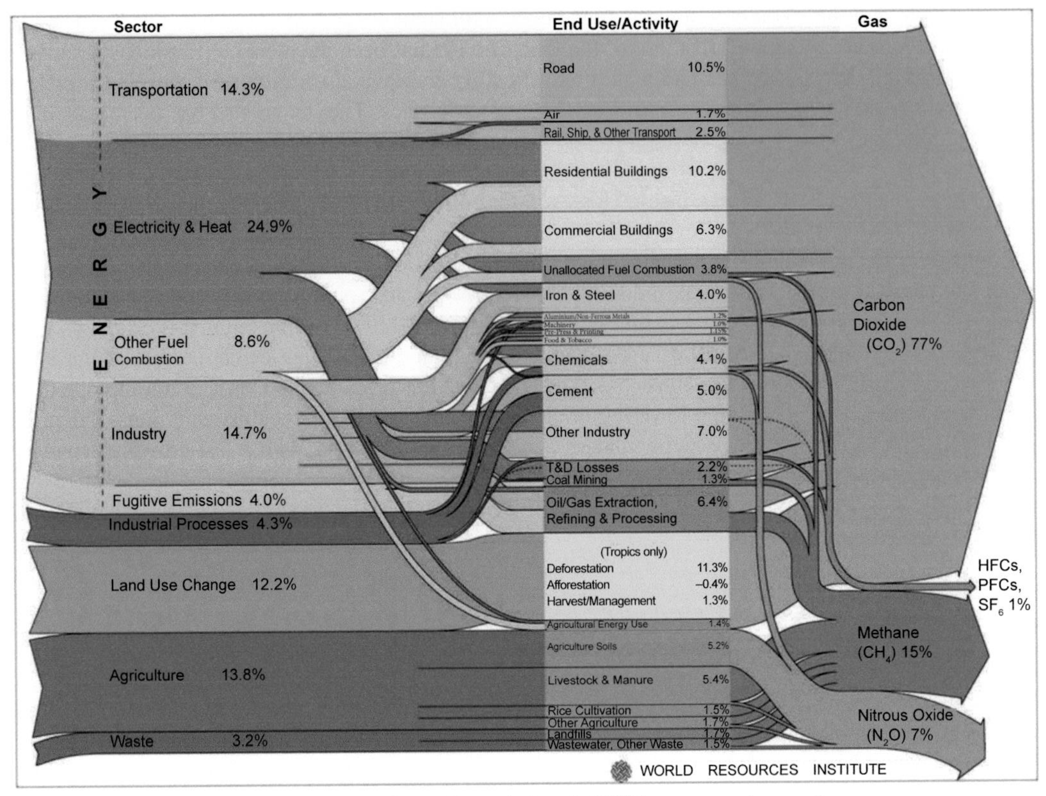

Figure 8. Global Greenhouse Gas emissions in 2005 by sector, end use and gas.

warming implications of six emission scenarios; Figure 9 shows the assumed emission scenarios in Gt CO_2 and Figure 10 shows the projected warming all the way to 2100, associated with these emission scenarios:

1) A business as usual case that assumes continued reliance on fossil fuels with no serious mitigation reductions for methane and nitrous oxide, the two other key GHGs.
2) A scenario consistent with the 2015 Paris Agreement where most countries "promised" significant but relatively modest emission reductions.
3) A serious CO_2 global emission reduction program with near term emission stabilization and 3% annual emission reductions, starting in **2035** and continuing for 65 years.
4) Scenario 3 above, with the addition of a complimentary CH_4 and N_2O emission reduction program, starting in 2035 as well.
5) Scenario 4 above, but with the addition of a major Direct Atmospheric Capture (DAC) program to remove CO_2 from the atmosphere, starting in **2030** and continuing for 70 years.
6) Scenario 5 above, but with CO_2, methane and N_2O mitigation starting ten years earlier, in **2025,** and maintained for 75 years.

As can be seen, with 3% annual CO_2 mitigation starting in 2035, it will be difficult to limit warming to below 2 ºC. The addition of CH_4 and N_2O mitigation significantly reduces the warming, but limiting warming to 2 ºC still appears unlikely. When one adds a major Direct Air Capture (DAC) component to the mitigation strategy it appears that limiting warming to 2 ºC may be achievable. Finally, if we start CO_2, CH_4 and N_2O emission reductions ten years earlier, in 2025, and again supplement with a major DAC complimentary program, further warming reduction is achieved, raising the probability that warming would be limited to 2 ºC. It is worth noting that none of these options appear to be able to limit warming to 1.5 ºC, a likely unattainable target. DAC technology involves the construction of massive chemical plants designed to remove CO_2 directly from the atmosphere.

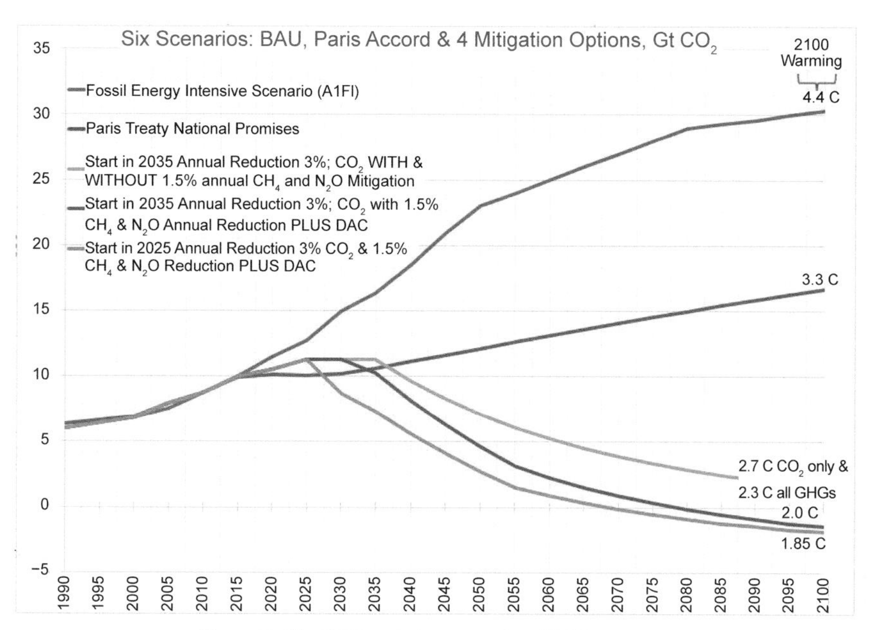

Figure 9. Six GHG emission scenarios, Gt CO_2 per year.

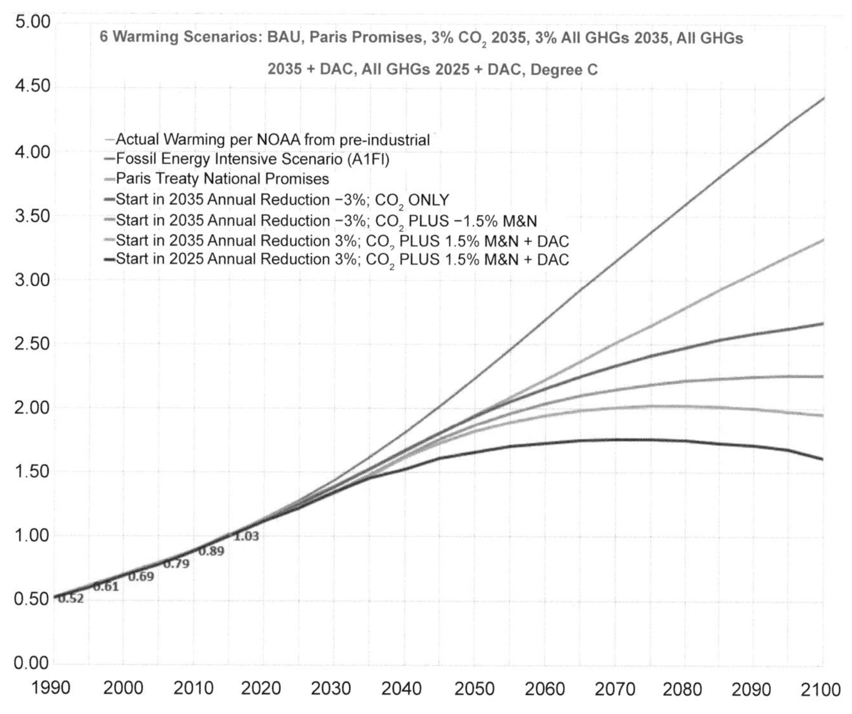

Figure 10. Projected warming associated with six GHG emission scenarios, Degree Celsius.

It should be noted that DAC technology is at a very early stage of development and implementation costs are likely to be very high per ton of carbon captured. Keith (2018) has analyzed a chemical sorbent process, aqueous sorbent KOH coupled to a calcium caustic sorbent recovery loop, and estimated capture costs at \$94 to \$232 per ton. Options 5 and 6 described above, assume DAC capture would start in 2030 with progressively greater annual removal quantities for a total of 37 Gt for the 70-year mitigation period.

Given this quantity, the estimated cost of such DAC capture would be between $1.3 to $2.5 trillion per year or $72 and $178 trillion over the 70-year period! This study did not include the required costs for permanent storage of the CO_2, probably in deep underground saline aquifers, which will likely add trillions of US$ to the cost of such an enterprise.

9. Emerging Technologies Will need to be Available and Extensively Utilized if Global Warming Target Levels Stand a Chance of Being Achieved

If the international community ultimately sets sufficiently aggressive emission requirements in order to minimize the worst consequences of climate change, affordable, practical low-carbon technologies would need to be commercially available within the next ten years. Of particular importance would be Carbon Capture and Storage (CCS) technologies, advanced nuclear generators, low-cost renewable generation with energy storage capability, efficient buildings and low-emission vehicles. Also, given the importance of methane and N_2O emissions (see Figures 10 and 11), the international community must agree on emission reductions for these pollutants as soon as possible as well. For methane, leakage from oil, gas and coal operations are particularly important. Agriculture operations are important sources for CO_2, methane and N_2O.

Figure 11, derived from IEA (2016), quantifies the amount of CO_2 that would have to be mitigated by technology, by sector, between now and 2050 in order to have a chance of limiting warming to 2 ºC. As can be seen, all sectors require major CO_2 reductions.

Tables 2 to 5 have been generated with the aim of summarizing for key technologies by sector: The current state of the art, issues that could limit near term utilization and research, development and demonstration priorities.

Note that for Table 2 a column has been added which quantifies what IEA (2016) has determined would be the potential mitigation impact of each power generation technology, in terms of Gt of CO_2 mitigated between now and 2050. It is worth noting that, in recent years, solar and wind technologies have seen their capital costs reduced substantially along with cost reductions in storage technologies needed if these technologies can continue to generate power when the sun doesn't shine and the wind doesn't blow. On the other hand, CCS technology development has stumbled. Princiotta (2017) summarized the state of the art of CCS technology as of 2016 and concluded that, despite an active demonstration program initiated 10 years ago, 43 projects have been shut down due to overruns yielding unacceptably high costs, degradation of power plant efficiency and serious capture and storage technical difficulties. Such technology is particularly important for relatively new coal and natural gas-fired power generators,

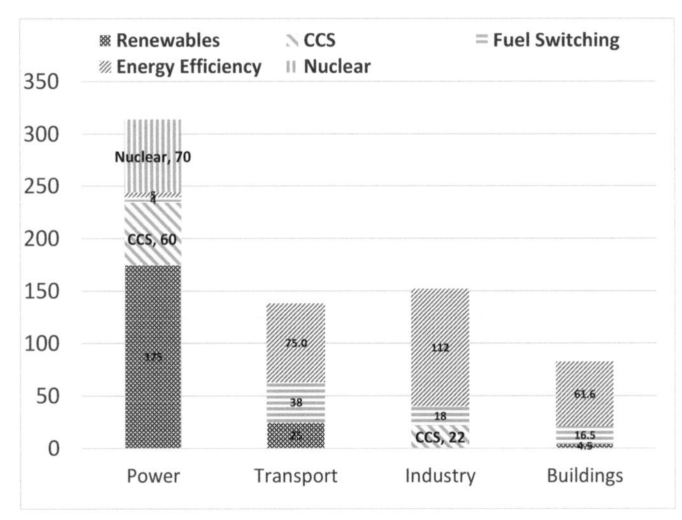

Figure 11. Cumulative Gt CO_2 reductions needed by sector per technology in 2050 in order to limit warming to 2 ºC.

Table 2. Power generation low-carbon technologies.

Technology	IEA 2050 carbon emission reduction, Gt; 2 °C goal	Current state of the art	Issues	Technology RD&D needs
Carbon Capture and Storage	3.7	Early commercialization for coal with many demos having cost overrun and operating issues	High capital costs, 20–30% conversion efficiency degradation, complexity and potential reliability concerns; Underground Storage: Cost, safety, efficacy and permanency issues	**High:** Demos on next generation technology on a variety of units burning coals & natural gas; enhanced Underground Storage program with long term demos evaluating large number of geological formations
Solar-Photovoltaic and Concentrating (renewable)	3.7	First generation commercial	Solar resource intermittent and variable, although costs have been reduced further efficiency/cost reductions needed	**High:** Research needed to develop and demo cells with higher efficiency, and lower capital costs; develop/commercialize affordable storage technology
Wind Power (renewable)	2.6	Commercial (on-shore)	Costs very dependent on strength of wind source, large turbines visually obtrusive, intermittent power source	**Medium:** Higher efficiencies, off-shore demonstrations. Affordable storage technology
Nuclear Power-advanced & next generation	2.6	Commercial BWR, PWR; Developmental: Generation III+ and IV: e.g., Pebble Bed Modular Reactor	Deployment targeted by 2030 with a focus on lower cost, minimal waste, enhanced safety and resistance to proliferation	**High:** Demonstrations of key advanced technologies with complimentary research on important issues; commercialization of fusion technology could be transformational, might be possible, mid to late century
Biomass as fuel gasified or co-fired with coal (renewable)	1.0	Early Commercial	Important to assess true renewability of biomass source, limited to 20% when co-fired with coal	**Medium:** Biomass/IGCC would enhance efficiency and CO_2 benefit; also, genetic engineering to enhance biomass plantations
Fuel Switching coal to gas	0.2	Commercial (w/o CCS)	Effectiveness of CCS on natural gas generators; CH4 emissions during hydro fracturing could reduce GHG reduction benefit	**High:** Hydro fracturing environmental mgt., CCS demos needed, especially in the U.S.
Smart Grids	Not calculated, but supports renewables	Early Commercial, with active research focused on next generation technologies	Telecommunications cost high, security concerns and questions regarding consumer acceptance/ participation	**High:** Enhanced smart grid modeling, reduce telecommunication cost component, demonstrate effectiveness in maximizing solar and wind power production in overall mix

Table 3. Mobile source technologies.

Technology	IEA number of light duty vehicles in 2050, millions	Current state of the art	Issues	Technology RD&D needs
Electric & Hybrid Gasoline and Diesel	91	Early commercial	For electric plugs-in, mileage (battery) limitations; charging durations and high purchase prices; benefits greater if power from low-carbon sources	**High:** Battery improvements in storage capability, cost and lifetimes important
Fuel Cell Electric Vehicle	27	Developmental	Fuel cell costs and fuel cell stack life; also, H_2 production and need for fueling infrastructure	**High:** Breakthrough R,D&D needed to develop competitive, long lived fuel cell stack; viable H_2 production and storage, with a focus on safety, needed
Ethanol from cellulosic biomass sources, e.g., wood	0	Developmental	Important to assess true renewability of biomass source; inability to convert wide range of biomass sources with competitive production costs	**High:** Breakthrough RD&D needed to develop economical technology capable of generating large quantities; especially critical for the aircraft industry
Biodiesel & other fuels from biomass; thermo chemical processes	0	Developmental	Important to assess true renewability of biomass source; inability to convert wide range of biomass sources with competitive production costs	**High:** Breakthrough RD&D needed to develop economical technology capable of generating large quantities; especially critical for the aircraft industry

Table 4. Industrial technologies.

Technology	Current state of the art	Issues	RD&D needs
CO_2 Capture and Storage (IEA ETP 2015 projects 1.6 Gt mitigation in 2050)	Early development	Applicability limited to large energy-intensive industries, including fuel transformation processes; key questions: Cost, safety, efficacy	**High:** Major program with long term demos evaluating large number of geological formations to evaluate efficacy, cost and safety
Motor Systems	Commercial	For most industries not a major cost; lack of expertise for some industries	**Medium:** Lower costs and higher efficiencies desirable
Enhanced energy efficiency: Existing basic material processes	Commercial	Developing countries can have low energy efficiency due to lack of incentive and/or expertise	**Low**
Steam systems (required for many industries)	Commercial	For most industries not a major cost; lack of expertise for some industries	**Low**
Materials/Product Efficiency	First generation: commercial	Little incentive to minimize the CO_2 "content" of materials and products; life cycle analyses required	**Medium:** Conduct life cycle analyses of key materials and products with the aim of minimizing CO_2 "content"
Cogeneration (combined heat and power)	Commercial	Limited by electric grid access that would allow the ability to feed electricity back to grid; also high capital costs	**Low**
Enhanced energy efficiency: New basic material processes	Developmental to Near-commercial depending on industry	New, innovative production processes require major RD&D and would need reasonable payback to replace more C intensive processes	**Medium/High:** Develop and demonstrate less carbon intensive production processes for key industries
Fuel Substitution in Basic Materials Production	Commercial	Natural gas substitution for oil and coal can be expensive	**Low**
Feedstock Substitution in key industries	Commercial	Biomass and bioplastics can substitute for petroleum feedstocks and products; however, cost high and availability low	**Medium:** Develop affordable substitute feedstocks and products based on biomass

Table 5. Building technologies.

Technology	IEA 2050 carbon emission reduction, Gt; 2 °C goal	Current state of the art	Issues	Technology RD&D priority and needs
Enhanced energy mgt. and high efficiency building envelope: Insulation, sealants, windows, etc.	2.5	Commercial	Lack of incentive, high initial costs, long building lifetime	**Low/medium** priority: Incremental improvements to lower cost and enhance performance
High efficiency building heating and cooling, including heat pumps	0.8	Commercial	Lack of incentive, high initial costs	**Low/medium** priority: Incremental improvements to lower cost and enhance performance
Solar heating and cooling	0.5	First generation commercial	High initial costs, availability of low-cost efficient biomass heating systems	**Medium:** Focus on development of advanced biomass stoves and solar heating technology in developing countries

since it is unlikely that such plants would be prematurely retired in favor of renewable or nuclear plants. The current generation of nuclear power generation is commercially available but is burdened with high capital costs, waste disposal issues and serious safety concerns. The March 2011 Fukushima Daichi nuclear disaster, caused by a severe tsunami, has been responsible for several countries reconsidering their commitment to future nuclear reactor construction.

Table 3 lists the key mobile source technologies projected to play an important emission reduction role between now and 2050. Included is a column based on IEA (2016) which projects the number of light duty vehicles on the road in 2050 for key low-carbon technologies. As can be seen, electric and hybrid electric gasoline and diesel vehicles are projected to be the most important. Their cost and, therefore, their market penetration will be heavily dependent on the costs and performance characteristics of the vehicle storage batteries. Also projected to be important are fuel cell vehicles. These vehicles are in the early stages of commercialization. Fuel cells generate electricity to power the motor, generally using oxygen from the air and compressed hydrogen. They are generally classified as zero emission vehicles since the only effluent is water, However, it is important to account for any carbon emissions that would be associated with the production of hydrogen. Biomass fueled vehicles are at early stages of development, but have the potential to be significant low-carbon mobile source options.

Table 4 summarizes key industrial low-carbon technologies. The technology with the greatest change chance of making the greatest impact in this sector is carbon capture and storage. This technology would be applicable to major industrial sources of CO_2, including cement, iron and steel, oil refining and pulp and paper operations. Unfortunately, as discussed in the power generation discussion above, serious cost, reliability and permanent storage issues. Clearly, an enhanced research, development and demonstration program utilizing the next generation CCS technologies is required.

Table 5 lists key low-carbon building technologies. Emission reduction in this sector depends less on the development of new technology and more on providing incentives to promote the use of state-of-the-art technology for retrofitting existing buildings and incorporating in new buildings. For this reason, priority for research, development and demonstration programs is lower than in the power, mobile source and industrial sectors.

10. Technology RD&D is Woefully Inadequate

Given the need for dramatic emission reductions required in all economic sectors, it is essential that global RD&D expenditures are adequate to ensure the availability of high performing low-cost technologies in the near term. Figure 12 summarizes IEA (2011) analysis of actual versus needed global energy technology

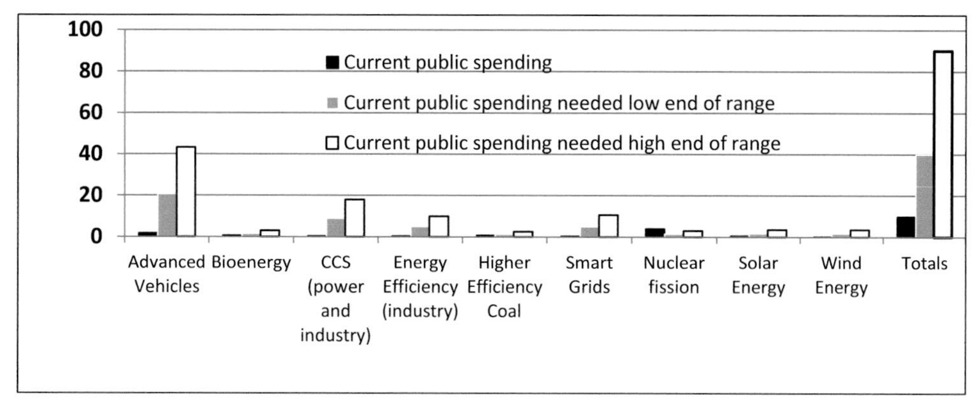

Figure 12. IEA analysis comparing actual low-carbon technology RD&D versus required spending, $ billions.

RD&D annual funding for key technologies consistent with reducing global emissions 50% by 2050. This analysis concludes that actual expenditures are a small fraction of what is required. The total required funding is estimated at $40–$90 billion per year, whereas actual spending for these key technologies is estimated to be only $10 billion annually. Although substantial, the $30–$80 billion annual funding gap is minuscule compared to the $1.7 trillion global military spending in 2017 (CNBC, 2018).

11. Geoengineering Options should be Studied

As previously discussed, meeting the global community's goal of limiting warming to 1.5–2 °C will be a monumental challenge. Given the massive and radical infrastructure change that will be needed in the near term, meeting such a target may not be possible. It has been suggested that geoengineering options could, at least in theory, buy us time to make the necessary infrastructure/technology changes to dramatically reduce global GHG emissions. They have been described as both a delaying tactic or as a possible "last resort" action to limit catastrophic climate change. Geoengineering measures attempt to compensate for GHG emissions via two fundamentally distinct approaches: (1) changing Earth's solar radiation balance by increasing reflectivity and (2) removing CO_2 from the atmosphere. Figure 13 mentions several of the most discussed geoengineering concepts in both categories. Table 6 compares the characteristics of these two concepts.

Note that two scenarios in Figures 9 and 10 assumed availability of commercial Direct Air Capture Technologies to compliment emission reduction efforts. DAC is sometimes referred to as the negative CO_2 emissions option. However, this approach, as well as all the options mentioned in Figure 13, are only

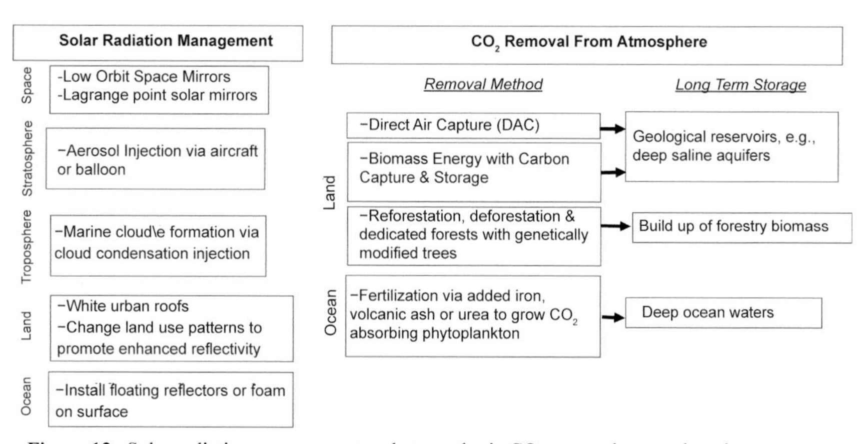

Figure 13. Solar radiation management and atmospheric CO_2 removal geoengineering concepts.

Table 6. Comparing solar radiation and atmospheric CO_2 removal concepts.

Characteristic	Carbon dioxide removal proposals	Solar radiation management proposals
Do they directly address the cause of GHG-induced climate change?	Yes: Such techniques act as negative CO_2 emitters	No: They utilize reflection of incoming solar radiation to compensate for heat added by GHGs; would not mitigate ocean acidification by CO_2
Can they intoduce novel risks?	No	Yes: Effects on stratspheric ozone levels and deleterious meterological impacts possible
How expensive?	Generally at least as expensive as emission reduction techniques	Although costs are very preliminary they appear relatively low re. emission control
How fast would results be realized?	Would take decades to realize significant results	Although still at conceptual stage, results could be realized within several years after installation
What level of international cooperation would be required?	Major cooperation involving multi-trillion US$ financial commitments needed	Less cooperation needed re. financial commitments but potential societal risks suggest international agreement before implementation needed
Key feasibility questions	Are the processes effective and affordable?	Are the processes effective with acceptable risks?

at a conceptual stage with serious performance, environmental impact and economic issues. Nevertheless, it is the author's view that, given the magnitude of the mitigation challenge, such approaches warrant serious feasibility evaluations, as soon as possible.

14. Conclusions

Humanity has dug itself a very deep hole. To limit the damage, it will take a concerted international effort, building on the 2015 Paris Accord, to aggressively reduce emissions in all sectors as soon as possible. Low-carbon technologies will of necessity play a crucial role in this process. It is the goal of this book to provide an assessment of the status and prospects of key low-carbon technologies.

References

CNBC. https://www.cnbc.com/2018/05/02/global-military-spend-rose-to-1-point-7-trillion-in-2017-arms-watchdog-says.

Cook, J. 2017. Impacts of Greenhouse Gas Emissions, Skeptical Science, https://www.beforetheflood.com/explore/the-deniers/fact-climate-change-is-very-very-dangerous/.

Climate Change Knowledge. 2014. http://www.climate-change-knowledge.org/uploads/GHG_pollution_schema.png.

Global Carbon Atlas. 2018. http://www.globalcarbonatlas.org/en/ CO_2-emissions.

International Energy Agency: Energy Technology Perspectives 2010 (IEA Publications, Paris, France, 2011).

International Energy Agency: Energy Technology Perspectives 2015 (IEA Publications, Paris, France, 2016).

Keith, D. 2018. A Process for Capturing CO_2 from the Atmosphere, Joule, Volume 2, August 2018.

Marcot, S. 2013. A reconstruction of regional and global temperature for the past 11,300 years, science. Science 08 Mar 2013: 339(6124): 1198–1201. DOI: 10.1126/science.1228026.

NASA GISS. 2019. Https://data.giss.nasa.gov/gistemp/graphs_v3/.

Princiotta, F. 2016. We are losing the climate change mitigation challenge, can we recover? MRS Energy & Sustainability, June, 2016, https://www.cambridge.org/core/journals/mrs-energy-and-sustainability/article/we-are-losing-the-climate-change-mitigation-challenge-is-it-too-late-to-recover/7EABEBD608FC3C651FE75931B19E7157/core-reader.

Skeptical Science. 2019. Global Warming at 4 Hiroshima Atomic Bombs Per Second https://4hiroshimas.com/.

Steffen, W. et al. 2018. Trajectories of the Earth System in the Anthropocene, 2018, Proceedings of the National Academy of Sciences of the USA.

USEPA. 2017. https://www.env-econ.net/2017/01/epa-separating-human-and-natural-influences-on-climate.html.

Warren, R. 2010. The Royal Society, The role of interactions in a world implementing adaptation and mitigation solutions to climate change. Philos. Trans. R. Soc. A 2011(369): 233. doi:10.1098/rsta.2010.0271 (published November 29, 2010).

World Resources Institute: World Greenhouse Gas Emissions in 2005 (2006). Available at: http://www.wri.org/resources/charts-graphs/us-greenhouse-gas-emissions-flow-chart (accessed April 9, 2019). Google Scholar.

Section 1

Biomass Sector

CHAPTER 1

Biomass as a Source for Heat, Power and Chemicals

*Kafarov, V** and *Rosso-Cerón, AM*

1. Introduction

Biofuels are those fuels obtained from biomass. Biomass is any type of organic matter that has its immediate origin in the biological process of living organisms, such as plants, or their metabolic waste (compost); the concept of biomass includes products of both vegetable and animal origins. This term has been accepted to name the group of energy products and materials of a renewable type that come from the organic raw material formed by biological route. Nowadays, different types of biomass can be differentiated.

Biofuels are alcohols, ethers, esters, and other chemical products that come from organic compounds of cellulosic base (biomass) obtained from wild plants or crops, which can replace, to a greater or lesser degree, the use of fuels destined for the production of electricity or for transportation.

The current biocomponents usually come from sugar, wheat, corn or oilseeds. The main objective of using these biofuels is to reduce the emission of greenhouse gases that overheat the earth's surface and accelerate climate change. Unlike the use of fossil hydrocarbons, the use of biomass for energy consumption reduces CO_2 emissions to the atmosphere and the associated climate change impacts.

Biofuels of biological origin can replace part of the consumption of traditional fossil fuels, such as oil, gas and coal. This type of fuel is usually in liquid form and is used to drive the combustion engines of land transport. The most developed and used biofuels are bioethanol and biodiesel.

In this way, biofuels appear as an alternative source of energy that can be used if hydrocarbon prices rise too high or over a long-term horizon in which they run out. A second purpose in their use is that they help in reducing CO_2 emissions and curbing global warming. However, the energy crops of corn, sugarcane, sorghum or soybeans imply an alternative use to food and this might again generate a great controversy (Herguedas et al., 2012).

Depending on the nature of the biomass, energy use and the desired biofuel, several methods could be taken into consideration in order to obtain biofuels: Mechanical processes (chipping, crushing and compaction), thermochemical procedures (combustion, pyrolysis, and gasification), biotechnological techniques (micro-bacterial and enzymatic) and extractive methods to obtain liquid, solid and gaseous fuels.

Carrera 27-Calle 9, Chemical Engineering Department, Industrial University of Santander, Bucaramanga, Colombia.
* Corresponding author: kafarov@uis.edu.co

Table 1. Type of biomass and characteristics.

Type of biomass	Characteristics
Primary biomass	Organic matter formed directly from photosynthetic beings. This group includes plant biomass, including agricultural and forestry waste.
Secondary biomass	Produced by heterotrophic beings who use primary biomass in their nutrition. They constitute fecal matter or meat of the animals.
Tertiary biomass	Produced by the beings that feed on secondary biomass, for example remains and droppings of carnivorous animals that feed on herbivores.
Natural biomass	Produced by wild ecosystems; 40% of the biomass produced in the Earth comes from the oceans.
Residual biomass	Can be extracted from the waste of human activities, such as agriculture and forestry.
Energy crops	Refers to any crop with the purpose of providing the biomass to produce biofuels.

Source: (Herguedas et al., 2012).

Table 2. Obtaining biofuels processes.

	Techniques	Products	Applications
Mechanic	Chipping Trituration Compaction/pellets*	Firewood Chips* Briquettes* Sawdust	Heating Electricity
Thermochemical	Pyrolysis Gasification	Carbon Oils Gasogen	Heating Electricity Transport Chemical industry
Biological	Fermentation Anaerobic digestion	Ethanol Biogas	Heating Electricity Transport Chemical industry
Extractive	Physical-chemical extraction	Oils Esters Hydrocarbons	Transport Chemical industry

Source: (Herguedas et al., 2012). * Pellets

2. Classification of Biofuels According to Biomass Feedstocks

2.1 First-generation biofuels

First-generation biofuels are made from the sugars and vegetable oils found in food crops using standard processing technologies. Its classification is shown below:

2.1.1 Bio-alcohols

These are alcohols of organic origin. They consist of two fundamental compounds: Ethanol and methanol. Ethanol presents better expectations in terms of use, the first is known as bioethanol. Ethanol is manufactured following a similar process to that of beer. The raw material is very varied: Cereals (corn, wheat and barley), tubers (cassava, sweet potato, potato and taro), and sucrose (beet, sugarcane, molasses and sweet sorghum) (Salinas-Callejas and Gasca Quezada, 2009). These energy compounds are transformed into sugars, and then converted to ethanol through alcoholic fermentation. It is used in mixtures with conventional gasoline to replace it as a fuel in larger or smaller proportions. Although it does not completely replace gasoline, it confers stability to the mixture and reduces volatility, which facilitates its daily use, storage and transport.

2.1.2 Bio-oils

These are obtained from oilseeds and fried vegetable oils (cooking oil). Conventional tests have been done to drive motors with supermarket oils, and they have been successful. For example, in a research carried out by the Faculty of Chemical Sciences of the National Autonomous University of Chiapas (Unach), they managed to create biodiesel from the transformation of fried cooking vegetable oil (Zhang et al., 2003). This would reduce fuel costs, double the useful life of the vehicles and, therefore, reduce emissions of carbon monoxide, sulfur, aromatic hydrocarbons and solid particles. Vegetable oil does not release contaminants such as sulfur dioxide.

With respect to the biodiesel, it is a liquid fuel that is obtained in a similar way, but in this case, part of the diesel fuel is replaced by various vegetable oils and oil crops from soybean, rapeseed,[1] palm, jatropha[2] and sunflower. Although these species are usually the most used in its production, it can be obtained from more than 300 plant species, depending on which is the most abundant in the country of origin (Zhang et al., 2003).

3. Disadvantages of First-generation Biofuels

3.1 Environmental

3.1.1 Food crisis

We can point out that an imminent disadvantage observed on first-generation biofuels has been the so-called "food crisis". Economist Don Mitchell, of the World Bank, estimated that the impact of the alternative use of food by biofuels implied a 70% increase in food prices. In the USA, the administration insisted on using corn to generate biofuels and dismissed its impact on the price of grain calculated at 5%; however, other estimations mention that the increase in the price of corn was 54% (Gerber et al., 2008).

The large increase in biofuel production in the USA and the European Union was supported by subsidies, mandates and preferential import tariffs, which has been an accelerated increase in food prices since 2002 (Valdés and Foster, 2002).

Without these policies, the production of biofuels would be lower, and the costs of food products would be smaller. Since biofuels are produced on the basis of food or compete for land, that can be used to produce food, they directly impact on the price of food by restricting the supply of cereals for food, or indirectly if the food is livestock inputs; which impacts the price of the meat and dairy product.

3.1.2 Water

The production of first-generation biofuels implies a high consumption of fresh water. The growth of ethanol production is directly related to the increase in the demand of fresh water to irrigate the fields. For each kilogram of cereal produced, 1m^3 of water is consumed. It has been estimated that the ethanol used in a car, for a route of 20,000 km, implies a water consumption equivalent to 100 people in Europe or 500 people in Africa. At the same time, the corn used to obtain the amount of ethanol for the aforementioned route, could feed 7 people for a whole year (Valdés and Foster, 2002).

3.1.3 Deforestation

Biofuel crops have several effects: Increases in cropland expansion; they take over a huge share in total cropland; they are mainly located in areas that today are occupied by intact ecosystems; and they increase

[1] The rapeseed (*Brassica napus*) is an oleaginous widely spread in the world, which produces edible oil of excellent quality and likewise is transformed into biodiesel.

[2] Jatropha, known as "*piñón de tempate*" or "jatropha", is a seed that contains an inedible oil that can be used directly to fuel lamps and combustion engines, or can be transformed into biodiesel.

CO_2 emissions from deforestation. Thus, converting intact ecosystems, such as tropical rainforests or open woodlands, which store large amounts of carbon and belong to the most diverse terrestrial ecosystems, counteracts global climate and biodiversity protection goals. For bioenergy to make a real net contribution to climate change mitigation, intact forests must be protected. However, currently, less than 3% of global agricultural land is used for cultivating biofuel crops and land use change associated with bioenergy represents only around 1% of the total emissions caused by land-use change globally, most of which are produced by changes in land use for food and fodder production, or other reasons (Popp et al., 2014).

Without considering co-emissions from deforestation, biodiversity issues, and impacts on food and water security, the biomass resource potential could deliver a considerable amount of the world's primary energy demand up to 2095 (Popp et al., 2014).

3.1.4 Social costs

The social costs are also strong. The areas are temperate and humid forests, meadows and pastures. Many of these areas have been the habitat of native communities linked to their peasant agriculture. There is not only a process of ecosystemic degradation, but also the displacement of these aborigines and their passage from backwardness to indigence. Furthermore, they would stop producing their self-consumption goods; besides, they would lose the use of reserves for large-scale business exploitation, they might receive lean wages that would expose them to possible famines in the face of rising prices or the shortage of food (Yirdaw et al., 2017).

3.1.5 High production costs

The production of biofuels still costs considerably more than the production of fossil fuels, even considering the strong increase in oil prices.

For the production, storage and transportation of biofuels, large quantities of inputs would be required (in addition to land and water) whose production and transport also would demand quantities of energy. It takes energy to plant, produce fertilizers or pesticides, harvest, transport and process the grains or plants to their final biofuel form. For example, the average production cost for ethanol today is $1.22 per gallon, which translates to $51.24 per barrel. Now, on an energy basis—given that ethanol has 67% of the energy content of a barrel of oil, that translates to $76.86 on a barrel-of-oil-equivalent basis (Lane, 2017).

4. Second-generation Biofuels

Second-generation biofuels, also known as advanced biofuels, are fuels that can be manufactured from various types of non-food biomass. They are made from different feedstocks, including lignocellulosic biomass or woody crops, agricultural waste, as well as dedicated non-food energy crops grown on marginal land unsuitable for crop production and, therefore, may require different technology to extract useful energy from them.

The second-generation biofuels are distinguished from the first-generation biofuels in two aspects: They are obtained from plants that do not have a food function, and they are produced with technological innovations that will allow them to be more ecological and advanced than the current ones. As they are obtained from non-food raw materials, they can be grown on marginal lands that are not used for growing food. Existing biomass industries and relevant conversion technologies must be considered when evaluating the suitability of developing biomass as feedstock for energy. In this sense, they allow greater diversification with new raw materials, new technologies and new final products, thus promoting agricultural, forestry and agroindustry development.

It has been found that biomass from cellulose can be a basic raw material in the production of second-generation biofuels. The cellulose biomass allows the generation of cellulosic bioethanol, so waste from sawmills can be used, and forestry can be reoriented and expanded to diversify the use of forests and protect them from their clearing for agricultural and livestock uses (Balan, 2004). The main resources of second-generation biofuels are:

4.1 Energy crops

An energy crop is a plant grown at a low-cost and low-maintenance harvest that is used to make biofuels, such as bioethanol, or combusted for its energy content to generate electricity and/or heat. Energy crops are generally categorized as woody or herbaceous plants; many of the latter are grasses of the family *Graminaceae*.

Commercial energy crops are typically densely planted, high-yielding crop species which are processed to biofuel and burnt to generate power. Woody crops, such as willow (Mola-Yudego and Arinsson, 2008) or poplar, are widely utilized, as well as temperate grasses, such as *Miscanthus* and *Pennisetum* purpureum (both known as elephant grass) (Hodsman et al., 2005). If carbohydrate content is desired for the production of biogas, whole crops such as maize, millet, white sweet clover, and many others can be made into silage and then converted into biogas (Masarovicova et al., 2009).

Through genetic modification and application of biotechnology, plants can be manipulated in order to obtain greater yields, high energy yields can also be realized with existing crops (Masarovicova et al., 2009).

4.2 Municipal solid waste

Municipal solid waste is formed by a very large range of materials, and total quantities of such waste are increasing. In the United Kingdom, recycling initiatives decrease the proportion of waste going straight for disposal, and the level of recycling is increasing each year. However, there remains significant opportunities to convert this waste to fuel via gasification or pyrolysis (Kretschmer et al., 2013).

4.3 Green waste

Green waste, such as forest residues or garden or park waste (Kretschmer et al., 2013) may be used to produce biofuel via different routes. For example, biogas captured from biodegradable green waste, and gasification or syngas for further processing to biofuels via catalytic processes. In this case, as with other biomass, problems of scale and variety of residues have yet to be overcome.

4.4 Black liquor

Worldwide, the pulp and paper industry currently process about 170 million tons of black liquor (measured as dry solids) per year, with a total energy content of about 2EJ (IEA Bioenergy, 2007), making black liquor a very significant biomass source.

A pulp mill that produces bleached Kraft pulp generates 1.7–1.8 tons of black liquor (measured as dry content) per tonne of pulp. Black liquor, thus, represents a potential energy source of 250–500 MW per mill. As modern Kraft pulp mills have a surplus of energy, they could become key suppliers of renewable fuels in the future energy system (IEA Bioenergy, 2007).

Today, black liquor is the most important source of energy from biomass in countries such as Sweden and Finland with a large pulp and paper industry. It is, thus, of great interest to convert the primary energy in the black liquor to an energy carrier of high value. Furthermore, in comparison with other potential biomass sources for chemicals production, black liquor has the great advantage that it is already partially processed and exists in a pumpable, liquid form (IEA Bioenergy, 2007).

4.5 "Drop-in" biofuels

Drop-in biofuels can be defined as "liquid bio-hydrocarbons that are functionally equivalent to petroleum fuels and are fully compatible with existing petroleum infrastructure" (Karatzoz et al., 2015).

There is considerable interest in developing advanced biofuels that can be readily integrated into the existing petroleum fuel infrastructure—i.e., "dropped-in"—particularly by sectors such as aviation, where there are no real alternatives to sustainably produced biofuels for low carbon emitting fuel sources. Drop-in biofuels should be fully fungible and compatible with the large existing "petroleum-based" infrastructure.

According to a July 2014 report published by the IEA Bioenergy Task 39, entitled "The Potential and Challenges of Drop-in Biofuels", there are several ways to produce drop-in biofuels that are functionally equivalent to petroleum-derived transportation fuel blend stocks (IRENA, 2017).

4.6 Environmental impact

4.6.1 Greenhouse gas emissions

Lignocellulosic biofuels reduce greenhouse gas emissions by 60–90% when compared with fossil petroleum (Borjesson et al., 2013), which is on par with the best of the current first-generation biofuels, where typical best values are currently 60–80%. In 2010, average savings of biofuels used within Europe was 60% (Lamers et al., 2012). In 2013, 70% of the biofuels used in Sweden reduced emissions by 66% or even higher (Westin and Forsbeng, 2014).

4.6.2 Advantages of second-generation biofuels

The advantages of second-generation biofuels are numerous:

- By having a greater variety of raw not edible materials, they do not compete with the food function, since they are not alternatives to food, although it may generate growth in the industry that uses vegetable fibers or wood.
- They can be planted in non-agricultural or livestock areas, particularly, they can diversify the use of forests and encourage forestry and stop deforestation. In some cases, they may be used to recover eroded land into hillsides or decertified areas and fix CO_2 through its root system.
- The water consumed is generated by the forests themselves due to their ecosystem function with the generation of rain.
- They do not require the massive use of agrochemicals (fertilizers, pesticides, water, land, etc.). The net ratio of energy produced will improve with respect to the current ones.
- They can use biomass from garbage, from industrial waste or from human consumption.
- Encourage technological development with diversification effects in the agro-industrial sector.
- They are highly efficient in reducing emissions of greenhouse gases, particularly CO and CO_2 from short to medium term.

In the long term, they can lower production costs compared to current biofuels.

4.6.3 Disadvantages of second-generation biofuels

One of the disadvantages faced by second generation fuels is the high costs they face because they are now at the marketing threshold due to their relatively high manufacturing cost. This means that second-generation biofuels cannot yet be produced economically on a large scale. Production costs for the pulp and ethanol process (alone) are higher than the prices of gasoline based on mineral oil and conventional bioethanol.

Production costs are uncertain and vary with the feedstock available and conversion process but are currently thought to be above USD 0.80/liter of gasoline equivalent (Sims et al., 2010). Even with oil prices remaining above 80 USD/bbl, second-generation biofuels will probably not become fully commercial nor enter the market for several years without significant additional government support.

Finally, there is no clear candidate for "best technology pathway" between the competing biochemical and thermo-chemical routes. The development and monitoring of several large-scale demonstration projects are essential to provide accurate comparative data (Sims et al., 2010).

5. Third-generation Biofuels or Algae Fuel

Its elaboration uses production methodologies like those of second generation but using bioenergy crops specially designed or adapted as a raw material to improve the conversion of biomass to biofuels. These improvements or adaptations frequently use molecular biology techniques, such as the development of trees with low lignin percentages, which would reduce costs and improve the production of ethanol, or the modification of corn to contain integrated celluloses.

Algae fuel, algal biofuel, or algal oil is an alternative to liquid fossil fuels, using microalgae as its source of energy-rich oils. Also, algae fuels are an alternative to commonly-known biofuel sources, such as corn and sugarcane (Scott et al., 2010; Darzins et al., 2010), since microalgae (unicellular photosynthetic microorganisms, living in saline or freshwater environments that convert to algal biomass) can be converted to bio-oil, bioethanol, bio-hydrogen and biomethane via thermochemical and biochemical methods. Several companies and government agencies are funding efforts to reduce capital and operating costs and make algae fuel production commercially viable (Oncel, 2013). Like fossil fuel, algae fuel releases CO_2 when burnt, but unlike fossil fuel, algae fuel and other biofuels only release CO_2 recently removed from the atmosphere via photosynthesis as the algae or plant grew. The energy crisis and the world food crisis have ignited interest in farming algae for making biodiesel and other biofuels using lands unsuitable for agriculture. Among algal fuels, the attractive characteristics are: they can be grown with minimal impact on fresh water resources (Jia et al., 2010), they can be produced using saline and wastewater, they also have a high flash point (Dinh et al., 2009), and are biodegradable and relatively harmless to the environment if spilled (Demirbas, 2011; Demirbas, 2009). Biofuels production costs can vary widely by feedstock, conversion process, scale of production and region. However, algae cost more per unit mass than other second-generation biofuel crops due to high capital and operating costs (Hodsman et al., 2005) but are claimed to yield between 10 and 100 times more fuel per unit area (Greenwall et al., 2009).

5.1 Nutrients and growth inputs

Nutrients can be provided through runoff water from adjacent land areas or by channeling the water from sewage/water treatment plants. Microalgae cultivation using sunlight energy can be carried out in open or covered ponds or closed photobioreactors. Algal cultures consist of a single or multiple specific strains optimized for producing the desired product. Water, necessary nutrients and CO_2 are supplied in a controlled way, while oxygen must be removed (Carlsson et al., 2007).

Nutrients like nitrogen (N), phosphorus (P), and potassium (K), are important for plant growth and are essential parts of fertilizers. Silica and iron, as well as several trace elements, may also be considered important nutrients, the lack of one can limit the growth or productivity in an area (Arumugam et al., 2013).

Although most algae grow at low rate when the water temperature gets lower, the biomass of algal communities can get large due to the absence of grazing organisms. The modest increases in water current velocity may also affect rates of algae growth since the rate of nutrient uptake and boundary layer diffusion increases with velocity (Greenwall et al., 2009).

5.1.1 Carbon dioxide

Bubbling CO_2 through algal cultivation systems can greatly increase productivity and yield (up to a saturation point). Typically, about 1.8 tons of CO_2 will be utilized per ton of algal biomass (dry) produced, though this varies with algae species (Moellering and Benning, 2009).

5.1.2 Nitrogen

This is a valuable substrate that can be utilized in algal growth. Various sources of nitrogen can be used as a nutrient for algae, with varying capacities. Nitrate was found to be the preferred source of nitrogen, regarding the amount of biomass grown. Urea is a readily available source that shows comparable results, making it an economical substitute for nitrogen source in a large-scale culturing of algae (Arumugam et al., 2013). Despite the clear increase in growth in comparison to a nitrogen-less medium, it has been shown that alterations in nitrogen levels affect lipid content within the algal cells. In one study (Moellering and Benning, 2009), nitrogen deprivation for 72 hours caused the total fatty acid content (on a per cell basis) to increase 2.4-fold. 65% of the total fatty acids were esterified to triacyl glycerides in oil bodies, when compared to the initial culture, indicating that the algal cells utilized *de novo* synthesis of fatty acids. It is vital for the lipid content in algal cells to be high enough, while maintaining adequate cell division times, so parameters that can maximize both are under investigation.

5.1.3 Wastewater

A possible nutrient source is wastewater from the treatment of sewage, agriculture or flood plain run-off, all currently major pollutants and health risks. However, this wastewater cannot feed algae directly and must first be processed by bacteria, through anaerobic digestion. If wastewater is not processed before it reaches the algae, it will contaminate the algae in the reactor and, at the very least, kill much of the desired algae strain. In biogas facilities, organic waste is often converted to a mixture of carbon dioxide, methane and organic fertilizer. Organic fertilizer that comes out of the digester is liquid, and nearly suitable for algae growth, but it must first be cleaned and sterilized (Pittman et al., 2011; Chong et al., 2000).

5.2 Cultivation

Microalgae cultivation using sunlight energy can be carried out in open or covered ponds or closed photobioreactors, based on tubular, flat plate or other designs. Closed systems are much more expensive than ponds, and present significant operating challenges, and due to gas exchange limitations, among others, cannot be scaled-up much beyond about a hundred square meters for an individual growth unit.

Algae grow much faster than food crops and can produce hundreds of times more oil per unit area than conventional crops, such as rapeseed, palms, soybeans, or jatropha (Atabani et al., 2012). Due to algae have a harvesting cycle of 1–10 days, their cultivation permits several harvests in a very short timeframe, a strategy differing from that associated with annual crops (Chisti, 2007). In addition, algae can be grown on lands unsuitable for terrestrial crops, including arid land and land with excessively saline soil, minimizing competition with agriculture (Schenk et al., 2008).

5.2.1 Closed-loop system

The lack of equipment and structures needed to begin growing algae in large quantities has inhibited widespread mass-production of algae for biofuel production. Maximum use of existing agriculture processes and hardware is the goal (Benemann et al., 2012).

Closed systems (not exposed to open air) avoid the problem of contamination by other organisms blown in by the air. The problem for a closed system is finding a cheap source of sterile CO_2.

5.2.2 Photobioreactors

Most companies pursuing algae as a source of biofuels pump nutrient-rich water through plastic or borosilicate glass tubes (called "bioreactors") that are exposed to sunlight (and so-called photobioreactors).

Running a photobioreactor is more difficult than using an open pond, and costlier, but may provide a higher level of control and productivity (Chisti, 2007). In addition, a photobioreactor can be integrated into a closed loop cogeneration system much more easily than ponds or other methods.

5.2.3 Open pond

This consists of simple in-ground ponds, which are often mixed by a paddle wheel. These systems have low power requirements, operating costs, and capital costs when compared to closed loop photo bioreactor systems (Lundquist et al., 2010). Nearly all commercial algae producers for high value algal products utilize open pond systems (Bannon and Adey, 2008).

5.2.4 Turf scrubber

The algae scrubber is a system designed primarily for cleaning nutrients and pollutants out of water using algal turfs. Algae surf scrubber mimics the algal turfs of a natural coral reef by taking in nutrient rich water from waste streams or natural water sources and pulsing it over a sloped surface. This surface is coated with a rough plastic membrane or a screen, which allows naturally occurring algal spores to settle and colonize the surface. Once the algae have been established, it can be harvested every 5–15 days, (Adey et al., 2001) and can produce 18 metric tons of algal biomass per hectare per year. In contrast to other methods, which focus primarily on a single high yielding species of algae, this method focuses on natural polycultures of algae. As such, the lipid content of the algae in a turf scrubber system is usually lower, which makes it more suitable for a fermented fuel product, such as ethanol, methane or butanol (Biddy et al., 2016). Conversely, the harvested algae could be treated with a hydrothermal liquefaction process, which would make biodiesel, gasoline and jet fuel production possible (Sheehan et al., 1998; Smith et al., 2010).

5.3 Environmental impact

In comparison with terrestrial-based biofuel crops, such as corn or soybeans, microalgal production results in a much less significant land footprint due to the higher oil productivity from the microalgae than all other oil crops (Herro, 2008). Algae can also be grown on marginal lands useless for ordinary crops and with low conservation value and can use water from salt aquifers that is not useful for agriculture or drinking (Bullis, 2007; Chaw's Environmental and Infrastructure Group, 2011). Algae can also grow on the surface of the ocean in bags or floating screens. Thus, microalgae could provide a source of clean energy with little impact on the provisioning of adequate food and water or the conservation of biodiversity. Algae cultivation also requires no external subsidies of insecticides or herbicides, removing any risk of generating associated pesticide waste streams. In addition, algal biofuels are much less toxic, and degrade far more readily than petroleum-based fuels. However, due to the flammable nature of any combustible fuel, there is potential for some environmental hazards if ignited or spilled, as may occur in a train derailment or a pipeline leak (Acién-Fernandez et al., 2012). This hazard is reduced compared to fossil fuels, due to the ability for algal biofuels to be produced in a much more localized manner, and due to the lower toxicity overall, but the hazard is still there, nonetheless. Therefore, algal biofuels should be treated in a similar manner to petroleum fuels in transportation and use, with enough safety measures in place always.

Studies have determined that replacing fossil fuels with renewable energy sources, such as biofuels, has the capacity to reduce CO_2 emissions by up to 80% (Hemaiswarya et al., 2012). An algae-based system could capture approximately 80% of the CO_2 emitted from a power plant when sunlight is available. Although this CO_2 will later be released into the atmosphere when the fuel is burned (Acién-Fernandez et al., 2012). The possibility of reducing total CO_2 emissions, therefore, lies in the prevention of the release of CO_2 from fossil fuels. Furthermore, compared to fuels like diesel and petroleum, and even compared to other sources of biofuels, the production and combustion of algal biofuel does not produce any sulfur oxides or nitrous oxides, and produces a reduced amount of carbon monoxide, unburned hydrocarbons, and reduced emissions of other harmful pollutants (Kumar et al., 2010). Due to the fact that terrestrial plant sources of biofuel production simply do not have the production capacity to satisfy the current energy requirements, microalgae may be the only option to approach complete replacement of fossil fuels (Milano et al., 2016).

Microalgae production also includes the ability to use saline waste or CO_2 waste streams as an energy source. This opens a new strategy to produce biofuel in conjunction with wastewater treatment, while being able to produce clean water as a byproduct (Kumar et al., 2010). When used in a microalgal bioreactor, harvested microalgae will capture significant quantities of organic compounds as well as heavy metal contaminants absorbed from wastewater streams that would otherwise be directly discharged into the surface and groundwater (Herro, 2008). Moreover, this process also allows the recovery of phosphorus from waste, which is an essential but scarce element in nature—the reserves of which are estimated to have been depleted in the last 50 years (Zivojnovich, 2010; Dixner, 2013; Blankenship et al., 2011).

5.4 Economic viability

There is clearly a demand for sustainable biofuel production, but whether a biofuel will be used ultimately depends not on sustainability but on cost efficiency. Therefore, research is focusing on cutting the cost of algal biofuel production to the point where it can compete with conventional petroleum (Chisti, 2007). The production of several products from algae has been mentioned as the most important factor for making algae production economically viable. Other factors are the improving of the solar energy to biomass conversion efficiency (currently 3%, but 5 to 7% is theoretically attainable) and making the oil extraction from the algae easier (Mata et al., 2010).

5.5 Use of byproducts

Many of the byproducts produced in the processing of microalgae can be used in various applications, many of which have a longer history of production than algal biofuel. Some of the products not used in the production of biofuel include natural dyes and pigments, antioxidants, and other high-value bioactive compounds (Pulz and Gross, 2004; Singh et al., 2005; Sporalore et al., 2006). These chemicals and excess biomass have found numerous uses in other industries. For example, the dyes and oils have found a place in cosmetics, commonly as thickening and water-binding agents (Tokusoglu and Uunal, 2003). Discoveries within the pharmaceutical industry include antibiotics and antifungals derived from microalgae, as well as natural health products, which have been growing in popularity over the past few decades. For example, Spirulina contains numerous polyunsaturated fats (Omega 3 and 6), amino acids, and vitamins (Vonshak, 1997) as well as pigments, such as beta-carotene and chlorophyll, that may be beneficial (Berla et al., 2013).

6. Fourth-generation Biofuels

Fourth-generation biofuels seek to adapt the raw material to improve efficiency in CO_2 capture and storage. They take advantage of synthetic biology of algae and cyanobacteria (Hays and Ducat, 2015), which is a young but strongly evolving research field. Synthetic biology is based on the design and construction of new biological parts, devices and systems, and the re-design of existing, natural biological systems for useful purposes. It is becoming possible to design a photosynthetic/non-photosynthetic chassis, either natural or synthetic, to produce high-quality biofuels (Scaife et al., 2015). For the first-, second- and third-generation biofuels, the raw material is either biomass or waste, both being results of "yesterday's photosynthesis" (yet not from fossil resources). While these biofuels often are very useful in a certain region or country, they are always limited by the availability of the corresponding organic raw material, i.e., the biomass, which limits their application on a global scale (Inganas and Sundstrom, 2016).

Fourth-generation biofuels will be based on raw materials that are essentially inexhaustible, cheap and widely available. Photosynthetic water splitting (water oxidation) into its constituents by solar energy can become a large contributor to fuel production on a global scale, both by artificial photosynthesis (Wijffels et al., 2013) and by direct solar biofuel production technologies. Not only the production of hydrogen but also the production of reduced carbon-based biofuels is possible by concomitant enhanced fixation of atmospheric CO_2 and innovative design of synthetic metabolic pathways for fuel production. The generation of "designer bacteria" with new useful properties requires revolutionary scientific breakthroughs in several areas of fundamental research.

Fourth-generation biofuels are produced (i) by designer photosynthetic microorganisms to produce photobiological solar fuels, (ii) by combining photovoltaics and microbial fuel production (electro biofuels) or (iii) by synthetic cell factories or synthetic organelles specifically tailored for production of desired high-value chemicals (production currently based on fossil fuels) and biofuels (Wijffels et al., 2013).

6.1 Designer microorganisms in production of solar biofuels

Key improvements in photon-to-fuel conversion efficiency as well as in the quality of the fuel will be based on generation of designer microorganisms. They produce a selected photobiological solar fuel, which is a non-fossil fuel made by direct solar energy conversion in photosynthetic microorganisms (algae or cyanobacteria), and are engineered, when necessary, to secrete the fuel. Photobiological solar fuel is made in photosynthetic cells from solar energy using only water, or water and CO_2 as raw materials, depending whether the produced fuel is hydrogen- or carbon-based fuel, respectively (Hays and Ducat, 2015). Knowledge based on powerful biochemical and biophysical insights in natural photosynthetic light harvesting, water splitting, electron transfer, hydrogenases and carbon metabolism is critical for the development of direct photobiological solar fuels. Microorganisms will be optimized for fuel production with synthetic biology approaches, metabolic engineering and organism design based on knowledge acquired by genomic research, molecular systems biology and extensive modelling research (Halfmann et al., 2014).

The most advanced research made towards direct photobiological solar fuel production is based on research with unicellular algae and cyanobacteria. Cyanobacteria are suitable as photosynthetic chassis for their well-developed genetic transformation technologies as well as for extensive knowledge on their light harvesting and electron transfer processes, on their metabolome and advanced modelling research of the entire cell. Cyanobacteria have been genetically engineered to produce various fuels and chemicals (e.g., H_2, ethanol, isobutanol, isoprene, lactic acid) (Savakis and Hellingwerf, 2015). Introduction of various fermentative metabolism pathways to cyanobacteria cells by synthetic biology approaches has made it possible to produce biofuels directly from solar energy and Calvin–Benson cycle intermediates (Rabaey and Rozendal, 2010). Furthermore, the efficient secretion of products from the cells will increase the production capacity.

Breakthroughs in direct photobiological solar fuel production call for further research in design of light-harvesting systems, modelling and simulation of biological reactions and systems as well as in development of synthetic biology tools and production systems. During the coming 10–20 years, it is expected that various photobiological solar fuels will gradually enter into the market (Hays and Ducat, 2015).

6.2 Electrobiofuels

It is possible, through synthetic biology approaches, to establish new-to-nature hybrid production organisms that use renewable electricity and carbon sources for the production of commodity chemicals and biofuels. A newly emerging field relies on the capability of certain microbes for direct electron capture from electrodes (e.g., from solar cells or any renewable electricity source) to assimilation of the reducing equivalents into metabolism, along with CO_2 utilization (Torella et al., 2015). The new advanced technologies based on the combination of photovoltaics and microbial electrosynthesis are called electrobiofuels. The microbial electrosynthesis is based on the concept of capturing the energy from the electrodes, i.e., via solar cells or any other renewable source of energy. With this system, energy from solar cells can be turned into storable energy sources (electrofuels). They allow renewables from all sources to be stored conveniently and efficiently as a liquid fuel.

Recently (Aro, 2016), a novel and scalable integrative bio electrochemical production system for isopropanol was demonstrated. The solar water-splitting catalyst, based on earth-abundant metals, was used to provide energy for growth of a bacterium, *Ralstonia eutropha*. In metabolically engineered *Ralstonia*, the energy from water splitting could be diverted for production of isopropanol. Authors claim the highest bio electrochemical fuel yield reported so far.

6.3 From current biorefineries to synthetic factories in production of solar biofuels

The design of synthetic factories and cell organelles for enhanced biofuel production is a future technology and is recently entering intense developmental phase at the basic level of biofuel research.

Likewise, the synthetic biology technology itself is still young and only a few truly synthetic examples have been realized by now (Cameron et al., 2014). For optimal solutions, one needs to go beyond the known possibilities offered by biology by speeding up evolution and screening a massive number of artificial biological combinations. Thus, automation, microfabrication and measurement technologies are an integral part of a successful synthetic biology platform. Most importantly, engineering principles should be used to guide modelling, design and standardization of the synthetic biological systems and the entire development process so that optimization can progress from the current trial-and-error situation to the design of truly programmable biofuel production systems (Cameron et al., 2014).

Synthetic biology, after reaching maturity as a technology, will give tools and concepts that will make biology fully engineerable, and thereby make it possible to take full advantage of the diversity, functionality and specificity that biology can offer. Such research and technology development programmers should be established for construction of synthetic factories and organelles towards efficient biofuel production (Biogas from Waste and Renewable Resources, 2011).

7. Types of Biofuels

7.1 Gaseous biofuels

7.1.1 Biogas

Biogas is understood as the mixture of gases that originates from the decomposition of organic matter, with the help of microorganisms and under anaerobic conditions. This process is called anaerobic digestion or biomethanization. The proportion of gas mixture depends on the digested substrate, being, in general, formed by 50–70% methane (CH_4), 30–40% carbon dioxide (CO_2), and small proportions of hydrogen sulfide (H_2S), nitrogen (N_2), hydrogen (H_2) and others. This biogas can be produced both naturally (buried organic waste), and artificially, in devices called biodigesters (Igoni Hilkia et al., 2009).

Its high content of methane gives it a high heat capacity (between 18.8 and 23.4 kcal/m^3), so, after being purified to eliminate water vapor and hydrogen sulfide, it is used in boilers, when its production is small, or in cogeneration engines for the generation of electricity and heat, when production is higher (Igoni Hilkia et al., 2009). In addition, if adequate biogas purification is carried out (up to 91–95% methane), it can also be used as fuel for vehicles, fuel cells or incorporated into the natural gas network (Harder and Witsch, 1942).

The process for the production of biogas is the anaerobic digestion, defined as the biological process in which the organic matter, in the absence of oxygen, is decomposed into gaseous products (biogas) such as CH_4, CO_2, H_2, H_2S, and in digestate, which is a mixture of mineral products (N, P, K, Ca, etc.), and compounds of difficult degradation (Harder and Witsch, 1942).

This process can be carried out, as has also been seen in the previous section, with raw materials from different origins, if they are liquid, have fermentable materials, and a composition and relatively stable concentrations.

Artificially, the process is carried out in a biodigester or organic waste digester, consisting of a closed, hermetic and impermeable container (reactor) that will contain the organic material to be fermented diluted in water.

7.1.1.1 Process stages

The four key stages of anaerobic digestion involve hydrolysis, acidogenesis, acetogenesis and methanogenesis. The overall process can be described by the chemical reaction, where organic material, such as glucose, is biochemically digested into carbon dioxide (CO_2) and methane (CH_4) by the anaerobic microorganisms.

$$C_6H_{12}O_6 \rightarrow 3CO_2 + 3CH_4$$

7.1.1.1.1 Hydrolysis

For the bacteria in anaerobic digesters to access the energy potential of the material, these chains must be broken down first into their smaller constituent parts. These constituent parts, or monomers, such as sugars, are readily available to other bacteria. The process of breaking these complex organic molecules and dissolving the smaller molecules into a solution is called hydrolysis. Therefore, hydrolysis of these high-molecular-weight polymeric components is the necessary first step in anaerobic digestion (Harder and Witsch, 1942). Through hydrolysis the complex organic molecules are broken down into simple sugars, amino acids, and fatty acids.

Acetate and hydrogen produced in the first stages can be used directly by methanogens. Other molecules, such as volatile fatty acids with a chain length greater than that of acetate must first be catabolized into compounds that can be directly used by methanogens (Igoni Hilkia et al., 2009).

7.1.1.1.2 Acidogenesis

The biological process of acidogenesis results in further breakdown of the remaining components by acidogenic (fermentative) bacteria. Here, volatile fatty acids are created, along with ammonia, carbon dioxide, and hydrogen sulfide, as well as other byproducts (Igoni Hilkia et al., 2009). The process of acidogenesis is like the way milk sours.

7.1.1.1.3 Acetogenesis

The third stage of anaerobic digestion is acetogenesis. Here, simple molecules created through the acidogenesis phase are further digested by acetogens to produce largely acetic acid, as well as carbon dioxide and hydrogen (Song et al., 2004).

7.1.1.1.4 Methanogenesis

The terminal stage of anaerobic digestion is the biological process of methanogenesis. Here, methanogens use the intermediate products of the preceding stages and convert them into methane, carbon dioxide, and water. These components make up most of the biogas emitted from the system. Methanogenesis is sensitive to both high- and low-pH and occurs between pH 6.5 and pH 8. The remaining, indigestible material that the microbes cannot use and any dead bacterial remains constitute the digestate (Jewell et al., 1993).

7.1.1.2 Configuration

Anaerobic digesters can be designed and engineered to operate using several different configurations and can be categorized into batch vs. continuous process mode, mesophilic vs. thermophilic temperature conditions, high vs. low portion of solids, and single stage vs. multistage processes. More initial build money and a larger volume of the batch digester are needed in order to handle the same amount of waste as a continuous process digester. Higher heat energy is demanded in a thermophilic system compared to a mesophilic system and has a larger gas output capacity and higher methane gas content. For solids content, low will handle up to 15% solid content. Above this level is considered high solids content and can also be known as dry digestion (Song et al., 2004).

7.1.1.3 Applications

7.1.1.3.1 Biogas upgrading

Raw biogas produced from digestion is roughly 60% methane and 29% CO_2 with trace elements of H_2S: Inadequate for use in machinery. The corrosive nature of H_2S alone is enough to destroy the mechanisms (Biomethanation, Vol II).

Methane in biogas can be concentrated via a biogas upgrader to the same standards as fossil natural gas, which itself must go through a cleaning process, and become biomethane. If the local gas network allows, the producer of the biogas may use their distribution networks. Gas must be very clean to reach pipeline quality and must be of the correct composition for the distribution network to accept. CO_2, water, hydrogen sulfide, and particulates must be removed if present (Biomethanation, Vol II).

7.1.1.3.2 Biogas gas-grid injection

Gas-grid injection is the injection of biogas into the methane grid (natural gas grid). Until the breakthrough of micro combined heat and power, two-thirds of all the energy produced by biogas power plants was lost (as heat). Using the grid to transport the gas to customers, the energy can be used for on-site generation, resulting in a reduction of losses in the transportation of energy. Typical energy losses in natural gas transmission systems range from 1% to 2%; in electricity transmission they range from 5% to 8% (Jewell et al., 1993).

7.1.1.3.3 Biogas in transport

If the biogas is concentrated and compressed, it can be used in vehicle transportation. Compressed biogas is becoming widely used in Sweden, Switzerland and Germany. A biogas-powered train (The Biogas Train Amanda) has been in service in Sweden since 2005. Biogas powers automobiles. In 1974, a British documentary film titled Sweet as a Nut detailed the biogas production process from pig manure and showed how it fueled a custom-adapted combustion engine (Richards et al., 1991). In 2007, an estimated 12,000 vehicles were being fueled with upgraded biogas worldwide, mostly in Europe (Dosta et al., 2007).

Biogas is part of the wet gas and condensing gas (or air) category that includes mist or fog in the gas stream. The mist or fog is predominately water vapor that condenses on the sides of pipes or stacks throughout the gas flow. Biogas environments include wastewater digesters, landfills and animal feeding operations (covered livestock lagoons).

7.1.1.3.4 Biogas generated heat/electricity

Biogas can be used as the fuel in the system of producing biogas from agricultural wastes and co-generating heat and electricity in a combined heat and power plant. Unlike the other green energy such as wind and solar, the biogas can be quickly accessed on demand. The global warming potential can also be greatly reduced when using biogas as the fuel instead of fossil fuel (Roubík et al., 2016).

However, the acidification and eutrophication potentials produced by biogas are 25 and 12 times higher, respectively, than the fossil fuel alternative. This impact can be reduced by using correct combination of feedstocks, covered storage for digesters and improved technique for retrieving escaped material. Overall, the results still suggest that using biogas can lead to a significant reduction in most impacts compared to fossil fuel alternatives. The balance between environmental damage and greenhouse gas emission should still be considered while assessing the system.

The three principal products of anaerobic digestion are biogas, digestate, and wastewater.

7.1.1.4.1 Biogas

Biogas is the ultimate waste product of the bacteria feeding off the input biodegradable feedstock (Dosta et al., 2007) (the methanogens stage of anaerobic digestion is performed by archaea, a micro-organism on a distinctly different branch of the phylogenetic tree of life to bacteria), and is mostly methane and carbon dioxide (Dosta et al., 2007), with a small amount of hydrogen and traces of hydrogen sulfide (As-produced, biogas also contains water vapor, with the fractional water vapor volume a function of biogas temperature) (Biomethanation, Vol II). Most of the biogas is produced during the middle of the digestion, after the bacterial population has grown, and tapers off as the putrescible material is exhausted. The gas is normally stored on top of the digester in an inflatable gas bubble or extracted and stored next to the facility in a gas holder.

7.1.1.4.2 Digestate

Digestate is the solid remnants of the original input material to the digesters that the microbes cannot use. It also consists of the mineralized remains of the dead bacteria from within the digesters. Digestate can come in three forms: Fibrous, liquor or a sludge-based combination of the two fractions. In two-stage systems, different forms of digestate come from different digestion tanks. In single-stage digestion systems, the two fractions will be combined and, if desired, separated by further processing (Biomethanation, Vol II).

7.1.1.4.3 Wastewater

The final output from anaerobic digestion systems is water, which originates both from the moisture content of the original waste that was treated, and water produced during the microbial reactions in the digestion systems. This water may be released from the dewatering of the digestate or may be implicitly separate.

The wastewater exiting the anaerobic digestion facility will typically have elevated levels of biochemical oxygen demand and chemical oxygen demand. These measures of the reactivity of the effluent indicate an ability to pollute. Some of this material is termed "hard chemical oxygen demand", meaning it cannot be accessed by the anaerobic bacteria for conversion into biogas. If this effluent were put directly into watercourses, it would negatively affect them by causing eutrophication. As such, further treatment of the wastewater is often required. This treatment will typically be an oxidation stage wherein air is passed through the water in a sequencing batch reactors or reverse osmosis unit (Sabiha-Hanim and Abd Halim, 2018).

7.1.2 Syngas

Syngas is an abbreviation for synthesis gas, which is a mixture of carbon monoxide, carbon dioxide and hydrogen. Syngas is produced by gasification of a carbon containing fuel to a gaseous product that has low heating value.

The efficiency of the gasification process varies between 70 and 80%, depending on the technology used and the fuel or the type of gasifying agent used. The rest of the energy is invested in the endothermic reactions of the process, in the heat losses of the reactors, in the drying, filtering and washing of the syngas.

In principle, the fate of the gas synthesis is usually the production of heat by direct combustion in a burner or the generation of electricity by means of a combustion motor or turbine. Currently, advanced gasification processes, based on fluidized bed systems, are the most promising for the generation of electricity, with high efficiency, based on combined cycles of gas turbine and steam cycle. For this purpose, it is very important to obtain clean gases (Sabiha-Hanim and Abd Halim, 2018).

7.1.2.1 The route to produce syngas is gasification

Regardless of the technology used in gasification, biomass goes through several stages, whose order and importance depends on the fuel and the elements of the gasifying agent used, but can be separated into three groups:

7.1.2.1.1 Pyrolysis or thermal decomposition

First, the biomass undergoes a drying and preheating process when entering the gasifier. Subsequently, and in the absence of oxygen, the fuel is broken down into a mixture of solid, liquid and gas. The originated is solid called "char" and the liquids, due to the presence of tars and condensable vapors, are called "tar".

7.1.2.1.2 Oxidation or combustion

This phase takes place when the gasifying agent is an oxidant such as oxygen or air, giving rise to a series of oxidation reactions, mainly exothermic. Through them, the necessary heat is generated so that the process is maintained.

7.1.2.1.3 Reduction or gasification

This phase contains the reactions in which the remaining solid becomes gas. They are fundamentally endothermic reactions. The oxidation and reduction stages can be considered together in a single stage of gasification in which all kinds of possible reactions take place between the char and the gas mixture present.

The different types of gasifiers can be classified as:

7.1.2.1.3.1 Fixed bed (mobile) or descending

This type of gasifier can be used with both air and O_2 and steam.

The biomass is introduced into the gasifier at the top and the ash extraction is usually carried out at the bottom of the bed. The rate of descent through the bed is regulated by the extraction of the ashes. This system has as an advantage, its simplicity and the relatively low investment costs; however, it is very sensitive to the mechanical characteristics of the fuel, requiring uniform and homogeneous fuel.

7.1.2.1.3.2 Updraft

The solid and the gas circulate in opposite directions. The solid descends slowly and the gasifying agents (air and oxygen and steam) circulate in an upward direction.

When the biomass descends, it is heated by the gas stream until it reaches the combustion zone where the maximum temperature is reached, suffering a subsequent cooling prior to the discharge of the ash. The regulation of the temperature is done by the injection of water vapor. In general, a fairly polluted gas is obtained since the low temperatures of the gases (250–500 ºC) do not allow the decomposition of oils, tars and gases formed (phenols, ammonia and H_2S). The synthetized gas comes out of top of the reactor (Sabiha-Hanim and Abd Halim, 2018).

7.1.2.1.3.3 Downdraft

The solid and the gas circulate in the same direction inside the gasifier. The biomass, which is introduced through the upper part, is subjected to a progressive increase in temperature, drying at the beginning and pyrolysis below. This temperature pattern is originated due to the high temperatures generated in the lower part of the reactor, through the partial combustion of the products that get there (gases, tars and coal). In this case, the synthetized gas is released through the lower part of the gasifier and the air supply is made through the middle part of the gasifier.

7.1.2.1.3.3 Crossflows

In this case, the oxidizing agent is introduced through one side of the reactor, the synthetized gas leaves the diametrically opposite side. This gasifier has certain advantages over the previous ones since it has lower starting times, it can operate with dry and wet fuels, and the temperature of the gas obtained is relatively high, so that the composition of the gas produced contains small amounts of H_2 and CH_4. As a disadvantage, the high tar content obtained in the process (Yahyaoui, 2018).

7.1.2.1.3.4 Fluidized bed

A fluidized bed gasifier is a technology that converts carbon-containing bio solids into syngas. There are two main types of fluidized bed systems:

7.1.2.1.3.4.1 Bubbling bed systems

In this kind of system, the bed material (which could be a mixture of inert particles such as sand along with finely ground biomass) rests on a distributor plate (either perforated or porous type) through which the fluidizing medium, that is, air is passed at a velocity of about five times that of minimum fluidization velocity. Typical temperature in the bed is about 700–900 ºC. The feed, which is finely grained biomass, is introduced just above the distributor plate. The biomass first undergoes pyrolysis in the hot bed above the distributor to form char and gaseous products due to devolatilization (Li, 2002). The char particles are lifted along with fluidizing air and undergo gasification in relatively upper portions of the bed. Due to contact with high temperature bed, the high molecular weight tar compounds formed are cracked; thus, reducing the net tar content of the producer gas to less than 1–3 g/Nm^3 (Li, 2002).

7.1.2.1.3.4.1 Circulating fluidized beds

They are an extension of the concept of bubbling bed fluidization. In this case, the velocity of the fluidizing air is much higher than the terminal settling velocity of the bed material. Thus, the entire bed material (biomass + inert material such as sand) is lifted by the fluidizing air. The exhaust of the gasifier is a relatively lean mixture of solids and gas. This exhaust is admitted into a cyclone separator where solids are disengaged from the gas and are returned to the bed through a down comer pipe. Depending upon the solids' concentration and size distribution, either single stage or multistage cyclones are employed. Circulation of the biomass particles is carried out until the particles are reduced in size due to combustion/gasification. An advantage that the beds design offers is that the gasifier can be operated at elevated pressures (Kishore, 2008).

In order to avoid biomass ash agglomeration problems, fluidized beds may be forced to run at lower than optimal temperatures, thus decreasing the combustion efficiency. Most of the large-scale facilities built in the last decade use fluidized bed designs.

7.2 Liquid biofuels

Liquid biofuels are all those fuels that are obtained from biomass and that can be used for any energy application, whether thermal, electrical or mechanical, and to feed boilers and internal combustion engines (Otto and diesel). This type of alternative fuel includes the so-called biofuels, which are liquid (and gaseous) biofuels used for transport, and bio liquids, which refer to liquid biofuels destined for energy uses other than transportation, including electricity production.

7.2.1 Bioethanol

Bioethanol can be produced from biomass by the hydrolysis and sugar fermentation processes. Lignocellulosic biomass contains a complex mixture of carbohydrate polymers (cellulose, hemi cellulose) on the fiber wall. In order to produce sugars from these biomasses, the biomass is pre-treated by physical, chemical (acids or ammoniac) or biological (enzymatic attack) routes in order to reduce the size of the feedstock and to open the plant structure. Once the raw material is divided, the cellulose and hemicellulose hydrolysis are carried out to obtain glucose and xylose, that is, convert the polysaccharides of matter into simple sugars (Sabiha-Hanim and Abd Halim, 2018). The lignin which is also present in the biomass is normally used as a fuel for the ethanol production plants boilers. There are three leading methods of hydrolyzing sugars from biomass:

7.2.1.1 Acid hydrolysis

Hydrolysis is carried out in an acidic medium by using H_2SO_4, HCl or sulfhydryl, either in concentrated or diluted form. The first one is carried out with temperatures close to the ambient. It also solubilizes the polysaccharides leaving as lignin residue. While the diluted solubilizes hemicellulose and leaves lignin and cellulose as a residue. Even so, a second stage can be done to solubilize the cellulose.

The Arkanol process works by adding 70–77% sulphuric acid to the biomass that has been dried to a 10% moisture content. The acid is added in the ratio of 1.25 acid to 1 biomass and the temperature is controlled to 50 °C. Water is then added to dilute the acid to 20–30% and the mixture is again heated to 100 °C for 1 hour. The gel produced from this mixture is then pressed to release an acid sugar mixture and a chromatographic column is used to separate the acid and sugar mixture (Abascal Fernandez, 2017).

7.2.1.2 Dilute acid hydrolysis

The dilute acid hydrolysis process is one of the oldest, simplest and most efficient methods of producing ethanol from biomass. Dilute acid is used to hydrolyze the biomass to sucrose. The first stage uses 0.7% sulphuric acid at 190 °C to hydrolyze the hemi cellulose present in the biomass. The second stage is optimized to yield the more resistant cellulose fraction. This is achieved by using 0.4% sulphuric acid at 215 °C. The liquid hydrolates are then neutralized and recovered from the process (Abascal Fernandez, 2017).

7.2.1.3 Enzymatic hydrolysis

This method is the most effective to produce ethanol from biomass. It consists of microbial degradation of lignocellulosic residues by enzymes, the most used being cellulose. This activity depends on the ability to absorb the substrate and form the active enzyme-substrate complex. This causes factors, such as physical, morphological, crystallinity and chemical structure configuration, as well as own characteristics of the enzyme, to affect the process. The main feature of this process is the significant energy and equipment savings compared to acid methods. The problem arises in some cases when they generate compound inhibitors and a detoxification of the currents is required to later ferment them. These inhibitory substances originate in the hydrolysis of different components, of the esterified organic acids of hemicellulose or solubilized phenolic derivatives of lignin as well as the degradation of soluble sugars. However, this process is very expensive and is still in its early stages of development.

The sugars released during enzymatic hydrolysis are fermented to carry out the production mainly of ethanol, carbon dioxide, butanol, organic acids, xylitol and furfural (Abascal Fernandez, 2017). These fermentations can be:

7.2.1.3.1 Fermentation in co-culture

This process performs the cultivation of two microorganisms in the same vessel. These microorganisms will carry out the fermentation of glucose and pentose for production of ethane. However, there are still obstacles to overcome to implement this process.

7.2.1.3.2 Sequential fermentation

This type of fermentation consists in carrying out continuous fermentation of the glucose and the xylose in the same container.

7.2.1.3.3 Separate hydrolysis and fermentation

These operations are carried out under optimal conditions of temperature and pH, due to the realization of both processes separately. Its main drawback is that the cellulose accumulation generated in hydrolysis inhibits cellulose activity.

7.2.1.3.4 Simultaneous saccharification and fermentation

The glucose obtained is rapidly transformed into ethanol. In this process, the number of reactors needed, and the number of enzymes needed is smaller. Also, the risk of generating inhibitors is reduced. The problem with this option is that a correlation between temperatures and temperature is required. Total process speed, because of the fermentation stage, is slower.

7.2.1.3.5 Separate saccharification and fermentation

The microorganisms produce cellulose. For this, anaerobic clostridia is used in order to produce cellulolytic enzymes that hydrolyze the substrate and the sugars generated are converted into ethanol. This option is less used due to by-product formation and the low tolerance of microorganisms to ethanol.

7.2.1.4 Other options

Another option of obtaining ethanol is by alcoholic fermentation of hexoses and pentoses. In addition to these two types of fermentations, yeasts and microorganisms.

Once the ethanol is obtained, a distillation will be carried out in order to rectify and purify ethanol. This process is carried out in a conventional manner in order to eliminate water present in the fuel, the processes are:

7.2.1.4.1 Fractional distillation process

The distillation process works by boiling the water and ethanol mixture. Since ethanol has a lower boiling point (78.3 °C) compared to that of water (100 °C), the ethanol turns into the vapor state before the water and can be condensed and separated (Gomis et al., 2007).

7.2.1.4.1 Dehydration

There are three dehydration processes to remove the water from an azeotropic ethanol/water mixture. The first process, used in many early fuel ethanol plants, is called azeotropic distillation and consists on adding benzene or cyclohexane to the mixture. When these components are added to the mixture, it forms a heterogeneous azeotropic mixture in vapor–liquid-liquid equilibrium, which, when distilled, produces anhydrous ethanol in the column bottom, and a vapor mixture of water, ethanol, and cyclohexane/benzene (Horn and Krupp, 2006).

When condensed, this becomes a two-phase liquid mixture. The heavier phase, poor in the entrained (benzene or cyclohexane), is stripped of the entrained and recycled to the feed—while the lighter phase, with condensate from the stripping, is recycled to the second column. Another early method, called extractive distillation, consists on adding a ternary component that increases ethanol's relative volatility. When the ternary mixture is distilled, it produces anhydrous ethanol on the top stream of the column (Lopes et al., 2016).

7.2.1.5 Post-production water issues

Bioethanol is hygroscopic, meaning it absorbs water vapor directly from the atmosphere. Because absorbed water dilutes the fuel value of the ethanol and may cause phase separation of ethanol-gasoline blends (which causes engine stall), containers of ethanol fuels must be kept tightly sealed. This high miscibility with water means that ethanol cannot be efficiently shipped through modern pipelines, like liquid hydrocarbons, over long distances (Lopes et al., 2016).

The fraction of water that an ethanol-gasoline fuel can contain without phase separation increases with the percentage of ethanol. For example, E30[3] can have up to about 2% water. If there is more than about 71% ethanol, the remainder can be any proportion of water or gasoline and phase separation does not occur. The fuel mileage declines with increased water content. The increased solubility of water with higher ethanol content permits E30 and hydrated ethanol to be put in the same tank since any combination of them always results in a single phase. Somewhat less water is tolerated at lower temperatures. For E10 it is about 0.5% v/v at 21 °C and decreases to about 0.23% v/v at −34 °C (Lopes et al., 2016).

[3] Ethanol fuel mixtures have "E" numbers which describe the percentage of ethanol fuel in the mixture by volume, for example, E30 is 30% anhydrous ethanol and 70% gasoline.

7.2.1.6 Environmental impact

7.2.1.6.1 Energy balance

All biomass goes through at least some of these steps: It needs to be grown, collected, dried, fermented, distilled, and burned. All these steps require resources and infrastructure. The total amount of energy input into the process in comparison to the energy released by burning the resulting ethanol fuel is known as the energy balance (or "energy returned or energy invested"). The energy balance for sugarcane ethanol produced in Brazil is more favorable, with one unit of fossil-fuel energy required to create 8 from the ethanol. Energy balance estimates are not easily produced; thus, numerous reports that have been generated are contradictory. For instance, a separate survey, reports (Nduka and Imoedemhe, 2015) that production of ethanol from sugarcane which requires a tropical climate to grow productively, returns from 8 to 9 units of energy for each unit expended, as compared to corn, which only returns about 1.34 units of fuel energy for each unit of energy expended.

7.2.1.6.2 Air pollution

Compared with conventional unleaded gasoline, ethanol is a particulate-free burning fuel source that combusts with oxygen to form CO_2, CO, water and aldehydes. The Clean Air Act requires the addition of oxygenates, such as ethanol, to reduce carbon monoxide emissions in the USA.

7.2.1.6.3 CO_2 production

The calculation about how much carbon dioxide is produced in the manufacture of bioethanol is a complex and inexact process and is highly dependent on the method by which the ethanol is produced, and the assumptions made in the calculation. A calculation should include:

- The cost of growing the feedstock
- The cost of transporting the feedstock to the factory
- The cost of processing the feedstock into bioethanol

Such a calculation may or may not consider the following effects:

- The cost of the change in land use of the area where the fuel feedstock is grown
- The cost of transportation of the bioethanol from the factory to its point of use
- The efficiency of the bioethanol compared with standard gasoline
- The amount of carbon dioxide produced at the tail pipe
- The benefits due to the production of useful by-products, such as cattle feed or electricity.

7.2.1.6.4 Land use

Cellulosic ethanol production is a new approach that may alleviate land use and related concerns. Cellulosic ethanol can be produced from any plant material, potentially doubling yields, to minimize conflict between food needs vs. fuel needs. Instead of utilizing only the starch by-products from grinding wheat and other crops, cellulosic ethanol production maximizes the use of all plant materials, including gluten. This approach would have a smaller carbon footprint because the amount of energy-intensive fertilizers and fungicides remain the same for higher output of usable material. The technology for producing cellulosic ethanol is currently in the commercialization stage (Lopes et al., 2016).

7.2.2 Biodiesel

Biodiesel is a biofuel that is obtained basically from seeds of oil plants. It can also be obtained from the used frying oils, microalgae and animal fats. In quantitative terms, it is estimated that, in 2017, the global production of vegetable oils was of 187 million tons, 37.6% (70.3 million tons) are provided by palm and palm kernel and 30% (55 million tons) by soybean. The remaining 32.5% are supplied by canola, sunflower, peanut and cottonseed oils (Colombo et al., 2017).

Biodiesel is a fuel composed of methyl esters of long-chain fatty acids obtained, as already mentioned, vegetable oils or animal fats, and characterized by its high energy density (37 MJ/kg) (Banatl et al., 2013). Natural vegetable oils cannot be used in conventional diesel engines without modifications. In order to obtain biodiesel from natural oils, a transesterification process is usually carried out, through which the properties of the original oil are transformed into others similar to those of diesel oil, for proper use as fuel. Below is a general outline of the production process of biodiesel from seeds of an oil species (vegetable oil) or from used vegetable oils and animal fats (Global agricultural information network, Oilseeds and production annual, 2017).

Pretreatment of seeds before extraction includes:

7.2.2.1 Cleaning

The seeds, when arriving at the industry, can contain undesired elements, such as earth, mud, stones, metal particles or others (ropes, rags, etc.). All these elements must be separated before the seed starts the process, since their presence can cause damage in the extraction process.

7.2.2.2 Drying

This task is important for the correct conservation of the seed during storage and for optimal operations to be carried out after extraction. In most cases, a maximum humidity of 8% is allowed (Eryilmaz et al., 2018).

7.2.2.3 Descaling

The seeds of small size, as well as flaxseed, rapeseed, sesame, etc., are handled without peeling, as it is a difficult and expensive operation. When working with larger seeds, it will be necessary to determine an optimum level of peeling, which will be the one in which the lowest loss of oil is valued, either in the portion of the film that adheres to the peel, or that absorbed in the peel, or the shell itself if it is not removed.

7.2.2.4 Grinding or milling

It has been shown that the extraction of oil from an oilseed, either by mechanical methods or by chemical methods, is carried out more quickly when the seed has been previously subjected to crushing or rolling. In the case of chemical extractions, grinding increases the contact surface between, for example, hexane and raw material, so it circulates much better through the bed of extraction.

7.2.2.5 Conditioning

The conditioning refers to the humidity index and the temperature that a seed must have to achieve greater efficiency during the extraction process. It has been determined that an oilseed that has a low moisture content (1–2%) releases oil with greater difficulty than when it has higher amounts of water (10%) (Eryilmaz et al., 2018).

On the other hand, the thermal treatments, in addition to regulating the humidity of the grains, increase their plasticity, increase the fluidity of the oil, coagulate the protein fractions, inactivate the enzymes and destroy the pathogenic microorganisms.

After, the extraction process can be done mechanically, by chemical solvents or by combining both.

7.2.2.6 Mechanical extraction

Mechanical extraction is usually done through an ejector press also called a screw or extruder. This press is a continuous mechanical extractor, where the oil is extracted from the raw material in a single step, with a high pressure.

Generally, a screw press consists of a continuous helical thread which is rotated concentrically inside a perforated static cylinder (referred to as a barrel or cage). The material is transported along the length of the barrel, and thanks to the increase in pressure, the oil is expelled and drained through small spaces or grooves that exist in the barrel.

The compressed material, or protein cake, is discharged at one end of the barrel and the oil and some solids are collected at the base of the press for further clarification, which will be done either by filtering or by decanting. The yields obtained from plant seeds with this type of extraction can be greater than 30% (Keneni and Marchetti, 2017).

7.2.2.7 Chemical extraction

The extraction of the oil is carried out with the help of organic solvents: Hexane, propane, supercritical fluids, etc. They have the ability to dissolve fatty substances. This type of extraction is carried out at temperatures of about 50 ºC with an 18:1 ratio of solvent to the amount of seed. The final yields of chemical extraction are of the order of 40% by weight (Eryilmaz et al., 2018).

7.2.2.8 Mixed extraction

It is obtained by combining the previous methods. First, a partial mechanical extraction is carried out leaving an oil residue of just over 20% by weight in the matrix, and then a chemical extraction is carried out, which increases the yield to almost 42% by weight.

The by-products of this phase of the production chain are the "expeller" in the case of mechanical extraction, the flours in the case of chemical extraction and the de-oiled flour in the case of mixed extraction. Due to the high contents of proteins, carbohydrates and fibers, these by-products are used, above all, in the feeding of animals, although the protein part of the waste can be valued for the production of thermal energy by direct combustion, with a lower calorific power of almost 21 MJ/kg (Eryilmaz et al., 2018).

The next stages are:

7.2.2.8 Refining

This phase is carried out to improve the quality of crude (vegetable) oils before subjecting them to the transesterification process to obtain biodiesel. First, a centrifuge, wherein the oil is separated from the solid waste, is made. Afterwards, they are treated with water or acid, eliminating the pigments, the waxes and the phosphates. Finally, a physical neutralization is carried out through high temperatures (240–260 ºC) and low pressure. It can also be carried out by chemical processes, for example, saponification with sodium hydroxide. In this way, the acidity of the oils is reduced (Evans, 2007).

7.2.2.9 Saponification with sodium hydroxide

This technique is used when the acidity of the used oil is less than or equal to 10%. The process requires temperatures between 60 and 80 ºC and later a centrifugation is performed that will separate the soaps. Finally, the degumming phase is carried out through water or acid.

7.2.2.10 Esterification of fatty acids

If we have acid values between 10% and 15%, the free fatty acids are converted to methyl ester by means of a reaction that is carried out at high temperatures (250–260 ºC), high pressures (0.5–0.6 MPa) and with acid catalysts (sulfuric or phosphoric acid) (Evans, 2007).

7.2.2.11 Transesterification

Although biodiesel could be obtained through processes such as pyrolysis, micro emulsions or esterification, the commercially used method is transesterification or alcoholics. Transesterification is classified into two trends: Catalytic and non-catalytic transesterification, which, in turn, consist of different variants.

The most used variant is the catalytic reaction in basic medium, valued for its high yields and for having moderate operating conditions of pressure and temperature. Therefore, and in general terms, the transesterification process can be described as the chemical reaction in which the contained glycerol in the oils is replaced by an alcohol in the presence of a catalyst.

The most commonly used alcohol in the production of biodiesel is methanol, although ethanol, isopropyl alcohol and butanol are also used. The choice of one type or another will depend on factors such as its water content, its price, the amount of alcohol needed for the reaction, the possibility of recovery and recycling, its pricing regime and environmental issues related to emissions to the atmosphere (Evans, 2007).

Catalysts can be of several types:

- Homogeneous acids: H_2SO_4, HCl, H_3PO_4, RSO_3
- Heterogeneous acids: Zeolites, Sulphonic Resins, SO_4/ZrO_2, WO_3/ZrO_2
- Heterogeneous basics: MgO, CaO, Na/NaOH/Al_2O_3
- Homogeneous basics: KOH, NaOH
- Enzymatic: Lipases (Candida, Penicillium, Pseudomonas)

Figure 1 illustrates the chemical reactions associated with transesterification. They consist of three reversible reactions, in which a triglyceride is consecutively converted into diglyceride, monoglyceride and glycerin, releasing in each of them one mole of methyl ester. Stoichiometrically, three moles of alcohol are required for each mole of triglyceride, in practice, however, an excess of alcohol is used to shift the balance towards greater ester formation.

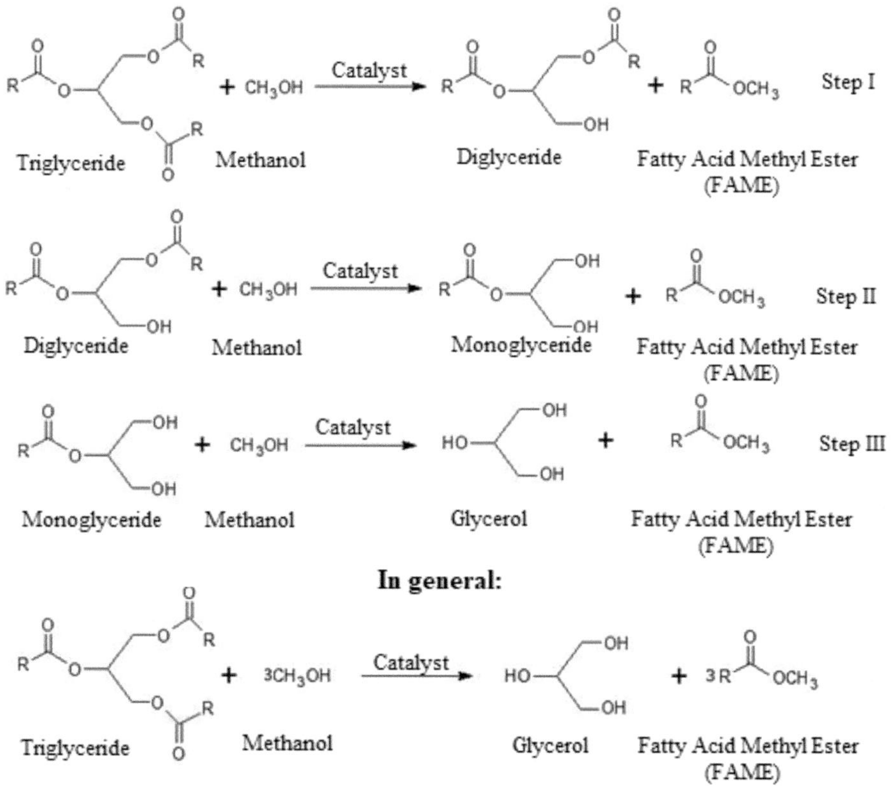

Figure 1. Transesterification reactions.

During this process, care must be taken since secondary reactions, such as saponification and neutralization of free fatty acids, and they would reduce the performance of the process. For this, glycerides and anhydrous alcohols must be counted (evaporating the excessive moisture from the oils before including them in the process) and with low proportions of free fatty acids.

7.2.3 Green diesel

Vegetable oil refining is a process to transform vegetable oil into biofuel by hydrocracking or hydrogenation. Hydrocracking breaks big molecules into smaller ones using hydrogen while hydrogenation adds hydrogen to molecules. These methods can be used for production of gasoline, diesel, propane, and another chemical feedstock. Diesel fuel produced from these sources is known as green diesel or renewable diesel.

7.2.3.1 Feedstock

Most of the plant and animal oils are vegetable oils, which are triglycerides, i.e., suitable for refining. Refinery feedstock includes canola, algae, jatropha, *salicornia*, palm oil and tallow. One type of algae, *Botryococcus braunii*, produces a different type of oil, known as a triterpene, which is transformed into alkanes by a different process.

7.2.3.2 Comparison to biodiesel

Based on its feedstock, green diesel could be classified as biodiesel; however, based on the processing technology and chemical formula, green diesel and biodiesel are different products. The chemical reaction commonly used to produce biodiesel is known as transesterification. Vegetable oil and alcohol are reacted, producing esters, or biodiesel and the coproduct, glycerol. When refining vegetable oil, no glycerol is produced, only fuels.

7.2.4 Straight vegetable oils

Straight unmodified edible vegetable oil is generally not used as fuel, but lower-quality oil has been used for this purpose. Used vegetable oil is increasingly being processed into biodiesel, or (more rarely) cleaned of water and particulates and then used as a fuel.

Vegetable oil can also be used in many older diesel engines that do not use common rail, unit injection or electronic diesel injection systems. Due to the design of the combustion chambers in indirect injection engines, these are the best engines for use with vegetable oil. This system allows the relatively larger oil molecules more time to burn. Some older engines, especially Mercedes, are driven experimentally by enthusiasts without any conversion, a handful of drivers have experienced limited success with earlier pre-"Pumpe Duse" engines and other similar engines with direct injection. Several companies, such as Elsbett or Wolf, have developed professional conversion kits and successfully installed hundreds of them over the past few decades (Kerry and Korpelshoek, 2007).

Oils and fats can be hydrogenated to give a diesel substitute. The resulting product is a straight-chain hydrocarbon with a high cetane number, low in aromatics and sulfur and does not contain oxygen. Hydrogenated oils can be blended with diesel in all proportions. They have several advantages over biodiesel, including good performance at low temperatures, no storage stability problems and no susceptibility to microbial attack (Kerry and Korpelshoek, 2007).

7.2.5 Bioethers

Bioethers (also referred to as fuel ethers or oxygenated fuels) are cost-effective compounds that act as octane rating enhancers. "Bioethers are produced by the reaction of reactive iso-olefins, such as iso-butylene, with bioethanol." Bioethers are created by wheat or sugar beet. They also enhance engine performance, while significantly reducing engine wear and toxic exhaust emissions. Although bioethers are likely to replace petroethers in the United Kingdom, it is highly unlikely they will become a fuel in and of itself due to the low energy density. Greatly reducing the amount of ground-level ozone emissions, they contribute to air quality (Kumar et al., 2014).

When it comes to transportation fuel, there are six ether additives: Dimethyl ether, diethyl ether, methyl tert-butyl ether, ethyl tert-butyl ether, tert-amyl methyl ether, and tert-amyl ethyl ether (Thomas et al., 2009).

The European Fuel Oxygenates Association identifies methyl tert-butyl ether and ethyl tert-butyl ether as the most commonly used ethers in fuel to replace lead. Ethers were introduced in Europe in the 1970s to replace the highly toxic compound. Although Europeans still use bioether additives, the USA no longer has an oxygenate requirement, therefore, bioethers are no longer used as the main fuel additive (Wessof, 2017).

7.2.6 Aviation biofuels

Aviation biofuel is a biofuel used for aircraft. It is considered by some to be the primary means by which the aviation industry can reduce its carbon footprint. After a multi-year technical review from aircraft makers, engine manufacturers and oil companies, biofuels were approved for commercial use in July 2011. Since then, some airlines have experimented with using biofuels on commercial flights (Wessof, 2017).

Currently, five aviation biofuel production pathways are approved for blending with fossil jet kerosene. However, only one—hydroprocessed esters and fatty acids synthetic paraffinic kerosene fuel —is currently technically mature and commercialized. Therefore, hydroprocessed esters and fatty acids synthetic paraffinic kerosene is anticipated to be the principal aviation biofuel used over the short to medium term.

Meeting 2% of annual jet fuel demand from international aviation with SAF could deliver the necessary cost reduction for a self-sustaining aviation biofuel market thereafter. Meeting such a level of demand requires increased HEFA-SPK production capacity. If met entirely by new facilities, approximately 20 refineries would be required. This could entail investment in the region of $10 billion. Although significant, this is relatively small compared to fossil fuel refinery investment of $60 billion in 2017 alone (Le Feuvre, 2019).

The focus of the industry has now turned to second generation sustainable biofuels (sustainable aviation fuels) that do not compete with food supplies nor are major consumers of prime agricultural land or fresh water. NASA has determined that 50% aviation biofuel mixture can cut air pollution caused by air traffic by 50–70% (Wessof, 2017).

Trials of using algae as biofuel were carried out by Lufthansa and Virgin Atlantic as early as 2008, although there is little evidence that using algae is a reasonable source for jet biofuels. By 2015, cultivation of fatty acid methyl esters and alkenones from the algae, Isochrysis, was under research as a possible jet biofuel feedstock (Wessof, 2017).

In 2017, there was little progress in producing jet fuel from algae, with a forecast that only 3 to 5% of fuel needs could be provided from algae by 2050. Further, algae companies that formed in the early 21st century as a base for an algae biofuel industry have either closed or changed their business development toward other commodities, such as cosmetics, animal feed, or specialized oil products (Wessof, 2017).

7.3 Solid biomass fuels

Solid biofuel is defined as any product derived from solid biomass that is likely to be used directly in energy conversion processes.

The most important are the primary type, made up of lignocellulosic materials from the agricultural or forestry sector and the processing industries that produce waste of that nature. The straw, the remains of vine pruning, olive and fruit trees, firewood, barks and the remains of pruning, clearings or clearings of the forest masses, are typical materials used to produce solid biofuels (Mohan et al., 2006).

Moreover, the husks of nuts and olive pits, the pomace obtained from the extraction of the oil in the mills and the remains of the industries of cork, wood and furniture, constitute a good raw material for the manufacture of solid biofuels. Another group of solid fuels, which require a transformation in order to be used as fuel, and which will be treated in this matter, are chips, charcoal, pellets and briquettes (Angin and Sensoz, 2014).

7.3.1.1 Chips

Chips are an organic material usually made up of wood and bark, which is obtained by fractionating the biomass of forest origin. Its use for energy purposes can be through direct combustion or can be used as raw material to manufacture other biofuels (pellets and briquettes). They can come from four types of sources:

- Forest and agricultural residues
- Energy woody crops
- First and second transformation forest industries
- Waste from agro-food industries

The physical-chemical properties are:

7.3.1.1.1 Flammability variables

These variables refer to the parameters that most influence the combustion process, such as the coefficient of thermal conductivity, the flammability, the times and temperatures at which combustion starts, etc.

7.3.1.1.2 Heat power

This feature is closely linked to the nature of the splinter and the place where the combustion will take place. It will depend, therefore, on the calorific power of the raw material, which varies depending on the chemical composition. The higher the calorific value, the higher the calorific power.

7.3.1.1.3 Chipping process

In general, the chipping process consists of three phases:

- Phase in the field: Chipping takes place *in situ*, feeding the chipper of logs and/or pruning remains.
- Transport phase: Transport of the chip to the consumer center or to the storage and/or chipping plant. The transport is carried out in containers of different sizes.
- Plant phase: Splintering and sorting of the chips in the plant is carried out.

7.3.1.2 Charcoal

Charcoal is a solid fuel that has a very high carbon content so its calorific value is much higher than that of wood (can reach up to 35 MJ/kg). It is produced thanks to incomplete combustion (up to temperatures of 400 to 700 ºC), wood and other plant residues and is difficult to alter, as well as not being affected by fungi and xylophagous insects. Charcoal is a source of renewable energy, which increases its value as a fuel (Garcia-Maraver, 2013).

There are different types of charcoal and they are especially appreciated for domestic use (mainly in barbecues). Its use is widespread in countries in process of industrialization. They are usually bagged and tend to be from different sources (oak charcoal, scorpionfish charcoal or high-performance coconut fiber charcoal).

The physical properties are:

7.3.1.2.1 Shape, size and appearance

The charcoal has very varied forms, depending on whether the wood of origin is chopped or not. Its color is bright black, due to the woody pyro liquids that are deposited on the surface of the wood when the charcoal is made. The size of this solid biofuel is like that of the original firewood, although it can be fractioned to make transportation more comfortable. The wood that is used as raw material usually has a diameter of between 5 and 50 cm, cutting it if it is bigger. Previous packaging, a homogenization of the coal through sieves, is carried out (Garcia-Maraver, 2013).

7.3.1.2.2 Density and specific surface

It is conditioned by the matter of origin, but the average values of apparent density range between 0.17 kg/dm^3 and 0.5 kg/dm^3. In general, hardwoods have higher densities than coniferous woods. In addition, we can obtain higher densities of carbon if the raw material is subjected to pressure during carbonization (Kumar et al., 2008).

The chemical properties are:

7.3.1.2.3 Chemical composition

The main composition of the charcoal will depend on the raw material used and of the method and temperature of carbonization, but in general, it is formed by carbon, hydrogen, oxygen and traces of nitrogen. The higher the carbon content of the original material, the higher the calorific value of the charcoal, which will be much greater than the rest of the solid biofuels. In some cases, the ash content limits the use for industrial purposes.

7.3.1.2.4 Heating power

As already mentioned in previous sections, the calorific value of charcoal is higher than that of wood, wood and bark, chips, pellets and briquettes, and can reach up to 35 MJ/kg. This parameter is strongly linked to the temperature used in charcoal, since when it is higher, the percentage of carbon, and therefore, the calorific value will be higher, but the yield decreases (i.e., the amount of carbon that is formed per kg from the original firewood) (Kumar et al., 2008).

7.3.1.2.5 Charring process

Pyrolysis or carbonization is the process of heating organic materials present in the biomass in the absence of air. Generally, the term pyrolysis is used when said process focuses on obtaining gases and oils, and carbonization, when the process is aimed at obtaining solid products.

As is the case with charcoal, although strictly speaking, carbonization is a type of extreme and anhydrous pyrolysis of solid materials.

Through the carbonization process, solid, liquid and gaseous products are obtained (depending on the temperature reached), which can be used directly as fuel or can be processed to obtain chemicals and other types of fuels.

Most pyrolysis reactions occur below about 800 K (527 °C). The gaseous fraction, called pyrogenic vapor, is composed mainly of CO_2 and CO, and there are also small proportions of CH_4 and H_2. The proportion of products obtained (carbonized, tar, water and gas) depend on the source material, although, in general terms, it can be stated that if the raw material has a high lignin content, higher yields are obtained (Kumar et al., 2008).

The general process of carbonization of biomass consists of three phases:

- Dehydration of the raw material: The wood is dried with a temperature close to 100 ºC until achieving a moisture content equal to 0. Subsequently, the temperature continues to rise until reaching 280 ºC. The energy of this stage comes from the partial combustion of the wood, so it is an endothermic reaction, which absorbs energy.
- Fragmentation of the wood: When the wood is dry and acquires temperatures close to 300 ºC it begins to split spontaneously, producing coal and water vapor, acetic acid and more complex chemical compounds (in the form of tars and non-condensable gases such as H_2, CO and CO_2). Since

the supply of air in the system is necessary to partially burn the wood, there will also be nitrogen in these gases. This reaction is exothermic and continues until only the carbonized residue remains charcoal (Comins et al., 2016).

- Removal and decompression of tars: The process, spontaneously, stops when the temperature reaches about 400 °C, but this coal still contains too many tar residues along with the ashes of the original wood. Therefore, a heating is carried out until reaching temperatures of about 600 ºC, which increases the fixed carbon proportions, eliminating and further decomposing the tars. With these temperatures a percentage of fixed carbon close to 90% and 10% of volatile materials is reached, with a yield of carbon over dry mass exceeding 30% (Comins et al., 2016).

7.3.1.3 Pellets and briquettes

The final objective of the processes of pelletizing and briquetting consists of obtaining a final product of greater density than the initial products, hence the name of densification. The products obtained, having higher density, will be transported occupying a smaller volume (equal weight) than firewood and chips, and will, therefore, be easier to handle.

Pellets and briquettes are used as fuel, having as an advantage that they can be dosed by automatic systems, so there is a possibility to use them in large facilities and in the industrial sector. In addition to the residential sector replacing firewood.

The physical properties are:

7.3.1.3.1 Shape and size and appearance (color and brightness)

The pellets can have different forms, although their commercialization is almost always in cylindrical form. Their most usual diameters oscillate between 0.7 and 2 cm and their length of 1 to 7 cm. Their appearance depends on the raw material with which they have been manufactured, ranging from light brown to blackish tones, which are obtained from different wastes related to wood, or green, made with alfalfa. There are European standards that classify the use of pellets depending on their composition, see (Alakangas, 2011).

Briquettes can be manufactured from many types of compacted biomass: Forest biomass, from waste wood factories (door, furniture, sawmill, etc., factories), industrial waste biomass, residual biomass urban, charcoal, etc.

Most briquettes are cylindrical, although we can also find quadrangular, rectangular, hexagonal, octagonal, etc. Likewise, they can be solid or hollow. Their diameters exceed 5 cm, the most common being between 7.5 cm and 9 cm. The length is variable, ranging between 50 and 80 cm and is proportional to the diameter (Garcia-Maraver, 2013).

The color of the briquettes, like that of the pellets, will vary depending on their composition. In addition, if the briquette is coated with paraffin on the outside, it will present a lighter color (yellowish-white). The most demanded briquette aspect is one that looks like wood.

7.3.1.3.2 Density

The density of the pellets will depend on the used raw material. For example, if the raw material comes from hardwood tree species, in general it is denser than that from conifers. This parameter also depends on the pressure that has been used during the manufacturing, design and handling process.

Two density values should be considered:

The real density and the apparent density.

- The real density is the quotient between the real mass and the actual volume of the pellets.
- The approximate density of the briquettes is calculated by means of laboratory tests in which their mass (on a scale) and their volume (by geometric calculations) are measured. If a more accurate estimate is needed, the so-called "water displacement" method can also be used, which consists in weighing a certain amount of the briquettes and stuffing them in silicone.

7.3.1.3.3 Humidity

The humidity of the pellet will depend on the way in which the product has been supplied. In the manufacture of pellets, it is usually based on dry particles with humidity below 12%, which are dried even more in the pelletizing process (link to the process).

In the manufacture of briquettes, dry particles which have a percentage of humidity less than 12% are usually used, although the manufacturing process can reduce this value to 8–10%, reaching the recommended humidity. If the biomass that forms the briquette has not been dried properly, the water is trapped in its insides and during the combustion, it tends to break the briquette when leaving in the form of water vapor (Garcia-Maraver, 2013).

During the pelletizing and briquetting processes, a warming on the outer side surface causes a bakelisation, which causes a thin plastic film of blackish color that prevents the entry of water inside the product and, therefore, avoids the increasing of humidity.

In the case of briquettes, their main properties are (Comins et al., 2016):

- Humidity < 10%
- PCI > 16.9 MJ/kg (4.7 kWh/kg)
- Density around 1,000 kg/m^3
- Ash content < 0.7%

There are European standards that classify the use of briquettes depending on their composition, see (Alakangas, 2011).

7.3.1.3.4 Densification process

Within the processes of physical transformation of products (in the great majority of the cases of lignocellulosic nature), the compaction of them to obtain densified combustible products usable as energy substitutes stands out.

The compaction of the biomass is achieved by applying a high pressure (above 100 bar) with rollers on a perforated matrix, through which the material is passed.

In general, it is necessary to have previously conditioned the waste through some of the stages of physical transformation (chipping, drying, grinding) so the final cost of the product obtained is higher than the rest of the waste transformed into fuels (chips, sawdust, chips, etc.). The three most important activities that have to be carried out to obtain briquettes and pellets are:

- Preparation of the raw material in appropriate dimensions
- Drying of the raw material
- Briquetting or pelletizing of the raw material

The main stages of transformation of the residual biomass that have to be carried out to obtain pellets and briquettes are chipping, drying, milling, sieving and densification (pelletizing and briquetting).

7.3.2 Chipping

This set of operations is essential when using waste material (forest lignocellulosic), both in direct applications and for the manufacture of densified elements. The first granulometric reduction is achieved with chipping, which allows to obtain chips (chips) with a maximum particle size that makes it possible to handle, store, load and transport waste (Garcia-Olivares, 2015).

In general, you can talk about 3 types of chipping equipment:

- Mobile chippers: Driven by forestry tractors. They can reach places of difficult access, but yields are usually low.
- Self-propelled chippers: They have their own traction system. They can move more quickly.
- Fixed or semi-fixed chippers: They are used in places where the size of the operation and the high volumes handled make this type of facilities amortizable.

7.3.3 Drying

Normally, the residual biomass has a high moisture content (above 100% on a dry basis), which requires prior conditioning for later use for energy purposes. We have two types of drying (Garcia-Olivares, 2015):

- Natural drying: Based on taking advantage of environmental conditions to facilitate.
- Dehydration of waste.
- Forced drying: Done when the humidity obtained with natural drying is not adequate for the processing of the material, or the necessary conditions for its realization are not available. In this case, the most commonly used equipment is classified into direct dryers (heat transfer through direct contact between wet material and hot gases) and indirect dryers (heat transfer carried out through a retaining wall).

7.3.4 Grinding and sieving

Subsequently, it is necessary to perform a grinding of the material to achieve a greater homogeneity and smaller particle size. Generally, hammer mills are usually used for densification applications, mainly due to their lower maintenance, compared to knife equipment. At this point, undesirable materials such as metal elements, stones, sand, large pieces, etc., are eliminated, as they can cause serious problems in grinding facilities. Large-gauge pieces and stones are removed at the beginning of the flow diagram by vibrating sieves or slanted screens.

Pelletizing and briquetting consist of:

7.3.5 Pelletizing

The pelletizing process consists of the compaction of lignocellulosic material under certain conditions (granulometry less than 2 cm and humidity below 12% on dry basis) to obtain cylinders with a diameter between 7 and 25 mm. The length of the pellet, as already indicated, is variable, although the most abundant oscillate between 3.5 and 6.5 cm (Garcia-Olivares, 2015).

For its production, it is necessary to have a base product with a humidity between 8–15% (preferably $< 12\%$), and a particle size of the order of 5 mm. The compaction of the biomass should be 1 to 6 times its volume and the extrusion holes of the matrix should have a cylindrical shape, with diameters between 0.5 and 2 cm and 1 to 3 cm in length. The density of the pellets is usually between 1,000–1,200 kg/m^3, although, when it is distributed in bulk, the bulk density is usually of the order of 800 kg/m^3 (Garcia-Olivares, 2015).

7.3.6 Briquetting

As previously mentioned, the briquettes are formed by pressed materials that are used as fuel. When lignocellulosic elements, such as wood, are used for its manufacture, the temperatures released during the pressing phase produce a softening of the lignin, so that after cooling it functions as a binder of the particles. The manufacture of briquettes is very similar to that of pellets, although the resulting product has larger dimensions: Diameter > 5 cm and length between 50 and 80 cm (Garcia-Olivares, 2015).

7.3.7 Use of additives

In general, no additives are used in the process of obtaining pellets and briquettes.

Although the additives can favor the process providing greater cohesion to the final product, thus improving its resistance to knocking, their use makes the process more expensive and can result in contaminants. This last aspect is problematic if the pellets are used in domestic chimneys where combustion must be clean.

Therefore, in the pelletizing process, chemical products are not usually used but simply pressure and steam, although it is also possible to find a reduced percentage of biological additives. In the case of briquettes, the agglomerating action provided by lignin makes it possible not to add any type of additive (resins or waxes).

References

Abascal Fernandez, R. 2017. Study of the bioethanol production from different types of lignocellulosic biomass. Matrix of reactions and optimization. Thesis. Escuela Politécnica de Ingeniería de Minas y Energía. Universidad de Cantabria-España.

Acién Fernández, F.G., González-López, C.V., Fernández Sevilla, J.M. and Molina Grima, E. 2012. Conversion of CO_2 into biomass by microalgae: How realistic a contribution may it be to significant CO_2 removal. Applied Microbiology and Biotechnology 96: 577–586.

Adey, W.H., Kangas, P.C. and Mulbry, W. 2011. Algal turf scrubbing: Cleaning surface waters with solar energy while producing a biofuel. BioScience 61: 434–441.

Advances in Biochemical Engineering/Biotechnology, ISSN 0724-6145. 2003. Springer.

Advances in Biochemical Engineering/Biotechnology. Biomethanation II. Vol. 82. Ed. Springer, 2005, Netherlands.

Alakangas, E. 2011. European standards for fuel specification and classes of solid biofuels. Solid Biofuels for Energy—A Lower Greenhouse Gas Alternative 28. 10.1007/978-1-84996-393-0_2.

Anderson, G. 2004. Seawater composition. On line: <http://www.marinebio.net/marinescience/02ocean/swcomposition.htm>.

Angin, D. and Şensöz, S. 2014. Effect of pyrolysis temperature on chemical and surface properties of biochar of rapeseed (*Brassica napus* L.). International Journal of Phytoremediation 16. 10.1080/15226514.2013.856842.

Aro, E. 2016. From first generation biofuels to advanced solar biofuels. Ambio 1: 24–31.

Arumugam, M., Agarwal, A., Arya, M.C. and Ahmed, Z. 2013. Influence of nitrogen sources on biomass productivity of microalgae Scenedesmus bijugatus. Bioresource Technology 131: 246–249.

Atabani, A.E., Silitonga, A.S., Badruddin, I.A., Mahlia, T.M.I., Masjuki, H.H. and Mekhilef, S. 2012. A comprehensive review on biodiesel as an alternative energy resource and its characteristics. Renewable and Sustainable Energy Reviews 16: 2070–2093.

Balan, V. 2014. Current challenges in commercially producing biofuels from lignocellulosic biomass. ISRN Biotechnol. 463074.

Banat, F., Pal, P., Jwaied, N. and Al-Rabadi, A. 2013. Extraction of olive oil from olive cake using soxhlet apparatus. American Journal of Oil and Chemical Technologies 18: 2326–6570.

Bannon, J. and Adey, W. 2008. Algal turf scrubbers: Cleaning water while capturing solar energy for biofuel production. Proceedings of the Fourth Environmental Physics Conference 10: 19–23.

Benemann, J., Ian, W. and Tryg, L. 2012. Life cycle assessment for microalgae oil production. Disruptive Science and Technology 1: 68–78.

Berla, B.M., Saha, R., Immethun, C.M., Maranas, C.D., Moon, T.S. and Pakrasi, H.B. 2013. Synthetic biology of cyanobacteria: unique challenges and opportunities. Frontiers in Microbiology 4: 246.

Biddy, M., Davis, R., Jones, S. and Zhu, Y. 2016. Whole Algae Hydrothermal Liquefaction Technology Pathway. National Renewable Energy Laboratory.

Biogas from Waste and Renewable Resources. An introduction. pp. 57–78. *In*: Deublein, D. and A. Steinhauser (eds.). 2011. Wiley-VCH Verlag GmbH & Co. KGaA. 10.1002/9783527632794.

Blankenship, R.E. et al. 2011. Comparing photosynthetic and photovoltaic efficiencies and recognizing the potential for improvement. Science (New York, N.Y.) 332: 805–9.

Börjesson, P., Lundgren, J., Ahlgren, S. and Nyström, I. 2013. Dagens Och Framtidens Hållbara Biodrivmedel.

Bullis, K. 2007. Algae-based fuels set to bloom-oil from microorganisms could help ease the nation's energy woes. MIT Technology Review. On line: www.technologyreview.com/s/407268/algae-based-fuels-set-to-bloom/.

Carlsson, A.S., van Beilen, J.B., Miller, R. and Clayton, D. 2007. Micro and macro-algae: Utility for industrial applications. Bowles, D. (ed.). Outputs from the EPOBIO: Realizing the Economic Potential of Sustainable Resources—Bioproducts from Non-food Crops Project. UK: CNAP. University of York.

Carriquiry, M.A., Du, X. and Timilsina, G.R. 2011. Second generation biofuels: Economics and policies. Energy Policy 39: 4222–4234.

Chisti, Y. 2007. Biodiesel from microalgae. Biotechnology Advances 25: 294–306.

Chong, A.M.Y., Wong, Y.S. and Tam, N.F.Y. 2000. Performance of different microalgal species in removing nickel and zinc from industrial wastewater. Chemosphere 41: 251–7.

Colombo, C.A., Díaz, B.G., Chorfi Berton, L.H. and Ferrari, R. 2017. Macauba: A promising tropical palm for the production of vegetable oil. OCL 25. 10.1051/ocl/2017038.

Cornell, C.B. 2008. First algae biodiesel plant goes. On line: <gas2.org/2008/03/29/first-algae-biodiesel-plant-goes-online-april-1-2008>.

Cosmin, S. et al. 2016. Testing model for assessment of lignocellulose-based pellets. Wood Research 61: 331–340.
Darzins, A., Pienkos, P. and Edye, L. 2010. Current status and potential for algal biofuels production. A Report to IEA Bioenergy Task 39.
Demirbas, A. 2011. Biodiesel from oilgae, biofixation of carbon dioxide by microalgae: A solution to pollution problems. Applied Energy 88: 3541–3547.
Demirbas, A.H. 2009. Inexpensive oil and fats feedstocks for production of biodiesel. Energy Education Science and Technology Part A: Energy Science and Research 23: 1–13.
Dinh, L.T.T., Guo, Y. and Mannan, M.S. 2009. Sustainability evaluation of biodiesel production using multicriteria decision-making. Environmental Progress & Sustainable Energy 28: 38–46.
Dixner, C. 2013. Application of algal turf scrubber technique to remove nutrient from a eutrophic reservoir in the Jiulong River watershed, Southeast China. International Summer Water Resources Research School. Lund University. Online: <http://www.tvrl.lth.se/fileadmin/tvrl/files/vvrf05/CharlottaD_Application_of_Algal_Turf_Scrubber_Technique.pdf>.
Dosta, J., Galí, A., Macé, S. and Mata-Álvarez, J. 2007. Modelling a sequencing batch reactor to treat the supernatant from anaerobic digestion of the organic fraction of municipal solid waste. Journal of Chemical Technology & Biotechnology 82: 158–64.
Eryilmaz, T., Aksoy, F., Aksoy, L., Bayrakceken, H., Aysal, F.-E., Sahin, S. and Yesilyurt, M.K. 2018. Process optimization for biodiesel production from neutralized waste cooking oil and the effect of this biodiesel on engine performance. Ciencia, Tecnologia y Futuro 8: 121–127.
Evans, G. 2007. International biofuels strategy project. The national non-food crops centre. Technology Status Report, NNFCC <http://www.globalbioenergy.org/uploads/media/0711_NNFCC___Liquid_Transport_Biofuels_Technology_Status_Report.pdf >.
Garcia-Maraver, A. 2013. Optimization of the palletization process of agricultural waste originating from olive farms for their application in domestic boilers. Universidad de Granada. ISBN 978-84-9028-596-1.
García-Olivares, A. 2015. Substitutability of electricity and renewable materials for fossil fuels in a post-carbon economy. Energies 8: 13308–13343.
Gerber, N., Eckert, M. and Breuer, T. 2008. The impacts of biofuel production on food prices: A review. ZEF—Discussion Papers on Development Policy Bonn. Ed. Zentrum für Entwicklungsforschung (ZEF), Germany.
Global agricultural information network Oilseeds and Products Annual. 2017. < https://www.fas.usda.gov/databases/global-agricultural-information-network-gain >.
Gomis, V., Pedraza, R., Francés, O., Font, A. and Asensi, J. 2007. Dehydration of ethanol using azeotropic distillation with isooctane. Industrial & Engineering Chemistry Research—Ind. Eng. Chem. Res. 46. 10.1021/ie0616343.
Greenwell, H.C., Laurens, L.M.L., Shields, R.J., Lovitt, R.W. and Flynn, K.J. 2009. Placing microalgae on the biofuel's priority list: A review of the technological challenges. Journal of the Royal Society Interface 7: 703–726.
Halfmann, C., Gu, L., Gibbons, W. and Zhou, R. 2014. Genetically engineering cyanobacteria to convert CO_2, water, and light into the long-chain hydrocarbon farnesene. Applied Microbiology and Biotechnology 98: 9869–9877.
Harder, R. and von Witsch, H. 1942. Bericht über versuche zur fettsynthese mittels autotropher. Microorganismen Forschungsdienst Sonderheft 16: 270–275.
Harder, R. and von Witsch, H. 1942. Die Massenkultur von Diatomeen. Berichte der Deutschen Botanischen Gesellschaft 60: 146–152.
Hays, S.G. and Ducat, D.C. 2015. Engineering cyanobacteria as photosynthetic feedstock factories. Photosynthesis Research 123: 285–295.
Hemaiswarya, S., Raja, R., Carvalho, I.S., Ravikumar, R., Zambare, V. and Barh, D. 2012. An Indian scenario on renewable and sustainable energy sources with emphasis on algae. Applied Microbiology and Biotechnology 96: 1125–1135.
Herro, A. 2008. Better than corn? Algae set to beat out other biofuel feedstocks 21. Worldwatch Institute. Online < www.researchgate.net/publication/296236728_Better_than_corn_Algae_set_to_beat_out_other_biofuel_feedstocks>.
Hodsman, L., Smallwood, M. and Williams, D. 2005. European Parliament's committee on Agriculture and Rural Development. The promotion of non-food crops. Policy Department Structural and Cohesion Policies. The National Non-Food Crops Centre, United Kingdom.
Horn, W. and Krupp, F. 2006. Earth: The Sequel: The Race to Reinvent Energy and Stop Global Warming 85: 20–56.
IEA Bioenergy. 2007. Summary and conclusions from the IEA bioenergy ExCo54 workshop. Black Liquor Gasification. Online: < https://www.ieabioenergy.com/wp-content/uploads/2013/10/Black-Liquor-Gasification-summary-and-conclusions1.pdf>.
Igoni Hilkia, A., Abowei, M.F.N., Ayotamuno, M.J. and Eze, C.L. 2009. Comparative evaluation of batch and continuous anaerobic digesters in biogas production from municipal solid waste using mathematical models. Agricultural Engineering International: CIGR Journal. ISSN 1682–1130.
Inganäs, O. and Sundström, V. 2016. Solar energy for electricity and fuels. Ambio 45: 15–23.
Institute for Environment and Sustainability (Joint Research Centre). 2011. Well-to-wheels analysis of future automotive fuels and powertrains in the European context WELL-to-WHEELS Report. Online: <publications.europa.eu/en/publication-detail/-/publication/708b39fd-4cc9-4456-9b63-6d1536240202/language-en>.
IRENA. 2017. Biofuels for aviation: Technology brief. International Renewable Energy Agency, Abu Dhabi.

Jewell, W., Cummings, R. and Richards, B. 1993. Methane fermentation of energy crops: Maximum conversion kinetics and *in situ* biogas purification. Biomass and Bioenergy 5: 261–278.

Jia, Y., Xu, M., Zhang, X., Hu, Q., Sommerfeld, M. and Chen, Y.S. 2010. Life-cycle analysis on biodiesel production from microalgae: Water footprint and nutrients balance. Bioresource Technology 10: 159–65.

Karatzos, S., McMillan, J.D. and Saddle, J.N. 2015. Summary of IEA Bioenergy Task 39 Report-the Potential and Challenges of Drop-In Biofuels, IEA Bioenergy.

Keneni, Y.G. and Marchetti, J.M. 2017. Oil extraction from plant seeds for biodiesel production. AIMS Energy 5: 316–340.

Kerry, R. and Korpelshoek, M. 2007. Bioethers impact on the gasoline pool. Digital Refining. Online: <www.digitalrefining.com/article/1000210,Bioethers_impact_on_the_gasoline_pool.html#.XXZ8QyhKi70>.

Kishore, V.V.N. 2008. Renewable Energy Engineering & Technology: A Knowledge Compendium TERI Press, New Delhi.

Kretschmer, B., Allen, B., Kieve, D. and Smith, C. 2013. Shifting away from conventional biofuels: Sustainable alternatives for the use of biomass in the UK transport sector. An IEEP discussion paper produced for ActionAid. Institute for European Environmental Policy (IEEP): London.

Kumar, A., Lijun, W., Yuris, A.D., David, J. and Milford, H. 2008. Thermogravimetric characterisation of corn stover as gasification and pyrolysis feedstock. Biomass & Bioenergy 32: 460–467. 10.1016/j.biombioe.2007.11.004.

Kumar, A., Ergas, S., Yuan, X., Sahu, A., Zhang, Q., Dewulf, J., Malcata, F.X. and Van Langenhove, H. 2010. Enhanced CO_2 fixation and biofuel production via microalgae: Recent developments and future directions. Trends in Biotechnology 28: 371–380.

Kumar, S.M., Bhaskar, T., Jain, A.K., Singal, S.K. and Garg, M.O. 2014. Bio-Ethers as Transportation Fuel: A Review. Indian Institute of Petroleum Dehradun.

Lamers, P., Junginger, M., Hamelinck, C. and Faaij, A. 2012. Developments in international solid biofuel trade—an analysis of volumes, policies, and market factors. Renewable and Sustainable Energy Reviews 16: 3176–3199.

Lane, J. 2017. Ethanol and biodiesel: dropping below the production cost of fossil fuels? Biofuels Digest. Online: <https://www.biofuelsdigest.com/bdigest/2017/05/18/ethanol-and-biodiesel-dropping-below-the-production-cost-of-fossil-fuels/ >.

Le Feuvre, P. 2019. Commentary: Are aviation biofuels ready for takeoff? IEA Energy Analyst. Online: https://www.iea.org/newsroom/news/2019/march/are-aviation-biofuels-ready-for-take-off.html.

Li, X. 2002. Biomass gasification in circulating fluidized bed (Ph.D. dissertation), University of British Columbia, Vancouver, Canada.

Lopes, M.L. et al. 2016. Ethanol production in Brazil: A bridge between science and industry. Brazilian Journal of Microbiology 47: 64–76.

Lundquist, T., Woertz, C., Quinn, N. and Benemann, J. 2010. A realistic technology and engineering assessment of algae biofuel production. Energy Biosciences Institute University of California, Berkeley, California.

Masarovičová, E., Kráľová, K. and Peško, M. 2009. Energetic plants—cost and benefit. Ecological Chemistry and Engineering 16: 263–276.

Mata, T.M., Martins, A.N.A. and Caetano, N.S. 2010. Microalgae for biodiesel production and other applications: A review. Renewable and Sustainable Energy Reviews 14: 217–232.

Moellering, E.R. and Benning, C. 2009. RNA interference silencing of a major lipid droplet protein affects lipid droplet size in *Chlamydomonas reinhardtii*. Eukaryotic Cell 9: 97–106.

Mohan, D., Pittman, Jr., C.U. and Philip, H.S. 2006. Pyrolysis of wood/biomass for bio-oil: A critical review. Energy Fuels 20: 848–889.

Mola-Yudego, B. and Aronsson, P. 2008. Yield models for commercial willow biomass plantations in Sweden. Biomass and Bioenergy 32: 829–837.

Milano, J., Ong, H.C., Masjuki, H.H. and Chong, W.T. 2015. Microalgae biofuels as an alternative to fossil fuel for power generation. Renewable and Sustainable Energy Reviews 58: 180–197.

Nduka, V. and Imoedemhe, G. 2015. Plant Design for Production of Fuel Grade Ethanol< www.academia.edu/11493222/PLANT_DESIGN_FOR_PRODUCTION_OF_FUEL_GRADE_ETHANOL>.

Oncel, S.S. 2013. Microalgae for a macroenergy world. Renewable and Sustainable Energy Reviews 26: 241–264.

Pittman, J.K., Dean, A.P. and Osundeko, O. 2011. The potential of sustainable algal biofuel production using wastewater resources. Bioresource Technology 102: 17–25.

Popp, J., Lakner, Z., Harangi-Rákos, M. and Fári, M. 2014. The effect of bioenergy expansion: Food, energy, and environment. Renewable and Sustainable Energy Reviews 32: 559–578.

Pulz, O. and Gross, W. 2004. Valuable products from biotechnology of microalgae. Applied Microbiology and Biotechnology 65: 635–648.

Rabaey, K. and Rozendal, R.A. 2010. Microbial electrosynthesis: Revisiting the electrical route for microbial production. Nature Reviews Microbiology 8: 706–716.

Refaat, A.A. 2010. Different techniques for the production of biodiesel from waste vegetable oil. Int. J. Environ. Sci. Tech. 7: 183–213.

Richards, B., Cummings, R., White, T. and Jewell, W. 1991. Methods for kinetic analysis of methane fermentation in high solids biomass digesters. Biomass and Bioenergy 1: 65–73.

Roubík, H., Mazancová, J., Banout, J. and Verner, V. 2016. Addressing problems at small-scale biogas plants: A case study from central Vietnam. Journal of Cleaner Production 112: 2784–2792.

Sabiha-Hanim, S. and Abd Halim, N.A. 2018. Sugarcane bagasse pretreatment methods for ethanol production. Fuel Ethanol Production from Sugarcane. DOI: 10.5772/intechopen.81656.

Salinas-Callejas, E. and Gasca Quezada, V. 2009. The biofuels. El Cotidiano 157: 75–82. Universidad Autónoma Metropolitana Unidad Azcapotzalco Distrito Federal, México.

Savakis, P. and Hellingwerf, K.J. 2015. Engineering cyanobacteria for direct biofuel production from CO_2. Current Opinion in Biotechnology 33: 8–14.

Scaife, M.A., Nguyen, G.T.D.T., Rico, J., Lambert, D., Helliwell, K.E. and Smith, A.G. 2015. Establishing Chlamydomonas reinhardtii as an industrial biotechnology host. The Plant Journal 82: 532–546.

Schenk, P.M., Thomas-Hall, S.R., Stephens, E., Marx, U.C., Mussgnug, J.H., Posten, C., Kruse, O. and Hankamer, B. 2008. Second generation biofuels: High-efficiency microalgae for biodiesel production. Bioenergy Research 1: 20–43.

Scott, S.A., Davey, M.P., Dennis, J.S., Horst, I., Howe, C.J., Lea-Smith, D.J. and Smith, A.G. 2010. Biodiesel from algae: Challenges and prospects. Current Opinion in Biotechnology 21: 277–286.

Shaw's Environmental and Infrastructure Group. Large Volume Ethanol Spills–Environmental Impacts and Response Options, 2011. < https://www.mass.gov/files/documents/2016/09/us/ethanol-spill-impacts-and-response-7-11_44776_56452.pdf>.

Sheehan, J., Dunahay, T., Benemann, J. and Roessler, P. 1998. A Look Back at the U.S. Department of Energy's Aquatic Species Program: Biodiesel from Algae. U.S. Department of Energy's Office of Fuels Development. Colorado, United States.

Sims, R.E.H., Mabee, W., Sanddler, J.N. and Taylos, M. 2012. An overview of second-generation biofuel technologies. Bioresource Technology 101: 1570–1580.

Singh, S., Kate, B.N. and Banerjee, U.C. 2005. Bioactive compounds from cyanobacteria and microalgae: An overview. Critical Reviews in Biotechnology 25: 73–95.

Smith, V.H., Sturm, B.S.M., Denoyelles, F.J. and Billings, S.A. 2010. The ecology of algal biodiesel production. Trends in Ecology & Evolution 25: 301–309.

Song, Y.C., Kwon, S.J. and Woo, J.H. 2004. Mesophilic and thermophilic temperature co-phase anaerobic digestion compared with single-stage mesophilic- and thermophilic digestion of sewage sludge. Water Res. 38: 1653–62.

Sporalore, P., Joannis-Cassan, C., Duran, E. and Isambert, A. 2006. Commercial applications of microalgae. Journal of Bioscience and Bioengineering 101: 87–96.

Thomas, F., McGowan, M.L., Brown, W.S., Bulpitt, J.L. and Walsh Jr., J.L. 2009. Biomass and Alternate Fuel Systems: An Engineering and Economic Guide. pp. 280. ISBN 978-0-470-41028-8. Wiley.

Tokuşoglu, O. and Uunal, M.K. 2003. Biomass nutrient profiles of three microalgae: spirulina platensis, chlorella vulgaris, and isochrisis galbana. Journal of Food Science 68: 1144–1148.

Torella, J.P., Gagliardi, C.J., Chen, J.S., Bediako, D.K., Colon, B., Way, J.C., Silver, P.A. and Nocera, D.G. 2015. Efficient solar-to-fuels production from a hybrid microbial-water-splitting catalyst system. Proceedings of the National Academy of Sciences USA 112: 2337–2342.

Valdés, A. and Foster, W. 2002. Reflections on the policy implications of agricultural price distortions and price transmission for producers in developing and transition economies. OECD Global Forum on Agriculture: Agricultural Trade Reform, Adjustment and Poverty, May 23–24, Paris, France.

Vonshak, A. 1997. Spirulina platensis (Arthrospira): Physiology, Cell-biology and Biotechnology. Taylor & Francis. London, England.

Wessof, E. 2017. Hard lessons from the great algae biofuel bubble. Greentech Media. Online: < www.greentechmedia.com/articles/read/lessons-from-the-great-algae-biofuel-bubble >.

Westin, P. and Forsbeng, M. 2014. Hållbara biodrivmedel och flytande biobränslen under. Online: < energimyndigheten.a-w2m.se/Test.ashx?ResourceId=3066>.

Wijffels, R.H., Kruse, O. and Hellingwerf, K.J. 2013. Potential of industrial biotechnology with cyanobacteria and eukaryotic microalgae. Current Opinion in Biotechnology 24: 405–413.

Yahyaoui, I. 2018. Advances in Renewable Energies and Power Technologies: Volume 2: Biomass, Fuel Cells, Geothermal Engines, and Smart Grids. *In*: Elseviers (eds.). Oxford, United Kingdom.

Yirdaw, E., Tigabu, M. and Monge, A. 2017. Rehabilitation of degraded dryland ecosystems—Review. Silva Fennica 51. https://doi.org/10.14214/sf.1673.

Zhang, Y., Dub, M.A., McLean, D.D. and Kates, M. 2003. Biodiesel production from waste cooking oil: 1. Process design and technological assessment. Bioresource Technology 89: 1–16.

Zivojnovich, M.J. 2010. Algae based water treatment systems—cost-effective nutrient pollution control and for point and nonpoint source applications. HydroMetia Water Treatment Technologies conference, Ocala, Fl, Feb. 16, 2010.

CHAPTER 2

From Sugarcane to Bioethanol

The Brazilian Experience

Daroda, RJ,[1,*] *Cunha, VS*[1] and *Brandi, HS*[2]

1. Introduction

1.1 Ethanol: from Henry Ford's prediction to Brazilian reality

The use of ethanol as an engine fuel became popular at the end of the 18th and beginning of the 19th century. In 1826 a prototype internal combustion engine was tested using a blend of alcohol and turpentine. The first engine running with alcohol was invented by Nicholas August Otto and is still known as Otto engine (Kitman, 2000). In 1896, Henry Ford projected the first "farm alcohol" fuel-powered car, the "Quadricycle" (Goettemoeller, 2007). At the same time, in an interview first published in the Christian Science Monitor, as a visionary he predicted: "The fuel of the future," he said, "is going to come from fruit like that sumac out by the road or from apples, weeds, sawdust—almost anything. There's fuel in every bit of vegetable matter that can be fermented. There's enough alcohol in one year's yield of an acre of potatoes to drive the machinery necessary to cultivate the field for a hundred years" (Kovarik, 1998). The 1908 Model T, the famous car produced by Ford Motors Company, was fueled by alcohol, gasoline or any combination of the two, and Ford's vision remains current to this day (Goettemoeller, 2007; Sukhdev, 2012).

The discovery of petroleum in Pennsylvania in 1859, and the development of derived products as a gasoline at lower price some years later, interrupted the use of ethanol as an engine fuel.

During the First and Second World War, many countries were forced to use ethanol due to oil shortages. Today, bioethanol is undoubtedly the main biofuel used in Otto engines.

Implementing an integrated infrastructure for production, distribution and consumption at large scale, Brazil became the pioneer in adopting a policy to encourage the use of bioethanol as a fuel.

1.2 The Brazilian bioethanol timeline

Since the sugarcane cycle, during colonial times, the sugar and alcohol sector has had an important role in the Brazilian social-economics system, when the Brazilian production of sugarcane has been

[1] Instituto Nacional de Metrologia, Qualidade e Tecnologia – INMETRO.
Email: vscunha@inmetro.gov.br
[2] Sociedade Brasileira de Metrologia – SBM.
Email: hsbrandi43@gmail.com
* Corresponding author: rjdaroda@inmetro.gov.br

implemented based in the plantation regime, using intensive slave manpower. After the end of slavery in 1888, there was no immediate adoption of wage labour for the freed slaves, leading to the transformation of agricultural crops areas into pastureland. Consequently, the small farmer was forced to sell his properties and become an employee or migrant to the cities, reflecting until today the situation of the Brazilian rural worker (Gorren, 2009). This situation has persisted over the years, with jobs in the sugar-alcohol sector being seasonal, low-paid and with precarious working conditions. On the other hand, biofuel production generates benefits for developing countries, such as the generation of a large number of jobs, especially for workers in rural areas with low levels of education. An example is the investment cost of US$ 11,000, required to create an agroindustrial job in Brazil, that is roughly 20 times lower than in the petrochemical industry (Gorren, 2009). Today, the impact of the "Programa Nacional do Álcool (Proálcool)", National Alcohol Program, is well known. When the oil price soared, Brazil was strongly dependent of imported oil, with nearly 80% of its consumption resulting in an important imbalance in the trade balance. In 1972, the spending corresponded to US$ 469 million, while in 1974, the value increased to US$ 2,840 million, representing 32.2% of all goods imported by the country (Scandiffio, 2005; Vieira Filho, 2004). Thus, in order to rebalance the commercial trade by decreasing the oil consumption, the Brazilian government established the use of a blend of anhydrous bioethanol with gasoline, starting with 4.5% and increasing yearly to 20%. In 1975, a government decree established the "Proálcool" (Barros, 2007), a program of investment in research to develop the production of sugarcane in order to increase productivity, quality and the sugar content in the sugarcane so as to encourage the use of bioethanol as motor fuel and to promote the development of new engine technologies to use only hydrated bioethanol. "Proálcool" improved the ethanol production process (fermentation, distillation and purification) and developed new engine technologies, initially to use a blend of high percentage of ethanol with gasoline and later to run on pure anhydrous ethanol. The program involved the agricultural sector, universities, public and private research centers and cars manufacturers (Original Equipment Manufacturer – OEM) (Cortez, 2016). From 1975 to 1979, during the first four years of "Proálcool", the efforts for the production of bioethanol from sugar cane were concentrated on producing anhydrous ethanol for use as a blend with gasoline. In 1979 and 1980, the oil crisis increased and the oil price had tripled compared to 1975, intensifying the Brazilian trade imbalance. Through tax incentives and subsidies to bioethanol, the OEMs (vehicle manufacturers) began developing new engines running only with bioethanol produced from sugarcane, a raw material readily available in Brazil. In 1980, the OEMs started to sell cars with engines powered by 100% hydrated bioethanol (BNDES/CGEE, 2008). The sugarcane plantation and the production of bioethanol were rapidly expanded. The growth of hydrated bioethanol consumption was so fast that in 1985, nearly 96% of the cars produced in Brazil were running on hydrated bioethanol produced from sugarcane (de Andrade, 2009).

Figure 1 shows the cars produced in Brazil changing from gasoline engine technology to the pure hydrated bioethanol engine technology encouraged by the "Proálcool" program.

During 1985, there was a change in the international scenario, characterized by falling oil prices and the recovering of the sugar price in the international market. At this time, sugar began to represent higher profits to farmers as well as better results in the trade balance, therefore, sugar production began receiving more government incentives. The result was a reduction in the supply of bioethanol and the consequent price escalation. With the fluctuation in bioethanol supply, consumer confidence in cars powered with hydrated bioethanol was affected, initiating a strong reversal in the acquisition and production of new cars. As can be observed in Figure 1, 1989 shows an increase in the production of cars driven with a blend of gasoline and anhydrous bioethanol and a decrease of pure hydrated bioethanol cars. While in 1985, the production of hydrated bioethanol cars represented more than 90% of the total, in 1990, this percentage dropped to only 11% (BNDES/CGEE, 2008).

The Brazilian policy related to sugar production has been extremely dependent on the international quotation of this commodity, causing permanent oscillations of bioethanol prices in the domestic market, leading the OEMs to accelerate the development of flex-fuel technology engines that allowed the use of any blend of bioethanol and gasoline. In 2003, with the new flex technology tested and approved, the OEMs launched the new flex-cars, significantly increasing the consumption of hydrated bioethanol

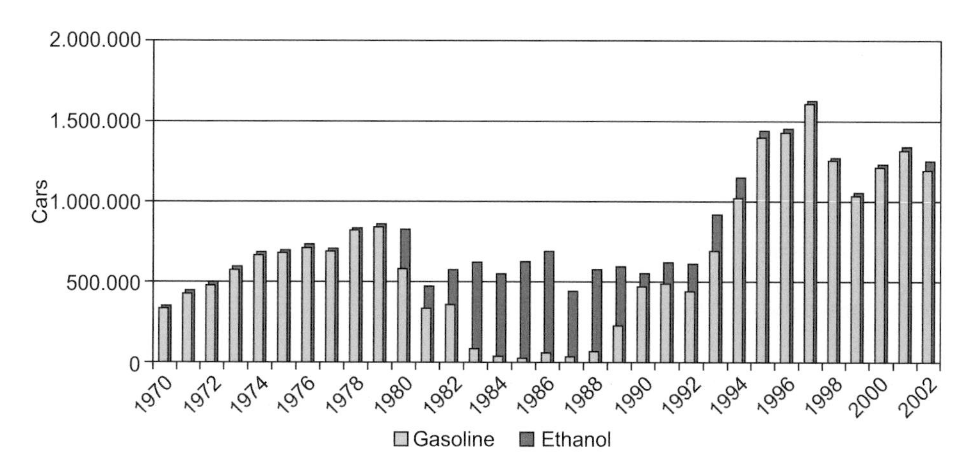

Figure 1. Yearly car production in Brazil from 1970 to 2002 (Source: Anuário da Indústria Automobilística Brasileira – ANFAVEA) (ANFAVEA, 2017).

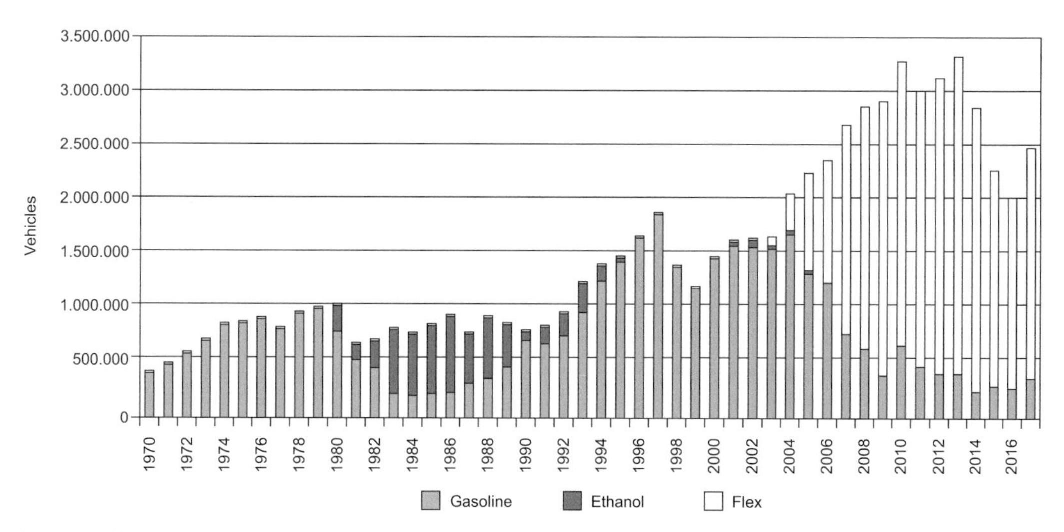

Figure 2. Car production in Brazil with different technologies: The evolution of bioethanol use as an engine fuel in Brazil-gasoline blended with anhydrous bioethanol, pure hydrated bioethanol and flex (any proportion of bioethanol and gasoline) (Source: Anuário da Indústria Automobilística Brasileira 2018 – Brazilian Automotive Industry Yearbook) (ANFAVEA, 2018).

(ANFAVEA, 2017). Figure 2 shows how the introduction of new technologies of engines in Brazil affected car production. In 2003, the OEMs started selling flex-cars that offered the owner the choice of using a blend of fuel, ranging from 100% gasoline (a gasoline blend with anhydrous bioethanol) to 100% of hydrated bioethanol. Ethanol and gasoline have a difference in calorific value. Ethanol has 70% of gasoline's calorific value, on average, which means roughly 30% higher consumption than gasoline (Barros, 2007; Antoniosi, 2016).

This same figure shows that more than 85% of the cars produced in Brazil, in 2017, were flex-cars. The 15% non-flex gasoline cars still sold in the country represented the sum of cars imported from different countries and brands (68%) and cars manufactured in Brazil (32%) (ANFAVEA, 2017). In the last 60 years the content of anhydrous bioethanol blended with gasoline ranged from 5% to 27%. The present bioethanol concentration of 27% has been legally established in 2015 (MAPA, 2019).

1.3 Raw materials for bioethanol production

The purpose of this chapter is not to do an exhaustive review of the production methods and raw materials for bioethanol production, but to focus on the comparison of the figures of beet, corn and sugarcane bioethanol production and present an overall view of the experience of Brazilian bioethanol use and production. The position of some players in the competition between biofuel and food crops, land use and other related issues is not covered.

The ethanol produced by biological fermentation of materials containing sugar and starch is known to be used as a beverage since the ancient civilizations and maintained through the ages (Kitman, 2000). Evidence of ethanol has been found in residue in Chinese pottery from thousands of years ago.

As mentioned in sub-section 1.1, the use of ethanol as a fuel began with a blend of ethanol and turpentine in an internal combustion engine developed by Morey. Pure corn-based ethanol as a fuel was used in the first car produced by Ford (Liu, 2012).

At the beginning of the 20th century, in Brazil, sugarcane ethanol produced from sugar was first used as a fuel for lighting lamps, and around 1920, with the increase the number of automobiles in Brazil, as an engine fuel blended with gasoline. Today, ethanol obtained from corn and sugarcane represents more than 85% of the world production. Although they are currently the main raw materials in ethanol production, other sources are also used. All are based on sugars and starch: Sugar beet, grape, cassava, sweet sorghum, potato and wheat, among others.

Intensive research and development projects in Industries, Universities and Research Centers with high investment made it possible to produce ethanol from cellulose and hemicellulose present in any biomass. This result gives another dimension to ethanol production. This new technology made it possible to use any biomass as the raw material, as is the case of sugar cane in Brazil, where the bagasse and straw can be used, increasing the ethanol production yield by a significant percentage. The same occurs in the case of corn with the use of straw.

Although the production processes are different in the stages of preparation and purification of the raw materials based on sugar, starch or cellulose/hemicellulose, its main production units of fermentation and distillation are the same.

2. Bioethanol Market

2.1 The origins of bioethanol in Brazil

As we mentioned in the subsection 1.3, ethanol can now be produced from any biomass source, but it can also be produced in a petrochemical process as a byproduct of the petroleum refining process. There is a consensus in calling the ethanol obtained from biomass "bioethanol" and the ethanol produced from petroleum "fossil ethanol".

In Brazil, all ethanol is produced from sugar cane. This raw material was introduced in Brazil between the year 1532 and 1548 by Martin Afonso de Souza, with the arrival of the Portuguese in Brazil. Sugar cane, although not native to the country, adapted well to the new environment and displayed high productivity, mainly due to the climate and soil. Sugar was a commodity that was highly valued and appreciated by the European nobles, this triggered a rapid expansion of sugarcane cultivation and growth in sugar production, which lead to it quickly becoming the main product of export to Europe (de Moura Filho, 2003; Monteiro, 2010; Frawley, 2016; Tavora, 2011).

The experience of the Portuguese in producing "bagaçeira", a spirit made from grapes, is probably the origin of the new product from sugar cane, the "cachaça". This spirit obtained by the fermentation followed by distillation of the sugarcane juice immediately had high acceptance and demand in the colony, affecting the consumption of the Portuguese "bagaçeira" and directly upsetting the Portuguese economical interest at that time. Some mills at that time practically stopped the production of sugar in order to dedicate themselves to the production of the spirit. In 1649, the growing demand for brandy and the reduction of sugar production led the Portuguese crown to temporarily ban the production of the spirit in Brazil. This act lead to the population revolting, resulting in the authorization to produce "cachaça", however, high taxes were imposed on the production and consumption of the spirit in Brazil (SEBRAE, 2019).

In 1650, the production of sugar from beet grew in Europe. It began to compete in the continent with the Brazilian sugar market, the largest at that time. At the same period, sugar cane is spreading through the Antilles, small tropical Caribbean islands, and Cuba, all European colonies. The production in Antilles starts to supply Europe and Cuba to supply the United States of America. The sugarcane-based economy started to decline with the defeat and expulsion of the Dutch from the northeast of Brazil, in the middle of the XVII century. Dutch businessmen moved to the Antilles to produce sugar and by controlling the transport and the commerce, could offer the product to Europe at lower price, establishing an enormous crisis in the Brazilian sugarcane economy that was stressed when the country economy turned the attention to the gold mines in the state of Minas Gerais. The economic cycle of sugarcane ended with the change of the Brazilian capital from Salvador, in Bahia, to Rio de Janeiro, in the southeast of Brazil [SUA PESQUISA.com, 2019]. The use of sugar to produce spirit or as a fuel in lighting lamps was a theme of interest during the first "International Exhibition of Alcohol Products and Equipment" and the "Congress of Industrial Applications of Alcohol" that took place in Rio de Janeiro in 1903. The lighting lamps burning ethanol, installed in Rio de Janeiro during the event, had an enormous success due to the clean burning, free of smoke, as opposed to the whale oil or kerosene lamps used in Europe. Later, in the 20th century, the crash of the American stock market in 1929 shook the world sugar market, and Brazil as a major exporter suffered a severe blow. The crisis that shook the sector stimulated the need to develop new uses of sugarcane that started to be considered since experiments of Nikolas Otto and Henry Ford, showing the feasibility of their use as a fuel in engines (Furtado, 2007; Goldemberg, 2004; da Silva Alves, 2009; Kirchof de Brum, 2016). From 1975, with the "Proálcool" program (Cortez, 2016), large investments were made by setting up a production, storage and distribution infrastructure to reach nearly 5 billion gallons per year. Bioethanol was consolidated as a fuel for vehicles, starting with the addition of anhydrous bioethanol in gasoline and resulted in the flex-cars offered to the consumer in 2003 (Goldemberg, 2004). Figure 3 shows the evolution of the flex-car market.

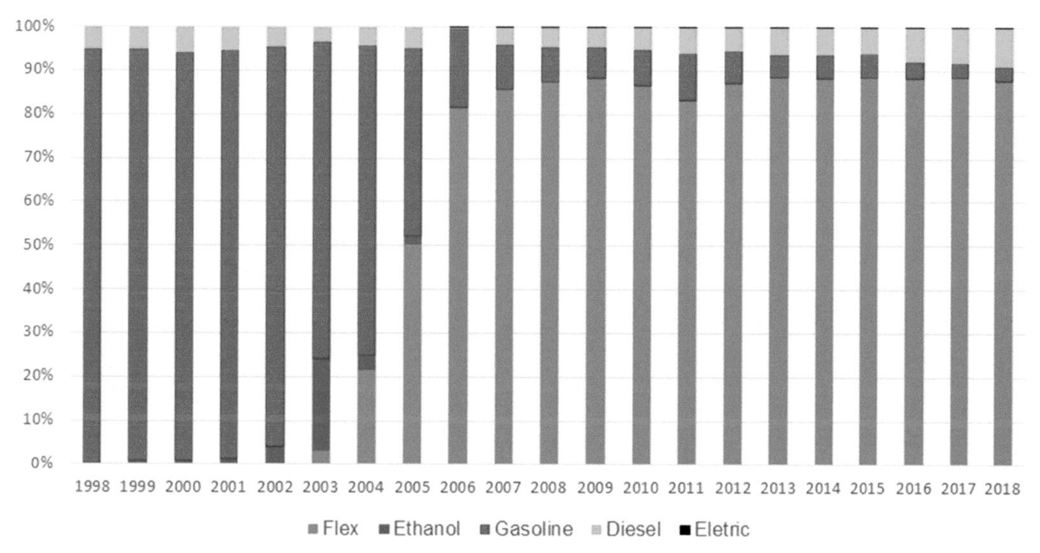

Figure 3. Evolution of licensing flex-fuel cars in Brazil: Cars and Light commercials (Data Source: Anuário da Indústria Automobilística Brasileira, 2019) [ANFAVEA, 2019].

2.2 Sugarcane and the bioethanol market

Although the production of cars in Brazil is predominantly flex favoring the consumption of bioethanol, in the last 10 years, the sale of gasoline exceeds the sale of ethanol. The consumers' choice is regulated by the price of the two commodities. Figure 4 shows that, in the years of high gasoline prices or of high production of ethanol lowering its pump prices, the sales consumed volume of gasoline and ethanol remained very similar.

Bioethanol is now produced from various biomass sources. The main sources are: Sugarcane in Brazil being the main producer; corn in USA as the main producer; sugar beet in some countries in Europe. Bioethanol from sweet sorghum, grape, cassava, potato and cellulose are produced in small units, some as pilot or even experimental units.

Figure 5 shows Brazil as the world's largest producer of bioethanol until 2006, when the United States became the world leader in bioethanol production and Brazil the second, a rank that still remains today.

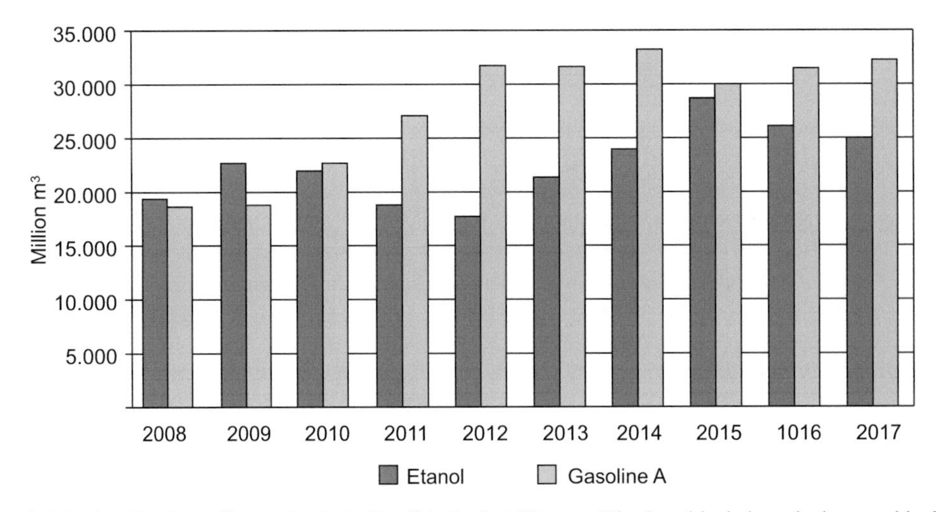

Figure 4. Bioethanol and gasoline total sales in Brazil in the last 10 years (Bioethanol includes anhydrous and hydrated) (Source: Agência Nacional de Petróleo, Gás Natural e Combustível – ANP – graphic 4.9) (ANP, 2018).

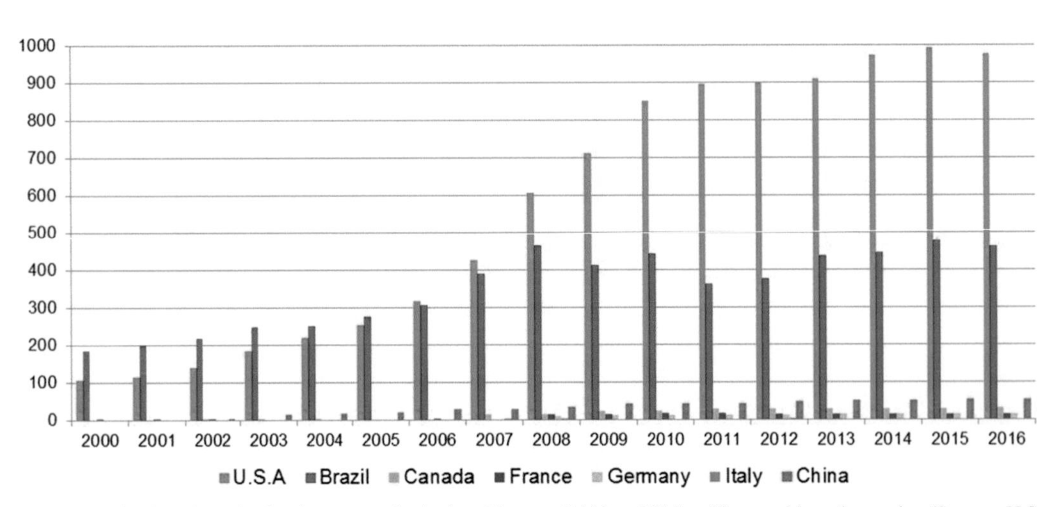

Figure 5. Bioethanol production by country in the last 17 years (2000 to 2016) – Thousand barrels per day (Source: U.S. Energy Information Administration – EIA, International Energy Statistics) (EIA Beta, 2018).

Figure 6 shows that a similar situation occurs in the ethanol trade between the two countries. Before 2010, the USA was a net importer of bioethanol. From 2010, the growth of production transformed the country into a net exporter. Since then, Brazil became the main American bioethanol destination. In 2017, the USA had a net export of 1,220 million gallons of bioethanol (RFA, 2018) and Brazil received 33% of this amount.

This inversion is related to the high sugar price volatility. When the sugar price falls in the international market, the ethanol production increases and the imports decrease. The reversal process also happens and, in this case, the imports increases to fulfill the Brazilian internal demand.

In 2017, both countries contributed nearly 85% of world bioethanol production. However their relative contribution had strongly changed. While the USA contribution had increased to 58.4% of the total global production, Brazil accounted for 26.1%. European countries participated with only 5%. Figure 7 shows the bioethanol fuel consumption of the seven largest consumers, in 2016.

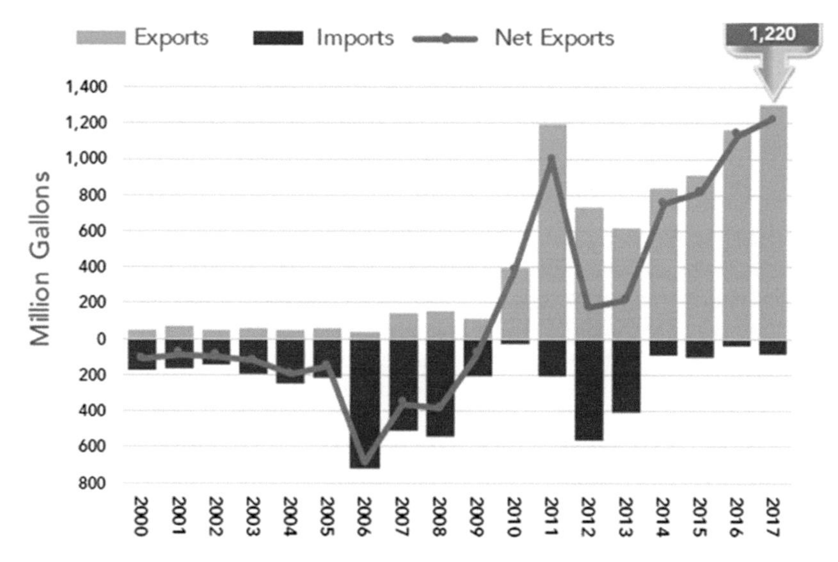

Figure 6. USA Bioethanol trade with Brazil from 2000 to 2017. Before 2010 Brazil was a bioethanol exporter. From 2010 to date, Brazil has become an importer from the USA (Source: U.S. Department of Commerce, U.S. Census Bureau, Foreign Trade Statistics) (RFA, 2018).

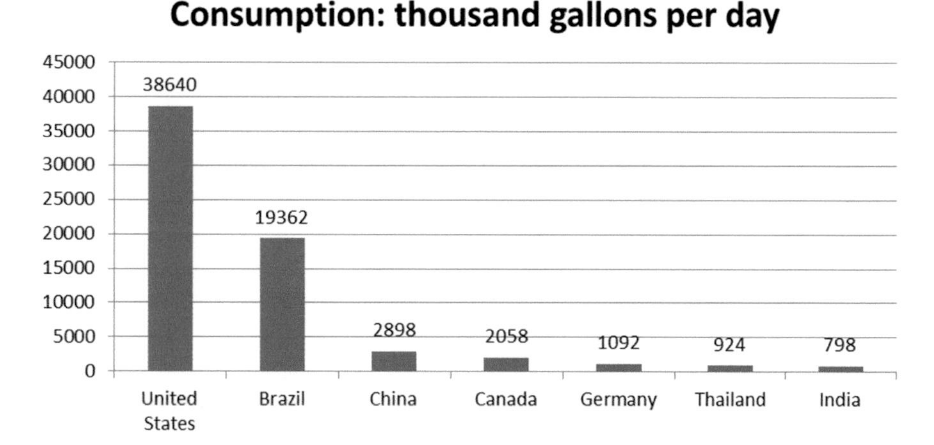

Figure 7. Consumption of bioethanol as an engine fuel for the first seven ranked countries Data adapted. Converted from barrels per day to gallons per day (1bb = 42 gal) (Knoema, 2018).

3. Brazilian Ethanol Production

Figure 8 shows two different behaviors of the world bioethanol production. From 1980 to 2002, the world bioethanol production presented a stability characterized by equilibrium between supply and demand, while from 2002 to 2012, there was constant growth, with the production slightly surpassing the demand. The need to reduce greenhouse gas emissions explains the growth in the world production and demand of bioethanol. In fact, it is almost a consensus among the scientific community that the expanded use of fossil fuels without actions to mitigate greenhouse gas emissions has global climate implications. The need to reduce greenhouse gas emissions explains the growth of the world production and demand for bioethanol (Guevara, 2017).

The success of the "Proálcool" program turned Brazil into a global reference in the use of biomass. In the initial phases of the use of ethanol as a fuel in Brazil, bioethanol was obtained by introducing new processes in the already operating sugarcane mills, built to produce sugar. Nowadays, all the mills in Brazil operate to produce both sugar and bioethanol, changing the main final product according to the market demand. Although in the years of the "Proálcool" program the bioethanol production and consumption in Brazil was not related to climate concerns but to the soaring of crude oil prices, today, the global efforts to reduce greenhouse gas emissions stimulated the crescent use of bioethanol fuel as the best alternative to gasoline, in Brazil. This trend brings a risk to the Brazilian ethanol production strongly based on sugarcane, especially considering that the sugarcane harvest occurs during the rainy season and that the yield of ethanol per ton of sugarcane can decrease more than 10%, depending on the volume of rain. To reduce this threat, the productivity of ethanol from other raw materials like corn, sweet sorghum and beet is being subjected to field experiments in Brazil. Another aspect to be considered is the comparative advantages related to stocking, in particular for corn, making the bioethanol production possible during the whole year (EPE–MME, 2018).

Although more expensive and less sustainable than to sugarcane, corn is the main raw material used to produce ethanol in the USA. Others sources of biomass with small production volume, like sugar beet in the EU and grape and sweet sorghum in other countries, are also used, showing different productivity and yields. The productivity and yield for sugarcane, corn and sugar beet are shown in Table 1 (Manochio, 2014).

Table 1 shows that the production of sugarcane corresponds to 5400 to 10800 (l/ha) and corn to 3450 to 4600 (l/ha), favoring the performance of sugarcane concerning the production of bioethanol per cultivate area (Manochio, 2014). The Brazilian average yield per ton of sugarcane is 71 kg of sugar, 42 liters of bioethanol or 11.5 tons of total recoverable sugars per hectare. Each ton of sugarcane has a potential energy equivalent to 1.2 barrels of fossil oil. This high productivity results from favorable

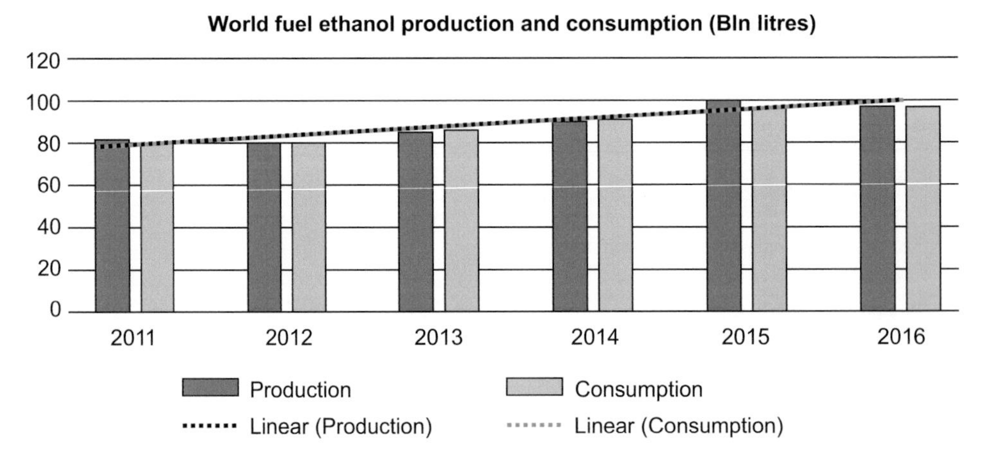

Figure 8. World Fuel Ethanol Production and Consumption from 2011 to 2016 (Source: International Sugar Organization – ISO 2019) (ISO, 2019).

Table 1. Productivity and yield for sugarcane, corn and sugar beet (Manochio, 2014).

Raw material	Productivity (ton/ha)	Productivity (l/t)	Productivity (l/ha)
Sugarcane	60–120	90	5400 to 10800
Corn	7.5–10	370–460	3450 to 4600
Sugar beet	50–100	100	2–1

climate conditions and appropriate soil for cultivation of sugarcane. The above figures concerning the yield of sugar and bioethanol per ton are obtained using one third of sugar cane, only. The other two thirds correspond to bagasse and straw, which are used as a fuel in boilers to produce energy as steam for the processing of sugar and bioethanol. The majority of this steam is used to generate electricity, which is partially consumed in the plant, and the remainder sent to the grid (EPE–MME, 2007).

The process to produce bioethanol from sugarcane is simpler than from corn, since it follows the sugar production sequence. The sugarcane is cut into small pieces through pickles and shredders and sent to the mill in order to obtain the highest possible yield in broth—a solution containing the sugars. Bagasse, which is also obtained as a residue of the process, is sent to the boilers and used to produce steam for the electricity-generating turbine. The broth goes through a purification processes via chemical treatment and a set of filtrations before is used to produce either sugar or bioethanol. Then, the purified broth can follow two different processes. First, the process for sugar production through broth concentration by water evaporation followed by crystallization. Second is the bioethanol production where the broth is submitted to a fermentation process, producing a water-ethanol solution which is sent to the distillation towers in order to obtain the hydrated bioethanol. The anhydrous bioethanol is obtained by adding a dehydration step (de Azevedo, 2012; Andrietta, 2006).

There is an important difference between the corn and the sugarcane processes to produce bioethanol. To obtain the fermentable sugars to produce bioethanol from corn, there is a critical step called "starch hydrolysis". This is not necessary if sugarcane is the raw material, because the fermentable sugars are already available in the broth. This step increases the cost of bioethanol production from corn. Figure 9 shows a simplified process flow chart for sugarcane, sugar beet and corn.

Today, Brazil is the world largest producer of sugarcane, having developed several varieties of sugarcane, especially adapted to different climate conditions and type of soils as well as being plague resistant, that may maximize the production of sucrose and increase the competitiveness and advantages of sugarcane bioethanol. In the last few years, the increase of ethanol production per hectare was related to a higher productivity with the same land use (Novacana, 2019).

Brazil has about 350 million ha of arable land adequate to food production. A small surface area of 4.8 million ha, or 1.5%, of this land (corresponding to approximately 15% of the UK territory and 6.5% of France) is used for bioethanol production (Zuurbier, 2008; ÚNICA, 2019). The sugarcane harvest and sugar, ethanol and bioelectricity plants are mostly concentrated in South-Central Brazil.

In 2017, renewable energy sources had a share of 43.2% in the Brazilian energy matrix, with sugarcane—bagasse/straw (11.7%) and bioethanol (5.7%)—contributing 17.4%. Considering just electricity generation, the share of sugarcane is 7% with an installed capacity of 11,298 MW (BOLETIM/ ÚNICA, 2018; EPE, 2018).

Under various aspects, sugarcane presents important competitive advantages in comparison to ethanol produced from other raw materials, both in terms of its total utilization but also in terms of all the

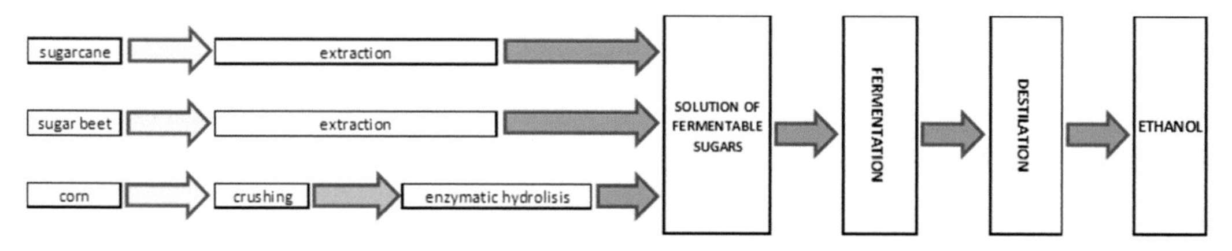

Figure 9. Simplified process flow chart for bioethanol production (source: Manochio, adapted) (Manochio, 2014).

three dimensions of sustainability, namely economics, environmental and social. The previous discussions were mainly focused in the economics advantages to produce bioethanol using sugarcane rather than other raw materials, demonstrating its competitiveness concerning economical sustainability. Concerning the environmental dimension, sugarcane is undoubtedly the best raw material for ethanol production. In fact, related to the capacity to reduce greenhouse gas emissions, the ratio between renewable energy and the energy involved in the process to produce bioethanol demonstrates a significant advantage for sugarcane bioethanol. On the social dimension, one of the most important benefits of bioethanol being used as fuel is that its production is mostly based on sugarcane, a sustainable source in Brazil, and its domestically-based crops and fabrication creates jobs, keeping low-skilled workers in the countryside, reducing the dependence on fossil fuel and increasing Brazilian energy independence. Table 2 presents the ratio renewable energy/energy involved to produce bioethanol (Macedo, 2007).

Table 2. The ratio of renewable energy/energy involved to produce bioethanol for different raw materials (Macedo, 2007).

Bioethanol raw material	Ratio
Sugarcane	8.9
Corn	1.3
Sugar beet	2.0
Sweet sorghum	4.0
Wheat	2.0

3.1 Sustainability of using bioethanol as a fuel

The results of Table 2 demonstrate that concerning the environmental aspects associated to greenhouse gas emissions, sugarcane is undoubtedly the best raw material to produce bioethanol, resulting in nearly 10 times more energy than the fossil fuel energy that is consumed in the process of production. Thus, the reduction of greenhouse gas emissions for bioethanol is roughly 90% while for corn is about 10% to 30% (Macedo, 2007). This significant result is obtained due to the complete utilization of sugarcane. As previously mentioned, today the sugars contained in the broth are converted into bioethanol and the bagasse and straw are used to generate thermal and electrical energy.

Brazil faces problems concerning poverty, widespread unemployment, lack of sanitation, low education level and lack of public infrastructure, common to the large majority of the developing countries. Nevertheless, in the past years, the country had an important participation in the global effort to adopt measures to mitigate the consequences of climate change. According to the International Energy Agency (IEA) Bioenergy, Countries Report, Brazil – 2018 Update, Brazilian actions "represent one of the largest undertakings by any single country to date, having reduced its emissions by 43.9% (GWP-100; IPCC SAR) in 2015 in relation to 2005 levels. Brazil is nevertheless willing to further enhance its contribution towards achieving the objectives of the convention (United Nations framework convention on climate change), in the context of sustainable development" (IEA, 2018). Among several targets intended to be adopted in the future, in the context of the present work, two deserve a few words:

i) Expanding the use of non-fossil fuel energy sources domestically, increasing the share of renewables (other than hydropower) in the power supply to at least 23% by 2030, including by raising the share of wind, biomass and solar.
ii) In the industry sector, promote new standards of clean technology and further enhance energy efficiency measures and low-carbon infrastructure.

Related to the first goal, it must be remarked that, although the numbers presented in this chapter are already comparatively quite in favor of sugarcane ethanol, another possible advantage that should be signalized is that only about 30% of the total sugarcane energy stored in the sucrose is used to produce sugar or bioethanol. The other 70%, consisting of straw and bagasse, is biomass containing cellulose (cellulosic biomass). Fuels derived from cellulosic biomass are generally called "second-generation" biofuels and offer an alternative to conventional energies with a large potential for economic and environmental impact. Presently, there is a strong research effort and investments focused on finding competitive and efficient processes to transform cellulosic biomass into bioethanol from sugarcane by turning the cellulose and hemicellulose into sugars using acid or enzymatic hydrolysis and these into bioethanol through the fermentation of the broth. The remaining lignin will continue to be used in the production of thermal and electrical energy. A semi industrial plant using bagasse and cane straw as raw material is in operation in Brazil using the DHR (Dedini Hidrólise Rápida – Rapid Hydrolysis Dedini) process developed by Dedini (LAMNET, 2004). The plant processes 2000 kg of bagasse and straw equivalent per hour for a daily production of 5000 liters of bioethanol.

According to the Plan for Energy Expansion (PDE) 5, Brazil has also two other commercial second-generation ethanol plants (Granbio and Raízen) and one experimental ("Centro de Tecnologia Canavieira" - CTC), with a nominal production capacity of 82, 42 and 3 million liters per year, respectively (IEA, 2018).

Undoubtedly, the introduction of bagasse and sugarcane straw as raw material for bioethanol production will give a new dimension to the sugar and alcohol industry in Brazil, contributing to the growing use of non-fossil fuel energy sources domestically, without the need to increase cultivated area, increasing competitiveness of this fuel (Soares, 2013; LAMNET, 2004).

Concerning the second target, it is a fact that stakeholders use documentary standards to optimize production and to implement health, consumer protection, security, quality, environment, sustainability and other public and private policies. They are also fundamental instruments for international trade. Because standards are used as a rule or guideline, it is a document constituted by technical criteria or precisely required specifications and to be credible and accepted by the market it must be issued by a well-recognized body. The International Standardization Organization (ISO) is one of the best examples of a standardization body (Sikdar, 2014). Measurement standards concede confidence to the process involving documentary standards and are the basis of the whole standardization process. In the case of biofuels, bioethanol and biodiesel, an important step to harmonize worldwide technical standards has been taken with the publication of "White Paper on Internationally Compatible Biofuel Standards" establishing the basis for the development of measurement and documentary standards, common to Brazil, the USA and the EU. During the next three years, the National Institute of Metrology, Quality and Technology (Inmetro), the Brazilian National Metrology Institute (NMI), the National Institute of Standards and Technology (NIST), the North-American NMI, and several European NMI and laboratories developed and began producing common Certified Reference Materials (CRM) for bioethanol and biodiesel. This accomplished a fundamental step toward the international standardization and the transformation of biofuels into a world commodity. This assures confidence in the analyses conducted by the industrial laboratories to promote new standards of clean technology and further enhance energy efficiency measures and low-carbon infrastructure (Brandi, 2011).

Finally, a third target for Brazil to attain the objectives of the Convention should be mentioned, because it relates the use of biofuel to the social dimension of sustainability. The target states that "in the transportation sector, further promote efficiency measures, and improve infrastructure for transport and public transportation in urban areas" (IEA, 2018).

These three targets highlight the role of bioethanol in the Brazilian policy to achieve the objective of United Nations Convention on climate change and to the goals of the 2030 Sustainable Development Agenda of the United Nations.

References

Andrietta, M.G.S., Steckelberg, C. and Roberto, S. 2006. Bioetanol – Brasil, 30 anos na vanguarda, Revista Multiciencia, n. 7, 2006; https://www.multiciencia.unicamp.br/artigos_07/a_02_7.pdf.

ANFAVEA (Associação Nacional dos Fabricantes de Veículos Automotores). 2017. Brazilian Automotive Industry Yearbook. http://www.virapagina.com.br/anfavea2017/ (Last accessed – June 2019).

ANFAVEA (Associação Nacional dos Fabricantes de Veículos Automotores). 2018. Brazilian Automotive Industry Yearbook, http://www.virapagina.com.br/anfavea2018/65. (Last accessed January 2019).

ANFAVEA (Associação Nacional dos Fabricantes de Veículos Automotores). 2019. Anuário da Indústria Automobilística Brasileira; 2.5—Licenciamento de autoveículos novos por combustível; p. 62; http://www.virapagina.com.br/anfavea2019/.

Antoniosi, L. and Maintinguer, S.I. 2016. A Evolução do Etanol Brasileiro: Do Proálcool aos Dias Atuais; Congresso de Inovação, Ciência e Tecnologia do IFSP – 2016, SP-Brasil.

Barros, R. 2007. Energia para um novo mundo. Rio de Janeiro: Monte Castelo Idéias, 160 p.

BNDES/CGEE. 2008. Banco Nacional de Desenvolvimento Econômico e Social (BNDES) e Centro de Gestão e Estudos Estratégicos (CGEE), 2008; Bioetanol de cana-de-açúcar: Energia para o desenvolvimento sustentável; organização BNDES e CGEE.—Rio de Janeiro: BNDES.

BOLETIM/ÚNICA. 2018. A Bioeletricidade da Cana em Números—2018, https://www.udop.com.br/index.php?item=noticias&cod=1174654.

Brandi, H.S., Daroda, R.J. and Souza, T.L. 2011. Standardization: An important tool in transforming biofuels into a commodity. Clean Technologies and Environmental Policy 13: 647–649.

Cortez, L.A.B. 2016. Universidades e empresas: 40 anos de ciência e tecnologia para o etanol brasileiro; 2016, Editora Edgard Blücher Ltda.

da Silva Alves, J. and Lima, R.C. 2009. Transmissão de Preços entre Mercados de Açúcar Espacialmente Separados no Brasil: Uma Análise de Co-Integração, Campo Grande, 25 a 28 de julho de 2009, Sociedade Brasileira de Economia, Administração e Sociologia Rural; http://www.sober.org.br/palestra/15/1036.pdf. http://www.agriculture.gov.au/SiteCollectionDocuments/abares/brazilian-sugar-industry.pdf.

de Andrade, E.T., de Carvalho, S.R.G. and de Souza, L.F. 2009. Programa do Proálcool e o Etanol no Brasil. ENGEVISTA 11(2): 127–136.

de Azevedo, M.S., de Oliveira Santos, R.V. and Magalhães, T.V. 2012. Produção de Etanol no Brasil. Bolsista de Valo 2(1): 151–154, http://essentiaeditora.iff.edu.br/index.php/BolsistaDeValor/article/view/2408.

de Moura Filho, H.P. 2003. 120 Anos de Produção Mundial de Açúcar: Comentário sobre Séries Estatísticas Tradicionais (1820–1940), https://www.researchgate.net/publication/4900142_120_anos_de_producao_mundial_de_acucar_comentario_sobre_series_estatisticas_tradicionais_1820-1940.

EIA Beta. 2018. US Energy Information Administration—EIA, International Energy Statistics. https://www.eia.gov/beta/international/data/browser/#/?pa=000003&c=00000212000g000400000004000000000000000000000002&ct=0&tl_id=79-A&vs=INTL.79-1-BRA-TBPD.A&cy=2015&vo=0&v=H&start=2000&end=2016 (Last accessed – June 2019).

EPE–MME. 2007. Plano Nacional de Energia 2030/Ministério de Minas e Energia; colaboração Empresa de Pesquisa Energética, Brasília: MME: EPE, 2007 O Mercado de Energia Elétrica: Evolução a Longo Prazo, p.159; http://epe.gov.br/sites-pt/publicacoes-dados-abertos/publicacoes/PublicacoesArquivos/publicacao-165/topico-173/PNE%202030%20-%20Proje%C3%A7%C3%B5es.pdf.

EPE–MME. 2018. Cenários de Oferta de Etanol e Demanda e Ciclo Otto 2018 – 2030; Ministério de Minas e Energia (MME) e Empresa de Pesquisa Energética (EPE), 2018 http://epe.gov.br/pt/publicacoes-dados-abertos/publicacoes/cenarios-oferta-etanol-e-demanda-ciclo-otto.

EPE. 2018. Balanço Energético Nacional 2018, Relatório Síntese – ano base 2017; Empresa de Pesquisa Energética – EPE, Rio de Janeiro, maio de 2018; www.epe.gov.br%2Fsites-pt%2Fpublicacoes-dadosabertos%2Fpublicacoes%2FPublicacoesArquivos%2F+publicacao303%2Ftopico397%2F+Relatório%2520Síntese%25202018-ab%25202017vff.pdf.

Frawley, N. 2016. The Brazilian Sugar Industry, Department of Agriculture and Water Resources, Agricultural Commodities report, JUNE QUARTER 2016 (Research by the Australian Bureau of Agricultural and Resource Economics and Sciences).

Furtado, C. 2007. Formação Econômica do Brasil, Companhia das Letras, São Paulo; http://www.afoiceeomartelo.com.br/posfsa/autores/Furtado,%20Celso/Celso%20Furtado%20-%20Forma%C3%A7%C3%A3o%20Econ%C3%B4mica%20do%20Brasil.pdf.

Goldemberg, J., Coelho, S.T., Nastari, P.M. and Lucon, O. 2004. Ethanol learning curve—the Brazilian experience. Biomass and Bioenergy 26: 301–304.

Gorren, R.C.G. 2009. Biocombustíveis aspectos sociais e econômicos: Comparação entre Brasil, Estados Unidos e Alemanha; http://www.iee.usp.br/producao/2009/Teses/1_Gorren_dissert.pdf.

Guevara, A.J.H., da Silva, O.R. and Hasegawa, H.L. 2017. Avaliação de Sustentabilidade da Produção de Etanol no Brasil: Um Modelo em Dinâmica de Sistemas. Brazilian Business Review 14(4): 435–447, Julho-Agosto, 2017; http://www.spell.org.br/documentos/buscaredicao/periodico/brazilian-business-review/idedicao/4719.

IEA. 2018. IEA Bioenergy, Countries Report, Brazil – 2018, Update, Bioenergy policies and status of implementation; https://www.ieabioenergy.com/wp-content/uploads/2018/10/CountryReport2018_Brazil_final.pdf.

ISO. 2019. International Sugar Organization – ISO, 2019; https://www.isosugar.org/sugarsector/ethanol.

Kirchof de Brum, A., de Moura, A.P., De Souza, C.C., Frainer, D.M. and dos Reis Neto, J.F. 2016. Matriz Energética Brasileira: Considerações Sobre A Configuração Institucional Na Competitividade Do Etanol; Revista SODEBRAS – Volume 11 N° 130 – OUTUBRO/ 2016; https://www.researchgate.net/publication/308046857.

Knoema. 2016. World Data Atlas, Fuel Ethanol Consumption, 2016; https://knoema.com/atlas/topics/Energy/Renewables/Fuel-ethanol-consumption.

Kovarik, B. 1998. Henry ford, charles kettering and the fuel of the future—ethanol history; bill kovarik, automotive history review. Spring 32: 7–27. http://www.ethyl.environmentalhistory.org/?page_id=56.

LAMNET. 2004. International Workshop on Bioenergy Policies, Technologies and Financing; 2004; 9th LAMNET Project Workshop; Ribeirão Preto, São Paulo, Brazil, 13–17 September 2004; https://www.bioenergy-lamnet.org/publications/source/bra2/Olivero.pdf.

Liu, K. and Rosentrater, K.A. 2012. Distillers Grains: Production, Properties, and Utilization. CRC Press.

Macedo, I.C. 2007. Situação atual e perspectivas do etanol, Núcleo Interdisciplinar de Planejamento Energético da Unicamp, Campinas (SP), ESTUDOS AVANÇADOS 21 (59), http://www.scielo.br/pdf/ea/v21n59/a11v2159.pdf.

Manochio, C. 2014. Produção de Bioetanol de cana-de-açúcar, milho e beterraba: Uma comparação dos indicadores tecnológicos, ambientais e econômicos; Trabalho de Conclusão de Curso (Graduação em Engenharia Química) – Universidade Federal de Alfenas–Campus de Poços de Caldas, MG; https://www.unifalmg.edu.br/engenhariaquimica/system/files/imce/TCC_2014_1/Carolina%20Manochio.pdf.

MAPA (Ministério da Agricultura, Pecuária e Abastecimento). 2019. Portaria MAPA Nº 75 DE 05/03/2015; https://www.legisweb.com.br/legislacao/?id=281775 (Last accessed: January 2019).

Monteiro, B.M. dos Santos. 2010. Composição Química de aguardentes de cana de açúcar obtidas por fermentação com diferentes cepas de levedura *saccharomyces cerevisiae*; Dissertação apresentada para obtenção do título de Mestre em Ciências; Escola Superior de Agricultura Luiz de Queiroz; Piracicaba, 2010.

Novacana. 2019. https://www.novacana.com/sustentabilidade/demanda-terras-aumento-producao-etanol-acucar.

RFA. 2018. Renewable Fuel Association–RFA, Ethanol Strong, Ethanol Industry Outlook, 2018; https://www.ethanolresponse.com/wp-content/uploads/2018/02/2018-RFA-Ethanol-Industry-Outlook.pdf.

Scandiffio, M.I.G. 2005. Análise prospectiva do álcool combustível no Brasil–Cenários 2004–2024; Campinas: Universidade Estadual de Campinas, Faculdade de Engenharia Mecânica (Tese de Doutorado).

SEBRAE. 2019. Serviço Brasileiro de Apoio às Micro e Pequenas Empresas (SEBRAE) Agronegócio, A História da Cachaça no Brasil, http://www.sebraemercados.com.br/a-historia-da-cachaca-no-brasil/(last accessed: April 2019).

Sikdar, S.K. and Brandi, H.S. 2014. How to Quantify Sustainability in Construction and Manufacturing and the Need for Standards, NIST-ASCE-ASME Sustainability Workshop, Rockville, MD, USA, 12-13/06/2014; Ayyub, B.M., Galloway, G.E. and Wright, R.N. (eds.). 2015. Proceedings of the Measurement Science for Sustainable Construction and Manufacturing Workshop, Volume I. Position Papers and Findings, University of Maryland Report to the National Institute of Standards and Technology, Office of Applied Economics NIST.GCR.15-986-1, Gaithersburg, MD. pp. 204–209, 2015; http://dx.doi.org/10.6028/NIST.GCR.15-986-1.

Soares, P.A. and Vaz Rossell, C.E. 2013. Conversão da Celulose pela tecnologia Organosolv, Núcleo de Análise Interdisciplinar de Políticas e Estratégias da Universidade de São Paulo – NAIPPE, Vol. 3 nova série; http://naippe.fm.usp.br/arquivos/livros/Livro_Naippe_Vol3.pdf.

suapesquisa.com. 2019. Ciclo do Açúcar no Brasil https://www.suapesquisa.com/historiadobrasil/ciclo_acucar.htm.

Sukhdev, P. 2012. Ford's Ethanol and Rockefeller's Gasoline: Who Won, Who Lost, and Why? July 15, 2012 http://corp2020.net/entries/general/ford%E2%80%99s-ethanol-and-rockefeller%E2%80%99s-gasoline-who-won-who-lost-and-why-), ("Ford Predicts Fuel from Vegetation," New York Times, 20 September 1925, 24).

Távora, F.L. 2011. História e Economia dos Biocombustíveis no Brasil, Textos para Discussão, Centro de Estudos da Consultoria do Senado Federal, http://www.senado.gov.br/conleg/centroaltosestudos1.html.

ÚNICA. 2019. (União da Indústria da Cana de Açúcar)—A Dimensão do Setor Sucroenergético; https://www.unica.com.br/wp-content/uploads/2019/06/A-Dimensao-do-Setor-Sucroenergetico.pdf.

Vieira Filho, J.E. 2004. Agroindústria canavieira—o momento oportuno para a discussão de um programa energético do álcool. Revista Confiança.

Zuurbier, P. and van de Vooren, J. 2008. Sugarcane Ethanol: Contributions to climate change mitigation and the environment, Wageningen Academic Publishers, The Netherlands, 2008.

CHAPTER 3

Biomass in Regional and Local Context

Michael Narodoslawsky

1. Introduction

Bio-resources offer the possibility to transform carbon dioxide, water, and nutrients, with the help of solar irradiation, into valuable material products. Primary terrestrial bio-resources include crops, forage and wood, which can be utilised directly as food, feed, construction material, energy sources or as a starting point of conversion chains that may lead to biofuels and a wide variety of chemicals. A coarse estimate on the base of FAO data puts current production of crops at roughly 6 billion t/yr[1] and global production of wood at 3.8 billion m^3/yr.[2] The wide range of use and, in particular, the central role for human nutrition leads to brisk demand and competition for bio-resources. Although they are renewable in the sense that, given sustainable care for forests, grass land and fields, they may be harvested again after every cultivation cycle, their annual yield is restricted by the finiteness of their basic resource, namely fertile land.

It is this bond to land that makes terrestrial bio-resources contextual. Solar irradiation as the driving force of plant growth requires area for its conversion to useful services, regardless if it is assimilation of water and carbon dioxide to organic compounds or its transformation to electricity by photovoltaic panels. So, all resources and services based on solar irradiation are inherently de-centralized and area bound. Bio-resources are no exceptions to this rule but are particularly inefficient resources when assessed according to their conversion rate of solar energy to useful energy content. According to Zhu et al. (2008), maximum theoretical conversion rates of solar irradiation into bio-resources are between 4.6% for C-3 plants like wheat, and 6% for C-4 plants like maize. Practical conversion rates are lower, usually around 50% of these theoretical values. This compares to conversion rates of 15–20%[3] for PV panels and even higher efficiencies for thermal solar collectors.

The low conversion rate of solar irradiation into bio-resources clearly assigns them their role within efforts to manage carbon emissions. A range of environmental Life Cycle Assessments (LCA) based on either Global Warming Potential (GWP), e.g., Schlömer et al. (2014), or on aggregated assessment methods, like the Sustainable Process Index (SPI), see Kettl et al. (2011a), show consistently that all renewables-based energy technologies perform substantially better in ecological terms than all fossil-based technologies, but that bio-energy performs worse than all other renewables-based competitors. Bio-resources are, however, material products with a much broader spectrum of applications than just providing energy services. This spectrum ranges from providing food and feed to industrial bio-products,

Institute for Process and Particle Engineering, Graz University of Technology, Inffeldgasse 13/3.
Email: narodoslawsky@tugraz.at

[1] Based on FAO data for 2014 to 2017, http://www.fao.org/faostat/en/#data/QC [last accessed June 2019].

[2] Based on FAO data for 2017, http://www.fao.org/forestry/statistics/80938/en/ [last accessed June 2019].

[3] See https://news.energysage.com/what-are-the-most-efficient-solar-panels-on-the-market/ [last accessed April 2019].

like paper, timber and a large variety of bio-chemicals to storable energy carriers. Rational use of bio-resources, therefore, favours their use as food, feed and material products. Their role within energy provision should be focussed on applications that require material energy carriers, such as bio-fuels for transport or for providing heat or electricity when no other renewable energy can be used.

The area to capture solar irradiation is, however, only one factor in the generation of bio-resources. Yields of bio-resources are critically dependent on soil quality and climate. Water and nutrients are also indispensable resources whose availability and quality are critically dependent on the spatial context, as are soil quality and climate. Together these factors define what crops may possibly be cultivated in a certain region.

Finally, cultivated land must be seen within its cultural, economic and social context: Yield of crops is critically dependent on the know-how of farmers as well as their economic ability to purchase farm equipment and auxiliary production means, such as fertilisers and pest control. This explains why there is a wide range in yields per hectare and year for crops in different locations, even for the same crop, e.g., the yield for maize varies from 6.6 t/ha.yr for Albania to 9.9 t/ha.yr for Austria and 11.1 t/ha.yr for the U.S.[4] The regional economic and industrial structure will also influence what crops are cultivated and how they are utilised. Infrastructure such as roads, railways, transport grids for gas and electricity as well as heat distribution grids are further essential spatial factors to market bio-resources and services and products derived from them. This chapter will discuss how utilisation technologies for bio-resources are influenced by their spatial context, what properties are critically important for optimal value chains based on bio-resources and what guides the structure of regional utilisation technology networks.

2. Influences of Spatial Context on Bio-resources based Technologies

The dependence of plants on solar irradiation makes bio-resources inherently de-central resources. This has considerable impact on logistics of value chains, size and structure of utilisation technologies. Contrary to fossil resources, which are point resources, bio-resources have a "first and last mile problem": They have to be collected from large areas. Besides that, the requirement to keep the land fertile entails that nutrients removed must be replenished. Since much of the nutrients contained in bio-resources may not be required (or even wanted) in the end product, wastes and by-products along the value chain holding valuable nutrients should be returned to land in an appropriate manner in order to reduce environmental pressure. This requires a by-product and waste distribution logistics in addition to the collection effort for bio-resources.

From the point of view of utilisation technologies, the first question to be answered is about the size of the land necessary to provide the resources for a given product volume P in t/yr, see equation (1). The most important factors defining this size A_R in ha are the yield of the bio-resource Y_R in t/ha.yr, the content of the compound utilised in the technology in the bio-resource C_R in percent of fresh weight and the fraction of the land available to support the industrial production f_{al} in percent of the total area. All these factors are dependent on the spatial context. Besides this, the technological efficiency of converting the compound from the bio-resource into the product e_t in percent of the input is required for calculating this area.

$$A_R = \frac{P.C_R}{Y_R.f_{al}.e_t} \tag{1}$$

Besides the area necessary to support industrial production, other data are also needed in order to determine the structure of the value chain. This applies in particular to the properties relevant to logistics, like humidity and transport density.

Bio-resources are always humid. Humidity means increased transport efforts in all cases, as the dead weight of water has to be shipped along with the valuable content of bio-resources. Moreover, the higher

[4] Data from FAO statistics for 2017, see http://www.fao.org/faostat/en/#data/QC [last accessed April 2019].

their humidity, the more susceptible they are to bio-degradation. This either requires increased and often costly efforts to store bio-resources or to adapt their conditioning or utilisation to the harvest period by operating industrial plants in campaign mode in order to minimise the storage time of fresh material. Conditioning of humid bio-resources usually requires drying to low humidity, a step that consumes a considerable amount of energy. Campaign mode operation entails larger plants than continuous operation, as the whole annual production of the bio-resource has to be treated within the harvesting period, which is usually only a fraction of the year.

Transport density of bio-resources influences the efficiency of collection transport. Vehicles have a maximum mass load M_m in t as well as a maximum transport volume V_m in m³. The critical transport density ρ_c in t/m³ is the quotient of these two numbers, see equation (2):

$$\rho_c = \frac{M_m}{V_m} \tag{2}$$

If the transport density of the load is larger than ρ_c, the transport is limited by mass. If it is, however, less than the critical transport density, volume becomes limiting. In this case, the ratio between payload and vehicle dead weight decreases, meaning that transport energy per ton of transported good increases, making the transport more expensive and causing higher environmental pressure. Table 1 lists important properties of selected bio-resources, according to Narodoslawsky (2019). Data for light fuel oil are also included in this table for comparison.

It can be seen from this table that bio-resources have considerable moisture content and lower heating values than light fuel oil while many have also low densities. This applies, in particular, to harvest residues. The effect of these properties can be seen when calculating the energy density of bio-resources: It ranges from 750 MJ/m³ for loose straw to 12,000 MJ/m³ for corn to 14,800 MJ/m³ for rape seed. This

Table 1. Properties of selected bio-resources (crops and harvest residues).

Bio-resource	Main Ingredient (MI)	MI content [% FM*]	Moisture content [% FM*]	Yield [t FM*/ha.yr]	Heating value [MJ/kg]	Transport density [kg/m³]
Corn	Starch	70	10–25	5–15	15	700–800
Wheat	Starch	70	10–25	2.5–7	15	700–760
Rape seed	Oil	40	6–10	1.3–4.2	26.4	500–560
Palm fruit	Oil	20	70	4–20	n.a.	610–660
Sugar beet	Sugar	15–20	75–80	40–95	2.7**	600
Sugar cane	Sugar	10–15	75	40–130	7	200–400
Soy bean	Protein/oil	34/18	10	2.9	23.2	700
Grass silage	Protein/ cellulose	1.2–2–5/n.a.	75–80	15–25	2.6**	400–600
Liquid manure	Nitrogen fert./ carbohydrates	0.3–0.9/ 2–4	90–95	–	0.7**	1,000
Miscanthus	Cellulose	32–42	12–20	8–17	14.2	120
Wheat straw	Cellulose	36–54	15	1.5–4.2	15	20–60/ 70–140****
Wood (forest)	Cellulose	29–32	35–42	3.5–6	6.8	300***/ 750–1,200
Corn cobs	Cellulose	39	15–30	1.2–2.2	10.4–13	130–220
Light fuel oil	Hydrocarbons	100	0	n.a.	42.7	840

* Fresh mass, as harvested from the field, before conditioning processes like drying.

** Represented as heating value of bio-methane generated by digestion.

*** Chips.

**** Compacted.

compares to almost 35,900 MJ/m³ for light fuel oil, revealing an inherent logistical disadvantage of bio-resources.

Table 1 also reveals limits on yields of the compounds used in industrial processes. For sugar plants, sugar yields can reach 16 t/ha.yr (sugar beet) to 19.5 t/ha.yr (sugar cane). Starch yields of corn (which has, by far, the highest yield of starch crops) can reach 10.5 t/ha.yr. Cellulose yields of forests are around 2 t/ha.yr and oil yields of oil seeds may reach 1.7 t/ha.yr for rape seed and up to 4 t/ha.yr for palm fruit.

Yields of harvest residues are generally lower and they also have low transport densities. This becomes even more important when the usual means of transport are analysed according to their critical densities in to equation (2) and the energy they require per ton kilometer. Table 2 lists these data. Transport energy rises for goods with densities below ρ_c as volume becomes limiting and the ratio of payload to dead weight decreases.

From Table 2, it becomes obvious that collection transport with tractors requires a relatively large amount of transport energy. This becomes more pronounced the lower the transport density of the good to be transported is, a problem that particularly arises with harvest residues like straw. Transport over longer distances will become more efficient when bio-resources are conditioned close to the point of harvest, either by drying (which reduces weight), compacting (which increases transport density) or pre-separation (which reduces the mass to be transported further). As a rule of the thumb, this becomes more important the lower the quality and the higher humidity of the bio-resource is. High quality, dry bio-resources, such as log wood, wheat, corn and rape seed, may easily be transported over long distances from collection points close to fields. This is exemplified by the global trade and transport of corn from the USA, soy bean from the USA and South America or the trade and transport of wood from Russia and Ukraine to European pulp and paper factories. Lower quality resources like low grade wood, grass, miscanthus and harvest residues should be utilised or conditioned close to harvest areas or points of emergence.

Until now, we have only looked at single resources. As a matter of fact, bio-resources are always the root of complex utilisation networks. Even in regions that are characterised by monoculture farming harvest residues and/or by-products from conditioning of bio-resources always arise. In many cases harvest residues and by-products have very different logistic parameters than main crops and local/regional closing of nutrient cycles is required to retain fertility of land. This means that, given the properties of low-grade bio-resources and the imperative to use the limited conversion rate of solar irradiation into biomass fully, bio-resource utilisation requires a combination of de-central and central technologies.

The situation becomes even more complex in regions where diverse bio-resources are cultivated and where stock farming is part of the agricultural profile. In such regions, not only does the supply of high quality bio-resources diversify, but the emergence of harvest residues, waste bio-resources and by-products from conditioning and bio-resource based technologies also proliferate. The challenge in such regions is to combine technologies that are necessary on the regional level to either condition bio-

Table 2. Critical transport densities and transport energy per t.km for selected means of transport.

Means of transport	ρ_c **in t/m³**	**Transport energy in kJ/t.km**				
		Corn	**Wood logs**	**Wood chips**	**Straw bales**	**Straw loose**
Tractor*	0.72	1.33	1.33	1.91	3.31	6.42
Lorry**	0.30	1.38	1.38	1.38	1.84	2.89
Rail (electric)***	0.39	0.18	0.18	0.20	0.30	0.51
River ship**	0.92	0.18	0.18	0.26	0.55	1.09

* Tractor data: http://deutzfahr.at/fileadmin/Bilder/Prospekte/Agrofarm_TTV_Profiline.pdf [April 2019]. Diesel consumption for 75 PS tractors from: Gastinger G.: Untersuchung des Kraftstoffverbrauchs in der 75 kW Traktorenklasse mit einem leistungsverzweigten und lastschaltbaren Getriebe, Master Thesis, University of Life Sciences Vienna, 2011. Available from http://epub.boku.ac.at/obvbokhs/content/titleinfo/1127129 [April 2019].

** Borken J., Patyk A., Reinhardt G.A.: Basisdaten für ökologische Bilanzierungen: Einsatz von Nutzfahrzeugen in Transport, Landwirtschaft und Bergbau, Vieweg & Teuber, Wiedbaden, 1999.

*** Data from: http://www.forschungsinformationssystem.de/servlet/is/342234/ [January 2017].

resources for further processing and transport or utilise (especially low-grade resources) them to cover the local/regional demand and ensure efficient nutrient management to retain fertility. This requires the "bio-refinery" approach, first introduced by Carlsson (1983), meaning a multi-input/multi-product technological scheme,[5] discussed elsewhere in this book in the contribution of Jhuma Sadhukhan. It may however be stated here that bio-refineries must be seen as technology networks that may comprise central and de-central steps, see Ecker et al. (2012), depending on logistic parameters of the resources and intermediate products (platforms) involved and taking local/regional demand into consideration.

Looking at the de-central aspect of such technology networks in particular, two factors are of specific interest: Liquid by-products and integration of energy in local/regional systems. The former is easily explained. Liquid by-products from stock farming, like manure, or from energy provision technologies, like digestate, or from the food industry, like whey, are characterised by very low concentrations of valuable compounds. Narodoslawsky (2019) argues, based on the data from Tables 1 and 2, that transporting manure for 5.3 km with either tractor or lorry would require 1% of the energy contained in the transported manure. Transporting such resources over long distances (without conditioning) is therefore economically and, given the fact that transport is still largely using fossil fuel, ecologically infeasible. They must, therefore, be utilised locally. In the case of resources that are mostly used as fertilisers, such as manure and digestate, it must also be considered that their dispersion on land is limited by agricultural and environmental considerations, depending on local circumstances. Given that their "last mile" is usually covered by tractor transport, this requires careful adaptation of location and size of processes where these resources emerge.

The integration of bio-based conditioning and utilisation technologies into local/regional energy systems requires a more detailed analysis. Bio-resources may be converted into different energy carriers, like bio-methane, bio-ethanol, bio-diesel, wood or straw pellets or bio-char. All these bio-energy carriers require technological treatment of the original bio-resource and the size and location of the site of this treatment depends critically on the logistic parameters of the original bio-resource. Again, as a rule of the thumb, the lower the quality of the bio-resource and the higher the humidity is, the smaller the size of the conversion technology and the more de-centralized the production. In terms of distribution, all liquid and solid energy carriers are fit for long distance transport, gaseous energy carriers may be transported over long distances if they are produced close to a gas grid.

Many low-grade bio-resources, however, may only be converted to energy and, given their disadvantageous logistic properties, this should be realised close to the point of their emergence. This is true for resources like low grade wood, straw, silage and manure. In general, there are two possible ways of utilising these resources: Dry resources (wood, straw, corn cobs, etc.), may be utilised thermally by technologies like combustion, gasification or pyrolysis, humid or liquid resources (silage, manure, etc.), may be digested. Combustion processes may render either heat or heat and power in combined heat and power (CHP) systems, in the case of gasification linked to gas cleaning and catalytic up-grading gaseous or liquid hydrocarbons (Fischer-Tropsch or FT fuel) may be obtained while pyrolysis renders liquid pyrolysis oil and bio-char. Digestion will produce bio-gas, which can either be combusted in a CHP system or cleaned to render bio-methane and used as gaseous bio-fuel or injected into gas grids. Liquid and solid energy carriers, like FT bio-fuel and pyrolysis oil or bio-char, can be transported easily over long distances. Bio-methane may also be transported over long distances if injected into existing gas grids. Electricity generated in CHP units may also be transported over long distances in existing grids. These products are typical export products from regional bio-resource utilisation systems.

The case is, however, different for heat, which is a natural by-product from CHP units and a main product from (usually small) bio-heat plants. Heat distribution is linked to considerable losses. These losses are dependent on the linear heat density (i.e., the cumulated heat load of customers divided by the length of the distribution line) and the temperature level in the distribution system. As a rule of thumb,

[5] The International Energy Agency (IEA) dedicated its Task 42 to the definition of biorefinery systems. The results of this Task are available at https://subsites.wur.nl/en/ieabiorefinery/Publications.htm [Dec. 2017]. The current text follows the logic laid down in IEA Task 42.

these losses decrease as the heat load per unit length of the distribution line increases and the temperature in the distribution system decreases. For more detailed information, see Nussbaumer and Thalmann (2014). As a result of the considerable losses of heat distribution systems, heat is a product that can only travel short distances, a few kilometres at most.

The consequences of this fact are decisive for the contextual nature of bio-resource utilisation. Heat is an almost inevitable (by-) product of almost any bio-resource based value chain and, given its short distribution range, its utilisation shapes optimal local and regional bio-resource exploitation systems; therefore, in order to utilise bio-resources fully, there must be an equilibrium of local heat sources and heat consumption. In addition to that, sustainability of bio-resource utilisation requires that nutrients be recycled locally as much as possible. These requirements shape the utilisation of bio-resources in their spatial context.

3. Rules for Establishment of Spatially Adapted Bio-Resource Utilisation

The properties of bio-resources, the characteristics of utilisation technologies and the requirement to keep the basic resource fertile land productive allows some generic rules for using these contextual resources to be formulated. Such rules are especially important for mobilising low-grade biomass, such as agricultural residues. Following an analysis of global agricultural production by Bakker (2013), such residues may amount to more than 80% of the current global crop production. Many of these residues, along with a considerable amount of by-products along the value chains of bio-resources, are under-utilised and usually have no application in the food sector. The rules given here have been laid down by Narodoslawsky (2016) and will here be discussed particularly with the spatial context focus in mind.

3.1 Use bio-resources fully within ecological limits

The relatively low conversion rate of solar irradiation into useful bio-resources, which is limited by the assimilation capacity of plants as well as the strong competition for these resources from the food, energy and chemical sector, necessitates the highest possible efficiency in their use. This, in turn, calls for fully utilising the resources within the limits posed by the often concurring need to retain land fertility. As a first step, it is, therefore, necessary to establish the rate of withdrawal of biomass from the land in question. This rate is critically dependent on the climatic situation and the state and quality of the soil as well as on the possibility to replenish nutrients by re-integrating by-products and waste flows from utilisation like manure, digestate or ashes. Within these limits, all biomass grown should be utilised. This includes harvest residues, which are predestined for energetic and industrial use as they usually have no competition from the food sector. This emphasises the importance of the concept of bio-refineries that are capable of utilising low-grade bio-resources.

The need for efficiency, however, applies also to the whole value chain based on bio-resources. Waste from these value chains represents an under-utilisation of the natural income represented in the original bio-resource and often an ecological problem if it is not sensibly integrated into the biosphere. In the current utilisation of food, almost a third of the original bio-resources end up as bio-waste along the whole food value chain. This amounts to a considerable scientific and technological challenge of finding technologies that can handle by-products and waste in innovative bio-refinery systems.

3.2 "Bio-refinerise" existing bio-resource based sectors

Many large industrial sectors, like food, timber, pulp and paper, fats and oils as well as existing bio-energy technologies, are already dependent on bio-resources. In their current way of operation, they are focussed on producing a particular service, product or class of products, generating wastes and by-products to be discarded or used by other sectors.

These bio-based plants have the advantage that they have already mastered the logistical challenges of collecting their bio-resource. They are, however, usually not able to utilise other bio-resources available in their spatial context and seldom able cover demand for other goods and services except for their particular product in their respective local/regional context. With their core know-how, infrastructure and functioning logistic for a particular bio-resource, they may become prime sites for more complex bio-resources, covering a broader portfolio of contextual resources and generating a more diverse range of products.

This will benefit the industry in question as well as the spatial context the industrial plants are in. Advantages for industry include economic stability due to serving different markets and better utilisation of existing capacities. Advantages for the local and regional context are a higher value added by utilising more of the available bio-resources and the provision of more goods and (energy) services by a local supplier. Besides that, increased activity at existing sites leads to more jobs and a general boost to the local/regional economy.

Care must be taken that increased removal of biomass is balanced by sensible management of nutrient cycles. Appropriate technologies to condition by-products and bio-waste from these bio-refineries for re-integration into ecosphere or agricultural activities (e.g., by providing feed or fertiliser) and an appropriate distribution logistic for these goods must be established.

A good example is the "Green Bio-Refinery" concept, as realised in pilot scale in Utzenaich, reported by Ecker et al. (2012). In this case, silage that used to be digested to provide bio-gas was pressed and lactic acid and a high value amino-acid product were separated from the juice. The remaining juice and the press cake were then digested and the digestate was returned to fields. This scheme allows the value gained from the primary resource to be increased by broadening the product portfolio.

3.3 Favour material goods over energy services

As the quest to de-carbonise the global economy progresses, bio-resources offer a well-established pathway to generate material goods. Furthermore, value added increases with the length of a value chain. This means that, to utilise just the energy content of a bio-resource is economically less attractive than converting it into material goods in longer value chains. A well-established example for this is the preferable use of wood in pulp and paper, giving rise to a long and diverse value chain rather than using higher quality wood as just an energy source.

When converting bio-resources into material goods, some heuristics may help in choosing the right bio-resource as the base for the right value chain. High quality crops that are not used for food or feed shall be reserved for processes generating high value products, such as pharmaceuticals or high quality bio-polymers. This is due to the fact that these processes involve considerable separation and polishing steps and that any additional input of inert material or contaminants with the raw material should be avoided.

Bio-resources of lower quality (grass, straw, low quality wood, etc.), that often also have disadvantageous logistic parameters, shall be utilised in de-central bio-refineries to produce platform materials that may be up-graded in central, large scale industrial sites. These platform materials range from simple pellets made from harvest residues or low-quality wood, to chemicals like lactic acid to pyrolysis oil and bio-char. What platform material (or combination of materials) shall be produced depends on the resource availability as well as the availability of existing sites of conventional bio-based sectors in the regional context.

Energy should be generated from the lowest quality bio-resource, such as waste wood, low grade harvest residues and manure, or other agricultural waste flows.

3.4 Retain as much material as possible in the region

This rule follows from the importance of retaining fertility of the land and the fact that most bio-waste or residues from bio-based processes are either very humid or have low transport densities. Replenishing the nutrients required in order to keep land fertile means that as much as possible from the harvested biomass that is not converted in tradeable products must be returned to the land. This is achieved by "rear guard technologies" that condition bio-waste flow for re-integration into agricultural and forestry operations, if possible while generating energy. Examples for such rear-guard technologies are digestion of liquid waste, combustion of dry solid waste and residues or composting of humid bio-waste. An appropriate distribution management and logistic system, taking the particular properties of the products of these technologies into account, must be established, too.

3.5 Use intersection points of distribution grids as sites

A major advantage that bio-resources bring to the table when they are used to generate energy is their storability. Being material energy carriers, they may be used "just in time" when energy is needed. This distinguishes them from most other renewable energy forms that are either cyclical (direct solar irradiation) or dependent on weather conditions (wind energy). The larger the share of solar and wind power in an electricity grid becomes, the more demand arises for energy forms that are able to cover the gap between demand and fluctuating provision. Storage of surplus electricity in batteries is one (costly) possibility to cover the gap, the use of reservoir power stations is another that is, however, restricted by the availability of suitable sites. The use of low-grade bio-resources as energy carriers for input to electricity provision technologies that can operate on demand is an attractive alternative to these possibilities to stabilise electricity grids.

Using bio-resources for energy generation involves thermal conversion that provides heat. If this heat is used to generate power, a part of it inevitably becomes available as lower temperature heat that can still be used to cover domestic or industrial heat demand.

In digestion and gasification, the bio-resource is first converted to a gaseous product. This product may either be cleaned (in the case of digestion) or converted (in the case of gasification) to bio-methane or combusted directly to again provide power and heat. Such tri-valent technologies are capable of providing three energy forms that may be distributed over grids: Gas, power and heat.

In order to fully exploit the energy contained in bio-resources, it is necessary to provide as many energy services as possible. This means to serve as many distribution grids as possible. These distribution grids have very different logistical characteristics. While electricity may be transported over long distances, the power grid has almost no capacity to store electricity. In contrast, heat, as mentioned earlier, may only be transported in the local context of a few kilometres, but may be easily and cheaply stored for days. The gas grid provides both long distance transport and, usually, large storage capacities. Putting bio-energy technologies at the intersection of grids allows for negotiation between them and usage of the full economic potential of bio-energy. Technologies providing heat and power (CHP technologies) should be operated in a way as to stabilise the power grid, providing electricity whenever there arises a gap between supply and demand. The heat inevitably provided as a couple product shall be stored and used when heat demand arises. Tri-valent technologies that may produce bio-methane should primarily inject bio-methane as long as there is no electricity demand and should also store heat to be used when heat demand arises.

As heat is most restricted in its distribution, bio-energy technologies can become hubs serving heat to other processes. Ideally, bio-energy sites utilise low grade bio-resources and wastes, meaning that they are typical de-centralized technologies. Heat, however, is a main input for many conditioning processes, such as drying. These technologies are also located close to harvest areas and are de-central. Combining bio-energy sites with these technologies, as well as with other industrial processes requiring heat, will improve the efficiency and economic potential of the use of bio-resources along the whole value chain considerably.

A good example is the bio-gas plant "Graskraft Reiterbach" in Eugendorf/A.[6] This bio-gas plant serves the regional electricity grid and a local heat grid when operating in CHP mode. If no electricity is

[6] For details see the country report summery 2015 of IEA Task 37 on http://task37.ieabioenergy.com/country-reports.html [last accessed June 2019].

demanded, bio-gas is separated and bio-methane is cleaned to natural gas standards. This bio-methane is either injected in the gas grid or sold as fuel in a bio-methane filling station for vehicles.

4. Planning Contextual Bio-Resource Utilisation

Bio-resource utilisation in the local and regional context requires a complex and strongly interlinked network of technologies, involving a large number of actors. Planning such technology networks can be aided by optimisation methods on the basis of Mixed Integer Linear Programming, described elsewhere in this book. Another approach involves combinatorial methods to optimise technology networks, as presented by Cabezas et al. (2015). In all these models, the strong interaction between utilisation pathways of different bio-resources in a given region, as well as the formative nature of the use of heat on local technology systems can be taken into account. This helps the large number of involved actors recognise interdependences between technologies, as well as between technologies and logistics. A region-centred method for designing and evaluating the ecological and economic parameters of such bio-resource utilisation systems is discussed by Kettl et al. (2011b).

5. Conclusion

Bio-resources transform solar radiation into material products, using carbon dioxide, water and nutrients in the process of assimilation. This process can only utilise a small fraction of the irradiation and, therefore, requires large areas of fertile land. This makes bio-resources into de-central resources that require sophisticated collection logistics. The kind of bio-resource that can be grown as well as the yield per area unit depend on local conditions of climate, fertility and cultivation know-how. This means that bio-resources are contextual resources.

Moreover, bio-resources are humid to varying degrees and often show low transport density. These aspects become more important the lower the quality of the resource is. While high quality bio-resources may easily be transported over long distances, the lower the quality of the resource, the closer to the point of harvest it must be utilised or at least conditioned for further transport or processes. This makes bio-resource utilisation also contextual, being shaped by the portfolio of resources available while shaping the social and economic structure of the location and region. Planning the most optimal utilisation system for bio-resources within a local and regional context requires that not only forestry, agricultural and technological considerations but also logistics, existing distribution grids for energy and the structure of the local and regional economic system be taken into account.

References

Bakker, R.R.C. 2013. Availability of lignocellulosic feedstocks for lactic acid production. Report 1391, Wageningen UR Food and Bio-based Research, Wageningen/NL.

Cabezas, H., Heckl, I., Bertok, B. and Friedler, F. 2015. Use the P-graph framework to design supply chains for sustainability. Chem. Engn. Progr. 111(1): 41–47.

Carlsson, R. 1983. Leaf protein concentrate from plant sources in temperate climate. pp. 52–80. *In*: Telek, L. and Graham, H.D. (eds.). Leaf Protein Concentrates, AVI, Westport.

Ecker, J., Schaffenberger, M., Koschuh, W., Mandl, M., Boechzelt, H.G., Schnitzer, H., Harasek, M. and Steinmüller, H. 2012. Green biorefinery upper austria-pilot plant operation. Separation and Purification Technology 96: 237–247.

Kettl, K.-H., Niemetz, N., Sandor, N. and Narodoslawsky, M. 2011a. Ecological impact of renewable resource-based energy technologies. Journal of Fundamentals of Renewable Energy and Applications (1): 1–5. DOI: 10.4303/jfrea/R101101.

Kettl, K.-H., Niemetz, N., Sandor, N., Eder, M., Heckl, I. and Narodoslawsky, M. 2011b. Regional optimizer (RegiOpt)–Sustainable energy technology network solutions for regions. Computer Aided Chemical Engineering 29: 1959–1963.

Narodoslawsky, M. 2016. Towards a sustainable balance of bio-resources use between energy, food and chemical feedstocks. Foundations and Trends® in Renewable Energy 1(2): 45–107.

Narodoslawsky, M. 2019. Bioresources and technologies. pp. 69–92. *In*: Krozer, Y. and Narodoslawsky, M. (eds.). Economics of Bioresources, Springer Nature, Heidelberg, in Press.

Nussbaumer, T. and Thalmann, S. 2014. Status Report on District Heating Systems in IEA Countries, prepared for the International Energy Agency IEA Bioenergy Task 32 and the Swiss Federal Office of Energy; verenum, Zürich; available from www.ieabioenergytask32.com [April 2019].

Schlömer, S., Bruckner, T., Fulton, L., Hertwich, E., McKinnon, A., Perczyk, D., Roy, J., Schaeffer, R., Sims, R., Smith, P. and Wiser, R. 2014. Annex III: Technology-specific cost and performance parameters. pp. 1329–1356. *In*: Edenhofer, O., Pichs-Madruga, R., Sokona, Y., Farahani, E., Kadner, S., Seyboth, K., Adler, A., Baum, I., Brunner, S., Eickemeier, P., Kriemann, B., Savolainen, J., Schlömer, S., von Stechow, C., Zwickel, T. and Minx, J.C. (eds.). Climate Change 2014: Mitigation of Climate Change. Contribution of Working Group III to the Fifth Assessment Report of the Intergovernmental Panel on Climate Change. Cambridge University Press, Cambridge, United Kingdom and New York, NY, USA.

Zhu, X.-G., Long, S.P. and Ort, D.R. 2008. What is the maximum efficiency with which photosynthesis can convert solar energy into biomass? Current Opinion in Biotechnology 19: 153–159.

CHAPTER 4

Prioritising Uses for Waste Biomass

A Case Study from British Columbia

Roland Clift,[1,2,*] *Xiaotao Bi,*[1] *Haoqi Wang*[1] and *Huimin Yun*[1]

1. Introduction: Biomass as an Energy Source

Before the general switch to fossil fuels that was a central feature of the industrial revolution, biomass was the main source of heat for industrial and domestic use and also, in the form of fodder for draught animals, of mechanical energy for agriculture and inland transport. Biomass was also a major source of chemicals, such as oils from seeds and fruits, and pharmaceuticals, in the form of herbal medications. The progressive legalisation of medicinal uses of herbal products, notably cannabis, is an example of the return to using vegetable materials for high-value products with tightly regulated supply chains (Clift, 2001). For bulk use of biomass, current interest centres mainly on the extent to which it can support a partial return from fossil carbon to the renewable carbon cycle (RCEP, 2004). However, the global human population is now an order of magnitude greater than that supported by biomass before the industrial revolution, and *per capita* energy consumption has increased by roughly another order of magnitude (e.g., Sørensen, 2012; Smil, 2017). The extent to which biomass can contribute to supporting the energy demand in the 21st century is, therefore, limited (Giampietro, 2009). Recognising that biomass is a limited resource introduces the further question of identifying those uses that bring the greatest benefit in managing carbon to mitigate greenhouse gas (GHG) emissions. This contribution explores how to identify the optimal uses of biomass by analysing a specific region and energy economy with conditions particularly favourable for biomass: The Canadian province of British Columbia (BC).

Biomass, particularly lignocellulosic ("woody") biomass, differs from most fossil fuels in that it is a distributed source with low energy density. Collection and transport are, therefore, more significant in the performance of systems for utilising biofuels than for conventional fossil fuels. In consequence, there are economic and environmental incentives for using the material locally or processing it into an energy-dense form before transporting it over long distances, in contrast to the approach, familiar in the petroleum and petrochemicals sector for example, of having large centralised facilities for processing and distribution. Given the relatively low energy content of biomass, there are also questions regarding how much it should be refined or processed before final use in order to keep a favourable energy balance over the fuel cycle (Clift and Mulugetta, 2007).

Figure 1 illustrates these concerns in simple energy balance terms, to show why the more the biomass is processed and refined, the greater the cost and energy losses are likely to be over the fuel supply

[1] Department of Chemical and Biological Engineering, The University of British Columbia, 2360 East Mall, Vancouver, British Columbia V6T 1Z3, Canada.
[2] Centre for Environment and Sustainability, University of Surrey, Guildford GU2 7XH, UK.
* Corresponding author: r.clift@surrey.ac.uk

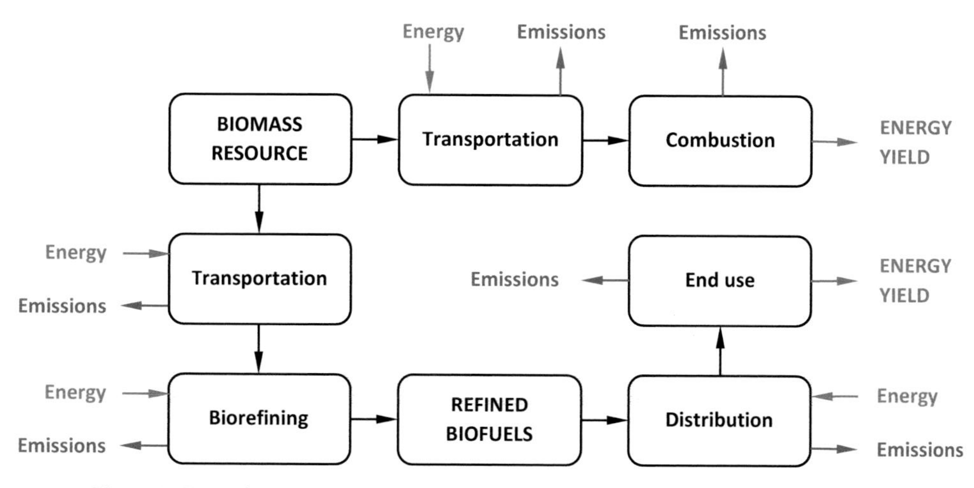

Figure 1. Energetic considerations in processing and using biomass, adapted from Clift (2012).

chain. Whatever the utilisation pathway, biomass used as an energy source is ultimately converted to combustion products; the energy available from combustion of the biomass is defined by its composition. If it is simply burned, as in the upper pathway in the diagram (Figure 1), the energy yield is the calorific value of the biomass, net of any energy inputs into cultivation, harvesting and transport. The lower pathway shows the case where the biomass is processed, for example into a liquid transport fuel. In this case, the processing plant is usually capital-intensive and therefore must be large to achieve economies of scale. Depending on the scale of the processing plant, the biomass may be drawn from a large area so that the transport distances and associated energy inputs are correspondingly large. Conversion of the biomass to refined products needs energy input and is also associated, as a thermodynamic inevitability, with energy (strictly, exergy) losses. Therefore, when the refined products are burned, for example in an internal combustion engine, the net energy release allowing for the energy inputs to transport and processing is inevitably smaller, and may be much smaller, than when the unprocessed biomass is simply burnt close to the point at which it arises.

The analysis in this contribution asks "What is the most beneficial use of the biomass resource?", rather than the question more commonly addressed: "How can this resource contribute to decarbonising a particular sector?" In the light of the increasing urgency to reduce GHG emissions to prevent a climate catastrophe (IPCC, 2018), the analysis considers technologies that are already deployable or are close to becoming deployable. The features of biomass use to be examined include quantifying the relative advantages of simple uses over "advanced" processing of biomass. In a carbon-constrained world, the objectives for biomass use must be to maximise life cycle energy yield or reduction in GHG emissions relative to fossil fuels, at the least cost; these objectives are not identical but are sufficiently aligned to give consistent prioritisation of the possible options for using biomass. Basing the analysis on a region relatively rich in biomass reveals the real constraints on the contribution that bioenergy can make to GHG abatement even under the most favourable circumstances.

2. British Columbia: A Suitable Case for Study

2.1 BC's energy system and climate aspirations

British Columbia (BC) is the westernmost province of Canada, between the Rocky Mountains and the Pacific Ocean with the U.S. states of Washington to the South and Alaska to the North West. BC has a number of broad characteristics that resemble other regions in the world, most obviously the Nordic countries: Relatively low overall population density but with most of the population—nearly 60% in the case of BC (Statistics Canada, 2019a)—concentrated in a few large- to medium-sized conurbations;

proportionately large forested areas, with large stocks and arisings of lignocellulosic biomass, particularly softwoods; and aspirations to be a leader in combatting climate change.

Currently, BC relies heavily on fossil fuels. Direct combustion of fossil fuels in the province generates 47.7 million tonnes (Mt) of CO_2 emissions, accounting for almost 80% of total provincial GHG emissions (Government of Canada, 2019), whilst upstream processing of fossil fuels contributes a further 13.6 Mt, i.e., most of the balance ((S&T)2 Consultants Inc., 2013). Total energy consumption in 2016 was estimated at 1165 PJ, of which 38% was supplied by Refined Petroleum Products (RPPs) used mainly in transport and 30% by natural gas, of which 42% (i.e., 12% of total energy use) was used for space and water heating in commercial and residential buildings (National Energy Board, 2019; Statistics Canada, 2019b). Thus, BC is typical of many developed societies: Reducing and decarbonising energy use in buildings is technologically unglamorous but is a key part of abating GHG emissions (Clift et al., 2006).

The average carbon intensity of electrical energy generated in the province is low—more than 90% of the electricity in BC is generated from renewable sources, primarily hydroelectricity—but there is some fossil generation, notably in small off-grid communities. However, the low carbon supply is fully committed, including for export to parts of the USA, so any substantial increase in demand in BC must be met by importing power, notably from the neighbouring province of Alberta where generation is dominated by fossil fuels so the carbon intensity of electrical energy is high. Conversely, expansion of low-carbon electricity generation in BC could offset imports of electrical energy with high carbon intensity. This raises methodological issues, discussed below, in assessing the environmental value of a new electricity source.

Canada ratified the Paris agreement in 2016, and committed to reducing GHG emissions by 30% below 2005 levels by 2030 (Environment and Climate Change Canada, 2017). BC has set more ambitious targets: To reduce emissions by 80% below 2007 levels by 2050 (British Columbia, 2016). It is important that these targets refer to GHG emissions within the country and province. This way of setting targets can have a perverse effect on selecting the best uses of bioenergy; this is explored in sections 4.6 and 5 below.

In support of its aspirations, BC was the first jurisdiction in North America to introduce a carbon tax, in 2008. It was accompanied by fiscal measures intended to make it revenue-neutral, so that it should not be seen as an additional tax. The tax was initially set at $10 (Cdn) per tonne of CO_2-equivalent but was subsequently raised in steps to reach $35/t in 2018 (i.e., nearly $130 per tonne of carbon), with plans for further increases. It is notable that BC's carbon tax is already at levels widely considered to be "politically unacceptable". For Canada as a whole, a carbon tax of $50 per tonne of CO_2-equivalent is projected (Government of Canada, 2017); this figure is used below as a reference value in assessing and prioritising possible uses of biomass.

In addition to the provincial carbon tax, RPPs are subject to various energy taxes. Bio-based fuels are not currently exempt from these taxes. However, the province's Renewable and Low Carbon Fuel Requirement (RLCF) currently requires a minimum "renewable" content of 5% in gasoline and 4% in diesel fuel, with plans to increase the RLCF to 15% by 2030 (British Columbia, 2016). Currently, most of the "renewable" fuel is imported from outside the province, but the expectation is that this will in time be provided from BC's own biomass resource. Tax on natural gas is much lower than on RRPs, because the government of BC sees natural gas as a source of "clean energy". However, development of Renewable Natural Gas (RNG) to displace fossil natural gas is promoted by a feed-in tariff.

Vancouver, BC's largest city, has announced a target of completely replacing fossil fuels with renewable energy by 2050 (City of Vancouver, 2015). The city's plans are based on a specific energy mix discussed in section 5 below.

2.2 Biomass resources in BC

Figure 2 summarizes the three major sources of waste in British Columbia, with their current uses or disposal routes.

The forestry sector in BC is the largest in Canada. It is estimated that 10.5 million tonnes dry matter (tDM) of woody biomass are currently available for energy production in BC every year, including unharvested

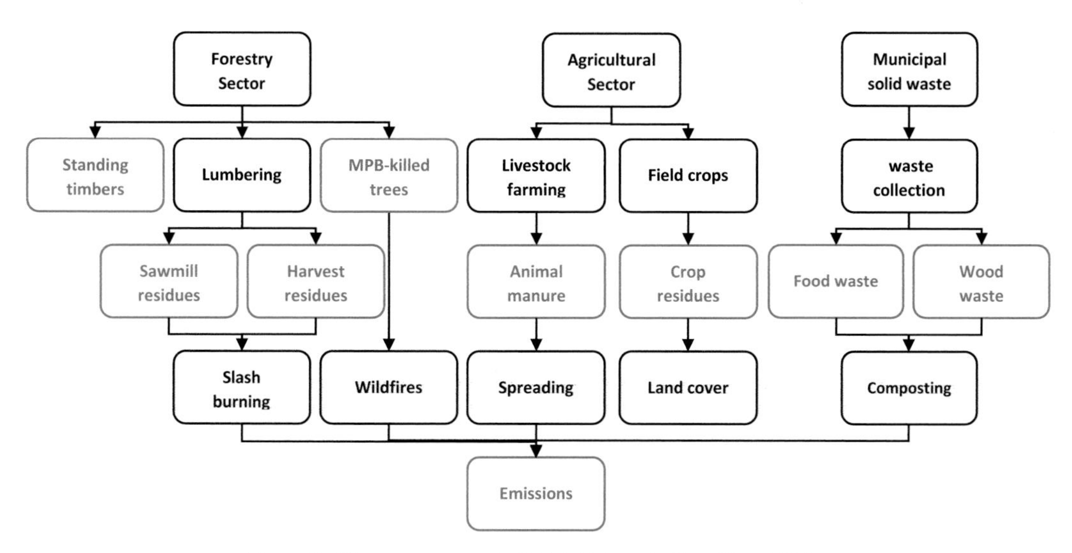

Figure 2. Biomass resources in British Columbia, from Wang (2019).

timbers within Allowable Annual Cut (AAC; i.e., the level of harvesting permitted by the province), trees killed by infestation of mountain pine beetle (MPB, a consequence of climate change), uncollected forest residues and sawmill residues (Wang, 2019). Of the forest and sawmill residues, 4.7 million tDM *per annum* are used in pulp, chip and pellet mills (British Columbia, 2019); the existing and potential trade in wood pellet fuel is discussed further in section 4.6 below. Exports of wood fuel amount to about 1.8 million tDM pa of pellets (British Columbia, 2018a) and 0.3 million tDM pa of chips (Natural Resources Canada, 2018), leaving 4.0 million tDM pa uncollected and potentially available for domestic use (Wang, 2019). In addition, at a rough estimate, about 4 million tDM of MPB-killed trees could be salvaged for energy production annually (Wang, 2019).

Forestry wastes are less costly than premium forestry resources and so represent potential feedstock for bioenergy. There are further incentives to harvest the wastes: Residues left *in situ* represent potential fuel for wildfires, so the province requires them to be burnt by uncontrolled open-air "slash burning" if they are not harvested. Thus, unused wastes are burnt to form carbon dioxide anyway, along with atmospheric pollutants with impacts on human health. This is significant in assessing the environmental implications of using the wastes for energy production; see section 3 below.

There are also unknown quantities of lignocellulosic waste generated by clearance of land for cultivation or construction and left *in situ*. At present, there is no market for this dispersed material, so it is also treated *in situ* by "slash burning" like other uncollected waste biomass. This raises the question, not explored here, of the potential use of mobile plants for chipping or pelleting of waste wood.

In addition, agriculture in BC generates about 7 Mt pa of animal manure and 0.45 Mt pa of crop residues (figures in wet tonnes with moisture content not specified). Roughly 0.7 Mt of food waste and 0.3 Mt of post-consumer wood waste *per annum* are also available (Wang, 2019). The quantities of animal manure, arising primarily from dairy and beefstock farming, are of particular interest: Disposal of this waste presents a problem, so there is an additional incentive to find beneficial uses.

3. Environmental Assessment

Assessing the environmental benefits and impacts of exploiting any source of energy requires consideration of the complete supply system for the fuel, its treatment and conversion, and delivery of usable energy, i.e., it requires Life Cycle Assessment (LCA) (Hammond et al., 2015). Because the focus is on assessing possible uses for a limited resource, the functional unit for the LCA (i.e., the basis on which alternatives are compared) is a mass of biomaterial.

For bioenergy crops, LCA is complicated by the need to consider changes in land use; this has generated heated discussion in LCA circles over two types of analysis termed *Attributional* and *Consequential* (e.g., Yang, 2016). However, the forms of biomass considered here are currently waste materials. This renders the assessment much more straightforward: the LCA can be carried out using an approach that is well established and incorporated in a number of software packages for assessing the environmental effects of alternative strategies for waste management allowing for recovery of energy and materials. The basic approach is shown in Figure 3. The waste to be used or treated exists regardless of the activities to manage it, so the processes generating the waste are common to all systems to use or manage the material and can, therefore, be omitted from comparative assessment. The starting point for the system is, therefore, the available and unused "waste" biomass. Even with this simplification, there is a methodological issue that affects the outcome of the assessment; it is discussed in section 4.5 below in connection with anaerobic digestion of food waste and animal manure.

Figure 3 shows the distinction between the *Foreground System*, comprising the operations whose selection or mode of operation is affected directly by decisions based on the study, and the *Background System*, i.e., other processes connected to the Foreground by exchange of energy and/or materials. The distinction is pragmatic, but a sufficient (but not necessary) condition for processes to be in the Background is that the exchange with the Foreground takes place through a homogeneous market. Ideally, the Foreground activities should be described by primary data, but the background can be described by generic data, such as those available in databases; in this work, GHGenius ((S&T)2 Consultants Inc., 2013) was used. The LCA must allow for the exchanges between the Foreground and Background, including energy recovery from the biomass "waste". It is assumed that the other functional outputs from the Background system are unchanged, so that the recovered energy displaces energy from other sources. The total inventory—i.e., resource inputs and environmental emissions—comprises the direct burdens associated with the Foreground plus the indirect burdens associated with Background production of materials and energy used in the Foreground minus the avoided burdens displaced from the Background by energy (and, where relevant, materials) recovered from the waste and from "slash-burning" of waste that would otherwise not be put to beneficial use. This modelling approach enables the assessment to account for the efficiency of the technology used in the Foreground to process the biomass and also to describe different uses for the energy recovered from the biomass.

In this work, the Foreground processes for forestry wastes include activities at the source (felling, skidding, etc.), pelletisation where relevant, and transport by rail or heavy diesel truck to the bioenergy conversion plant. Further details are given by Wang (2019). Transport distances have been based on the averages for BC but the overall results are not sensitive to this detail. Inputs to and emissions and transport from the conversion plant are specific to each technology. The focus here is primarily on carbon management, i.e., on life cycle GHG emissions, but human health impacts were also assessed and are a significant consideration in the hierarchy of preferred uses; see section 4.4 below.

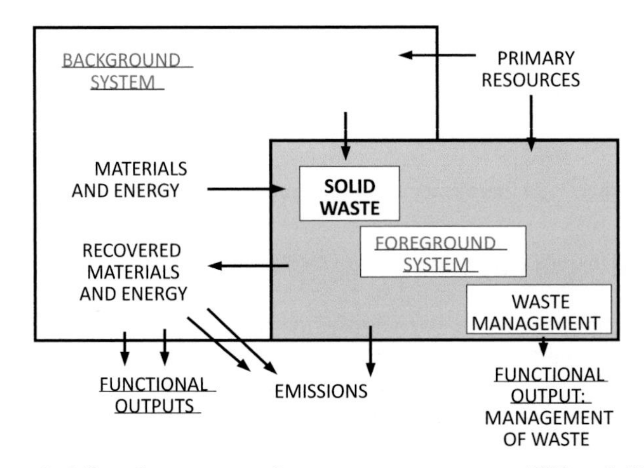

Figure 3. Life cycle assessment of waste management systems (Clift et al., 2000).

The climate-forcing impacts of emissions are evaluated in terms of CO_2-equivalents based on the values for Greenhouse Warming Potential (GWP) given in the IPCC Fifth Assessment Report using a time horizon of 100 years (IPCC, 2013). Following common convention, GWP values for a time horizon of 100 years are used, although the increasing urgency of the climate crisis argues for using a shorter horizon. Because carbon in the biomass forms part of the renewable carbon cycle, removed from the atmosphere by incorporation into biomass grown to replace that used, carbon dioxide emitted from combustion of biomass is conventionally treated as climate-neutral; i.e., it does not contribute to climate change (IPCC, 2011). This is common practice in the LCA of waste management (Astrup et al., 2015), and is further justified in the present study because the biomass will be burned anyway (see Figure 2) so that use as a traded fuel does not lead to additional carbon emissions. For carbon emitted in forms other than carbon dioxide, notably methane, the impact assessment allows for the different GWP. Emissions from other energy sources supplied from the Background and used in the Foreground are treated as non-renewable.

4. Uses of Biomass

4.1 Technologies and efficiencies

As noted above, in view of the urgency of the climate crisis, only technologies already deployable or considered close to deployable have been assessed. The overall energy conversion efficiencies cited in this section refer to net conversion of the higher heating value of the feedstock into that of the product in the conversion process, and are representative median values based on available literature. Conversion efficiency is the dominant parameter determining the ranking of technological alternatives (Wang, 2019; Wang et al., 2020), as expected from Figure 1. The LCA also allows for other processes in the supply chain, including transport; again, as expected from Figure 1, the results are sensitive to transport distances for forestry waste materials. The avoided burdens arising from the use of bioenergy are assessed for the specific conventional energy source replaced, mainly using the GHGenius database ((S&T)2 Consultants Inc., 2013).

4.1.1 Heat, power and cogeneration

Gasification of biomass with combustion of the product gas has been assessed as a relatively clean form of the simple combustion route (i.e., the upper route in Figure 1). The energy can be used for space and water heating in commercial, institutional and residential buildings via district heating; the efficiency for this use is taken as 73%. For heat output only, the biomass displaces natural gas, the most widely used fuel for heating in BC. Alternatively, the energy can be used to generate electric power; the results presented here are based on a conversion efficiency of 38%. For the initial assessment, the output is taken to replace the average supply in BC but the significance of this assumption is explored in section 4.5 below. The gasified biomass can also be used for Combined Heat and Power generation (CHP). The efficiency is taken here as 30% for electrical output and 50% for heat; i.e., a total overall efficiency of 80%.

4.1.2 Gasification to synthesis gas

Woody feedstock can also be gasified to produce synthesis gas (a mixture of H_2, CH_4, CO and CO_2), for subsequent conversion to an energy-carrier product. Energy inputs and losses are substantial (cf. Figure 1): The overall energy conversion efficiencies have been taken here as 54% for methanol production, 45% for ethanol and 58% for renewable natural gas (RNG).

4.1.3 Hydrothermal Liquefaction

Hydrothermal Liquefaction (HTL) converts woody feedstock into a mix of liquids that can be separated into gasoline, diesel, jet fuel and fuel oil. The energy supplied to the process and embodied in the other inputs is significant: The net energy efficiency of the process is less than 60% (Nie and Bi, 2018). HTL also produces a solid co-product containing carbon and nitrogen; the LCA includes avoided burdens, assessed assuming that it is used as a soil improver and displaces nitrogen fertiliser.

4.1.4 Anaerobic Digestion

Anaerobic Digestion (AD) can be used to convert food waste, animal manure and crop residues into biogas, typically containing 60% methane and 40% CO_2, with traces of other gases that impart the characteristic odour. Methane is a strong greenhouse gas, so fugitive emissions are a significant part of the GHG assessment; fugitive emissions are taken here as 2% of the total methane produced (Evangelisti et al. 2015). The indirect burdens (cf. Figure 3) associated with electricity and heat consumption in feed preparation and digestion are estimated as fractions of the energy content of the biogas: 11% for electricity and 13% for heat (Berglund and Börjesson, 2006). Overall energy efficiency depends on the use of the biogas. If it is used close to the AD unit, for heat, electrical power or CHP, it may not be necessary to separate the CO_2 from the methane before combustion. However, using the methane as renewable natural gas, for example for distribution via a gas grid, requires separation and purification processes that lower the net energy efficiency and raise the cost. AD also yields a digestate residue rich in soil nutrients. Avoided burdens have been calculated assuming that the digestate is used to replace synthetic fertilisers (Evangelisti et al. 2014).

4.2 Assessment of costs and benefits

To assess the economic viability of different applications, the minimum selling price (MSP) of energy products was estimated as the price at which the Net Present Value of a new plant is zero after 20 years, allowing for a real rate of return of 10% and inflation at 2% pa (Wang, 2019; Wang et al., 2020). Comparison of the MSPs with current commodity prices gives an indication of which technologies may be viable. However, this comparison is specific to the current BC context. It is more informative to show the results of the environmental and economic assessment in terms of possible contribution to GHG abatement and the associated cost.

Figure 4 shows the reduction in GHG emissions estimated as potentially available and the associated extra cost (not allowing for carbon tax), for the different combinations of feedstock and technology. The gradient of the line from the origin to the point representing each waste stream and technology represents the associated marginal GHG abatement cost per tonne of CO_2-equivalent. For comparison, lines are shown for the currently projected Canada-wide carbon tax of $50 per tonne and for a tax at double this level, i.e., $100 per tonne. Points with negative costs represent technologies that are economically viable even in the absence of a carbon price. Points below the $50 line identify technologies that would be attractive with carbon price at that level. Technologies represented by points above the $100 line would require measures in addition to the carbon tax before they would be pursued.

The Marginal Abatement Costs (MAC) indicated by Figure 4 are broadly consistent with other studies (Wang, 2019). The minimum values are generally lower than values suggested by other studies, but this is to be expected given that the majority of studies have investigated wood chips and whole logs, whereas the waste materials considered here tend to have shorter supply chains. The maximum values obtained in this work, for whole wood logs, are consistent with literature data. A systematic sensitivity analysis shows that wide variations in the most significant variables do not change the ranking of options for use of the available biowastes that emerges from Figure 4. Overall, the priority order appears to be robust.

4.3 Prioritisation of uses

Figure 4 shows that District Heating (DH) fuelled by wood waste, crop residues and forestry waste has negative GHG reduction cost; i.e., it should be viable even in the absence of a carbon price. This confirms the conclusion from other studies (e.g., RCEP, 2004; Tagliaferri, 2018) that District Heating (with or without CHP—see below) should be the priority use for woody biomass because of its cost-effectiveness

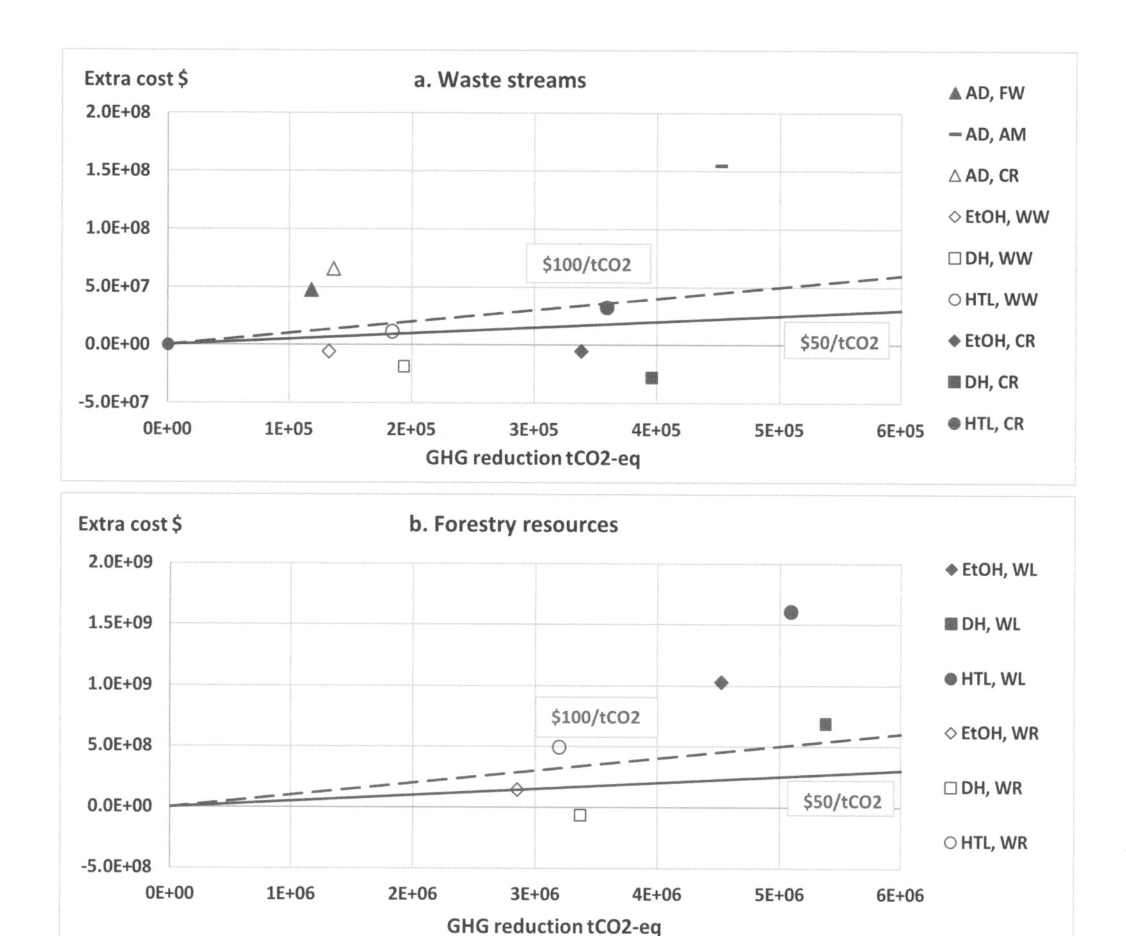

Figure 4. Total GHG reductions and costs (Wang et al., 2020).

Abscissa shows the total GHG reduction achievable by using all the available biomass in the way specified; ordinate shows the associated extra cost in Canadian dollars. Note difference in scales between Figures 4a and 4b.
Applications: AD – Anaerobic Digestion; DH – District Heating; EtOH – Ethanol produced via gasification and synthesis gas; HTL – Hydrothermal Liquefaction.
Feedstocks: AM – Animal Manure; CR – Crop residues; FW – Food Waste; WL – Wood Logs; WR – Wood Residues; WW – Wood Waste.

in GHG abatement. The alternative of converting the biomass to RNG for use in heating is less efficient and more expensive (Wang et al., 2020). District Heating fuelled by wood logs offers a much larger contribution to GHG abatement but is more expensive: To be viable, it would require a carbon price greater than $100 per tonne. Given the concentration of the population of BC in urban centres, District Heating is technologically viable. However, there are barriers to the adoption of biomass-fired District Heating, discussed in section 4.4 below.

Anaerobic Digestion (AD) appears to be associated with high costs, but this is a more complex case, discussed further in section 4.5 below: The possible contribution to GHG abatement is arguably underestimated in Figure 4, so the abatement cost is overestimated. There are further reasons to promote AD: It represents a beneficial use of wastes that currently present disposal problems.

Figure 5 shows the extent of GHG abatement achievable as a function of abatement cost, derived from the results in Wang et al. (2020); it can be interpreted as showing which technologies become viable at different levels of carbon tax. It shows that using all the waste biomass available in BC could reduce GHG emissions from the province by more than 10 million tonnes per year, representing more than 15%

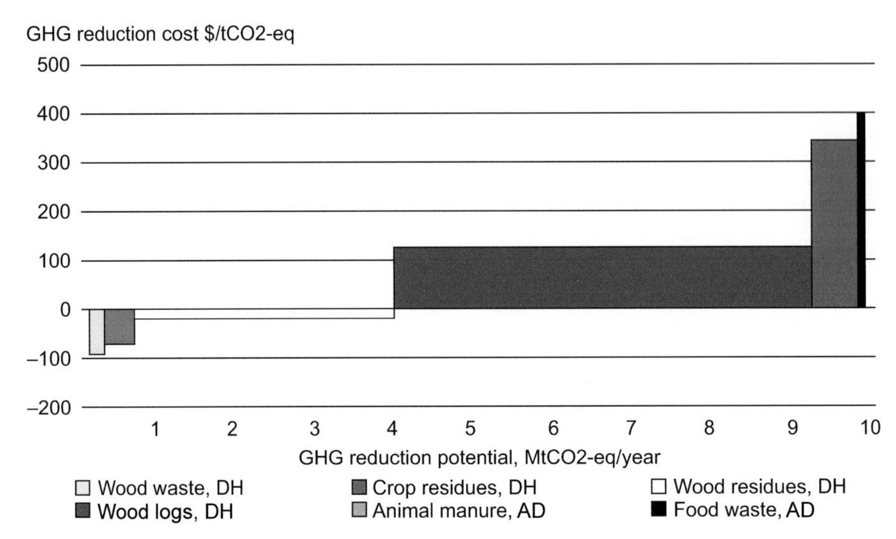

Figure 5. Potential for GHG Abatement and Marginal Costs for Different Fuels and Uses: DH - District Heating; AD – Anaerobic Digestion.

of current emissions. This confirms that energy from biowastes has an essential role in decarbonising the energy system. It is also notable that the uses foreseen in Figure 5 are all within current demand; i.e., Figure 5 represents the priority domestic uses for all the available resources shown in Figure 2 and these demands would use all the biomass available.

Hydrothermal Liquefaction (HTL) appears in Figure 4 as the most expensive option for lignocellulosic feedstocks: HTL of wood logs is associated with an abatement cost in excess of $300 per tonne, although this is still lower than the simple abatement cost for AD. HTL of wood waste (WW) and wood residues (WR) is less expensive but the quantities available are too small to make a significant contribution to decarbonisation. Gasification to produce simple energy carriers, such as ethanol or methanol, particularly from crop residues, appears to be more viable, and could compete with the uses shown in Figure 5. Methanol production is not shown in Figure 4 but the conversion efficiency is greater than for ethanol. In view of the current interest in methanol as a potential marine fuel (e.g., Andersson and Salazar, 2015), this appears to be another possible use with more realistic potential than liquid fuels produced by HTL.

4.4 District heating

Figure 5 demonstrates the GHG and economic benefits and potential scope of using waste biomass as a fuel for District Heating. However, adverse impacts on human health must also be considered at this point. They arise primarily from final combustion, with other stages in the life cycle being much less significant. Therefore, human health impacts are relevant in decisions that lie outside the scope of conventional LCA (Clift et al., 2000): Siting and scale rather than selection of generic technology.

Rabl and Spadaro (2000), amongst others, have shown that the human health impacts of emissions from combustion plants, specifically particulates and oxides of nitrogen and sulphur, can be at least as significant as the contributions to global climate change. However, these emissions are specific to any combustion plant; amongst other factors, they depend on the gasification and combustion process and on the plant's gas cleaning system. In the terms of Figure 3, this means that the combustion plant must be treated as part of the Foreground and described by data specific to the particular plant. Furthermore, in contrast to usual practice in LCA, which bases assessment of health impacts on simple generic exposure models (see e.g., Spadaro and Rabl, 1999), assessment of health impacts of operations in or close to heavily populated areas also needs to consider local conditions. The precise location and even the timing of operation of the combustion plant can affect human exposure and, hence, the impacts of emissions

(Petrov et al., 2015). As a specific example, Petrov et al. (2017) assessed the health impacts caused by emissions from a real biomass-fired district heating demonstration plant in BC, located at the University of British Columbia in an outlying suburb of Vancouver. The analysis was based on detailed operating data. The plant is typical of small- and medium-sized installations, with abatement of particulate emissions but no NO_x suppression. Compared to natural gas, the plant results in significantly increased harm to human health, primarily due to NO_x emissions.

The results of Petrov et al. (2017) illustrate the point that, to avoid introducing significant health effects, DH plants without highly effective gas cleaning must be located away from population centres in order to avoid human exposure, for example to provide heating and, if CHP plants are used, some electrical power in small remote communities. For larger population centres, high efficiency gas cleaning (including NO_x suppression) is essential, implying that plants must be large enough for this to be economically feasible (Guest et al., 2011). In centralised biomass-fired DH plants, Selective Catalytic Reduction can reduce NO_x emissions by 80% whilst electrostatic precipitation can collect 90% of the particulate matter (Pa et al., 2011). Technologies to achieve significantly cleaner gas emissions are already available (Seville, 1997) and have been deployed in specific waste-to-energy and biomass fired plants (e.g., Tagliaferri et al., 2015; Tagliaferri et al., 2018). With efficient gas cleaning, conversion from heating plant at the scale of individual buildings to centralised heating plant can improve air quality; this has been one of the reasons for the continuing popularity of biomass-fired District Heating in the Nordic countries (EU, 2018; IEA, 2019). Concerns over local air pollution have prevented the development of at least one District Heating plant in BC (Lee, 2015) but there are signs of revived interest (BC, 2018b). To exploit this use of biomass to the maximum extent possible will probably require a high-profile demonstration plant with lower emissions than the existing demonstration plants in BC, as a route to greater public acceptance.

The potential constraints on the use of biomass for District Heating mean that some feedstock is available for uses other than those within BC, shown in Figure 5. However, this does not mean that material is available for expensive uses, such as HTL to liquid biofuels. Rather, it should mean that biomass will continue to be available for export to users outside the province. The international trade in wood pellets is assessed in this context in section 4.6.

4.5 Anaerobic digestion

Anaerobic Digestion also merits specific discussion. The way this type of LCA is formulated and the current pricing structure in BC both distort the assessment and make AD appear unattractive. Electricity production by AD has been assessed using avoided burdens (i.e., carbon credits) corresponding to the average low-carbon generating mix in BC. This makes AD for electrical power and CHP appear unattractive in BC: CHP emerges as more beneficial than generation of heat only if the power output displaces electricity with higher carbon intensity (Evangelisti et al., 2015; Tagliaferri et al., 2015). This is an example of a methodological issue, introduced in section 3, at the heart of the distinction between attributional and consequential LCA (Yang, 2016): Whether a new source or demand should be assessed according to the average or marginal supply. There is an argument in this case that production of electricity from biowastes in BC would replace marginal imports with a high carbon intensity; this can change the position of AD in the ranking of possible technologies (Evangelisti et al., 2014). If the avoided burdens are assessed as displacing high-carbon generation, for example, imports to BC from the neighbouring province of Alberta, the GHG savings for AD to power CHP become much greater and the marginal abatement cost is correspondingly much smaller. However, the overall benefits may still be undervalued by current GHG accounting protocols that only recognise changes associated with emissions arising within the province: Any reductions in emissions from electricity generation would be credited to Alberta rather than to BC. This is one instance of a more general problem discussed in ensuing sections: That the current approach to accounting and attributing reductions can distort policies aimed at reducing GHG emissions.

The relative environmental benefits of producing biogas as a substitute for natural gas also depend on the way the avoided burdens from electricity production are estimated: If the electricity displaced by AD has a very low carbon intensity, purifying the biogas for use as RNG rather than using the raw gas to produce electricity can lead to greater GHG benefits (Tagliaferri et al., 2015). Set against this, the pricing structure for renewable natural gas in BC is complicated, analogous to the structure for pricing "green" electricity in Europe. Consumers can contract to use RNG, but for a premium price, currently $6.75 per GJ above the base price paid for natural gas. This additional premium is the source of the feed-in tariff available to RNG suppliers. However, gas prices in BC are in any case low, so production of RNG is not economically attractive and it is, therefore, not amongst the options considered in Figure 5.

Additional benefits can be realised from AD if it is integrated locally with other activities. As a notable example, using biogas from AD of animal manure to provide heat and CO_2 for greenhouse cultivation can yield very large GHG savings at low cost (Zhang et al., 2013; Zhang et al., 2015). This goes beyond the GHG savings considered in Figures 4 and 5.

4.6 Biomass exports: Wood pellets

In addition to the uses of biomass introduced in earlier sections, there is already a substantial industry in BC producing wood pellets for use as fuel. As noted in section 2.2, about 1.8 million tonnes of wood pellets and 0.3 million tonnes of chips are exported annually, much of it to Europe, although there is potential scope to expand exports to Asia (Yun, 2018). Canadian wood pellets compete mainly with pellets from the South Eastern USA. The pellets are used primarily for power generation, where they displace coal. Some generating plants in Europe changed from coal to wood-fuel in order to avoid closure under the decarbonisation programme of the European Union, and are considering implementing carbon capture and storage so that they can claim to represent a "negative emission" operation by sequestering carbon from biogenic sources.

Whether wood pellets should really be considered a "renewable" or low-carbon fuel has been questioned (e.g., Coath and Pape, 2013; Drouin, 2015). However, the criticisms centre on forestry practices: Possible deforestation, compounded by harvesting of slow-growing hard-woods. Whether or not these are valid criticisms of pellet production in the USA, they do not apply to the industry in BC, which is based on waste materials that would otherwise be unused. The relevant question for wood pellets from BC is whether the burdens associated with transport to the end-user are so large as to offset the avoided burdens from displacing the use of fossil fuel. Even for pellets shipped from Vancouver to Europe via Panama, the life cycle GHG emissions only amount to about 15% of the emissions avoided by replacing coal (Yun et al., 2020). Shipping accounts for about 70% of the life cycle emissions; i.e., it reduces the avoided emissions by about 11%. Over the whole life cycle, BC wood pellets used to replace coal for power generation in Europe reduce global GHG emissions by up to 0.9 kg of CO_2-eq. per kWh of electrical output (Yun et al., 2020). The marginal abatement cost over the life cycle is in the range $90 to $130 per tonne of CO_2-eq., comparable with the technologies represented in Figure 5 (Yun, 2018). From the BC perspective, the trade in wood pellets is profitable, generating a revenue without government support. It also eliminates the emissions that would otherwise arise from burning the wood waste. Therefore, within the province, it should be seen as a technology with a negative abatement cost.

In the present context, it is notable that this successful and environmentally beneficial industry is barely mentioned in BC's "Climate Leadership" plan (British Columbia, 2016) and is not identified for support. Under the current international protocols for GHG accounting, the "credit" for replacing fossil fuels by low-carbon biofuels is attributed to the user rather than the producer of the fuel. Therefore, exporting wood pellets does not contribute to meeting BC's climate targets, even though it does reduce global emissions of greenhouse gases and, thus, contributes to mitigating the global problem of climate change. The perverse consequences of the GHG protocols are explored further in the next section.

5. Discussion

The energy system in British Columbia, the abundance of waste biomass and the province's aspirations to be a leader in abating GHG emissions mean that the prospects for developing the bioenergy sector should be as good in BC as anywhere. However, the assessment presented here shows that even in BC the potential contribution of waste biomass in a low-carbon energy system is limited by the quantities of material available. Given that biomass is a constrained resource, the merits of alternative uses need to be assessed systematically: The question addressed must be "What are the best uses for biomass?" rather than "How can biomass contribute in a specific sector?"

Furthermore, the definition of "best uses" is itself problematic. The examples of wood pellets and CHP outlined here show how protocols to account for and attribute GHG emissions can lead to perverse conclusions and policies. The uses of biomass that appear attractive in meeting provincial and national targets are not necessarily those with the greatest global benefit. At a more general level, the perverse consequences extend as far as encouraging policies to reduce national or provincial GHG emissions but at the same time promoting export of carbon-intensive fuels (such as bitumen, a topic of notorious and intense current debate in Canada) for use elsewhere to contribute to other countries' GHG emissions.

Within the province, District Heating emerges clearly as the use for waste lignocellulosic biomass with the best greenhouse gas and economic performance. Converting biomass to hydrocarbon transport fuels by Hydrothermal Liquefaction is an intriguing technological challenge but is less beneficial or economically attractive: HTL technology could contribute to decarbonisation of transport but, on the basis of current performance projections, the associated cost will be much larger than the levels of support envisaged under current policies. Furthermore, the uses giving the greatest reduction in GHG emissions —District Heating and wood pellets exported for power generation—can account for the entire resource of waste biomass in BC. We, therefore, conclude that development efforts should first be concentrated on the less glamorous but probably more tractable challenge of reducing emissions from small- to medium-sized biomass-fired CHP and heating plants so that they can achieve the level of public acceptance seen in the Nordic countries. The tendency to focus on an exciting technological challenge rather than more mundane but more beneficial developments is not new and is not limited to BC (Clift and Mulugetta, 2007).

It was noted above that the city of Vancouver (in BC) has announced a target of eliminating dependence on fossil fuels by 2050. The target is to be met by following a trajectory including reducing energy consumption by about 35% and moving to an energy mix comprising 60% renewable electricity (mainly hydro), 15% district heating and cooling fuelled by biofuels, 14% biofuels for transport, 10% biomethane, and 1% hydrogen for niche uses in fuel cells (City of Vancouver, 2015). While the general aspiration is laudable, the rationale for this particular mix is not clear: targets and policies for decarbonising energy supply need to be based on the kind of analysis presented here, without making *a priori* assumptions over the best energy mix. The relationship between the energy systems of the city, the province and the country further illustrates the problems in reconciling plans at different spatial scales: Localising targets to the scale of a city raises the question of whether meeting the targets will involve transferring emissions elsewhere, for example, by using biomass that could otherwise be put to more beneficial use. Carbon management is a multi-scale "wicked" problem.

6. Conclusions

Bioenergy is indispensable to meeting British Columbia's decarbonisation targets. However, even in BC, biomass is a scarce resource, so uses should be prioritised, particularly given the immediacy of the global challenge to abate GHG emissions. Analysis of the cost and effectiveness of possible uses confirms the expectations from a simple thermodynamic analysis, that applications requiring the least processing and transport are cheapest and most effective and should, therefore, be prioritised. Use of biomass as fuel for district heating of commercial and residential buildings should be the highest priority, although some non-technical barriers to adoption must be overcome. Export of biomass to replace coal for power generation is also beneficial in reducing global GHG emissions, but the international protocols to account for emissions make domestic uses of the limited resource appear preferable. Anaerobic Digestion has a particular role as a beneficial way of treating problematic wastes, but is a further example which is undervalued by current carbon accounting protocols.

For liquid fuels produced from biomass, the more refined the fuel, the greater the cost and energy loss in producing it. Production of simple fuels, such as methanol or ethanol, may be viable if biomass supply exceeds the demands for which it can bring about the greatest GHG reductions. However, conversion of woody biomass into liquid hydrocarbon transport fuels by Hydrothermal Liquefaction (HTL) is associated with much larger GHG abatement costs. Major improvements in conversion efficiency and reductions in cost relative to the other possible uses will be needed before HTL can become a serious competitor in a carbon-constrained world where biomass is a limited resource.

References

Andersson, K. and Salazar, C.M. 2015. Methanol as a Marine Fuel. FCBI Energy Ltd., London, UK.

Astrup, T.F., Tonini, D., Turconi, R. and Boldrin, A. 2015. Life cycle assessment of thermal waste-to-energy technologies: Review and recommendations. Waste Management 37: 104–115.

Berglund, M. and Börjesson, P. 2006. Assessment of energy performance in the life-cycle of biogas production. Biomass and Bioenergy 30: 254–266.

British Columbia. 2016. Climate Leadership Plan. Victoria, BC, Canada.

British Columbia. 2018a. 2017 Economic State of the BC Forest Sector. Victoriia, BC, Canada.

British Columbia. 2018b. District Energy Sector in British Columbia. Obtained from http://www.biomassenergyresearch.ca/linked/districtheatinginbc.pdf.

British Columbia. 2019. Major Primary Timber Processing Facilities in British Columbia—2017. Victoria, BC, Canada.

City of Vancouver. 2015. Renewable City Strategy 2015–2050. Vancouver, BC, Canada.

Clift, R., Doig, A. and Finnveden, G. 2000. The application of life cycle assessment to integrated solid waste management: Part 1—methodology. Trans. IChemE B: Process Safety and Environmental Protection 78: 279–287.

Clift, R. 2001. Think global, shop local, roll your own. J. Ind. Ecol. 5: 7–9.

Clift, R., Sinclair, P. and Johnsson, F. 2006. Windows of opportunity. Parliamentary Monitor (section Energy), 141 (July/August), 47.

Clift, R. and Mulugetta, Y. 2007. A plea for common sense (and biomass). The Chemical Engineer October, 24–26.

Clift, R. 2012. An industrial ecology approach to the use of biomass waste. International Top-Level Forum on 'Approach for Achievement of Recycling Economy of Biomass Waste', Chinese Academy of Engineering, Beijing, China.

Coath, M. and Pape, S. 2013. Bioenergy: A Burning Issue. Royal Society for the Protection of Birds, London, UK.

Drouin, R. 2015. Wood pellets: green energy or new source of CO_2 emissions? Yale Environment 360, Jan. 22.

Environment and Climate Change Canada. 2017. Canadian Environmental Sustainability Indicators—Greenhouse Gas Emissions. Doi: 10.1016/B978-0-12-409548-9.05178-2.

European Union. 2018. Energy Statistical Pocketbook.

Evangelisti, S., Lettieri, P., Borello, D. and Clift, R. 2014. Life cycle assessment of energy from waste via anaerobic digestion: A UK case study. Waste Management 34: 226–237.

Evangelisti, S., Lettieri, P., Clift, R. and Borello, D. 2015. Distributed generation by energy from waste technology: A life cycle perspective. Process Safety and Environmental Protection 93: 161–172.

Giampietro, M. 2009. The Biofuel Delusion: The Fallacy of Large-scale Agro-biofuel Production. Earthscan, London, UK.

Government of Canada. 2017. Pan-Canadian Framework on Clean Growth and Climate Change. Ottawa, Canada.

Guest, G., Bright, R.M., Cherubini, F., Michelsen, O. and Strømman, A.H. 2011. Life cycle assessment of biomass-based combined heat and power plants-centralized versus decentralized deployment strategies. J. Ind. Ecol. 15: 908–921.

Hammond, G.P., Jones, C.I. and O'Grady, A. 2015. Environmental Life Cycle Assessment (LCA) of Energy Systems. Handbook of Clean Energy Systems, John Wiley & sons, Chichester, UK.

[IEA] International Energy Agency. 2018. World Energy Statistics. Retrieved from https://www.iea.org/statistics/?country=COSTARICA&year=2016&category=Energy%20supply&indicator=TPESbySource&mode=chart&dataTable=BALANCES, 2019.

Industrial Forestry Service Ltd. 2015. Wood-Based Biomass in British Columbia and its Potential for New Electricity Generation. Vancouver, BC, Canada.

[IPCC] Intergovernmental Panel on Climate Change. 2011. Special Report: Renewable Energy Sources and Climate Change; Annex II: Methodology. UN Environment Programme, Paris, France.

[IPCC] Intergovernmental Panel on Climate Change. 2013. Fifth Assessment Report, Working Group I: The Physical Science Basis. UN Environment Programme, Paris, France.

[IPCC] Intergovernmental Panel on Climate Change. 2018. Special Report: Global Warming of 1.5 ºC. UN Environment Programme, Paris, France.

Lee, M. 2015. Innovative approaches to low-carbon urban systems: a case study of vancouver's neighbourhood energy utility. Retrieved from https//www.policyalternatives.ca/sites/default/files/uploads/publications/BC%20Office/2015/02/CCPA-BC-NEU-Case-Study.pdf.

National Energy Board. 2019. Provincial and territorial energy profiles—British Columbia. Retrieved from https//www.neb-one.gc.ca/nrg/ntgrtd/mrkt/nrgsstmprfls/bc-eng.html.
Natural Resources Canada. 2018. Forest resources: Statistical data. Retrieved from https://cfs.nrcan.gc.ca/statsprofile/trade/ca.
Nie, Y. and Bi, X. 2018. Life-cycle assessment of transportation biofuels from hydrothermal liquefaction of forest residues in British Columbia. Biotechnol. for Biofuels 11: 1–14.
Pa, A., Bi, X. and Sokhansanj, S. 2011. A life cycle evaluation of wood pellet gasification for district heating in British Columbia. Bioresour. Technol. 102: 6167–6177.
Petrov, O., Bi, X. and Lau, A. 2015. Impact assessment of biomass-based district heating systems in densely populated communities. Part I: dynamic intake fraction methodology. Atmospheric Environment 115: 70–78.
Petrov, O., Bi, X. and Lau, A. 2017. Impact assessment of biomass-based district heating systems in densely populated communities. Part II: would the replacement of fossil fuels improve ambient air quality and human health. Atmospheric Environment 161: 191–199.
Rabl, A. and Spadaro, J.V. 2000. Public health impact of air pollution and implications for the energy system. Ann. Rev. of Energy and the Environment 25: 601–627.
[RCEP] Royal Commission on Environmental Pollution. 2004. Biomass as a Renewable Energy Source. The Stationery Office, London, UK.
(S&T)2 Consultants Inc., 2013. GHGenius Model 4.03 Volume 2: Data and Data Sources. Natural Resources Canada, Delta, BC, Canada.
Seville, J.P.K. (ed.). 1997. Gas Cleaning in Demanding Applications. Blackie Academic & Professional, London, UK.
Smil, V. 2017. Energy and Civilisation; A History. MIT Press, Cambridge, Mass, USA.
Spadaro, J.V. and Rabl, A. 1999. Estimates of real damage from air pollution: site dependence and simple impact indices for LCA. Int. J. LCA 4: 229–243.
Statistics Canada. 2019a. Population and Dwelling Count Highlight Tables, 2016 Census, Retrieved from https://www12.statcan.gc.ca/census-recensement/2016/dp-pd/hlt-fst/pd-pl/Table.cfm?Lang=Eng&T=801&SR=1&S=3&O=D&RPP=50&PR=59&CMA=0#tPopDwell.
Statistics Canada. 2019b. Table 25-10-0030-01: Supply and Demand of Primary and Secondary Energy in Natural Units. Retrieved from https//www150.statcan.gc.ca/t1/tbl1/en/tv.action?pid=2510003001.
Sørensen, B. 2012. A History of Energy: Northern Europe from the Stone Age to the Present Day. Taylor and Francis, London, UK.
Tagliaferri, C., Evangelisti, S., Clift, R., Lettieri, P., Chapman, C. and Taylor, R. 2015. Life cycle assessment of conventional and advanced two-stage energy-from-waste technologies for methane production. J. Cleaner Prod. 100: 212–223.
Tagliaferri, C., Evangelisti, S., Clift, R. and Lettieri, P. 2018. Life cycle assessment of a biomass CHP plant in UK: The heathrow energy centre case. Chem. Eng. Res. and Design 133: 210–221.
Wang, H. 2019. Environmental, Economic and Policy Analysis of Energy Production from Biomass Residues in British Columbia. Ph.D. Thesis, University of British Columbia, Vancouver, B.C.
Wang, H., Zhang, S., BI, X. and Clift, R. 2020. Greenhouse gas emission reduction potential and cost of bioenergy in British Columbia. Energy Policy 138: on-line.
Yang, Y. 2016. Two sides of the same coin: Consequential life cycle assessment based on the attributional framework. J. Cleaner Prod. 127: 274–281.
Yun, H. 2018. Wood Fuel from British Columbia: Multi-scale Assessment of the Economic, Energetic and Environmental Efficiencies of the Supply Chains of Conventional and Torrefied Wood Pellets. Ph.D. Thesis, University of British Columbia, Vancouver, B.C.
Yun, H., Clift, R. and Bi, X. 2020. Environmental and economic assessment of torrefied wood pellets from British Columbia. Energy Conversion and Management 208: on-line.
Zhang, S., Bi, X.T. and Clift, R. 2013. A life cycle assessment of integrated dairy farm-greenhouse systems in British Columbia. Bioresour. Technol. 160: 496–505.
Zhang, S., Bi, X.T. and Clift, R. 2015. Life cycle analysis of a biogas-centred integrated dairy farm-greenhouse system in British Columbia. Process Safety and Environmental Protection 93: 18–30.

CHAPTER 5

Industrial Oleochemicals from Used Cooking Oils (UCOs)

Sustainability Benefits and Challenges

Alvaro Orjuela

1. Introduction

Oil roasting, sautéing, and frying are ancient techniques that are frequently used for food processing and preparation in domestic and industrial environments worldwide. As the cooking oils and fats can reach high temperatures when heated (150–190 °C), contact with or immersion within hot frying oil accelerates the cooking process; in this case, the oil acts both as a heating medium and as lubricant. During this process, there is simultaneous mass and heat transfer among the oil, air and food, resulting in a complex set of physicochemical interactions that lead to the characteristic sensory profile of fried or roasted food. Due to the high temperature of the oil, the external surface of foodstuff dehydrates rapidly, losing water by steam evaporation in a bubbling action. After evaporation subsides, the characteristic porous crust of the fried food is obtained, while the internals are kept moist and soft, also acquiring a desirable odour, color, flavor and taste. This pleasing sensory profile is enhanced by chemical transformations such as the Maillard reaction, carbohydrate caramelization, protein gelatinization, oxidation, etc. Besides the convenience of fast cooking, the frying process also has a preventative action as a result of the thermal degradation of microorganisms, enzymes, and the reduction of the superficial water activity (Bordin et al., 2013).

Owing to the changes in human behavior after the industrial revolution, frying became a popular way to produce fast food and snacks (primarily potato chips), mainly in urban centers. At the end of the nineteenth century, large-scale open kettles were run under batch operation, doing the transition from domestic kitchens to garages, and subsequently to industrial factories. In 1929, the development of continuous fryers (Ferry's continuous potato chip cooker, J.D. Ferry Co.) started a new era in frying oil consumption in modern society; food frying systems were able to process from hundreds to thousands of kilograms per hour (Banks, 2007). This also boosted vegetable oil production with the consequent price reduction and higher accessibility for household users. Nowadays, the *per capita* world consumption of vegetable oils is around 21.2 kg/person/yr (OECD, 2020).

Department of Chemical and Environmental Engineering, Universidad Nacional de Colombia, 111321, Bogotá D.C., Colombia.

Food frying is generally performed using refined oils in order to avoid undesired off-flavors and to enhance oil thermal and chemical stability. These oils are mainly composed of triacyl glycerides of different fatty acids of an even number of carbon atoms, typically from C8 to C22, but particularly with higher content of C16 and C18. As presented in Table 1, the composition of refined oils and fats largely depends on their natural source, and they can contain saturated fatty acids and fatty chains with conjugated double bonds in their structure.

Table 1. Typical distribution of fatty acids in triglycerides of commercial oils and fats.

Glycerin backbone — Fatty acid chains

Oil or fat	Fatty acid type % wt. Saturated / Unsaturated	Carbon chain content in triglycerides, % wt.								Unsaturated C18 acids in triglycerides, % wt.			
		8	10	12	14	16	18	20	22	C18:0	C18:1	C18:2	C18:3
Palm	51 / 49				2	42	56			5	41	10	
Soybean	15 / 85					8	91			4	28	53	6
Rapeseed	7 / 93					4	93	1	2	1	56	26	10
Sunflower	8 / 92					4	93		1	4	84	5	
Palm kernel	86 / 14	4	5	50	15	7	18			2	15	1	
Peanut	19 / 89					10	84	3	3	4	55	25	
Cottonseed	27 / 73				1	15	84			6	76	2	
Coconut	90 / 10	8	7	48	17	9	10			2	7	1	
Olive oil	15 / 85					15	84	1		2	70	12	
Corn	13 / 87					13	87			3	31	52	1
Beef tallow	52 / 48				3	38	57			7	48	2	
Lard	41 / 59				1	27	70	2		11	44	11	4
Chicken fat	45 / 55				1	28	64	1		6	37	20	1
Butter	66 / 34		3	4	12	29	41			11	28	2	

2. Chemistry of Frying

A deep understanding of the chemical changes that occur during the frying process is fundamental in comprehending the nature of used cooking oils and defining the required upgrading processes, operating conditions, potential valorization routes, and their final uses. During oil frying, most chemical reactions occur by interaction with the carboxylic moiety and the unsaturations of the fatty acid chains (De Alzaa et al., 2018). Water from food and the evolved steam can hydrolyze the ester bonds, producing partial glycerides (i.e., monoglycerides (MG) and diglycerides (DG), free fatty acids (FFAs) and glycerol (G)). Due to hydrolysis, the acidity of the oil increases over time, thus, the FFA content in the oil is an important quality factor and a good indicator of the degree of use of a frying oil.

Simultaneously with esters hydrolysis, thermal and auto-oxidation reactions mainly occur at the double bonds of glycerides and FFAs. This leads to the formation of peroxides, and the subsequent rupture into volatile (e.g., hydrocarbons) and nonvolatile polar compounds (e.g., epoxides, alcohols, aldehydes, ketones, dicarboxylic acids, etc.). Therefore, while low peroxide and low volatiles contents are good indicators of the integrity of a new oil, they are insufficient to assess the quality of a used oil. This is because peroxides decompose and volatiles are removed when frying at high temperatures and for long periods, and when the oil has a high degree of reuse.

Finally, dimerization of triacyl glycerides, and polymerization of nonvolatile compounds are major decomposition processes during oil frying (Choe and Min, 2007). These are free radicals driven reactions, and the obtained dimers, oligomers and polymers can be very large molecules, reaching up to 1600 Daltons. These polymers with a high oxygen content accelerate a further degradation of the oil and are mainly responsible for the higher viscosity, the foaming tendency at high temperatures, the undesirable dark color, and the propensity to form resin-like residues. As the polymers might still contain some ester groups, a reduction in the acid-discounted saponification value and an increase in the unsaponifiable matter content in the used oil could be surrogate of polymers presence.

Other reactions can occur between the oil and the components contained in and extracted from the foodstuff (e.g., proteins, carbohydrates, fats, flavonoids, minerals, etc.). Thus, all the chemical changes during frying depend not only on the oil type, but also on the characteristics of the food (nature, composition, water content, size, shape, etc.), degree of reuse, air incorporation rates into the oil, frying temperature, kettle geometry, and the heating process (heating rate, surface area, surface temperature, operating time, etc.). Major physical and chemical changes during oil frying are summarized in Figure 1.

Besides the reduction of the nutritional value caused by the degradation of lipids and other components (e.g., antioxidants, vitamins), prolonged reuse of frying oils has a major impact in their edible character. Nitrogen-containing compounds extracted from food can react with oil degradation products to form highly toxic chemicals, such as acrylamides and heterocyclic amines. In addition, highly reactive oxygenated species formed during frying can decrease the radical scavenging capacity, causing oxidative stress (i.e., potential carcinogenicity, mutagenicity and genotoxicity). Furthermore, during the frying process, trans fatty acid chains, which are linked with a large variety of deleterious health effects (Bordin et al., 2013; Perumalla and Subramanyan, 2016), are formed. Another issue is the potential contamination with animal-derived proteins associated with different diseases (e.g., bovine spongiform encephalopathy, swine flu). Also, the potential absorption of bio-concentrated chemicals, such as persistent organic pollutants (POPs), dioxins, dioxin-like compounds, and polychlorinated biphenyls (PCBs), is of major concern. Finally, cyclic monomers formed during frying can enhance lipids intake and the subsequent over-accumulation within the organisms (Boatella and Codony, 2000; Sanli et al., 2011). For all these reasons, highly reused frying oil becomes a noxious product, so it cannot be indefinitely utilized for cooking purposes, nor incorporated into animal feed (i.e., without proper refining). Consequently, used cooking oils (UCOs) turn into problematic waste that requires proper disposal.

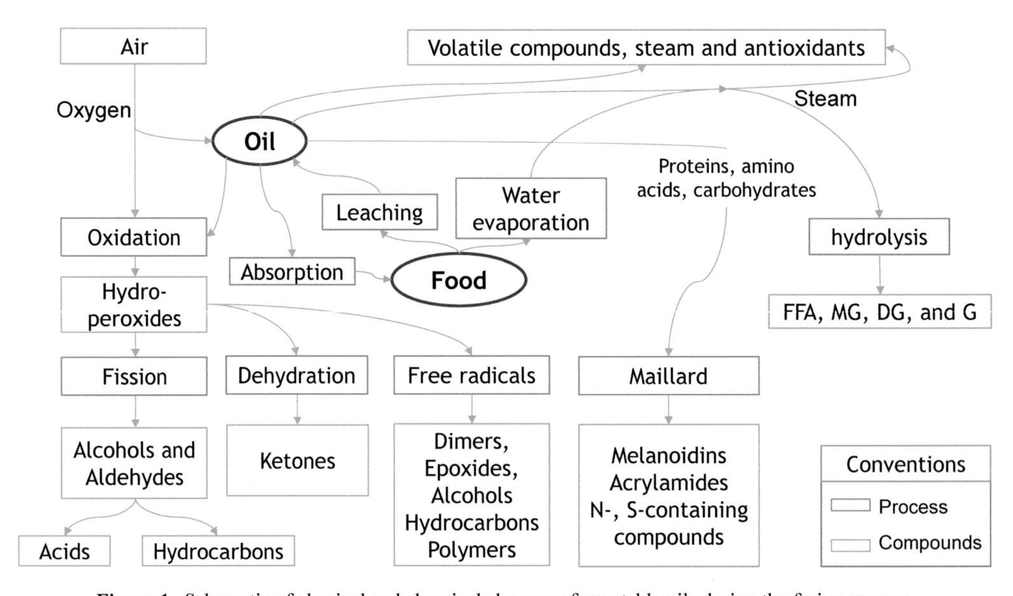

Figure 1. Schematic of physical and chemical changes of vegetable oils during the frying process.

3. Impacts of Waste Fats and Oils

UCOs management and disposal is a very challenging issue that has been exacerbated by population growth and higher consumption of oils and fats. In particular, consumption of vegetable oils has steadily increased around 3% per year over the last decade (OECD, 2020), with the consequent rise in waste oil generation, primarily in heavily populated areas. UCOs are generated at household, HORECA sites (HOtels, REstaurants, CAsino and catering), and at industrial facilities. While institutional and industrial UCOs management is somehow regulated and enforced, household disposal is a major problem for cities. Few countries have implemented laws for domestic UCOs management (e.g., Belgium), though most traditional household disposal is done by pouring UCOs through syphons or within the solid residues. In the first case, there is a cascade of negative impacts, including blockage of domestic and urban sewage pipes, sanitary sewer overflows, flooding during rainy seasons, damage of private/public infrastructure, proliferation of vectors (e.g., rats, cockroaches, mosquitos), bad odors, and increase in the operating costs of sewage and treatment plants, among others. When disposed of in landfills, UCOs leaching can affect the normal biodegradation processes, and can contribute to the increase of lixiviates and methane emissions. Sometimes, a more dramatic public health issue occurs, mainly in not well-regulated areas; used frying oil is illegally collected, filtered, bleached and redistributed as new oil among low-income populations.

Either from the sewage systems or from lixiviates, once UCOs reach surface or underground waters, they generate great environmental impacts on local and regional ecosystems: A supernatant lipid layer prevents water oxygenation, fresh water sand basins are contaminated with organic matter and noxious chemicals, oxygen depletion is caused by a higher chemical oxygen demand (COD), and eutrophication is boosted. While this could be considered an issue of underdeveloped and developing countries, this situation has already created problems in developed economies as well. For instance and as observed in Figure 2, in spite of the collection programs implemented in most European countries for the management UCOs, only about 6% of the domestic UCO generated is collected and properly disposed of (around 48 kt). This indicates that problems with sewage clogging are still common even in developed regions. This has prompted the implementation of household management policies that have been effective in the Netherlands, Belgium, Austria, and Sweden, where collected household UCOs represent 30 to 65% of the generated volume (Figure 2). Comparatively, the European HORECA and industrial sectors are highly regulated, and nearly 85% of UCOs generated from these sectors are collected and reused.

To aggravate the management problem, other waste fats, oils, and greases (FOG) are also generated, namely yellow, brown, and trap greases. While UCOs are essentially produced as a residue of food processing, yellow and brown greases are blends of vegetable oils and animal fats (e.g., lard, tallow, poultry, and fish) coming from food processing and animal rendering. Typically, a yellow grease contains from 2 to 15% wt. FFA; if the FFA content is higher, it is marketed at a discounted price, otherwise it can be blended with UCOs to meet yellow grease specifications. When the FFA content is above 35% wt., the waste is considered a brown grease (Canakci and Gerpen, 2001). Trap grease or black grease is obtained from grease collectors at industrial and HORECA locations, and from sewage interceptors and wastewater treatment plants. Normally, UCOs and yellow grease have found some secondary uses as energy sources or in the oleochemical industry (after suitable pretreatments). In contrast, brown grease and trap grease are not generally exploited because of the high content of impurities, the large degree of chemical and biochemical decomposition, and the presence of halogenated products from sanitization and cleaning chemicals (Ward, 2012). Brown and trap greases are normally disposed of in landfills, or used for biodigestion or composting. Typical market specifications of waste fats and oils are summarized in Table 2.

4. Production and Potential Supply of UCOs

Despite the large content of toxic chemicals and impurities, UCOs and yellow greases are mostly composed of triglycerides, partial glycerides and fatty acids. This turns them into potential feedstock for the oleochemical industry, similarly to crude or acidified vegetable oils, crude animal fats, and other

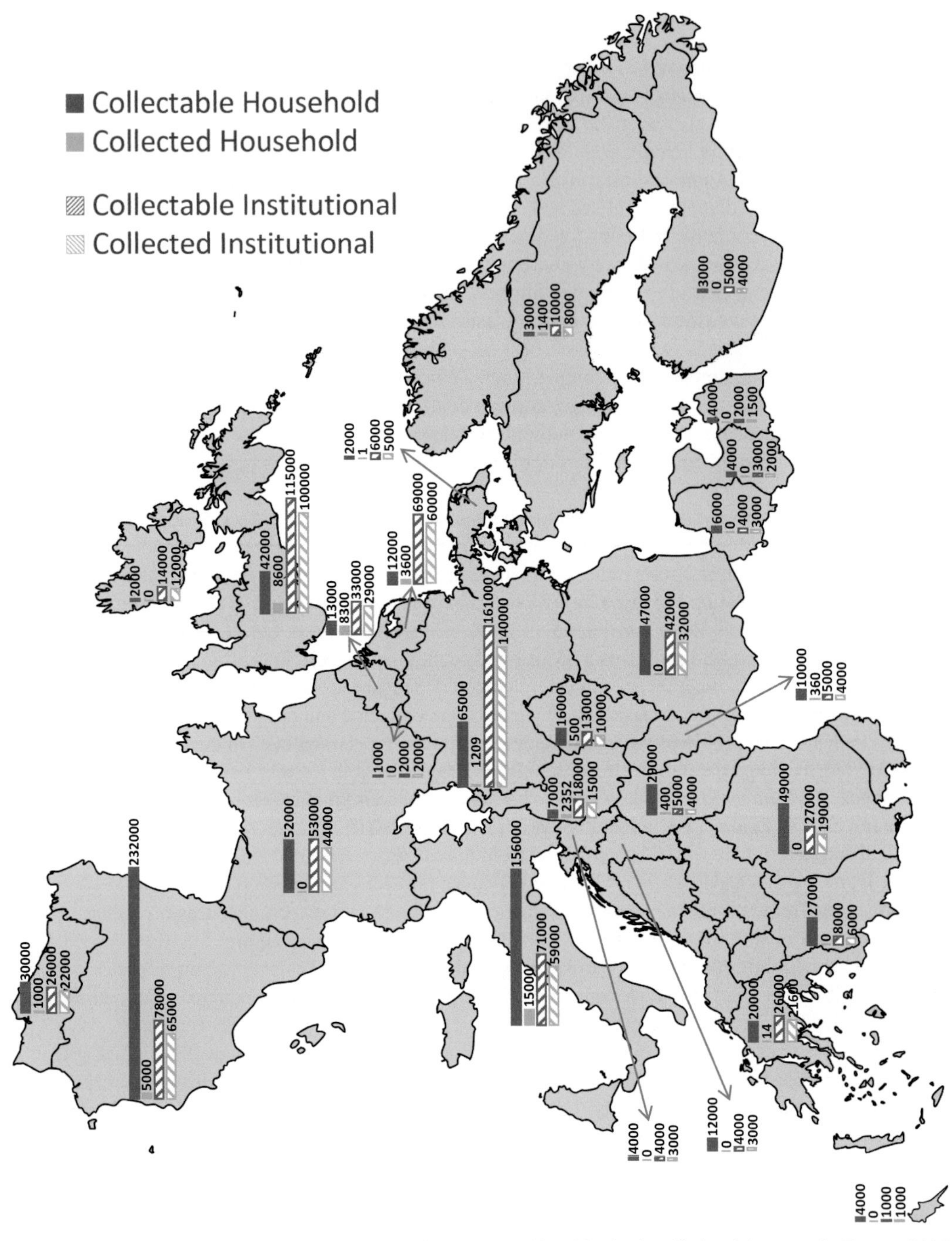

Figure 2. Collected and collectable UCOs (ton/yr.) from household and institutional/industrial sectors in Europe, 2016. (Data from Greenea, 2016a; Greenea, 2016c; Greenea, 2018).

non-edible oily residues (e.g., soapstocks and deodorizing distillates; Dumont and Narine, 2007). Thus, by implementing a circular economic model, UCOs could be effectively exploited as feedstock for the production of a variety of chemicals and fuels in biorefinery centers. Similarly to the traditional petroleum refineries, a suitable and optimized portfolio of derivatives (i.e., commodities, fine and specialty chemicals) would allow the benefits for a sustainable UCOs exploitation to be maximized.

Table 2. Typical market specifications of waste fat and oils (NRA, 2008; Ragauska et al., 2013; USDA, 2013; Greenea, 2016b; UGC, 2019).

Specification	**UCO**	**UCO Asia**	**Yellow grease**	**Brown grease**	**Trap grease**
Free Fatty Acids (FFA) % wt.	max 6	max 6	max 15	15–35	35–100
Moisture, Impurities and Unsaponifiables (MIU) max	3	3	3	3	none
Moisture % wt. max	1	1	1	1	none
Iodine index, cg I_2/g. min	75	50	75	none	none
Color (FAC)	39	39	39	45	none

In order to estimate the world production of UCOs and their potential as oleochemical feedstock, it is necessary to understand the vegetable oils market. According to recent data (OECD, 2020; Statista, 2019), the total world consumption of vegetable oils ranges between 203 and 210 million tons. Currently, around 67% of this volume is consumed for food applications, 20% for biofuels manufacture, and 13% for other uses (e.g., chemical feedstock). As observed in Figure 3a, the global oil consumption of vegetable oils has doubled in the last two decades, mostly driven by their use as food. Comparatively, world's population has increased nearly 24% in the same period (UN, 2019). This evidences a rapid increase of the *per capita* oil consumption worldwide, most probably due to more access to food, changes in food habits (e.g., more fast food and snacks), and more incorporation of oils as ingredient in processed foods. According to Figure 3b, palm, soybean, rapeseed and sunflower oils account for nearly 85% of the world's total oils consumption.

Due to the varying availability of crops and different culinary customs, consumption of the different types of oils changes among regions worldwide; for instance, soybean oil dominates in North America, palm oil in sub-Saharan countries and Asia, and rapeseed and sunflower in Europe (Ribau et al., 2018). Also, the share for food applications or for biofuels production changes significantly among countries. In Figure 4, the current distribution of vegetable oils consumption among continents according to their final use is presented. As observed, while most oil consumption in Africa, Asia and Oceania is related to food applications, the use for biofuels manufacturing represents nearly 30% of oil consumption in South America and Europe, and 20% in North America. It is also important to point out that vegetable oils are used in food applications not only for frying. They are important ingredients to enhance workability, texture, aeration, appearance, and conservation in processed food. Thus, their use as ingredients in consumer products (e.g., spreading margarines, sauces, ice creams, cookies, chocolate, etc.), represents a large share of oil consumption in the food industry. However, in general, there is no waste oil generation associated with these food applications. Future developments might enable the recovery of such lipids by processing and extraction of food waste.

As a consequence of the variability of oil consumption among regions, and also because of the different shares with respect to the final use, a proper estimation of UCOs generation as potential feedstock for the global oleochemical industry is difficult to assess. Nonetheless, as UCOs management has become a worldwide issue in recent years, several studies have estimated the potential production volumes of waste cooking oils at the local and regional level. Thus, based upon a review of the open literature and public statistics, Table 3 presents a comprehensive summary of the most current data on UCOs generation worldwide.

Data in Table 3 were calculated and reported on a *per capita* basis, taking into account that normalized data tend to change less over time than on a total basis. This allows projections to be made with less uncertainty and comparisons to be made among countries in the different regions. Figure 5 presents a comparison among vegetable oil consumption for food purposes and UCOs generation (i.e., where available) in countries from all continents. As expected, there is a large scatter among data, even from countries in the same region. Besides the uncertainty of collected data, this is a consequence of the different food consumption habits and the socio-economic conditions in the different regions. Despite the scatter and the low correlation ($R^2 = 0.18$), the slope of a simple linear regression of data in Figure 5 indicates that about 20% of vegetable oil used for food applications ends up as UCOs.

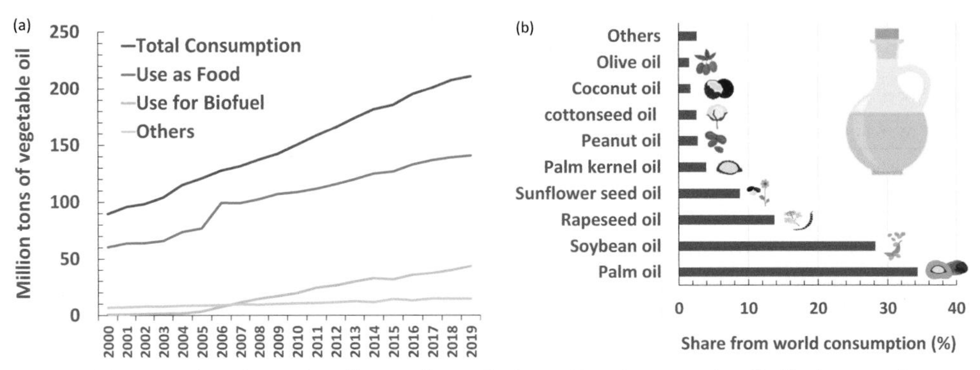

Figure 3. World consumption of vegetable oils according to final uses (a) and consumption distribution according to crop source (b) (Data from OECD, 2020; Statista, 2019).

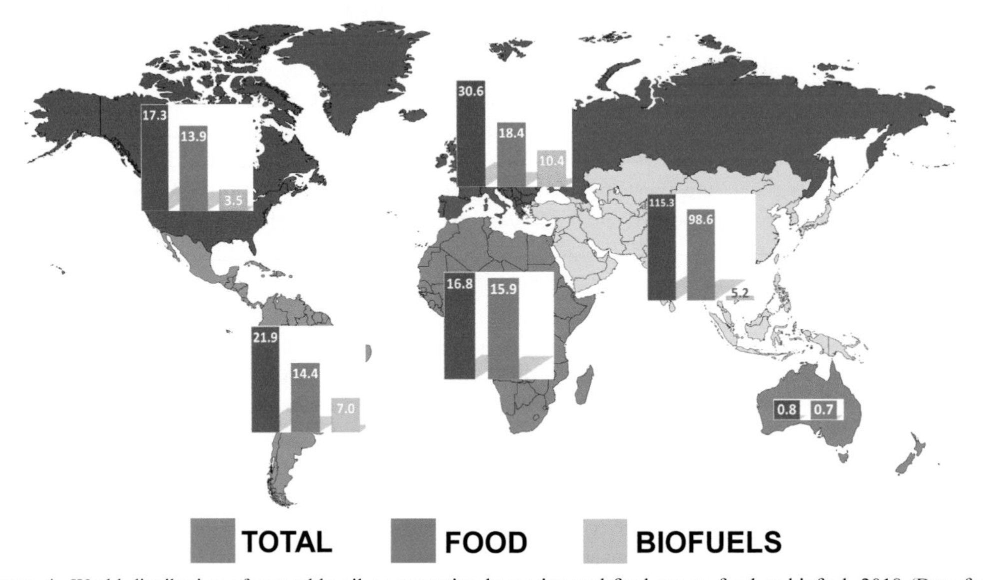

Figure 4. World distribution of vegetable oil consumption by region and final use as food or biofuel, 2018 (Data from OECD, 2020).

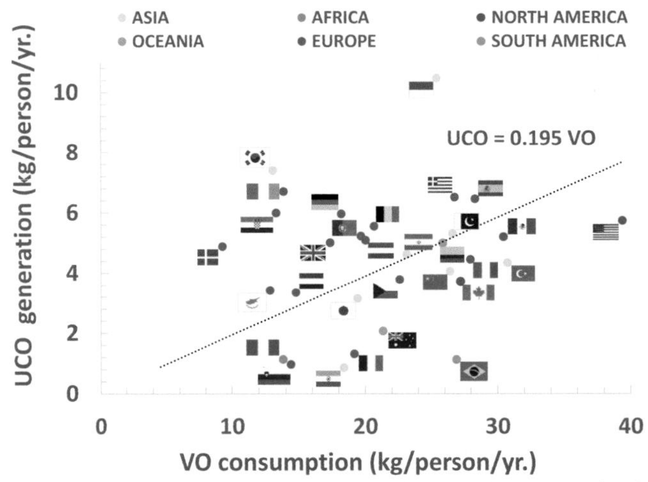

Figure 5. Distribution of *per capita* vegetable oil consumption and used cooking generation in different regions and countries.

Table 3. Annual *per capita* vegetable oil (VO) consumption and used cooking oil (UCO) generation among different regions of the world for 2017–2018.

Country	Population	VO consumption (kg/person/yr.)	UCO generation (kg/person/yr.)	Country	Population	VO consumption (kg/person/yr.)	UCO generation (kg/person/yr.)
ASIA				**EUROPE**			
China	1378665000	26.40	4.06	Austria	8747000	22.30	
India	1324171000	18.39	0.86	Belgium	11348000	20.66	5.55
Indonesia	263991379	25.41	10.47	Bulgaria	7128000	12.74	
Iran	81162788	23.12	4.62	Croatia	4171000	13.28	5.99
Israel	8547000	20.37		Cyprus	1170000	12.82	3.42
Japan	126995000	19.41	3.15	Czech Republic	10562000	22.58	3.79
Kazakhstan	18204499	18.19		Denmark	5731000	9.28	4.89
Korea	51246000	13.05	7.40	Estonia	1316000	8.36	
Kuwait	4053000	4.56		Finland	5495000	9.99	
Malaysia	31624264	30.43		France	66896000	19.16	1.32
Pakistan	197015955	26.58	5.32	Germany	82668000	18.20	5.96
Philippines	104918090	13.59		Greece	10747000	26.74	6.51
Saudi Arabia	32276000	20.91		Hungary	9818000	20.01	5.09
Thailand	69037513	18.07		Iceland	334000	8.68	
Turkey	80745020	30.73	4.35	Ireland	4773000	13.83	6.70
United Arab Emirates	9270000	19.75		Italy	60601000	27.95	4.46
Vietnam	95540800	15.40		Latvia	1960000	12.76	
				Lithuania	2872000	7.73	
AFRICA				Luxembourg	583000	11.32	
Egypt	97553151	23.64		Malta	437000	8.24	
Ethiopia	104957438	5.49		Montenegro	17018000	14.75	
Nigeria	190886311	13.79	1.13	Netherlands	17018000	14.75	3.35
South Africa	56717156	23.44		Norway	5233000	30.84	
				Poland	37948000	14.41	

Table 3 contd. ...

...Table 3 contd.

Country	Population	VO consumption (kg/person/yr.)	UCO generation (kg/person/yr.)	Country	Population	VO consumption (kg/person/yr.)	UCO generation (kg/person/yr.)
AMERICA				Portugal	10325000	19.66	5.23
Argentina	43847000	18.72		Romania	19705000	13.10	
Brazil	207653000	26.88	1.13	Russian Fed.	144342000	27.89	
Canada	36286000	27.21	3.72	Slovakia	5429000	11.51	
Chile	17910000	30.32		Slovenia	2065000	14.38	0.97
Colombia	49065615	25.85	5.00	Spain	46444000	28.31	6.46
Mexico	31624264	30.43	5.20	Sweden	9903000	16.89	
Paraguay	6811297	18.99		Switzerland	8372000	16.92	
USA	323128000	39.38	5.74	Ukraine	44222947	15.27	
				UK	65637000	17.35	5.01
OCEANIA							
Australia	24127000	21.34	2.07				
New Zealand	4693000	21.96					

Data from (Ghobadian, 2012; Sheinbaum et al., 2013; SAGE, 2015; Nelson and Searle, 2016; Da Silva et al., 2017; DARPRO, 2017; UN, 2017; Khan et al., 2018; Kharina et al., 2018; OECD, 2020; Ribau et al., 2018; Rincon et al., 2019).

Surprisingly, this matches a typical rule of thumb commonly used at the oleochemical sector for a rapid estimation of UCOs generation. Previous estimations based upon the total consumption of vegetable oils indicated that about 32% of the consumed oil is wasted as UCOs (Ribau, 2018). Taking this into account, it is estimated that the potential global supply of UCOs ranges between 41 and 67 million tons per year. According to plotted data of Figure 3a, this volume has the potential to supply a major fraction of the global amount of vegetables oils currently consumed by the oleochemical industry for biofuels and chemicals manufacturing, and also for other applications.

Taking into account that the food value of the UCOs has been already exploited, that they generate major environmental and social impacts when mismanaged, that there is a large, evenly distributed, and global supply, and that they are available at lower costs than traditional refined oils, UCOs have the potential to become a major alternative feedstock for the oleochemical industry. This becomes more attractive when considering that public policies are being implemented in different countries in order to promote the use of waste-based feedstock. Favorable prices, capital investment subsidies, tax reduction or exemptions and double counting schemes, among others, are typical promotion policies for UCOs recollection and reuse (EBIA, 2015; Boutesteijn et al., 2017). In addition to the economic incentives, reutilization of UCOs as oleochemical raw materials is an environmentally friendly alternative since it reduces the life cycle impacts associated to consumption of fossil-based feedstock, landfills disposal, discharge in sewer systems and consumption of refined oils, among others (Ortner et al., 2016; Tsoutsos et al., 2016). Nevertheless, challenges with recollection and refining need to be addressed before realizing the potential benefits of UCOs harnessing.

5. Used Cooking Oil Collection and Pre-treatment

UCOs are highly heterogeneous materials; in addition to the aforementioned degradation products, they can contain food residues, suspended solids, moisture and other impurities. Typical ranges of some of the characteristic properties of UCOs from fast food restaurants are summarized in Figure 6. As observed, the heterogeneity and the content of impurities limit UCOs usefulness as oleochemical feedstock, and affect their long term stability. For this reason, suitable pre-treatment processes are required to stabilize and upgrade waste oils. Normally, UCOs are collected in tank trucks or within plastic containers. Those are transported to centralized facilities where they are weighed and characterized in order to define subsequent processing conditions and also to establish price-related properties (e.g., FFA, moisture, rancidity, etc.).

Initially, collected UCOs are screened through gross and fine meshes to remove large particles, and then heated in storage tanks (50 to 80 °C). At this condition, solid oils and fats (i.e., highly saturated lipids) melt, enhancing flow properties and allowing water separation by decanting. If the FFA content

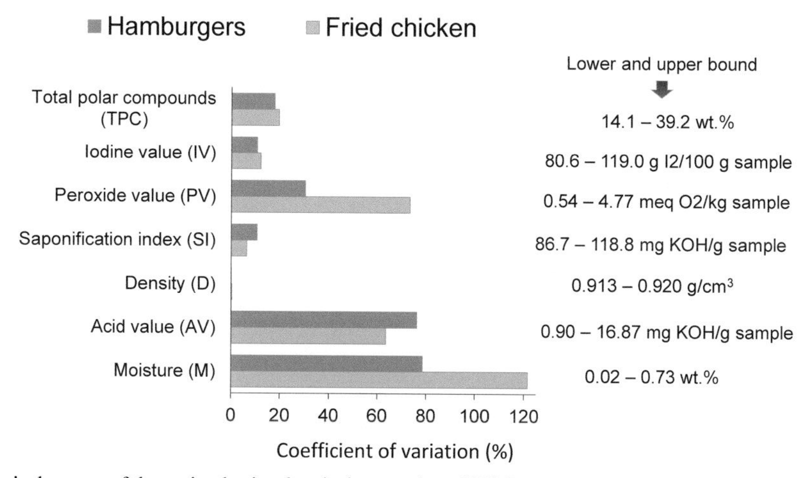

Figure 6. Typical ranges of the main physicochemical properties of UCOs generated in fast food restaurants (hamburgers and fried chicken) (Rincon et al., 2019).

is above the required specifications, blending with other low FFA UCOs can be done; alternatively, neutralization with a base can be accomplished. When used in biodiesel production, FFA can be processed by esterification before the transesterification in alkaline conditions. After FFA reduction, the oil can be subjected to vacuum to remove humidity and remaining volatiles, and then treated with adsorbents (e.g., bleaching earths, active clays) to remove unsaponifiable and nonvolatile matter, and also as a bleaching method. Afterwards, the dry oil is filtered to remove the spent adsorbent, and characterized to ensure market specifications before distribution. In general, after the pre-treatment process, 80 to 90 % wt. yield of refined UCOs is obtained with respect to the collected amount. Resulting waste water, fatty slurries, and solid residues need to be treated before final disposal or used as composting material. A recent study (Torres, 2019) carried out an inventory analysis of a UCOs small-scale batch processing facility, reporting the specific utilities consumptions per ton of UCO. According with this study, electricity requirement borders 6 kW-h per ton of UCO, energy consumption in a natural gas boiler is around 720 MJ per ton of UCO, and water use is around 0.33 m^3 per ton of UCO.

6. UCOs Valorization Alternatives

Once purified, UCOs could be used as chemical feedstock for a variety of oleochemical derivatives. Nevertheless, the potential valorization routes are largely dependent on the specific physicochemical characteristics of the available waste oils and fats. Properties related with the chemical composition and reactivity of the acylglycerides in UCOs (i.e., saponification index, acid value, iodine index, unsaponifiable matter) and the sensory profile (i.e., odour and color) need to be assessed in order to define feasible oleochemical targets. Thus, highly heterogeneous and dark colored UCOs are better suited for biofuels production, thermochemical transformations, biodigestion, and some drop-in applications (substrate for fermentation, additives in asphalt, concrete and rubber, etc.). In some other cases, highly functionalized (i.e., high iodine index) and more clear UCOs would be suitable raw materials for applications in which unsaturations play a major role, for instance in the synthesis of resins, epoxides, polyols, and some chemical specialties (e.g., azelaic and pelargonic acids and their esters). Similarly, UCOs with high content of saturated C12–C14 fatty acids (e.g., coconut oil derived UCOs) could be used for fatty alcohols and detergents production (Hill, 2000; Decker, 2006). In contrast, those would not be preferred for biodiesel production, as the C12–C14 methyl esters are more soluble in water and they can capture a high amount of moisture during handling and storage. Similarly, UCOs mostly composed of highly saturated C16 and C18 acyl chains would not be ideal as biodiesel feedstock; methyl esters of saturated fatty acids are solid, reducing the cloud point of the fuel and affecting its cold flow performance. An attempt to capture a comprehensive list of potential oleochemical derivatives that could use UCO as building block, within existing or new biorefineries, is presented in Figure 7.

Typically, most oleochemical processes at the industrial scale (> 90%) are performed by reacting the carboxylic groups of fats and oils. Comparatively, less than 10% of the industrial transformations are done on the alkyl chains or on the unsaturations (Hill, 2000). In either case, if chemically transformed, UCOs could be converted into a large variety of derivatives of commercial interest, entering into the global oleochemical market currently valued at $21.76 billion (AMERI, 2018). Within this market, there are largely consumed and low value-added commodities (e.g., biofuels, fatty acids, alcohol, esters), as well as niche and high value-added fine and specialty chemicals (e.g., surfactants, polyols, drying oils, plasticizers, lubricants, etc.). As a way to identify potential derivatives of commercial interest, Table 4 summarizes the added market value of some basic and functional oleochemicals with respect to the current value of UCOs (i.e., 620 USD/ton; Greenea, 2019a). Added value was calculated as follows:

$$VA = \frac{m_{Der.} \times f_{Der.}^{UCO} \times \$_{Der.}}{m_{UCO} \times \$_{UCO}}$$

Here, $m_{Der.}$ is the weight of the oleochemical derivative obtained from a given weight of UCO (m_{UCO}) according to the reaction stoichiometry. The $f_{Der.}^{UCO}$ value accounts for the weight fraction of the oleochemicals coming from UCO, and $\$_{UCO}$ and $\$_{Der.}$ correspond to the market prices of UCOs

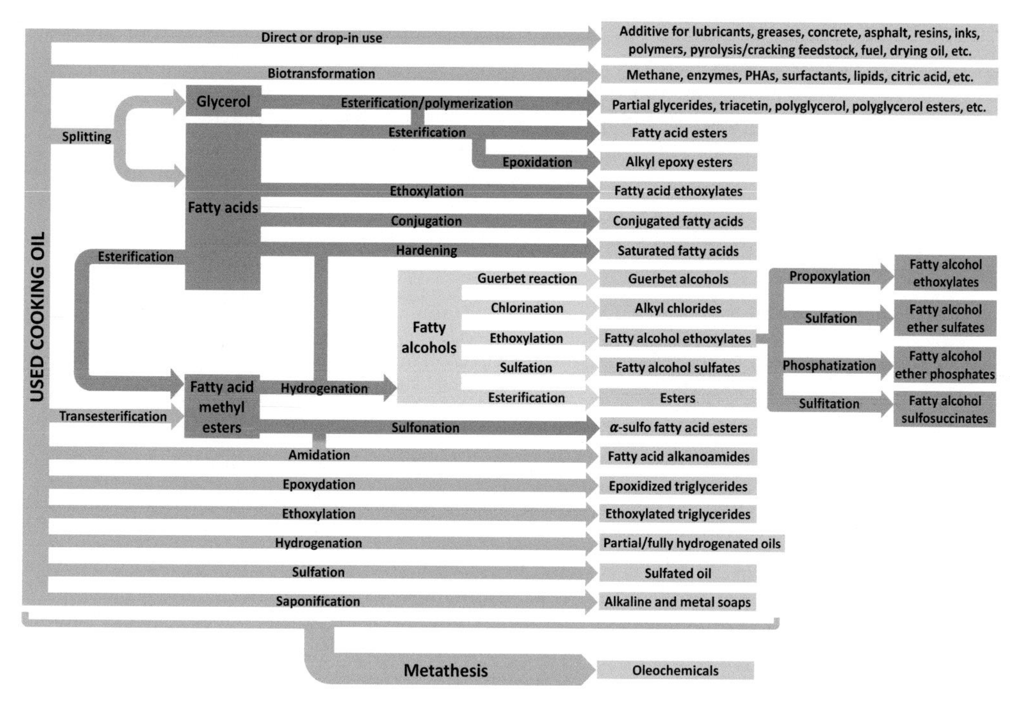

Figure 7. Potential valorization alternatives for used cooking oils (Adapted and updated from Brackman et al., 1984).

Table 4. Value added of potential UCOs derivatives (Costs from Alibaba, 2019; Greenea, 2019; Oleoline, 2018).

Oleochemical derivatives	Bulk price USD/t	Source	Stoichiometric yield (t/t_{UCO})[§]	Fraction of UCO (*f*) in the product	Added value with respect to UCOs
UCOs	620	Greenea	1.00	1.000	1.00
Biodiesel	952	Greenea	1.01	0.887	1.37
Fatty acids C16	1010	Oleoline	0.96	0.938	1.47
Fatty alcohol C16/C18	1350	Oleoline	0.90	0.965	1.90
Isopropyl palmitate	1600	Oleoline	1.10	0.802	2.27
Glyceryl palmitate	1500	Alibaba	1.21	0.997	2.92
Polyglyceryl-3-ester	2000	Alibaba	1.76	0.498	2.82
Epoxidized Oils	2000	Alibaba	1.00	0.941	3.06
Polyester-polyol (EG)	4000	Alibaba	1.18	0.803	6.13
Metallic soap (Co)	5000	Alibaba	1.05	0.848	7.18
Sucrose palmitate	5000	Alibaba	2.13	0.418	7.18

[§] Calculated based upon a saponification index of 195 mgKOH/g, and FFA content of max. 5% (acidity as palmitic acid), and an iodine value of 1 g I_2/g.

and the corresponding oleochemical, respectively. Although some derivatives from this table might look promising at first glance, it is important to consider the capital and operating costs, and the complexity of the process at the industrial scale. For instance, biodiesel can be produced in stirred tank reactors at low temperatures and pressures, whereas the manufacturing of fatty alcohols requires hydrogen supply, expensive catalysts, and the process operates under energy intensive conditions (i.e., high pressures and temperatures). Similarly, epoxides can be readily obtained in stirred reactors operating at low temperatures;

however, safety measures and strict control must be implemented in order to deal with the inherent risk of working with hazardous agents (e.g., hydrogen peroxide, ozone) in a highly exothermic process. Regarding specialty chemicals, such as sucrose fatty esters and other surfactants, the high value-added comes to the price of highly energy- and materials-intensive production and downstream processes. So, all these not easily foreseen processing complexities would involve high capital investment and operating costs.

7. UCOs as Energy Source

A basic and straightforward to reuse UCOs is to exploit their caloric value; they can be directly burned for heat and power generation, or transformed into a variety of biofuels. Currently, there are several processes industrially implemented for the production UCOs derived biofuels, namely biodiesel (i.e., fatty acid methyl esters), green diesel (i.e., hydrotreated vegetable oil, HVO), green fuel (i.e., obtained by feeding UCOs into existing fluidized catalytic crackers), and bio oil (i.e., liquid fraction from pyrolysis) (Singhabhandhu and Tuzuka, 2010a; Singhabhandhu and Tuzuka 2010b; Ward, 2012; Zhang et al., 2012; EBIA, 2015; Ortner, 2016; Yuste, 2016; Boutesteijn, 2017; Ribau et al., 2018; Lombardi et al., 2018; Sonthalia and Kumar, 2019). UCOs have been also used for the production of other biofuels, such as biogas (Alqaralleh et al., 2016; Ortner et al., 2016; Carnevale et al., 2017; He et al., 2018), and solid blends (Xiong et al., 2019), however, these are intended as small-scale valorization alternatives. Table 5 summarizes the heat of combustion of most typical UCOs derived biofuels, together with those of crude vegetable oils and fossil diesel. These values give an idea of the energy potential of the different bio-based resources.

Table 5. Comparison of the heating value of different fuels.

UCOs	Heat of combustion, MJ/kg
Used cooking oil	37.3–39.4
Crude palm oil	41.3
Crude soybean oil	41.9
Palm oil biodiesel	37.9
Soybean oil biodiesel	38.1
Pyrolysed palm oil	38.9
Pyrolysed soybean oil	39.5
Hydrotreated vegetable oil	44.0
Fossil diesel	46.6

Data from Thompson, 1981; De Oliveira et al., 2006; Singhabhandhu and Tezuka, 2010a; Lombardi et al., 2018.

7.1 UCOs for heat and power

As observed in the reported data in Table 5, the heating value of UCOs is similar to that of other oleochemical-based materials, but requiring far less processing than the other biofuels. Nonetheless, the physicochemical properties of UCOs (e.g., flow properties, solid contents, oxidation potential, moisture content, etc.), limit their potential as fuel for combustion engines, so they would be better suited for steam boilers and heaters. In this case, burning is applied as a UCOs abatement method, obtaining heat and energy as valuable byproducts (Singhabhandhu et al., 2010a; Singhabhandhu et al., 2010b; Zhang et al., 2012; Lombardi et al., 2018). The use of UCOs for heat and power has the advantage of reducing global warming potential (GWP, 100 years' time span). Recent studies indicate that this alternative generates a similar amount of greenhouse gases to that of UCOs derived biodiesel, but far less than biodigestion (Ortner et al., 2016). Despite the potential of UCOs as fuel, major challenges arise when dealing with the content of suspended solids (i.e., solid fatty acid chains, resins, polymers, etc.), and the reliability of injectors in the combustion chambers. Also, as the flash temperature of UCOs is quite high, incomplete combustion, fumes, and flying ashes are common, thus, solid particles can accumulate on the heat transfer surfaces, reducing energy recovery efficiency. Additionally, combustion chambers need to be optimally operated in order to avoid negative impacts of gas emissions at the local level (i.e., particular matter, odorous volatile organic compounds, sulfur- or nitrogen-containing compounds, etc.).

7.1.1 Biodiesel

The production of fatty acid methyl esters is by far the most widely-used valorization alternative at the industrial scale for UCOs harnessing. This is a result of the public policies implemented in developed countries since the beginning of the century. The first and second stages of the renewable energy directive (RED I and II) and the fuel quality directive from the European Union, and the Renewable Fuel Standards mandates from USA, have become major driving forces for UCOs collection and harnessing worldwide (Weaver and Howell, 2018; Greenea, 2019). According to recent estimations, 90% of total collected UCOs in Europe (Figure 2) are transformed into biodiesel, whereas the remaining 10% is used in the oleochemical industry (i.e., soaps, candles, etc.) (Toop et al., 2013). Currently, from the total biodiesel production in Europe (ca. 12.6 million tons, UFOP, 2018), and in the USA (8.7 million tons, Weaver et al., 2018) around 15% is obtained from UCOs. Regarding total world production (ca. 34 million tons) about 10% corresponds to waste-based biodiesel (UFOP, 2018).

The technical literature on the production of UCOs-based biodiesel is already very extensive and widely covered. At this point, it would be even possible to do a review of the dozens of reviews already published on this topic, only in the last decade. Therefore, there is no need to rewrite most of what have been covered in fair detail elsewhere. Interested readers are welcome to check the variety of compilations describing most current advances on biodiesel production using different raw materials (Kulkarni and Dalai, 2006; Bacovsky et al., 2007; Banerjee and Chakraborty, 2009; Knothe and Steidley, 2009; Abbaszaadeh et al., 2012; Oh et al., 2012; Kiss and Bildea, 2012; Santacesaria et al., 2012; Ward, 2012; Mazubert et al., 2013; Yaokob et al., 2013; Aranisola et al., 2014; Zhang et al., 2014; Haas et al., 2016; Rathore et al., 2016; Verma and Sharma, 2016; Weaver and Howell, 2018; Fonseca et al., 2019; Günaya et al., 2019, Thoai et al., 2019). Also, there are very comprehensive publications dealing with the assessment of the economic (Zhang et al., 2003; Nas and Berktav, 2007; Eguchi et al., 2015; Gebremariam and Marcheti, 2018; Chrysikou et al., 2019), environmental (Zhang et al., 2012; Eguchi et al., 2015; Ortner et al., 2016; Tsoutsos et al., 2016; Behrends, 2018, Lombardi et al., 2018; Chrysikou et al., 2019), and social dimensions (Zhang et al., 2014; Boutesteijn et al., 2017; Hajjari et al., 2017) of UCOs-based biodiesel. Nevertheless, it is important to mention the most important key elements of using UCOs as raw material for biodiesel production. The main chemical routes used in the most common industrial processes (i.e., Agrartechnik, Axens, BDI, CD Technology, CMB, DSmet Ballestra, Energea, Lurgi; Bacovsky et al., 2007) are described in Figure 8, together with some typical process flow diagrams.

Because of the large amount of FFAs, a two-step process is generally required when using UCOs as feedstock. Initially, esterification of FFA with methanol is carried out using a homogeneous or heterogeneous acid catalyst. Alternatively, FFAs can be initially removed by vacuum stripping similarly to the deodorization process carried out during traditional crude vegetable oils refining. Once acidity is treated, transesterification of acylglycerides with additional methanol is performed using an alkaline catalyst. In a different approach, esterification and transesterification can be run simultaneously using a homogeneous or heterogeneous acid catalyst; in this case, higher operating temperatures (i.e., high pressures) are required in order to achieve high conversions. As the extents of esterification and transesterification reactions are limited by the chemical equilibrium, multiple reaction-separation stages are generally required in both processes in order to remove glycerol. This allows high conversions to be achieved and the required biodiesel specifications regarding the low content of partial glycerides to be fulfilled. A slightly different approach for the exploitation of high FFA feedstock or aqueous oil solutions (e.g., algal oil) has been also implemented at the industrial scale; UCOs are subjected to complete hydrolysis into fatty acids, followed by a straightforward esterification with methanol using an acid

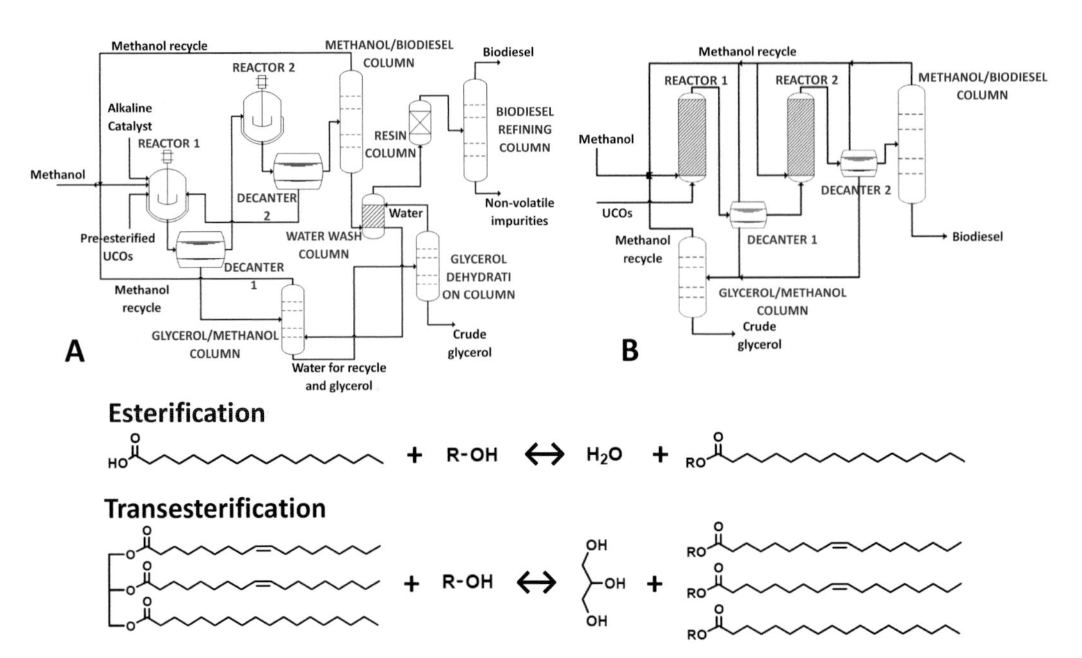

Figure 8. Simplified schematic of the most common industrial scale processes for UCOs-based biodiesel (Bacovsky et al., 2007; Abbaszaadeh et al., 2012). (A) Acid catalyzed esterification of FFA followed by alkaline transesterification (e.g., Lurgi). (B) Simultaneous acid catalyzed esterification and transesterification (e.g., Axens).

catalyst within a reactive separation unit (Kusdiana and Saka, 2004; Kiss and Bildea, 2012; Mazubert et al., 2013; Song et al., 2016).

Generally, in all current industrial processes, there are major challenges that jeopardize biodiesel production from UCOs, even under subsidized schemes. Biodiesel is a low added value commodity, so feedstock and operating costs represent a large fraction of biodiesel selling price (> 80%; Gebremariam and Marcheti, 2018). Then, even if the UCOs raw material has a lower cost than the typical vegetable oils used for biodiesel production, process instabilities caused by impurities present in the rather heterogeneous UCOs feedstock can occur. As a consequence, reprocessing or intensive refining would be required when using UCOs as feedstock, causing serious economic impacts. Also, a large amount of waste is generated in the downstream processes, namely waste water, neutralization salts, deactivated catalysts, bleaching earths, etc. Additionally, the associated water use footprint is very high. Finally, a large amount of crude glycerol with high content of impurities is produced. In spite of its potential as a chemical building block, the current demand for crude glycerol is very limited, due to the presence of salts (e.g., sulfates, chlorides) and the high purification costs required to obtain USP grade glycerol.

7.1.2 Hydrotreated vegetable oil

Due to the high oxygen content, the presence of polymer residues, and the labile nature of ester groups, there are several drawbacks to the use of biodiesel as fuel. Fatty acid methyl esters are susceptible to hydrolysis, external and auto-oxidation, and biochemical degradation. Also, the content of solid esters from saturated fatty acids and the presence of resinous solids negatively affect cold flow properties; this leads to the formation of precipitates that block fuel lines and accumulate in tanks. Moreover, fatty acids are corrosive even at low concentrations. Most of these drawbacks have been overcome by incorporating UCOs as raw materials of hydrotreatment processes (Bezergianni and Kalogianni, 2009; Luna et al., 2014; Zhao et al., 2017; Sonthalia et al., 2019). Hydrotreated vegetable oil (HVO) and some other non-

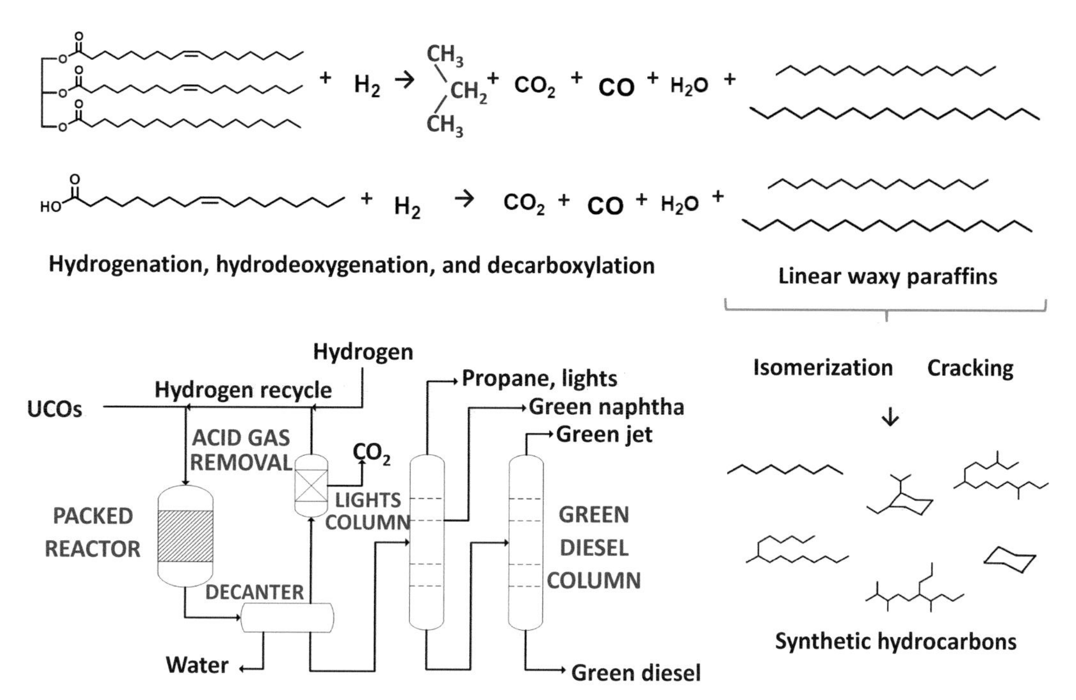

Figure 9. Simplified schematic of reactions and process flow diagram in a UCOs hydrotreating process.

ester biofuels (e.g., pyrolised UCOs) are more diesel-like biofuels, so they are more stable and more compatible with current fuel infrastructure, and they can even improve diesel properties (e.g., higher cetane index). A simplified schematic of the reported chemistry (Sonthalia and Kumar, 2019) and a conceptual process flow diagram for UCOs hydrotreatment is presented in Figure 9.

The HVO process involves hydrogenation of fatty acids unsaturations, and competing decarboxylation and hydrodeoxygenation of the carboxylic moiety. Generally, the process operates with alumina-supported metallic catalysts (e.g., Ru, Ni, Pd, Co, Mo, Pt) under high pressures (6–20 MPa) and temperatures (300–450 °C). Simultaneously with hydrogenation and hydrogenolysis, short chain and branched hydrocarbons are produced by series-parallel cracking and isomerization, generating a broad range of molecules. The main products of reaction are propane, carbon dioxide, water, and a variety of hydrocarbons. Hence, the process not only generates diesel-like molecules, but also a whole range of so called "hydrobiofuels" for different applications, namely LPG, jet fuel and naphtha. While a broad and tunable portfolio of fuels would be attractive from a biorefinery perspective, the side generation of CO, CO_2 and water makes the yields lower than those obtained in the biodiesel route.

As UCOs hydrotreatment can be readily implemented on existing refineries, investment costs and construction time can be reduced. In these facilities, it is possible to do UCOs hydrotreatment separately, or to do simultaneous co-processing with petroleum fractions. Current industrial frontrunners in HVO production are NESTE Oil with the NExBTL process, Honeywell/UOP with the Ecofining process, and the H-Bio from Petrobras. However, as observed in Table 6, other players are entering the market, and several plants are under construction or planned in the coming future. Currently, the world production capacity of HVO is around five million tonnes, and it already represents around 17% of the biodiesel market in Europe. The fact that UCOs-based HVOs generate less life cycle greenhouse gas emissions than other biofuels (Figure 10) might become a major driving force to boost its production within the portfolio of different alternative biofuels. Thus, it is expected that the HVO production capacity would triple during the coming decade (Greenea, 2019).

Table 6. Current and planned HVO units around the world (Greenea, 2017).

Company	Location	Capacity (Million Ton)	Notes
AltAir fuels	USA	125000	
Diamond green diesel	USA	500000	Expansion to 800000 MT
REG	USA	250000	
Emerald biofuels	USA	280000	Under construction
Petrobras	Brazil	230000	
Total	France	500000	Under construction
CEPSA	Spain	180000	Co-processing
Repsol	Spain	60000	Co-processing
ENI	Italy	550000	
ENI	Italy	580000	Under construction
Preem	Sweden	180000	Co-processing
UPM	Finland	100000	
NESTE	Netherlands	1000000	
NESTE	Finland	260000	
NESTE	Finland	260000	
NESTE	Singapore	1000000	
SB Petrixo	U.A.E.	400000	Under construction
Sinopec	China	20000	

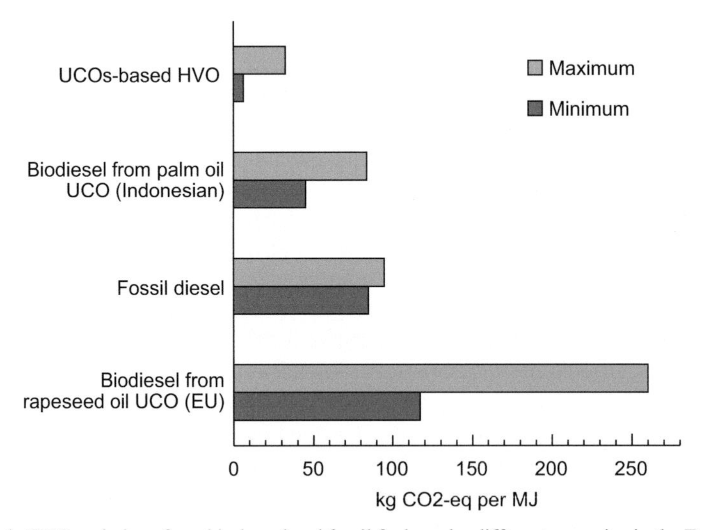

Figure 10. Life cycle GHG emissions from bio-based and fossil fuels under different scenarios in the European union (Data from Behrends, 2018).

8. Current use of UCOs as Drop-in Functional Chemical or as Oleochemical Feedstock

With the increasing global awareness on the environmental impacts of pollution, a variety of innovations are being developed for the exploitation of large volumes of waste materials; this also applies for UCOs. As mentioned earlier, waste cooking oils can be directly used as ingredients in different applications, or transformed into a variety of oleochemical derivatives. The direct use of crude fats and oil as candle fuels or as soap feedstock has been recognized since ancient times. These applications are suitable for UCOs exploitation in small businesses (Félix et al., 2014), and pretreatment is required, taking into account that appealing sensory profiles are required in those consumer products.

Recent reports indicate that UCOs have been directly used as drop-in crosslinking and bonding agents for a variety of final products, such as asphalt, resins, inks, concrete, and building blocks, among others (Sun et al., 2016; Jaya et al., 2018; Wang et al., 2018; Chandrasekar et al., 2016; Olufemi et al., 2018). In this case, the unsaturations in the fatty chains suffer complex physical and chemical oxypolymerization reactions, binding with some of the components of the composite end products. Also, the saturated chains in UCOs form a protective film that is resistant to water and alkali, but permeable to oxygen. This film allows a continuous oxygen intake, enhancing oxidation, crosslinking and, consequently, the drying effect.

As feedstock, UCOs con provide the nonpolar moieties required in surfactant molecules. This has been validated in the synthesis of zwitterionic surfactants (Zhang et al., 2015), methyl ester sulfonates (Jin et al., 2016; Permadani et al., 2018; Junior et al., 2019), sophorolipids and rhamnolipid by biological pathways (Maddikeri et al., 2015; Lan et al., 2015; Ozdal et al., 2017; Sharma et al., 2019). Also, another increasingly explored valorization alternative is epoxidation. The double bonds of fatty acid chains can form oxyrane rings by reaction with oxidizing agents such as ozone and hydrogen peroxide. This route enables epoxides that can be directly used as plasticizers, drying oils, and as ingredient in epoxy resins, among other potential applications to be obtained. Subsequently, epoxides can be transformed in a large variety of polyols for the manufacture of polyesters, resins and polyurethanes (Paul et al., 2018; Salleh et al., 2018; Suzuki et al., 2018). Likewise, UCOs have been used in the production of fine and specialty chemicals, such as plasticizers, lubricants, drilling muds, biobased polymers, lipases, graphene, etc. (Dumont and Narine, 2007; Wang et al., 2014; Li and Wang, 2015; Panadare and Rathod, 2015; Zhang et al., 2015; Borugadda and Goud, 2016; Kamilah et al., 2018; Mamat et al., 2018; Paul et al., 2018; Zheng et al., 2018; Lopes et al., 2019; Sarnoa et al., 2019).

9. Concluding Remarks

As with many other food waste streams, UCOs management represents a major issue for the sustainability of modern society. Particularly, the increasing consumption of oils and fats in food preparation, the large amount of derived residues, and the consequent negative social, environmental and economic impacts of their mishandling have become major driving forces to pursue sustainable ways for UCOs collection and reuse. The estimations here presented indicate that UCOs global generation can potentially supply the current oils and fats demand of the entire oleochemical industry. Moreover, this available volume could be even higher if novel waste reduction methods are implemented (e.g., solvent extraction of oils and fats from food waste or from trap greases), enabling the implementation of circular economy models within the oleochemical industry.

While most current applications are energy-related and they will certainly represent the major end use of collected UCOs in the next years, it is expected that the waste oils will increasingly be incorporated as feedstock in the oleochemical industry in the future. Initially, those applications where heterogeneity and unseemly sensory properties are not a limiting factor (e.g., tire rubber additive, asphalt, concrete, substrate for bioprocess, etc.), could welcome the use of re-greening bio-based ingredients. Also, some current oleochemical processes that normally handle colored and smelly raw materials or products (e.g., oxidation, sulfonation, sulfation) might be able to tune their facilities to incorporate certain types of UCOs. Later, it would be possible to see a gradual incorporation of pretreated or specifically selected UCOs (e.g., high iodine index) in chemical processes where they would provide advantageous properties, for instance in the production of epoxides, polyols, metallic soaps, non-isocyanate polyurethanes, biopolymers, etc.

Despite their potential, major limiting factors need to be overcome for the implementation of sustainable UCOs exploitation alternatives. As the major fraction of UCOs are generated by households, sustainable recollection programs must be implemented through promotion policies and law enforcement in order to realize the required supply as oleochemical feedstock. Thus, there is a need to involve the participation of different stakeholders (governments, public and private institutions and investors, oleochemical and food industry, academia, society, etc.), for the implementation of standardized and regulated practices for UCOs management and transformation. This might involve, for example, the promotion and implementation of conscious practices for UCOs management and disposal (e.g., oil-supply-UCOs-collection contracts, standardized degrees of reuse, source separation systems, etc.). These kinds of practices will help in reducing the heterogeneity issues, ensuring the availability of suitable feedstock for a variety of broad portfolio of oleochemical products.

Acknowledgments

This work has been partially funded by the Royal Academy of Engineering under the grant IAPP18-19\65, and the project entitled: *Valorization of Urban Used Cooking Oils by transformation into value-added oleochemicals. Study Case for Bogota, Colombia.*

References

Abbaszaadeh, A., Ghobadian, B., Omidkhah, M. and Najafi, G. 2012. Current biodiesel production technologies: A comparative review. Energy Conversion and Management 63: 138–148.

Alibaba. 2018. Chemicals. Available at https://www.alibaba.com/trade/(accessed May 3, 2019).

Alqaralleh, R.M., Kennedy, K., Delatolla, R. and Sartaj, M. 2016. Thermophilic and hyperthermophilic co-digestion of waste activated sludge and fat, oil and grease: Evaluating and modeling methane production. J. Environ. Manag. 183(Part 3): 551–561.

AMERI. Global Oleochemicals Market Outlook to 2025. Ameri Research Inc. Available at: https://www.ameriresearch.com/product/oleochemicals-market/(accessed May 3, 2019).

Aransiola, E., Ojumu, T., Oyekola, O., Madzimbamuto, T. and Ikhu-Omoregbe, D. 2014. A review of current technology for biodiesel production: State of the art. Biomass and Bioenergy 61: 276–297.

Bacovsky, D., Körbitz, W., Mittelbach, M. and Wörgetter, M. 2007. Biodiesel production: Technology and European providers. IEA Task 39. Report T39-B6.

Banerjee, A. and Chakraborty, R. 2009. Parametric sensitivity in transesterification of waste cooking oil for biodiesel production—A review. Resources, Conservation and Recycling 53(9): 490–497.

Banks, D. 2007. Industrial frying. pp. 291–304. *In*: Erickson, M. (ed.). Deep Frying (Second Edition) Chemistry, Nutrition, and Practical Applications. AOCS Press.

Behrends, F.J. 2018. Greenhouse gas footprint of biodiesel production from used cooking oils. Doctoral dissertation. Utrecht University.

Bezergianni, S. and Kalogianni, A. 2009. Hydrocracking of used cooking oil for biofuels production. Bioresource Technology 100: 3927–3932.

Boatella, J. and Codony, R. 2000. Recycled cooking oils: Assessment of risks for public health. European Parliament. Luxembourg. Available at: http://nehrc.nhri.org.tw/toxic/news/1030905-3.pdf (accessed May 3, 2019).

Bordin, K., Kunitake, M., Aracava, K. and Trindade, C. 2013. Changes in food caused by deep fat frying—A review. Arch. Latinoam Nutr. 63: 5–13.

Borugadda, V. and Goud, V. 2016. Improved thermo-oxidative stability of structurally modified waste cooking oil methyl esters for bio-lubricant application. J. Clean. Prod. 112: 4515–24.

Boutesteijn, C., Drabik, D. and Venus, T.J. 2017. The interaction between EU biofuel policy and first- and second-generation biodiesel production. Industrial Crops and Products 106(1): 124–129.

Brackman, B., Knaut, J. and Wallscheid, P. 1984. Oleochemicals. Henkel KGaA, Dusseldorf. Germany.

Canakci, M. and Gerpen, J.V. 2001. Biodiesel production from oils and fats with high free fatty acids. Trans. ASAE 44(6): 1429–1436.

Carnevale, E., Molari, G. and Vittuari, M. 2017. Used cooking oils in the biogas chain: A technical and economic assessment. Energies 10(2): 192.

Chandrasekar, M., Kavitha, S., Snekha, G. and Vinothini, A. 2016. Experimental investigation on usage of waste cooking oil (WCO) in concrete making and adopting innovative curing method. IJERT 5: 146–149.

Choe, E. and Min, D. 2007. Chemistry of deep-fat frying oils. J. Food Sci. 72(5): R77–R86. DOI: 10.1111/j.1750-3841.2007.00352.x.

Chrysikou, L., Dagonikou, V., Dimitriadis, A. and Bezergianni, S. 2019. Waste cooking oils exploitation targeting EU 2020 diesel fuel production: Environmental and economic benefits. Journal of Cleaner Production 219: 566–575.

Da Silva, A., Werderits, D., Leal, G. and da Silva, C. 2017. The potential of waste cooking oil as supply for the Brazilian biodiesel chain. Renewable and Sustainable Energy Reviews 72: 246–253.

DARPRO. 2017. Maximizing the value of cooking oil. Darling ingredients. Available at: https://d10k7k7mywg42z.cloudfront.net/assets/58b6ee1ba0b5dd4be900cd84/Todd_Mathes_Maximizing_the_Value_of_Used_Cooking_Oil.pdf (accessed May 3, 2019).

De Alzaa, F., Guillaume, C. and Ravetti, L. 2018. Evaluation of chemical and physical changes in different commercial oils during heating. Acta Scientific Nutritional Health 2(6): 2–11.

De Oliveira, E., Quirino, R., Suarez, P. and Prado, A. 2006. Heats of combustion of biofuels obtained by pyrolysis and by transesterification and of biofuel/diesel blends. Thermochimica Acta 450: 87–90.

Decker, M. 2006. Raw materials and processes in oleochemistry. Available at: http://www.abiosus.org/docs/1.3_VortragEmdenDierker.pdf (accessed May 3, 2019).

Dumont, M. and Narine, S. 2007. Soapstock and deodorizer distillates from North American vegetable oils: Review on their characterization, extraction and utilization. Food Research International 40: 957–974.

EBIA. 2015. Transformation of Used Cooking Oil into biodiesel: From waste to resource. European Biomass Industry Association. Available at: https://www.eubren.com/UCO_to_Biodiesel_2030_01.pdf (accessed May 3, 2019).

Eguchi, S., Kagawa, S. and Okamoto, S. 2015 Environmental and economic performance of a biodiesel plant using waste cooking oil. Journal of Cleaner Production 101: 245–250.

Félix, S., Araújo, A., Pires, A. and Sousa, A. 2017. Soap production: A green prospective. Waste Manage. 66: 190–195.

Fonseca, J., Teleken, J., Almeida, V. and Da Silva, C. 2019. Biodiesel from waste frying oils: Methods of production and purification. Energy Conversion and Management 184: 205–218.

Gebremariam, S. and Marchetti, J. 2018. Economics of biodiesel production: Review. Energy Conversion and Management 168: 74–84.

Ghobadian, B. 2012. Liquid biofuels potential and outlook in Iran. Renewable and Sustainable Energy Reviews 16: 4379–4384.

Greenea. 2016a. Waste-based feedstock and biofuels market in Europe. 2016. Available at https://www.greenea.com/wp-content/uploads/2016/11/Argus-2016.pdf (accessed May 3, 2019).

Greenea. 2016b. Used Cooking Oil technical data sheet. Available at: http://www.greenea.com/wp-content/uploads/2016/07/UCO.pdf (accessed May 3, 2019).

Greenea. 2016c. Analysis of the current development of household UCO collection systems in the EU. Available at https://theicct.org/sites/default/files/publications/Greenea%20Report%20Household%20UCO%20Collection%20in%20the%20EU_ICCT_20160629.pdf (accessed July 13, 2019).

Greenea. 2017. New players join the HVO game. Available at https://www.greenea.com/publication/new-players-join-the-hvo-game/(accessed May 3, 2019).

Greenea. 2018. And do you recycle your used cooking oil at home? Available at https://www.greenea.com/wp-content/uploads/2017/03/Greenea-article-UCO-household-collection-2017.pdf (accessed May 3, 2019).

Greenea. 2019a. Waste-based market performance (2019). Available at: http://www.greenea.com/en/market-analysis (accessed May 3, 2019).

Greenea. 2019b. EU Biodiesel Market Outlook 2019. Available at: http://www.greenea.com/wp-content/uploads/2019/01/Greenea-market-outlook-2019-V2.pdf (accessed May 3, 2019).

Günay, M., Türker, L. and Tapan, N. 2019. Significant parameters and technological advancements in biodiesel production systems. Fuel 250: 27–41.

Haas, M., Wyatt, V., Moreau, R., Cairncross, R. and Hums, M. 2016. Exceptionally low quality feedstocks for biodiesel production: Brown/Trap Grease. Available at http://www.biodieselsustainability.com/wp-content/uploads/2016/04/5._haas_trapgrease.pdf (accessed May 3, 2019).

Hajjari, M., Tabatabaei, M., Aghbashlo, M. and Ghanavati, H. 2017. A review on the prospects of sustainable biodiesel production: A global scenario with an emphasis on waste-oil biodiesel utilization. Renewable and Sustainable Energy Reviews 72: 445–464.

He, J., Wang, X., Yin, X., Li, Q., Li, X., Zhang, Y. and Deng, Y. 2018. Insights into biomethane production and microbial community succession during semi-continuous anaerobic digestion of waste cooking oil under different organic loading rates. AMB Expr. 8(92): 1–11.

Hill, K. 2000. Fats and oils as oleochemical raw materials. Pure Appl. Chem. 72(7): 1255–1264.

Jaya, R., Lopa, R., Hassan, N., Yaacob, H., Ali, M., Hamid, N. and Abdullah, M. 2018. Performance of Waste Cooking Oil on Aged Asphalt Mixture. ICCEE 2018. E3S Web of Conferences 65, 02002.

Jin, Y., Tian, S., Guo, J., Ren, X., Li, X. and Gao, S. 2016. Synthesis, characterization and exploratory application of anionic surfactant fatty acid methyl ester sulfonate from waste cooking oil. Journal of Surfactants and Detergents 19(3): 467–475.

Junior, G., Ibadurrohman, M. and Slamet, S. 2019. Synthesis of eco-friendly liquid detergent from waste cooking oil and ZnO nanoparticles. AIP Conference Proceedings 2085: 020075.

Kamilah, H., Al-Gheethi, A., Yang, T. and Sudesh, K. 2018. The use of palm oil-based waste cooking oil to enhance the production of polyhydroxybutyrate [P(3HB)] by Cupriavidus necator H16 strain. Arab. J. Sci. Eng. 43: 3453–3463.

Khan, H., Ali, C., Iqbal, T., Yasin, S., Sulaiman, M., Mahmood, H., Rashid, H., Pasha, M. and Mu, B. 2018. Current scenario and potential of biodiesel production from waste cooking oil in Pakistan: An overview. Chinese Journal of Chemical Engineering. In press.

Kharina, A., Searle, S., Rachmadini, D., Kurniawan, A. and Prionggo, A. 2018. The potential economic, health and greenhouse gas benefits of incorporating used cooking oil into Indonesias's biodiesel. International council on clean transport. Available at: https://www.theicct.org/sites/default/files/publications/UCO_Biodiesel_Indonesia_20180919.pdf (accessed May 3, 2019).

Kiss, A. and Bildea, C. 2012. A review of biodiesel production by integrated reactive separation technologies. J. Chem. Technol. Biotechnol. 87(7): 861–879.

Knothe, G. and Steidley, K. 2009. A comparison of used cooking oils: A very heterogeneous feedstock for biodiesel. Bioresour. Technol. 100: 5796–801.

Kulkarni, M. and Dalai, A. 2006. Waste cooking oils—an economical source for biodiesel: A review. Ind. Eng. Chem. Res. 45: 2901–2913.

Kusdiana, D. and Saka, S. 2004. Two-step preparation for catalyst-free biodiesel fuel production—hydrolysis and methyl esterification. Appl. Biochem. Biotechnol. 113(16): 781–791.

Lan, G., Fan, Q., Liu, Y., Chen, C., Li, G., Liu, Y. and Yin, X. 2015. Rhamnolipid production from waste cooking oil using Pseudomonas SWP-4. Biochemical Engineering Journal 101: 44–54.

Li, W. and Wang, X. 2015. Bio-lubricants derived from waste cooking oil with improved oxidation stability and low-temperature properties. J. Oleo. Sci. 64: 367–74.

Lombardi, L., Mendecka, B. and Carnevale, E. 2018. Comparative life cycle assessment of alternative strategies for energy recovery from used cooking oil. Journal of Environmental Management 216: 235–245.

Lopes, M., Miranda, S., Alves, J., Pereira, A. and Belo, I. 2019. Waste cooking oils as feedstock for Lipase and Lipid-Rich biomass production. Eur. J. Lipid Sci. Technol. 121: 1800188.

Luna, D., Calero, J., Sancho, E., Luna, C., Posadillo, A., Bautista, F., Romero, A., Berbel, J. and Verdugo, C. 2014. Technological challenges for the production of biodiesel in arid lands. Journal of Arid Environments 102: 127–138.

Maddikeri, G., Gogate, P. and Pandit, A. 2015 Improved synthesis of sophorolipids from waste cooking oil using fed batch approach in the presence of ultrasound. Chemical Engineering Journal 263: 479–487.

Mamat, R., Hamzah, F., Hashim, A., Abdullah, S., Alrokayan, S., Safiay, M., Asli, A. and Rusop, M. 2018. Influence of volume variety of waste cooking palm oil as carbon source on graphene growth through double thermal chemical vapor deposition. 2018 IEEE International Conference on Semiconductor Electronics (ICSE).

Mazubert, A., Poux, M. and Aubin, J. 2013. Intensified processes for FAME production from waste cooking oil: A technological review. Chemical Engineering Journal 233: 201–223.

Nas, B. and Berktay, A. 2007. Energy potential of biodiesel generated from waste cooking oil: An environmental approach. Energy Sources, Part B: Economics, Planning, and Policy 2(1): 63–71.

Nelson, B. and Searle, S. 2016. Projected availability of fats, oils, and greases in the U.S. Working paper 2016-15. International council on clean transport. Available at: https://www.theicct.org/sites/default/files/publications/Biodiesel%20 Availability_ICCT_20160707.pdf (accessed May 3, 2019).

NRA. 2008. Pocket information manual. A buyer's guide to rendered products. National Renderers Association, Inc. Alexandria, Virginia, USA. Available at: http://assets.nationalrenderers.org/pocket_information_manual.pdf (accessed May 3, 2019).

OECD. OECD-FAO Agricultural Outlook 2018–2027. Available at: https://stats.oecd.org (accessed 28 Feb., 2020).

Oh, P., Nang, H., Chen, J., Chong, M. and Choo, Y. 2012. A review on conventional technologies and emerging process intensification (PI) methods for biodiesel production. Renewable and Sustainable Energy Reviews 16: 5131–5145.

Oleoline. 2018. Oleochemical market report. Available at https://www.oleoline.com/products/Oleochemical-Market-Report-4.html (accessed May 3, 2019).

Olufemi, J., Napiah, M., Ibrahim, K. and Raduan, M. 2018. Evaluation of Waste Cooking Oil as Sustainable Binder for Building Blocks. ICCEE 2018. E3S Web of Conferences 65, 05003.

Ortner, M., Müller, W., Schneider, I. and Bockreis, A. 2016. Environmental assessment of three different utilization paths of waste cooking oil from households. Resources, Conservation and Recycling 106: 59–67.

Ozdal, M., Gurkok, S. and Ozdal, O. 2017. Optimization of rhamnolipid production by Pseudomonas aeruginosa OG1 using waste frying oil and chicken feather peptone. Biotech. 7(2): 117.

Panadare, D. and Rathod, V. 2015. Applications of waste cooking oil other than biodiesel: A review. Iran J. Chem. Eng. 12: 55–76.

Paul, A., Borugadda, V., Bhalerao, M. and Goud, V. 2018. *In situ* epoxidation of waste soybean cooking oil for synthesis of biolubricant basestock: A Process Parameter Optimization and Comparison with RSM, ANN, and GA. Can. J. Chem. Eng. 96: 1451–1461.

Permadani, R., Ibadurrohman, M. and Slamet. 2018. Utilization of waste cooking oil as raw material for synthesis of Methyl Ester Sulfonates (MES) surfactant. IOP Conf. Series: Earth and Environmental Science 105: 012036.

Perumalla, R. and Subramanyan, R. 2016. Evaluation of the deleterious health effects of consumption of repeatedly heated vegetable oil. Toxicology Reports 3: 636–643.

Ragauska, A., Pu, Y. and Ragauskas, A. 2013. Biodiesel from grease interceptor to gas tank. Energy Science and Engineering 1(1): 42–52.

Rathore, V., Newalkar, B. and Badoni, R. 2016. Processing of vegetable oil for biofuel production through conventional and non-conventional routes. Energy for Sustainable Development 31: 24–49.

Ribau, M., Nogueira, R. and Miguel, L. 2018. Quantitative assessment of the valorisation of used cooking oils in 23 countries. Waste Manage. 78: 611–20.

Rincón , L., Cadavid, J. and Orjuela, A. 2019. Used cooking oils as potential oleochemical feedstock for urban biorefineries—Study case in Bogota, Colombia. Waste Management 88: 200–210.

SAGE Global. 2015. Cal-biofuel. University of Calabar. http://www.sageglobal.org/new_layout/wp-content/uploads/2017/12/SAGE-Nigeria-SEB-World-Cup-Champion-2015-Calabar.pdf (accessed May 3, 2019).

Salleh, W., Tahir, S. and Mohamed, S. 2018. Synthesis of waste cooking oil-based polyurethane for solid polymer electrolyte. Polym. Bull. 75: 109–120.

Sanli, H., Canakci, M. and Alptekin, E. 2011. Characterization of waste frying oils obtained from different facilities. World renewable energy congress. May 8–23, Linköping, Sweden.

Santacesaria, E., Martinez, G., Di Serio, M. and Tesser, R. 2012. Main technologies in biodiesel production: State of the art and future challenges. Catalysis Today 195: 2–13.

Sarnoa, M., Iuliano, M. and Cirillo, C. 2019. Optimized procedure for the preparation of an enzymatic nanocatalyst to produce a bio-lubricant from waste cooking oil. Chem. Eng. J. In press.

Sharma, S., Datta, P., Kumar, B., Tiwari, P. and Pandey, L.M. 2019. Production of novel rhamnolipids via biodegradation of waste cooking oil using Pseudomonas aeruginosa MTCC7815. Biodegradation.

Sheinbaum, C., Calderon, A. and Ramirez, M. 2013. Potential of biodiesel from waste cooking oil in Mexico. Biomass and Bioenergy 56: 230–238.

Singhabhandhu, A. and Tezuka, T. 2010a. Prospective framework for collection and exploitation of waste cooking oil as feedstock for energy conversion. Energy 35: 1839–1847.

Singhabhandhu, A. and Tezuka, T. 2010b. The waste-to-energy framework for integrated multi-waste utilization: Waste cooking oil, waste lubricating oil, and waste plastics. Energy 35: 2544–2551.

Song, C., Liu, Q., Ji, N., Deng, S., Zhao, D., Li, S. and Kitamura, Y. 2016. Evaluation of hydrolysis–esterification biodiesel production from wet microalgae. Bioresource Technology 214: 747–754.

Sonthalia, A. and Kumar, N. 2019. Hydroprocessed vegetable oil as a fuel for transportation sector: A review. Journal of the Energy Institute 92: 1–17.

Statista. 2019. Consumption of vegetable oils worldwide from 2013/14 to 2018/2019. Available at: https://www.statista.com/statistics/263937/vegetable-oils-global-consumption/(accessed Feb. 28, 2020).

Sun, Z., Yi, J., Huang, Y., Feng, D. and Guo, C. 2016. Properties of asphalt binder modified by bio-oil derived from waste cooking oil. Constr. Build. Mater. 102: 496–504.

Suzuki, A., Botelho, B., Oliveira, L. and Franca, A. 2018. Sustainable synthesis of epoxidized waste cooking oil and its application as a plasticizer for polyvinyl chloride films. Eur. Polym. J. 99: 142–149.

Thoai, D., Tongurai, C., Prasertsit, K. and Kumar, A. 2019. Review on biodiesel production by two-step catalytic conversion. Biocatalysis and Agricultural Biotechnology 18: 101023.

Thompson, J. 1981. Final report on waste vegetable oil as a fuel extender. United States Army.

Toop, G., Alberici, S., Spoettle, M. and van Steen, H. 2013. Trends in the UCO market. Available at https://assets.publishing.service.gov.uk/government/uploads/system/uploads/attachment_data/file/307119/trends-uco-market.pdf(accessed May 3, 2019).

Torres, P. 2019. Life cycle assessment of the exploitation and valorization of used cooking oil in Bogotá as oleochemical feedstock (In Spanish). Master's Dissertation. National University of Colombia, 2019.

Tsoutsos, T.D., Tournaki, S., Paraíba, O. and Kaminaris, S.D. 2016. The Used Cooking Oil-to-biodiesel chain in Europe assessment of best practices and environmental performance. Renewable and Sustainable Energy Reviews 54: 74–83.

UFOP. 2018. UFOP Report on Global Market Supply 2017/2018. Available at: https://www.ufop.de/files/3515/1515/2657/UFOP_Report_on_Global_Market_Supply_2017-2018.pdf (accessed May 3, 2019).

UGC. 2019. Fats, oils & grease. Procurement, supply & management. Available at https://ugcinc.com/trade-markets/commodities-2/(accessed May 3, 2019).

UN. World Population Prospects 2017. Available at https://population.un.org/wpp/DataQuery/(accessed May 3, 2019).

USDA. 2013. USDA Commodity requirements. BOT2 Bulk oil and tallow. United States department of agriculture. Available at: https://www.fsa.usda.gov/Assets/USDA-FSA-Public/usdafiles/Comm-Operations/procurement-and-sales/export/pdfs/bot2.pdf (accessed May 3, 2019).

Verma, P. and Sharma, M. 2016. Review of process parameters for biodiesel production from different feedstocks. Renewable and Sustainable Energy Reviews 62: 1063–1071.

Wang, C., Xue, L., Xie, W., You, Z. and Yang, X. 2018. Laboratory investigation on chemical and rheological properties of bio-asphalt binders incorporating waste cooking oil. Constr. Build. Mater. 167: 348–358.

Wang, E., Ma, X., Tang, S., Yan, R., Wang, Y., Riley, W. and Reaney, M. 2014. Synthesis and oxidative stability of trimethylolpropane fatty acid triester as a biolubricant base oil from waste cooking oil. Biomass Bioenergy 66: 371–378.

Ward, P. 2012. Brown and black grease suitability for incorporation into feeds and suitability for biofuels. Journal of Food Protection 75(4): 731–737.
Weaver, J. and Howell, S. 2018. Biodiesel industry overview & technical update. Available at: https://biodiesel.org/docs/default-source/ffs-basics/biodiesel-industry-and-technical-overview.pdf?sfvrsn=20 (accessed May 3, 2019).
Xiong, Y., Miao, W., Wang, N., Chen, H., Wang, X., Wang, J., Tan, Q. and Chen, S. 2019. Solid alcohol based on waste cooking oil: Synthesis, properties, micromorphology and simultaneous synthesis of biodiesel. Waste Management 85: 295–303.
Yaakob, Z., Mohammad, M., Alherbawi, M., Alam, Z. and Sopian, K. 2013. Overview of the production of biodiesel from Waste cooking oil. Renewable and Sustainable Energy Reviews 18: 184–193.
Yuste, R. 2016. Biofuels production in conventional oil refineries through bio-oil co-processing. Available at https://www.repsol.com/imagenes/global/en/Yuste%20R%20-%20Biofuels%20production%20in%20conventional%20oil%20refineries%20through%20bio-oil%20co-processing_tcm14-58321.pdf (accessed May 3, 2019).
Zhang, H., Wang, Q. and Mortimer, S. 2012. Waste cooking oil as an energy resource: Review of Chinese policies. Renewable and Sustainable Energy Reviews 16: 5225–5231.
Zhang, H., Ozturk, U., Wang, Q. and Zhao, Z. 2014. Biodiesel produced by waste cooking oil: Review of recycling modes in China, the US and Japan. Renewable and Sustainable Energy Reviews 38: 677–685.
Zhang, Q., Cai, B., Xu, W., Gang, H., Liu, J., Yang, S. and Mu, B. 2015. The rebirth of waste cooking oil to novel bio-based surfactants. Sci. Rep. 5: 9971.
Zhang, Y., Dubé, M., McLean, D. and Kates, M. 2003. Biodiesel production from waste cooking oil: 2. Economic assessment and sensitivity analysis. Bioresource Technology 90: 229–240.
Zhao, X., Wei, L., Cheng, S. and Julson, J. 2017. Review of heterogeneous catalysts for catalytically upgrading vegetable oils into hydrocarbon biofuels. Catalysts 7(83): 1–25.
Zheng, T., Wu, Z., Xie, Q., Fang, J., Hu, Y. and Lu, M. 2018. Structural modification of waste cooking oil methyl esters as cleaner plasticizer to substitute toxic dioctyl phthalate. J. Clean. Prod. 186: 1021–1030.

CHAPTER 6

Advances in Carbon Capture through Thermochemical Conversion of Biomass

Sonal K Thengane

1. Introduction

Climate change and energy crisis are amongst the alarming issues of the world that are inter-related to each other. In 2015, world leaders had pledged at the UN climate change conference in Paris to reduce the greenhouse gas (GHG) emissions by 40–70% in the next four decades compared to year 2010 data (Aalbers and Bollen, 2017). The use of renewable energy sources, especially biomass, is expected to play a significant role in attaining this objective. With a share of 10%, biomass is the fourth most important energy source, after oil, coal and natural gas. Owing to the climate change policies, this share is expected to increase further, as biomass-based technology is an option that can often work as carbon neutral or sometimes carbon negative. The process of photosynthesis captures about 140 TW of energy, which is approximately 0.08% of the total incident solar energy on earth (Goldemberg and Johansson, 2004). In spite of such small percentage, the total volume of biomass produced is about 10 times more than our present energy demand. Nearly 100 billion tonnes of carbon are converted to biomass every year (Abbasi and Abbasi, 2010). Though these figures look attractive, there are some serious limitations on the extent of utilization of biomass for producing energy.

The burning, combustion or decomposition of biomass releases that carbon to the atmosphere which it had recently captured from the atmosphere during photosynthesis. Hence, there is no net addition of carbon or CO_2. However, the burning of fossil fuels results in net addition of CO_2 in the atmosphere as they are derived from biological matter that is millions of years old. This carbon neutral nature of biomass as an energy source has created great interest to utilize it in different ways as a substitute for fossil fuels. Figure 1 shows the net carbon balance for different energy alternatives, including fossil fuels and renewables (Quader and Ahmed, 2017; Thengane and Bandyopadhyay, 2020). Recent studies have used integrated assessment models to develop mitigation strategies in order to achieve the Paris Agreement's targets. The default pathways show an early peak in emissions, followed by rapid emission reductions and, finally, a period of net negative emissions (Fuss et al., 2014). These net negative emissions refer to active removal of carbon dioxide from the atmosphere, achieved by introducing new carbon sinks on a large scale (Anderson and Peters, 2016). The advantages of using negative emission technologies as part of a mitigation strategy are that they can: (1) somewhat dampen the need for quick near-term emission reductions and (2) compensate emissions from hard to moderate sectors (Vaughan et al., 2018).

Department of Mechanical Engineering, Massachusetts Institute of Technology, 77 Massachusetts Ave, Room 3-339F, Cambridge, MA 02139, USA.

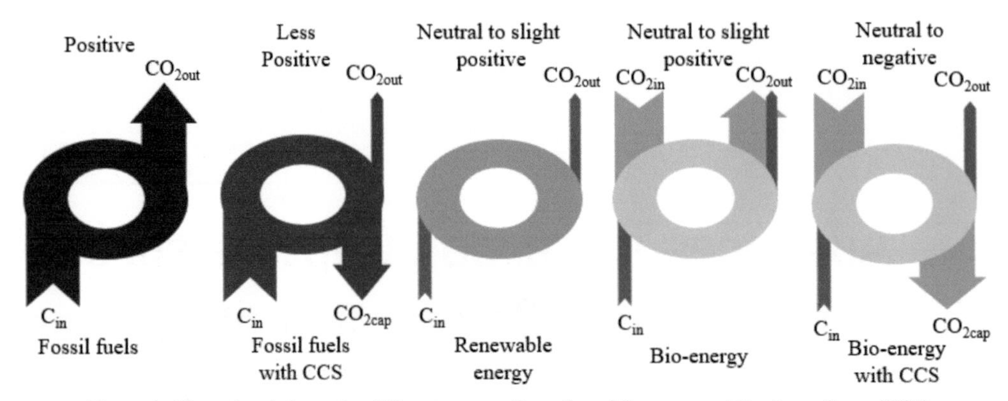

Figure 1. Net carbon balance for different energy alternatives (Thengane and Bandyopadhyay, 2020).

Carbon capture and storage (CCS) is considered as a key technology amongst the greenhouse gas (GHG) emission reduction options, in addition to energy savings and renewable energy technologies, to attain the stringent climate targets. The captured CO_2 can be stored in depleted oil and gas wells, inaccessible coal seams, saline aquifers, and other geological structures that can act as reservoirs (Thengane et al., 2019). CCS, often linked with fossil fuel-based processes, can be combined with biomass-based processes too, sometimes referred to as bio-CCS (Koornneef et al., 2012) or bioenergy with CCS (BECCS) (Azar et al., 2013). BECCS consists of multiple components and stages: biomass feedstock and collection, conversion of the biomass feedstock into energy, production of heat, electricity, or fuels, and capture and sequestration of the carbon resulting from using that energy (NAS, 2018). Figure 2 shows the schematic of a BECCS supply chain, starting from CO_2 in the atmosphere to the end applications and storage of captured CO_2. Bui et al. (2018) recently evaluated a 500 MW pulverized fuel BECCS plant in terms of energy efficiency, carbon intensity and pollutant emissions. They observed the strong dependence of carbon negativity of the technology on plant efficiency, and predicted the energy efficiency of 38% with heat recovery.

In general, biomass can be grass, plants, trees, wood, and several residues, as well as purposely grown food and energy crops. The use of biomass as fuel may lead to competition with other uses of biomass, such as food, paper, fibreboard, furniture, and as a feedstock for some other industries. Hence, it is important to define the kind of biomass that is being targeted for conversion and utilization. The waste biomass or residual biomass or simply 'biomass' that we would refer to in this chapter includes various agricultural residues, crop wastes, forestry waste, leaf litter, sawmill waste, food waste, some components

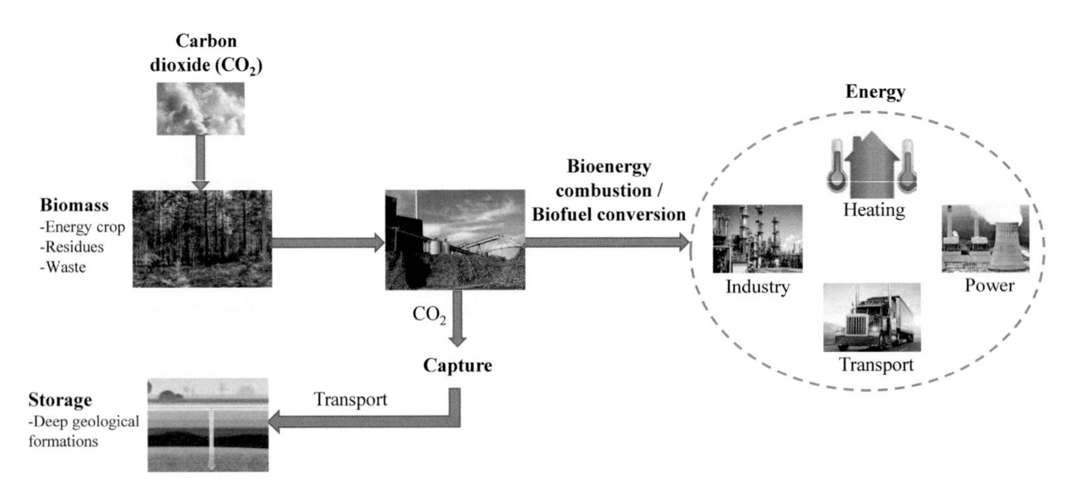

Figure 2. Schematic of BECCS supply chain (Kemper, 2015).

of municipal solid waste, and any other unused biomass which would otherwise decompose. The two major pathways of biomass conversion are thermo-chemical (e.g., combustion, pyrolysis, torrefaction, gasification) and biochemical (fermentation, bio-methanation). However, thermochemical methods have much lower conversion time, can convert the entire biomass without rejecting any component, and are less sensitive to feedstock, unlike biochemical methods (Bhaskar and Steele, 2015).

This chapter discusses the advances in different thermochemical conversion processes for various biomass wastes, with respect to carbon capture, and the environmental impacts associated with each process. Combustion of biomass being the primary mode of utilization has been the main focus in understanding the applicable carbon capture and storage (CCS) approaches. Most of the CCS approaches for biomass combustion are also applicable to gasification of biomass. The new approaches based on chemical looping for carbon capture in biomass combustion and gasification processes are subsequently discussed. The other modes of biomass conversion and utilization, such as pyrolysis and torrefaction, are also discussed and compared with respect to their carbon capture potential.

2. Thermochemical Conversion of Biomass

Thermochemical processing of biomass involves heating of biomass to different temperatures in the presence of differing concentrations of oxygen in order to produce a range of products, such as heat, fuel or chemicals. Any kind of biomass may be used to extract energy or derive one or other fuel/chemical from it. However, one biomass may provide better quality product at lesser costs than others, depending on the process and operating conditions. Hence, it is important to study the performance of different kinds of biomass in different processes and a varied set of operating conditions. This justifies the high number of research and review articles being published in the areas of combustion, gasification and pyrolysis (Akhtar et al., 2018). Torrefaction, or low-temperature pyrolysis, is a relatively new approach, yet the number of publications in this area is also rising steeply (Ribeiro et al., 2018). The following sub-sections discuss the processes of combustion, gasification, pyrolysis and torrefaction, with respect to their carbon capture potential and approaches.

2.1 Combustion

The use of biomass in earlier times was based on open fires having low efficiencies between 5–10%, but this has significantly improved over the last century. The combustion of biomass and biomass-derived fuels has the potential to partially or substantially substitute fossil fuels, such as coal, oil and natural gas. Industrialized economies are moving towards employing biomass combustion as one of the major options for heat and electricity production. Table 1 shows the broad applications of biomass primarily in four sectors, with the highest usage in heating and cooking, and the lowest usage in transportation. Though biomass is a renewable source, there are environmental emissions happening when it is used as a fuel, especially for combustion. WHO report (2014) estimated about 1.5 million human deaths per annum as a result of this type of pollution.

The three stages in general biomass combustion are drying, volatiles release and burning, and char combustion as shown in Figure 3. Drying requires heat to evaporate moisture, and the rate of drying depends on the particle size and temperature. Next, devolatilization takes place and pyrolysis gases are released, for which oxygen is needed for combustion. The char combustion stage requires oxygen, releases

Table 1. Sector-wise applications of biomass (Vakkilainen et al., 2013).

Description	Energy unit (Gtoe)
Domestic cooking and heating; Heating of commercial premises and large buildings	1
Industrial steam raising-waste biomass, medium size power generation/CHP/district heating units	0.19
Large electrical power plants	0.17
Transportation	0.08

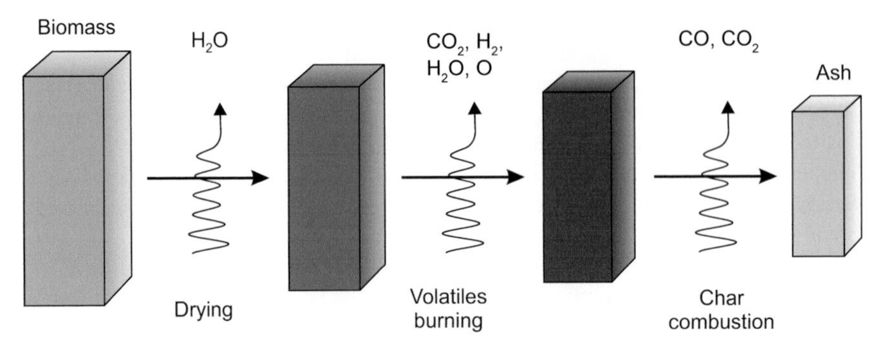

Figure 3. Simple combustion mechanism for biomass.

heat, reduces particle size, and leaves residual ash. It can be seen that the combustion of carbonaceous fuel, such as biomass, results in the formation of carbon dioxide and water as the major products. The increase in CO_2 concentrations in the atmosphere is related to climate change and the considerable efforts have been devoted to mitigating these emissions.

For large scale solid biomass utilization, most of the combustion technologies are based on those developed for coal (Spliethoff, 2010). Common technologies involve combustors of different types, such as fixed bed (< 5 MW_{th}), moving or travelling grate (up to 100 MW_{th}), fluidized bed (up to 500 MW_e), in addition to suspension firing combustion and co-firing with fossil fuels (up to 900 MW_e) (Jones et al., 2014). Then, there are several small scale processes involving combustion, such as open fires, cookstoves, and small boilers (< 10 kW_{th}) which do not have the flue gas treatment unit present in larger scale processes and, hence, emit lot of particulates and unburnt volatiles into the atmosphere. Hence, several design improvements, such as improved air/fuel mixing and secondary air circulation, have been implemented in order to improve the thermal efficiency of the smaller systems used for cooking and boiling water. The common industrial applications (< 100 MW_{th}) employ either travelling bed combustion or fluidized bed combustion. Travelling grate combustors involve a continuously moving grate and are widely used in incinerators and for biomass combustion of various feedstock sizes. Fluidized bed combustors involve a combustion chamber into which air introduced via a perforated plate keeps the char in fluidized state. Biomass typically has higher percentage of volatiles and, hence, most of the combustion takes place above the bed. For sustained fluidization, biomass is usually used in the form of pellets of fine particle sizes, depending on the fluidization requirements.

2.1.1 CCS for biomass combustion

Figure 4 shows the common processes in practice for carbon capture employed in biomass combustion processes. The properties of flue gas, such as CO_2 concentration, temperature and pressure, are the most effective factors for selection of suitable process for CO_2 separation. Bulk absorption and adsorption processes are usually the best suitable processes for CO_2 separation for flue gases which usually have high temperature (about 100 ºC), low pressure (1 bar), and low CO_2 concentration (< 15% mol). The industrial plants have preferred absorption over adsorption due to the simplicity of the absorption process, although R&D efforts to prepare adsorbents with high selectivity and capacity are still ongoing (Songolzadeh et al., 2014). For coal-based plants, co-firing biomass with coal has been looked upon as an option to reduce emissions. Retrofitting conventional coal units with post-combustion CCS can certainly lower carbon emissions, but it is constrained by a technical limit for the biomass co-firing ratio, which consequently limits the carbon mitigation potential. The present biomass share in biomass/coal co-firing plants is usually below 5% and rarely exceeds 10% on a continuous basis, although 20% co-firing is technically feasible. In contrast, coal-bioenergy with carbon capture and storage (CBECCS) technology can operate not only at high biomass ratios but can achieve zero lifecycle CO_2 emissions with a biomass ratio as low as 35% (Lu et al., 2019). Other approaches, such as cryogenic separation and membrane processes, are efficient for gas streams with high CO_2 concentration and, hence, are economically

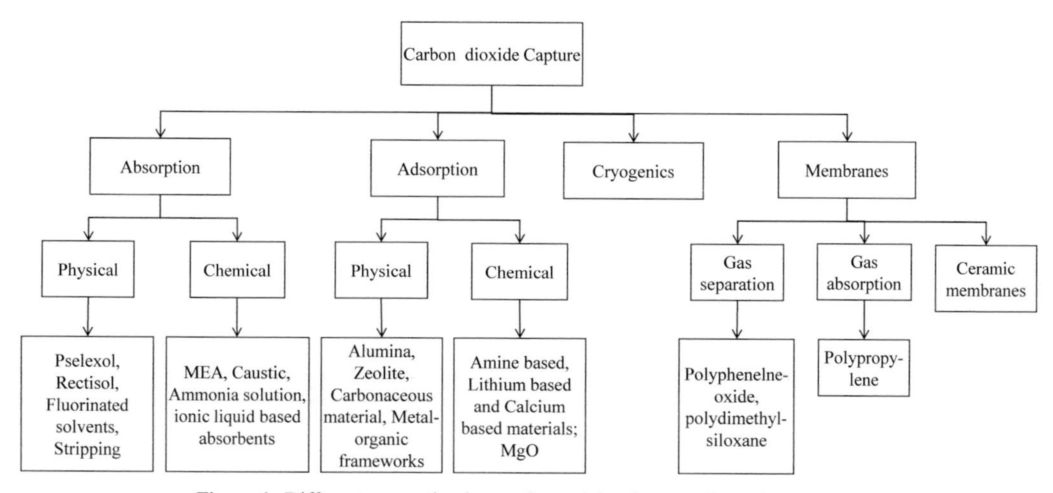

Figure 4. Different approaches in practice and development for carbon capture.

efficient for pre-combustion capture. The membrane-based approach is a continuous, steady-state process driven by pressure difference and, hence, suitable for high pressure streams with high CO_2 concentration (> 10 vol%). Different kinds of membranes have their own benefits and limitations. For example, inorganic membranes have higher thermal and chemical stability but lower selectivity than polymeric membranes. Due to operating problems and high compression costs, membrane separation is not suitable for post-combustion capture, but can be suitable for producing oxygen-enriched streams from air, in oxy-fuel combustion systems (Songolzadeh et al., 2014).

Figure 5 summarizes the different types of carbon dioxide capture processes that can be used during technologies using thermochemical conversion of biomass. The three major approaches for carbon capture and storage (CCS) are pre-combustion capture, oxy-fuel process, and post-combustion capture. There are some advanced approaches based on the concept of chemical looping which are discussed in next section. In pre-combustion capture, fuel reacts with oxygen/air/steam to produce synthesis gas (syngas), and the produced carbon monoxide reacts with steam in a catalytic shift converter to give CO_2 and more hydrogen. This is most commonly used in integrated gasification combined cycle (IGCC) power plants. CO_2 could then be separated by pressure-swing-adsorption, cryogenic distillation or chemical absorption process, resulting in a hydrogen-rich stream. The capture of CO_2 from the flue gas released after the combustion of biomass is called "post-combustion capture". Several technologies, such as absorption, adsorption, membranes and cryogenic technologies, exist for post combustion capture of CO_2 (Williams et al., 2012). Some of these, such as absorption using an amine solvent, have been commercially developed and implemented at industrial scales but are found to be suitable for only low CO_2 concentrations in flue gas. Also, the amine based solvents require a large amount of energy to regenerate in the solvent stripper, sometimes as high as 80% of the total energy of the process. Hence, the research efforts in post combustion capture are mainly focused on developing new solvents, membranes, process integration and pilot plant.

The third common category, which has been proposed as a strong contender for carbon capture technologies, is oxyfuel combustion. Nearly pure oxygen is used for combustion instead of ambient air, thereby forming mostly CO_2 and H_2O, which can be easily separated by condensing water. Some processes involve sending recycled flue gas with oxygen which further ensures a flue gas with higher CO_2 concentrations. Major benefits of this approach are high CO_2 concentration in output stream (above 80% v/v), high flame temperature, and easy separation of exhaust gases. Gaseous pollutant formation and SO_x and NO_x emissions are usually lower during oxy-fuel combustion. In one of the studies (Sher et al., 2018), when woody and non-woody biomass were tested in a 20 kW_{th} fluidized bed combustor, it was found that the concentration of oxygen in the oxidizing medium has a significant effect on temperature profiles and gas emissions (NO_x and CO). It needed oxygen concentration of about 30%, unlike the 21%

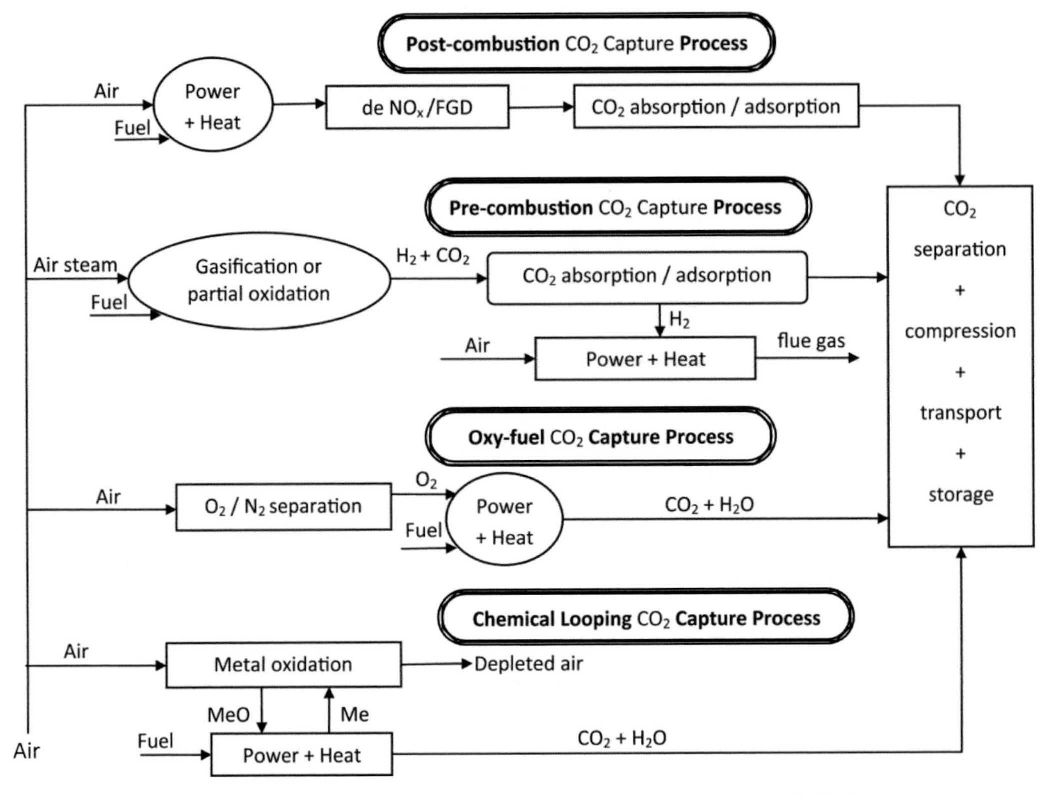

Figure 5. Different types of carbon dioxide capture processes (Leung et al., 2014).

in normal air, to maintain the same temperature, and above 25% oxygen to reduce CO and NO_x emissions. The oxyfuel combustion approach looks promising for coal but the works reported on co-firing of coal and biomass have noted a drop in the ignition temperature and rise in the proportion of biomass (Dincer et al., 2015). However, the fuel burnout improves with oxy-firing blends of coal and biomass (Arias et al., 2008). The recent projects investigating oxyfuel combustion for biomass co-firing observed that an increase in biomass co-firing ratio led to higher carbon savings but lower net energy efficiency (Falano et al., 2015). Another possible variant of oxy-fuel combustion uses steam mixed with oxygen to moderate the flame temperature (Seepana and Jayanti, 2010). The advantage of this variant is that there is no need to recirculate flue gas. Also, the flue gas, after condensation of steam, consists primarily of CO_2, which can be directly compressed and sequestrated. For biomass combustion, this may be advantageous due to the high oxygen content in the fuel. The common limitations for this technology are high capital cost and large electric power requirement to separate oxygen from air.

To summarize, biomass combustion in the presence of air results in the production of flue gas mainly containing N_2, CO_2 and H_2O. For such cases, a post-combustion separation system to capture CO_2 is required. Oxyfuel combustion needs pure O_2 alone or with some other oxidative medium as the gasifying agent, but obtaining pure oxygen is energy-intensive and expensive. Cryogenic separation is also expensive due to the energy requirement for refrigeration system. Hence, there is a need to develop a continuous process that can produce H_2-enriched gas from biomass with in-process CO_2 capture and sorbent regeneration.

2.1.2 Chemical looping for carbon capture

The concept of chemical looping combustion (CLC) involves direct use of solid fuels (coal, biomass, etc.) as solids are considerably more abundant and less expensive than natural gas. The fuel is physically mixed with the oxygen carrier (i.e., metal oxides) in the fuel reactor. The multiple reactors in parallel for oxidation and reduction reactions facilitate inherent separation of CO_2 from combustion product gas through the use of solid oxygen carrier. Separation of CO_2 do not need any additional energy. Several review articles have been published on the process and applications of CLC (Adanez et al., 2012; Fang et al., 2009; Hossain and de Lasa, 2008; Lyngfelt, 2014; Moghtaderi, 2012) with some of them also highlighting chemical looping reforming (Adanez et al., 2012) and chemical looping gasification (Acharya et al., 2009; Lahijani et al., 2015). The CLC process has been implemented at different scales, ranging from 10–140 kW_{th} for the combustion of fuels, such as natural gas or syngas, using oxides of nickel, copper, iron, and others as oxygen carriers. The application of chemical looping in gasification started with the motivation to increase the hydrogen content of the producer gas, and eliminate CO_2 within the gasifier. Fluidized bed reactors (FBR) are generally preferred for biomass gasification as they ensure efficient gas-solid interaction, and enhance CO_2 capture using an appropriate sorbent. Acharya et al. (Acharya et al., 2009) had proposed a continuous H_2 production process from agricultural biomass using FBR with *in situ* CO_2 capture and catalyst regeneration. Table 2 lists the summary of different chemical looping processes for CO_2 capture employed for the purposes of combustion and hydrogen generation.

Amongst different solid oxygen carriers employed for biomass gasification, quicklime (CaO) is one of the earliest options explored in 1970s. The *in situ* CO_2 capture using calcium oxide (CaO), a cost-effective catalyst favoring H_2 formation, during the steam reforming of biomass appears to be highly attractive and promising process (Sikarwar et al., 2016). Figure 6 shows the CaO-assisted chemical looping gasification (CLG) with two reactors, gasifier and regenerator. In the gasifier, steam reforming of biomass takes place in the presence of CaO, which captures CO_2 as $CaCO_3$ via the carbonation reaction. $CaCO_3$ particles are transferred to the other reactor for calcining back to CaO, with the production of a pure CO_2 stream which can be sent for storage. This CaO is recycled back to the gasifier, along with the heat of calcination, in order to contribute to endothermic reactions in the gasifier. The elimination of CO_2 during biomass gasification shifts the equilibrium of product gas and enhances hydrogen yield.

Table 2. Summary of chemical looping processes for CO_2 capture (Adanez et al., 2012).

Aim	Primary fuel	Process	Features
Combustion	Gas	CLC	- Gaseous fuels combustion with oxygen-carriers
	Solid	Syngas-CLC	- Previous gasification of solid fuel - Oxygen requirement for gasification
	Solid	iG-CLC	- Gasification of the solid fuel inside the fuel-reactor - Low cost oxygen-carriers are desirable
	Solid	CLOU	- Use of oxygen-carriers with gaseous O_2 release properties - Rapid conversion of the solid fuel H_2 production
Hydrogen production	Solid	SCL (Syngas-CL)	- H_2 produced by oxidation with steam of the oxygen-carrier - Previous gasification of solid fuel with O_2 - Reducer, oxidiser, and combustor are needed
	Solid	CDCL (Coal direct CL)	- H_2 produced by oxidation with steam of the oxygen-carrier - Coal and O_2 are fed to the reducer reactor - Reducer, oxidiser, and combustor are needed
	Gas	SR-CLC (Steam reforming-CLC)	- Steam reforming in usual tubular reactors - Energy requirements for SR supplied by CLC fuelled by tail gas
	Gas	a-CLR (autothermal-CLR)	- Partial oxidation of fuel with oxygen carriers - Process can be fit to produce pure N_2 stream and the desired CO/H_2 ratio
	Gas	CLH (Chemical looping hydrogen)	- H_2 produced by oxidation with steam of the oxygen-carrier - Three reactors are needed (Fuel Reactor, Air Reactor, and Steam Reactor)

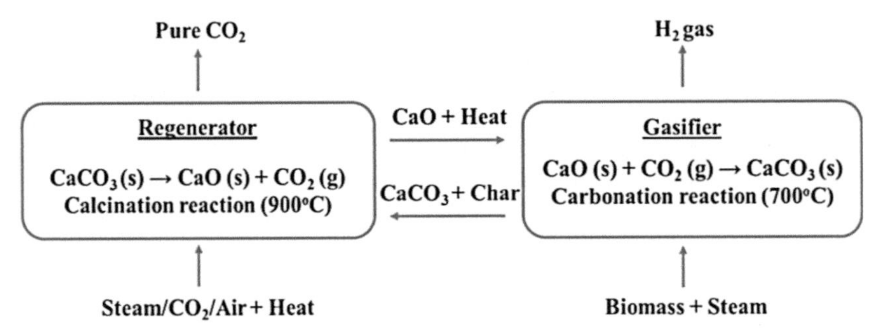

Figure 6. CaO looping biomass gasification using sorption enhanced reforming.

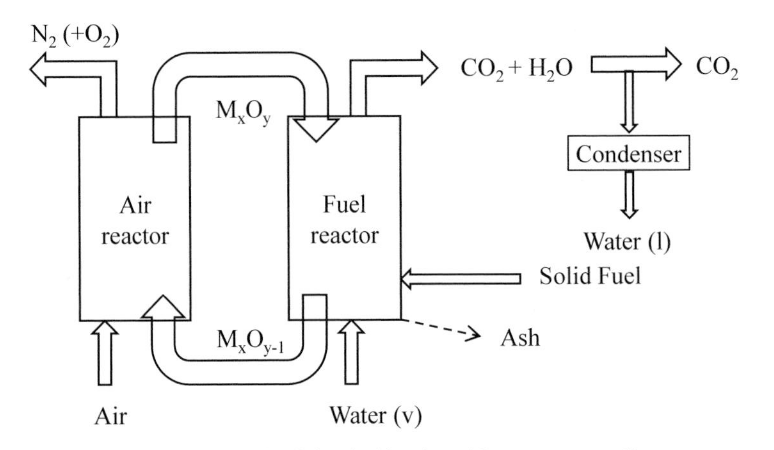

Figure 7. Schematic of chemical looping with oxygen uncoupling.

The role played by CaO as sorbent, tar cracker and heat carrier in FBR also results in high H_2 yield and conversion efficiency.

The *in situ* Gasification CLC (iG-CLC) process involves reaction of oxygen carrier with the gasification products of the solid fuel (e.g., biomass) generated inside the fuel reactor. The main process issues are the slow gasification process and the incomplete combustion of gases released in the fuel reactor. As a solution, an alternative process, chemical looping with oxygen uncoupling (CLOU), was proposed. This process is based on using oxygen carrier materials which release gaseous oxygen and, thereby, allow the solid fuel to burn with the gas phase oxygen (Mattisson et al., 2009). The reduced oxygen carriers could be regenerated at high temperatures. Figure 7 shows a schematic diagram of a CLOU system, with an air reactor, fuel reactor, and a condenser to separate water from CO_2. The experimental studies on CLOU have reported a similar conversion rate of biomass char to that obtained with lignite type of coal. Adánez-Rubio et al. (2014) achieved the complete combustion of biomass char and 100% CO_2 capture efficiency at fuel reactor temperature of 935 °C using a low solid fuel reactor inventory. The rate of char conversion rate of biomass in the CLOU process was 3 to 4 times higher than that corresponding to the iG-CLC process at temperatures above 900 °C (Adánez-Rubio et al., 2014). Further research and development is going on in the area of chemical looping-based carbon capture, especially for biomass conversion and utilization of biomass derived products.

As discussed, there are different processes for carbon capture that can be employed in biomass conversion processes, similar to coal-based processes. Some studies have compared these processes on the basis of different parameters using different approaches. Zhou et al. (2014) compared different CO_2 capture systems using life cycle analysis tool and found that the pre-combustion and oxy-fuel technologies performed better than post-combustion in several environmental impact categories, and required less coal to produce the same amount of electricity. Leung et al. (2014), in their review study, concluded that the

Table 3. Advantages and disadvantages of different carbon capture processes (Leung et al., 2014).

Carbon capture process	Advantages	Disadvantages
Post-combustion capture	Technology more mature than other alternatives; can easily retrofit into existing plants	Low CO_2 concentration affects the capture efficiency
Pre-combustion capture	High CO_2 concentration enhances sorption efficiency; fully developed technology, commercially deployed in some industrial sectors; opportunity for retrofit to existing plant	Heat transfer and efficiency decay issues for use of hydrogen-rich gas turbine fuel; high parasitic power requirement for sorbent regeneration; high costs for current sorption systems
Oxyfuel combustion	Very high CO_2 concentration enhances absorption efficiency; mature air separation technologies available; reduced volume of gas to be treated, hence, required smaller boiler and other equipment	High efficiency drop and energy penalty; cryogenic O_2 production is expensive; corrosion problem may arise
Chemical looping combustion	CO_2 is the main combustion product, which remains unmixed with N_2, thus avoiding energy intensive air separation	Inadequate large scale operation experience; Process still under development

selection of specific CO_2 capture technology heavily depends on the type of the plant and the fuel. For example, for gas-fired power plants, post-combustion capture technology is preferred due to its lower cost. Table 3 shows the advantages and disadvantages of the various carbon capture processes discussed so far.

2.2 *Gasification*

Gasification is a thermochemical conversion process involving the partial oxidation of solid fuels, such as coal or biomass, to generate producer gas that could be used for making fuels, chemicals, and generating power. In addition to biomass, such as agricultural residue, some components of municipal solid waste (e.g., yard or garden waste, paper, cardboard) could be used in pellet or particle form in fixed or fluidized bed gasifiers (Thengane, 2018). The transportation, handling and storage costs for residual biomass, especially dry leaves litter and crop residue, are higher due to their low density (in the range of 50–100 kg/m^3). Hence, it is essential to densify this waste through compaction- and compression-based processes, such as pelletization or briquetting, which would also enhance its volumetric energy density (Tumuluru et al., 2011). The briquettes and pellets made from biomass, such as agricultural residue, have found applications ranging from small cookstoves to large gasifiers in power plants. The inherent lignin and moisture content of most biomass act as a binder during the densification process, therefore, external binders are not required (Kaliyan and Morey, 2010). However, some pre-processing steps, such as drying, shredding, sand and soil removal, need to be undertaken before densifying such biomass. The pelletization process at pilot scale for biomass, such as agro residue and garden waste, has been explained in detail by Pradhan et al. (2018).

As a technology, biomass gasification offers multiple benefits of bioenergy in terms of producing heat, power and biofuels for useful applications. Dinca et al. (2018) recently studied biomass gasification with CO_2 capture technology using an aqueous 30 wt% monoethanolamine (MEA) solution to convert the treated syngas into electricity in a combined cycle power plant. The overall efficiency of the plant was reduced from 50.9% to 45.8% in normal case and to 47.03% in the case with the heat recovery steam generator unit which recovered the heat from syngas to reduce the heat requirement for the chemical solvent regeneration. For a biomass gasification-based combined heat and power (CHP) plant, carbon capture can be applied to either the combined flue gas streams originating from the combustion reactor and the gas engine; or the syngas or producer gas stream generated from gasifier. Though the conventional amine process is preferred when capturing carbon from flue gas (Ahn et al., 2013), it may not be the preferred option for a syngas stream. The syngas stream has a significantly lower volumetric

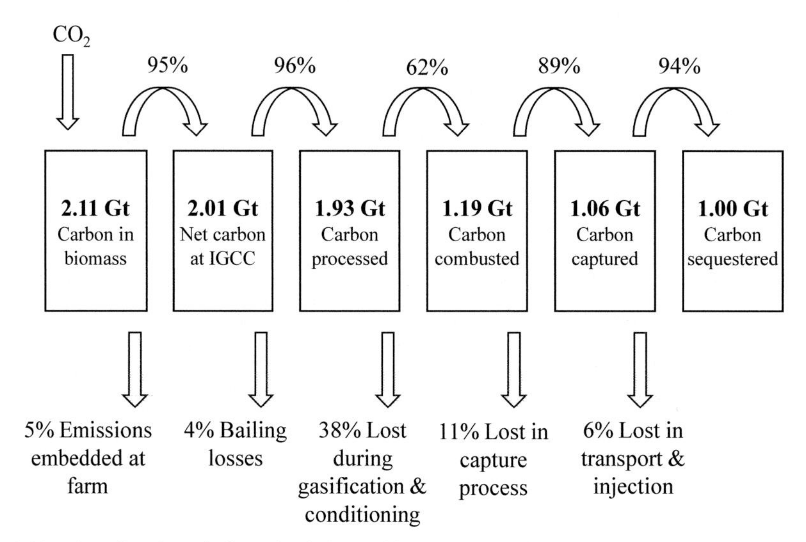

Figure 8. BECCS carbon flow for a dedicated switchgrass biomass IGCC plant (Kemper, 2015; Rhodes and Keith, 2005).

flowrate, approximately 11% of the flue gas, and a higher CO_2 mole fraction of 33 vol% in comparison to 13 vol% in the flue gas, and may be fit for an adsorption-based capture approach. For a 10 MW_{th} biomass-fuelled CHP plant, Oreggioni et al. (2015) compared the carbon capture using adsorption with that using conventional amine process, and demonstrated by process simulation the adsorptive capture unit to be more economical than conventional amine capture unit in all aspects. Figure 8 shows the BECCS carbon flow for a dedicated switchgrass biomass IGCC plant which implies that storage of 1Gt carbon (3.67 $GtCO_2$) requires fixation of at least 2.1 Gt carbon (7.7 $GtCO_2$).

Gasification of biomass using air or steam is commonly practiced, depending on the objective of the application. Unlike combustion, however, biomass gasification and pyrolysis do not need oxygen in pure form for reactions. Hence, these processes provide more options to contribute towards mitigating CO_2 emissions. The efforts have also been focused on using flue gas, CO_2, and other solid oxygen carriers as individual or combined gasifying agents to promote carbon capture. Lahijani et al. (2015) reviewed several CO_2-biomass gasification studies and observed the process to be chemically controlled at lower particle sizes where the diffusional effects are not prominent. Ahmed and Gupta (2011) analyzed the steam and CO_2 gasification of woodchips char at 900 °C and inferred that the partial pressure of the gasifying agent did not affect the reaction rate of the process. This indicated that both CO_2 and steam gasification processes were not controlled by the adsorption step. One of the studies used biomass gasification to capture CO_2 from an internal combustion engine and use it in gasifier (Sandeep et al., 2011). Sadhwani et al. (2016) compared oxygen gasification with CO_2 gasification and observed the lower char yield and, hence, higher carbon conversion in oxygen gasification due to the presence of free oxygen available for combustion reactions. The temperature had a profound effect on biomass gasification, affecting CO and H_2 which increase with a rise in temperature. At lower temperatures, pyrolysis is dominating, forming more hydrocarbons, which was the reason for higher peak HHV value for CO_2 gasification than air gasification. At higher temperatures, CO_2 gasification produces highly microporous char that greatly enhances CO_2 diffusion during the gasification step, leading to higher conversion. The primary reactions during the gasification step are char-CO_2 reaction (Boudouard), water gas reaction (char reforming), and free radical reactions leading to molecular hydrogen formation (Sadhwani et al., 2016).

Introducing carbon dioxide as a gasifying agent not only assists in carbon capture for biomass utilization but also contributes to the governing of the H_2/CO ratio. The increasing research on using biomass for producing chemicals has provided momentum in the use of Fischer Tropsch (FT) technology, which is highly dependant on the H_2/CO ratio. The syngas from a biomass gasification set-up and its subsequent transformation via FT synthesis can produce a wide range of synthetic hydrocarbons or clean sulfur-free liquid fuels. Thus, the processes of gasification and pyrolysis typically recover the chemical value of the waste, unlike combustion or incineration which recover mainly the energy value. In terms of thermodynamic efficiencies, extracting chemical value is always advantageous over extracting thermal energy (Thengane et al., 2016). The generated chemical products or gases could either be used as fuel to generate energy or as secondary feedstocks (char) for subsequent fuel production (Gumisiriza et al., 2017).

2.3 Pyrolysis

Pyrolysis involves thermal decomposition of biomass in absence of air, producing mainly liquid oil and charcoal, unlike gasification which needs some oxygen and generates producer gas (Axelsson and Kvarnström, 2010). The process can occur at a wide range of temperatures between 200–900 °C in inert atmosphere with or without catalyst, and is a preferred approach to convert waste plastics and biomass residues to valuable oils and chemical intermediates (Suriapparao and Vinu, 2015). For commercial scale processes, the majority of the product formed is liquid oil (calorific value 35–40 MJ/kg) as it is easier to transport liquid fuel than gas or solid fuels. Also, the flue gases produced per kg of waste are less in pyrolysis than in gasification and incineration. Though the direct use of pyrolysis oils in engines is still facing lot of issues and undergoing development, the vapors from pyrolysis have been successfully used in gasification to make syngas (Hornung et al., 2011). Figure 9 shows the concept of low-temperature (< 450 °C) pyrolysis bio-energy with carbon capture through biochar sequestration. About 50% of the pyrolyzed biomass is converted into biochar and can be returned to soil; the remaining 50% is converted into biofuel which can be used for energy application.

Earlier pyrolysis processes focused primarily on maximizing liquid biofuel production but new technologies allow for carbon sequestration in addition to production of sustainable energy from biomass. The process produces solid biochar and energy in the form of heat, steam, electricity or liquid fuels. The biochar carbon sequestration refers to the capture and subsequent storage of carbon, avoiding the emissions that would have occurred had the biomass been left to decompose (Mulabagal et al., 2017). Based on the carbon content of biomass feedstock, it can be assumed that, for every ton of biochar applied to the soil, 0.6–0.8 tonne of carbon (2.2–2.9 tonne of CO_2 equivalent) can be sequestered (Galinato et al., 2011). Biochar is a carbon-rich co-product containing nutrients from biomass ash, such as potassium (K), phosphorous (P), magnesium (Mg) and calcium (Ca). It has the potential to act as a soil quality enhancer by increasing soil organic matter, water retention, and soil biological activity, thereby resulting in better crop productivity (Hawkins et al., 2007). The biochar obtained during the process of pyrolysis may achieve negative greenhouse gas emissions when used in soil abatement. However, this is achieved at the

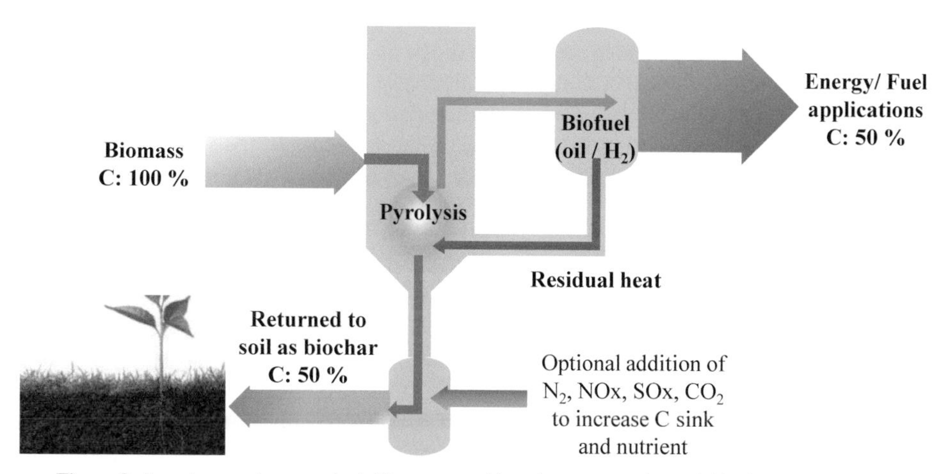

Figure 9. Low-temperature pyrolysis bio-energy with carbon capture through biochar sequestration.

expense of lower energy efficiency and higher impacts in the other assessed categories when compared to direct biomass combustion. Hence, the most favorable use of biochar is char co-firing, substituting fossil coal, even assuming high long-term stability of the char (Peters et al., 2015). However, for low-quality soils, biochar application to obtain high crop yield would show a more favorable performance in terms of global warming (Steiner et al., 2013).

The other developments in the technology are similar to those in the gasification area. For example, CO_2 as a pyrolysing medium enhances CO production by cracking of tars, and also improves the properties of biochar (Cho et al., 2015). Addition of CaO to biomass increases H_2 formation from reforming and char gasification reactions and lowers the onset temperatures of char decomposition and tar cracking (Widyawati et al., 2011). The pyrolyzers can produce more power than they consume, and can supply their own power, utilizing waste heat from the system. Hence, this technology could be deployed at different scales from small farm to industry without the need for existing energy infrastructure. The deployment of new biochar and bio-energy systems could create economic opportunities for local communities through the creation of new businesses that develop to support its infrastructure. These could be suppliers of biomass residues, plant operators and workers, manufacturer and distribution of co-products, and related agricultural application services.

2.4 Torrefaction

Torrefaction, sometimes called mild pyrolysis, is a thermal pre-treatment technology employing temperatures lower than 300 ºC to upgrade biomass to a product retaining about 70–80% of its mass and 80–90% of its energy. The process performance depends on different parameters, such as biomass characteristics, reaction temperature, heating rate, residence time, and the reaction medium. Recent studies on the topic are mainly concerned with investigating torrefaction chemistry, equipment performance and design, and elucidation of supply chain impacts for the technology (Chen et al., 2015; Eseyin et al., 2015; Felfli et al., 2005). This indicates that there is a long way to go in understanding torrefaction and making it a commercially feasible technology. However, because of the substantial amount of biochar it can produce, the potential of technology in carbon capture and storage appears high. Being an upgrading treatment, torrefaction can become a key unit operation for a wide range of applications, including biomass storage and transport. The technology could have tremendous influence on other biomass densification and conversion processes, with some obvious benefits of torrefied biomass over raw biomass. Torrefied biomass in the form of pellets or briquettes retains up to 96% of its chemical energy and is resistant to biodegradation, making it a good substitute for coal/charcoal (Axelsson and Kvarnström, 2010). The pretreatment of biomass, with processes such as torrefaction, fast pyrolysis, and hydrothermal carbonization (HTC), produces a carbon-rich solid product which can be a better starting feed for gasification than raw biomass (Erlach et al., 2012). These processes raise the heating value of the biomass, increase the energy density, and improve its mechanical properties, such as grindability. However, these technologies are still in the research and development stage.

Biomass torrefaction in partial oxidative medium or air, is believed to reduce the operating costs and the emissions at the expense of slight loss in the solid yield (Chen et al. 2015; Kung et al. 2019). This motivated the researchers to explore exhaust or flue gases as a torrefaction medium. The boiler exhaust gases containing CO_2, O_2, and N_2 can be utilized for torrefaction of biomass, which will improve the thermal efficiency of the plant in addition to producing a good quality biochar (Thanapal et al., 2014). Figure 10 shows the variation in C, H, and O components of torrefied biomass under nitrogen and carbon dioxide environments at different temperatures. As expected, the H/C and O/C ratios decreased with increase in the temperature of torrefaction, however, at higher temperatures, carbonization was greater in the presence of nitrogen. This is mainly due to the oxidation of some carbon in the biomass in the presence of CO_2, also resulting in a decrease in mass yields. Sarvaramini et al. (2014) reported the possibility of trapping CO_2 evolved during biomass torrefaction using cheap, abundant and already size-reduced mining residues, resulting in a carbon-negative torrefaction process. The process of torrefaction as a prerequisite step for an efficient biomass/coal co-firing could, therefore, be further enhanced by curbing the overall process CO_2 emissions. The torrefaction

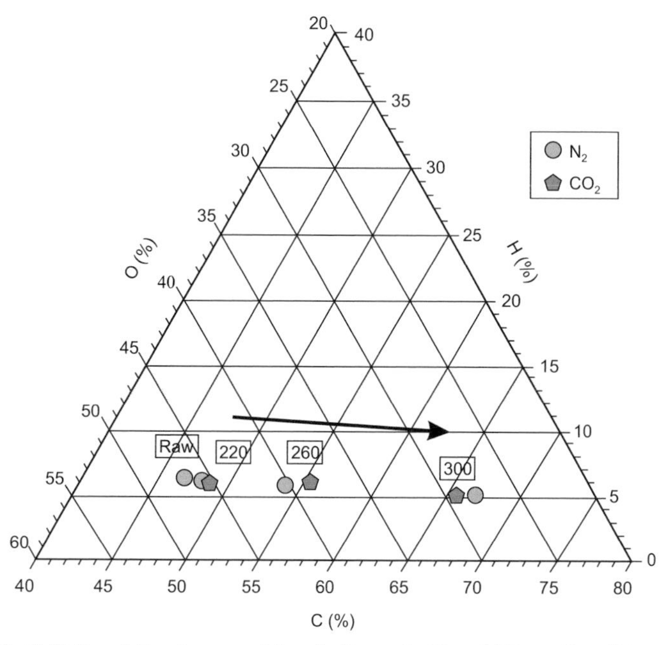

Figure 10. C, H, O variation for corncob torrefaction under N_2 and CO_2 medium (Li et al., 2018).

of biomass is still at an experimental level, with few pilots in operation; however, the upcoming results have generated a lot of interest from investors in the energy sector. Though torrefaction seems to be a promising technology for managing biomass residue and generating energy and value, the risk of the widespread consumption of biomass affecting water, soil and the food chain should not be ignored. Hence, the utilization of either waste biomass or less environmentally friendly, more abundant, and faster growing biomass should be targeted for this technology. Moreover, the application of torrefied biomass or biochar should be decided based on the value created, similar to that discussed in the pyrolysis section.

3. Comparison of Carbon Capture Potential

Figure 11 shows the overall carbon balance for different processes of biomass conversion. The options of natural decomposition, burning (i.e., combustion), and anaerobic digestion are also considered for the sake of comparison. The oil and gas products coming from each of the processes release carbon into the atmosphere quickly on combustion or end-use. On the other hand, the volatiles release process from biochar or char residue is extremely slow in standard atmospheric conditions. The process of pyrolysis and torrefaction appear to send 50% and 80%, respectivly, less carbon to the atmosphere in the form of carbon dioxide. These processes convert this portion of carbon into biochar, which can be either stored long term for energy purpose or be used in soil amendment. In either case, the time scale for which this carbon is trapped and prevented from going back to the atmosphere is very high compared to decomposition or burning of biomass. A recent perspective discusses this dual benefit of mitigating emissions and having energy security through the concept of "biochar mines" (Thengane and Bandyopadhyay, 2020). The figure also explains the role played by utilization of biomass as an energy source in the mitigation of CO_2 emissions without any additional CCS investments. The application of CCS technologies to these biomass conversion processes are expected to reduce the carbon sent back to atmosphere significantly. For example, the plants involving burning or combustion of biomass could be retrofitted with CCS options, as explained in the section on combustion, in order to reduce CO_2 emissions. Similarly, the advanced processes for CCS could be implemented for other biomass conversion processes in order to further reduce the carbon footprint of these processes.

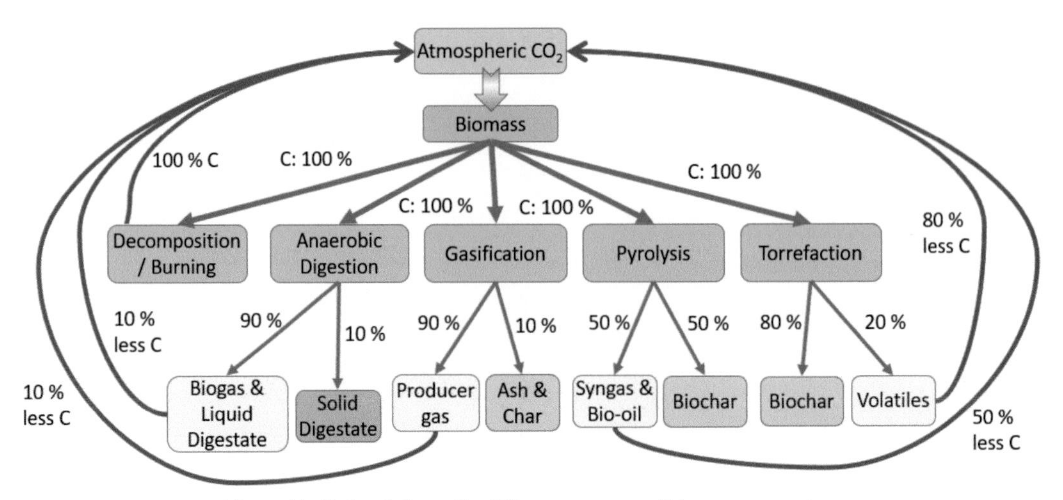

Figure 11. Carbon balance for different processes of biomass conversion.

When the biomass conversion technology is equipped with CCS, it is also termed as bio-CCS or BECCS, as discussed earlier. Some studies in literature have predicted the CO_2 mitigation potential and the opportunities for further improvement through the implementation of BECCS and its variants. The capture and storage of carbon in existing unit or plant comes at an additional cost and it is important to balance the economic investment with the benefits in environmental and the efficiency aspects of the plant. A recent study by Bui et al. (2018) evaluated opportunities to improve the performance of BECCS and presented three inferences: (i) increasing biomass co-firing proportion reduced SO_X emissions due to the decrease in fuel sulphur content; (ii) using a high performance solvent with heat recovery, the BECCS system could achieve an efficiency of 38% higher than the current fleet of coal-fired power plant with efficiencies ranging from 26 to 35%; (iii) to achieve greater carbon negativity (on a per MWh basis), a low efficiency system is more desirable due to increased consumption of biomass fuel, which results in more CO_2 being captured and permanently stored. Vaughan et al. (2018) proposed that the limits set by Paris Agreement are unachievable without large-scale use of BECCS, and attributed biomass residues as a major source of biomass as they are associated with fewer sustainability concerns than dedicated bioenergy crops. The study inferred the deployment rates for CCS in the biomass scenarios to be more challenging than in the case of fossil, renewable or nuclear technologies and, hence, will require ambitious policy interventions.

4. Conclusions

This chapter discussed different thermochemical conversion processes for biomass with respect to carbon capture and storage. For biomass combustion, different types of CCS approaches, such as pre-combustion, post-combustion and oxyfuel combustion, were presented, most of which are also applicable to biomass gasification. Chemical looping-based approaches for CCS look promising for biomass similar to coal but need more development. Based on only CCS potential, pyrolysis and torrefaction transferred 50% and 80%, respectively, less carbon to the atmosphere and, hence, are better in terms of carbon capture and storage without additional expense of energy. These percentages of CO_2 reduction are expected to increase further if CO_2 or flue gases are used as pyrolysing or torrefying medium. As part of CCS, the use of biochar as a soil amender is beneficial and recommended only for low-quality soils as it will improve the crop yield enormously. On the other hand, the biochar or torrefied biomass can prove to be a better feedstock for gasification or combustion applications, improving their efficiencies. Hence, the choice for utilization of biochar should be appropriately made depending on the context of application. The advantage of thermochemical processing is that it can convert nearly all the organic components of the biomass, unlike biochemical processing. However, the start-up and plant maintenance costs of thermochemical processes are high due to the demands of high temperature and preprocessing of feedstock. For efficient operation, thermochemical processing must be done on a large scale, which necessitates the transportation of biomass over long distances, resulting in an increase in cost. The processes such as drying, torrefaction, and densification can reduce the transportation costs and emissions but will need additional investment at the source. Fossil fuel-based energy may be utilized at several points in the course of conversion of biomass into fuels, including transportation. This may reduce the carbon neutrality of the process and, hence, it needs to be techno-economically optimized before being put into practice at a pilot or commercial scale. The development of negative emission technologies, such as Bio-CCS or BECCS, could reduce the atmospheric CO_2 concentration substantially, add flexibility to CO_2 reduction pathways with respect to time, and accelerate decarbonisation in the energy and transport sector. However, the impact of utilizing biomass in BECCS on the food system, land use, and other emissions, such as due to use of fertilizers, should not be ignored. The success of technologies utilizing biomass with the objective of mitigating climate change is yet to be determined at commercial level, and will totally depend on the perspective of the different stakeholders involved with respect to different technologies.

References

Aalbers, R. and Bollen, J. 2017. Biomass Energy with Storage can reduce Carbon Capture and costs of EU's Energy Roadmap with 15–75%. CPB Netherlands Bureau for Economic Policy Analysis, The Hague.

Abbasi, T. and Abbasi, S.A. 2010. Biomass energy and the environmental impacts associated with its production and utilization. Renewable and Sustainable Energy Reviews 14: 919–937.

Acharya, B., Dutta, A. and Basu, P. 2009. Chemical-looping gasification of biomass for hydrogen-enriched gas. Energy and Fuels 23: 5077–5083.

Adánez-Rubio, I., Abad, A., Gayán, P., De Diego, L.F., García-Labiano, F. and Adánez, J. 2014. Biomass combustion with CO_2 capture by chemical looping with oxygen uncoupling (CLOU). Fuel Processing Technology 124: 104–114.

Adanez, J., Abad, A., Garcia-labiano, F., Gayan, P. and de Diego, L.F. 2012. Progress in chemical-looping combustion and reforming technologies. Progress in Energy and Combustion Science 38(2): 215–282.

Ahmed, I.I. and Gupta, A.K. 2011. Kinetics of woodchips char gasification with steam and carbon dioxide. Applied Energy 88(5): 1613–1619.

Ahn, H., Luberti, M., Liu, Z. and Brandani, S. 2013. Process configuration studies of the amine capture process for coal-fired power plants. International Journal of Greenhouse Gas Control 16: 29–40.

Akhtar, A., Krepl, V. and Ivanova, T. 2018. A combined overview of combustion, pyrolysis, and gasification of biomass. Energy and Fuels 32: 7294–7318.

Anderson, K. and Peters, G. 2016. The trouble with negative emissions. Science 354(6309): 182–184.

Arias, B., Pevida, C., Rubiera, F. and Pis, J.J. 2008. Effect of biomass blending on coal ignition and burnout during oxy-fuel combustion. Fuel 87(12): 2753–2759.

Axelsson, C. and Kvarnström, T. 2010. Energy from municipal solid waste in Chennai, India—a feasibility study. Swedish University of Agricultural Sciences ISSN 1654-9392.

Azar, C., Johansson, D.JA. and Mattsson, N. 2013. Meeting global temperature targets—The role of bioenergy with carbon capture and storage. Environmental Research Letters 8: 034004.

Bhaskar, T. and Steele, P.H. 2015. Thermo-chemical conversion of biomass. Bioresource Technology 178: 1.

Bui, M., Fajardy, M. and Mac Dowell, N. 2018. Bio-energy with carbon capture and storage (BECCS): Opportunities for performance improvement. Fuel 213: 164–175.

Chen, W.H., Peng, J. and Bi, X.T. 2015. A state-of-the-art review of biomass torrefaction, densification and applications. Renewable and Sustainable Energy Reviews 44: 847–866.

Cho, D.W., Cho, S.H., Song, H. and Kwon, E.E. 2015. Carbon dioxide assisted sustainability enhancement of pyrolysis of waste biomass: A case study with spent coffee ground. Bioresource Technology 189: 1–6.

Dinca, C., Slavu, N., Cormoş, C.C. and Badea, A. 2018. CO_2 capture from syngas generated by a biomass gasification power plant with chemical absorption process. Energy 149: 925–936.

Dincer, I., Colpan, C.O., Kizilkan, O. and Ezan, M.A. 2015. Progress in Clean Energy Volume 2 Novel Systems and Applications. Springer (Vol. 1). Springer International Publishing Switzerland.

Erlach, B., Harder, B. and Tsatsaronis, G. 2012. Combined hydrothermal carbonization and gasification of biomass with carbon capture. Energy 45: 329–338.

Eseyin, A.E., Steele, P.H. and Pittman, C.U. 2015. Current trends in the production and applications of torrefied wood/ biomass—A review. BioResources 10(4): 8812–8858.

Falano, T., Thornley, P. and Roeder, M. 2015. Air/Oxy biomass combustion with carbon capture and storage. In European Biomass Conference (pp. 1–8). Vienna.

Fang, H., Haibin, L. and Zengli, Z. 2009. Advancements in development of chemical-looping combustion: A review. International Journal of Chemical Engineering, 2009(ii).

Felfli, F.F., Luengo, C.A. and Rocha, J.D. 2005. Torrefied briquettes: Technical and economic feasibility and perspectives in the Brazilian market. Energy for Sustainable Development 9(3): 23–29.

Fuss, S., Canadell, J.G., Peters, G.P., Tavoni, M., Andrew, R.M., Ciais, P., Jackson, R.B., Jones, C.D., Kraxner, F., Nakicenovic, N., Quere, C., Raupach, M.R., SHarifi, A., Smith, P. and Yamagata, Y. 2014. Betting on negative emissions. Nature Climate Change 4: 850–853.

Galinato, S.P., Yoder, J.K. and Granatstein, D. 2011. The economic value of biochar in crop production and carbon sequestration. Energy Policy 39(10): 6344–6350.

Goldemberg, J. and Johansson, T. 2004. World Energy Assessment: Overview 2004 Update. New York: United Nations Development Programme. New York.

Gumisiriza, R., Hawumba, J.F., Okure, M. and Hensel, O. 2017. Biomass waste-to-energy valorisation technologies: A review case for banana processing in Uganda. Biotechnology for Biofuels 10(11): 1–29.

Hawkins, R., Nilsson, J., Oglesby, R. and Day, D. 2007. Utilization of biomass pyrolysis for energy production, soil fertility and carbon sequestration. In UN Commission on Sustainable Development Partnerships Fair (pp. 1–8).

Hornung, A., Apfelbacher, A. and Sagi, S. 2011. Intermediate pyrolysis: A sustainable biomass-to-energy concept-biothermal valorisation of biomass (BtVB) process. Journal of Scientific and Industrial Research 70: 664–667.

Hossain, M.M. and de Lasa, H.I. 2008. Chemical-looping combustion (CLC) for inherent CO_2 separations—A review. Chemical Engineering Science 63(18): 4433–4451.

Household Air Pollution and Health. 2018. WHO Fact sheet no. 292, Geneva. https://www.who.int/news-room/fact-sheets/detail/household-air-pollution-and-health.

Jones, J.M., Lea-Langton, A.R., Ma, L., Pourkashanian, M. and Williams, A. 2014. Pollutants Generated by the Combustion of Solid Biomass Fuels. Leeds UK: Springer Briefs In Applied Sciences and Technology.

Kaliyan, N. and Morey, R.V. 2010. Natural binders and solid bridge type binding mechanisms in briquettes and pellets made from corn stover and switchgrass. Bioresource Technology 101: 1082–1090.

Kemper, J. 2015. Biomass and carbon dioxide capture and storage: A review. International Journal of Greenhouse Gas Control 40: 401–430.

Koornneef, J., van Breevoort, P., Hamelinck, C., Hendriks, C., Hoogwijk, M., Koop, K., Koper, M., Dixon, T. and Camps, A. 2012. Global potential for biomass and carbon dioxide capture, transport and storage up to 2050. International Journal of Greenhouse Gas Control 11: 117–132.

Kung, K.S., Shanbhogue, S., Slocum, A.H. and Ghoniem, A.F. 2019. A decentralized biomass torrefaction reactor concept. Part I: Multi-scale analysis and initial experimental validation. Biomass and Bioenergy 125: 196–203.

Lahijani, P., Alimuddin, Z., Mohammadi, M. and Rahman, A. 2015. Conversion of the greenhouse gas CO_2 to the fuel gas CO via the Boudouard reaction: A review. Renewable and Sustainable Energy Reviews 41: 615–632.

Leung, D.Y., Caramanna, G. and Maroto-Valer, M.M. 2014. An overview of current status of carbon dioxide capture and storage technologies. Renewable and Sustainable Energy Reviews 39: 426–443.

Li, S.X., Chen, C.Z., Li, M.F. and Xiao, X. 2018. Torrefaction of corncob to produce charcoal under nitrogen and carbon dioxide atmospheres. Bioresource Technology 249: 348–353.

Lu, X., Cao, L., Wang, H., Peng, W., Xing, J., Wang, S., Cai, S., Shen B., Yang, Q., Nielsen, C.P. and McElroy, M.B. 2019. Gasification of coal and biomass as a net carbon-negative power source for environment-friendly electricity generation in China. Proceedings of the National Academy of Sciences 116(17): 8206–8213.

Lyngfelt, A. 2014. Chemical-looping combustion of solid fuels—Status of development. Applied Energy 113: 1869–1873.

Mattisson, T., Lyngfelt, A. and Leion, H. 2009. Chemical-looping with oxygen uncoupling for combustion of solid fuels. International Journal of Greenhouse Gas Control 3(1): 11–19.

Moghtaderi, B. 2012. Review of the recent chemical looping process developments for novel energy and fuel applications. Energy and Fuels 26(1): 15–40.

Mulabagal, V., Baah, D., Egiebor, N. and Chen, W.Y. 2017. Biochar from biomass: a strategy for carbon dioxide sequestration, soil amendment, power generation, and CO_2 utilization. pp. 1938–1960. *In*: Handbook of Climate Change Mitigation and Adaptation (2nd ed.). Switzerland: Springer International Publishing.

National Academies of Sciences. 2018. Bioenergy with Carbon capture and storage approaches for carbon dioxide removal and reliable sequestration: Proceedings of a workshop—in Brief. pp. 1–24. *In*: National Academies of Sciences. Washington, DC: The National Academic Press.

Oreggioni, G.D., Brandani, S., Luberti, M., Baykan, Y., Friedrich, D. and Ahn, H. 2015. CO_2 capture from syngas by an adsorption process at a biomass gasification CHP plant: Its comparison with amine-based CO_2 capture. International Journal of Greenhouse Gas Control 35: 71–81.

Peters, J.F., Iribarren, D. and Dufour, J. 2015. Biomass pyrolysis for biochar or energy applications? A life cycle assessment. Environmental Science and Technology 49: 5195–5202.

Pradhan, P., Arora, A. and Mahajani, S.M. 2018. Pilot scale evaluation of fuel pellets production from garden waste biomass. Energy for Sustainable Development 43: 1–14.

Quader, M.A. and Ahmed, S. 2017. Bioenergy with carbon capture and storage (BECCS): Future prospects of Carbon-Negative Technologies. pp. 91–140. *In*: Clean Energy for Sustainable Development–Comparisons and Contrasts of New Approaches. Elsevier.

Rhodes, J.S. and Keith, D.W. 2005. Engineering economic analysis of biomass IGCC with carbon capture and storage. Biomass and Bioenergy 29: 440–450.

Ribeiro, J.M.C., Godina, R., Matias, J.C. de O. and Nunes, L.J.R. 2018. Future perspectives of biomass torrefaction: Review of the current state-of-the-art and research development. Sustainability 10: 1–17.

Sadhwani, N., Adhikari, S. and Eden, M.R. 2016. Biomass gasification using carbon dioxide: effect of temperature, CO_2/C ratio, and the study of reactions influencing the process. Industrial and Engineering Chemistry Research 55: 2883–2891.

Sandeep, K., Snehesh, S. and Dasappa, S. 2011. Carbon dioxide caputre through biomass gasification. 19th European Biomass Conference & Exhibition, (June): 1127–1133.

Sarvaramini, A., Assima, G.P., Beaudoin, G. and Larachi, F. 2014. Biomass torrefaction and CO_2 capture using mining wastes—A new approach for reducing greenhouse gas emissions of co-firing plants. Fuel 115: 749–757.

Seepana, S. and Jayanti, S. 2010. Steam-moderated oxy-fuel combustion. Energy Conversion and Management 51: 1981–1988.

Sher, F., Pans, M.A., Sun, C., Snape, C. and Liu, H. 2018. Oxy-fuel combustion study of biomass fuels in a 20 kWth fluidized bed combustor. Fuel 215: 778–786.

Sikarwar, V.S., Zhao, M., Clough, P., Yao, J., Zhong, X., Memon, M.Z., Shah, N., Anthony, E.J. and Fennell, P.S. 2016. An overview of advances in biomass gasification. Energy & Environmental Science 9(10): 2939–2977.

Songolzadeh, M., Soleimani, M., Takht Ravanchi, M. and Songolzadeh, R. 2014. Carbon dioxide separation from flue gases: A technological review emphasizing reduction in greenhouse gas emissions. The Scientific World Journal 2014: 1–34.

Spliethoff, H. 2010. Power Generation from Solid Fuels. USA: Springer.

Steiner, C., Teixeira, W.G., Lehmann, J. and Zech, W. 2013. Microbial response to charcoal amendments of highly weathered soils and amazonian dark earths in central amazonia—preliminary results. pp. 195–212. *In*: Amazonian Dark Earths: Explorations in Space and Time.

Suriapparao, D.V. and Vinu, R. 2015. Bio-oil production via catalytic microwave pyrolysis of model municipal solid waste component mixtures. RSC Advances 5: 57619–57631.

Thanapal, S.S., Chen, W., Annamalai, K., Carlin, N., Ansley, R.J. and Ranjan, D. 2014. Carbon dioxide torrefaction of woody biomass. Energy and Fuels 28: 1147–1157.

Thengane, S.K., Hoadley, A., Bhattacharya, S., Mitra, S. and Bandyopadhyay, S. 2016. Exergy efficiency improvement in hydrogen production process by recovery of chemical energy versus thermal energy. Clean Technologies and Environmental Policy 18: 1391–1404.

Thengane, S.K. 2018. Assessment of different technologies for managing yard waste using analytic hierarchy process. Process Integration and Optimization for Sustainability 3(2): 255–272.

Thengane, S.K., Tan, R.R., Foo, D.C. and Bandyopadhyay, S. 2019. A pinch-based approach for targeting carbon capture, utilization, and storage systems. Industrial & Engineering Chemistry Research 58(8): 3188–3198.

Thengane, S.K. and Bandyopadhyay, S. 2020. Biochar Mines: Panacea to climate change and energy crisis? Clean Technologies and Environmental Policy 22: 5–10.

Tumuluru, J.S., Wright, C.T., Hess, J.R. and Kenney, K.L. 2011. A review of biomass densification systems to develop uniform feedstock commodities for bioenergy application. Biofuels, Bioproducts and Biorefining 5: 683–707.

Vakkilainen, E., Kuparinen, K. and Heinimö, J. 2013. Large industrial users of energy biomass. IEA Bioenergy Task 40: 1–52.

Vaughan, N.E., Gough, C., Littleton, E.W., Mander, S., Welfle, A., Gernaat, D.E.H.J. and van Vuuren, D.P. 2018. Evaluating the use of biomass energy with carbon capture and storage in low emission scenarios. Environmental Research Letters 13: 044014.

Widyawati, M., Church, T.L., Florin, N.H. and Harris, A.T. 2011. Hydrogen synthesis from biomass pyrolysis with *in situ* carbon dioxide capture using calcium oxide. International Journal of Hydrogen Energy 36: 4800–4813.

Williams, A., Jones, J.M., Ma, L. and Pourkashanian, M. 2012. Pollutants from the combustion of solid biomass fuels. Progress in Energy and Combustion Science 38: 113–137.

Zhou, Q., Manuilova, A., Koiwanit, J., Piewkhaow, L., Wilson, M., Chan, C. and Tontiwachwuthikul, P. 2014. A comparative of life cycle assessment of post-combustion, pre-combustion and oxy-fuel CO_2 capture. Energy Procedia 63: 7452–7458.

CHAPTER 7

Phytowaste Processing

Josef Maroušek, Otakar Strunecký* and *Vojtěch Stehel*

1. Introduction

An excessive proportion of biowaste from the food industry is being routinely landfilled or combusted. In addition, about 30–40% of the mass of the municipal solid waste produced is represented by voluminous biowaste (equivalent to 170 kg per inhabitant annually, of which usually above 85% is of plant origin), whereas developed countries are well above the global average, the USA far ahead of others. More and more countries are concerned about the accelerating trend in biowaste generation, as this means investing in new landfills (3 to 8 USD m^{-3} on average) and increasing the environmental burden. For example, the EU has set obligations to minimize the amount of biodegradable waste deposited in landfills (Pubule et al., 2015). However, these targets have proved to be too ambitious, as Bulgaria, Croatia, Greece, Malta and Romania remained far behind their duties with > 80% of generated municipal solid waste still being landfilled. In some member states, this amount still exceeds 90% (Scarlat et al., 2018).

The efficient management of biowaste is one of the key conditions of the sustainable development of any country since the state of waste management, i.e., the environment, also affects tourism, agriculture, the health care system and the overall economy in general (Piippo et al., 2014). In addition to direct financial losses (average landfilling cost 40–90 USD t^{-1} biowaste), the failures in prevention and recycling in this area, among other factors, lead to a loss of materials and energy, unnecessary production of greenhouse gases and a wide range of negative environmental impacts, including threats to soil and water systems (Bovea et al., 2010). Effective biowaste management can also lead to nutrient recovery, such as phosphates and nitrogen, thereby lowering the nutrient loss and reducing the use of fossil minerals required for fertiliser production. Such nutrient losses were estimated by the European Commission to be in the tens of billions of €. Returning carbon (solid residues of pyrolysed phytowaste is known as biochar, the recommended application for conventional farming is 4 t ha^{-1} $year^{-1}$) and nutrients (originating from the phytowaste mass plus enhancement by liquid biowaste) back into the agricultural land instead of landfilling or combustion could significantly reduce the gap in the nutrient pool and sequester the additional atmospheric carbon, thereby potentially creating assumptions for sustainable cycles (Maroušek et al., 2019). Over a long period of time, the addition of biochar (average wholesale price estimated to be around 160 USD t^{-1}) and organic matter to soil improves its pH, cation-exchange capacity and other properties, such as water retention, resistance to erosion and nutrients leaching (Lehmann and Joseph, 2012). However, special attention must be given to microplastic contamination when processing municipal biowaste as microplastics could have various negative implications on human and animal health (Rillig et al., 2017).

The Institute of Technology and Business in České Budějovice, Okružní 517/10, České Budějovice, 37001, Czech Republic.
* Corresponding author: josef.marousek@gmail.com

Studies have shown that it is baseless to only transfer the organic waste into arable lands to achieve nutrient recycling without the proper separation and processing of the waste (Vaneeckhaute et al., 2012). From the agronomical point of view, two important biotechnological, phytowaste treatments include composting (biodegradation in the presence of oxygen—also known as aerobic fermentation) and biogas production (biodegradation in the absence or minimal levels of oxygen—also known as anaerobic fermentation). In addition, combustion and pyrolysis are also considered to play significant roles. However, in the case of pyrolysis, the unattractiveness of this technology may be due to its high investment (0.1–0.5 M USD on average) (Mardoyan and Braun, 2015). The composting market is also undeveloped since the quality standards are set too low and, despite the relatively low compost price (most often around 10 USD t^{-1}), farmers see the return on investment to be in the distant future (with regards to the current state of the market, the benefits from compost application can hardly cover the costs of the field application, not to mention the price of the compost) (Kolář et al., 2011). In the case of anaerobic fermentation for biogas production, it should be noted that the electricity generated from biowaste often receives a lower donation in comparison to the electricity produced from the purpose-grown phytomass, such as maize (Homolka et al., 2014). This has resulted in the excessive cultivation of these crops, leading to negative impacts, such as an increase in erosion (Mardoyan and Braun, 2015).

Environmentally responsible combustion of the organic matter requires high investments associated with separation (some 2 USD t^{-1}), pre–drying (on average 6 USD t^{-1}) and particularly the modifying of the combustion bed to minimize the formation of pollutants that may occur under improper combustion conditions (Demirbaş, 2005). Higher temperatures (preferably over 1,200 °C), and thus safer combustion conditions (resulting in lower levels of pollutants), may be achieved by co-combustion with coal, for example, but the resulting ash may be subsequently problematic in terms of application into the arable soil due to the presence of heavy metals (Sami et al., 2001).

With the exception of starch grains and juicy fruits, the mechanical and biotechnological utilization of different kinds of phytomass is evolutionarily hindered by its cell wall, made up of lignocellulose corpus. This component provides tensile strength to the plant and acts as a physical barrier to any environmental stressors. Lignocellulosic materials are mainly composed of three polymers, namely cellulose, hemicellulose, and lignin, which are tightly associated with one another. The relative composition of the three polymers varies with the type of the material and play an important role in determining the energy required for disintegration (Bajpai, 2016). Cellulose fibers (hundreds to millions of crystallized glucose molecules) resemble iron rods, hemicellulose appears as a binding barbed wire and lignin acts as ballast cement (Maroušek, 2015). The recalcitrant (resistance of the plant cell walls to deconstruction) lignocellulose structure makes the disintegration process energy intensive and prevents the hydrolytic enzymes from liberating more easily digestible (biodegradable) substances (Himmel et al., 2007). Consequently, the majority of up–to–date lignocellulose biorefineries or various biotechnologies have to manage large amounts of ballast organic matter. This increases the requirements on sub-processes, such as grinding, mixing, heating, and pumping, as well as the consumption of reactants, catalysts, and processing liquids (Maroušek et al., 2014). Provided that gate fees for landfilling as well as the taxation on fossil fuels does not correspond to its real environmental damage, the advanced biowaste processing methods are currently only profitable in the separation of bioactive substances for drug production, in the food and cosmetics industries, with limited applications in the chemical industry or subsidized products (Coady et al., 2017).

Pretreatment plays an important step in processing the lignocellulosic phytomass by modifying the recalcitrant material to be more amenable for further processes or reactions (Qian, 2014). Many phytomasses, particularly lignocellulose pretreatment procedures and their variant combinations, include shockwave treatment, steam explosion, acid hydrolysis, enzymatic hydrolysis, etc. (Singh et al., 2015). Disrupting the recalcitrant nature of the lignocellulosic biomass, thereby exposing the cellulose embedded in a polymeric matrix to enzymatic hydrolysis, forms a key goal of any phytomass pretreatment process (Himmel et al., 2007). Reports have shown that pretreatment can improve the sugar yields to over 90% of the theoretical yield for biomass, such as wood, grasses, and corn (Brodeur et al., 2011). However, the implementation of any pretreatment procedure means additional costs proportional to the level of phytomass disintegration (Maroušek et al., 2018) and these are profitable only for products with a high added value or products that are subsidized. Furthermore, the inherent, general disadvantages of pretreatment techniques (including the purely mechanical methods) include the risk of the possible formation of inhibitors, such as phenolics, quinones, furan aldehyde, levulinic acid, dihydroxy and dicarboxylic acids (Jönsson and Martín, 2016). If this problem could be addressed, it would be possible to successfully replace the purpose-grown phytomass for higher amounts of the waste phytomass, which is often hardly digestible. The benefits also include energy savings and the minimization (if not totally avoidance) of the use of expensive reactants, including rare catalysts (Sun and Cheng, 2002).

2. A Brief Overview of Processing Techniques

2.1 Maceration

It is generally known that when any material is disintegrated into smaller particles the surface area increases, resulting in the acceleration of the process with higher yields. As a side effect, the density of the material often increases, which impacts the subsequent material and energy flows. This rapid increase in energy consumption is seen with a slight decrease in the particle size (Himmel et al., 2007). When discussing the particle size, it is important to consider all the relationships between the hydromodule, dry mass content, feed rate, costs, including the consumed energy related to the process conditions, and all the necessary phytomass properties and specificities (Maroušek, 2014). For instance, in the case of knife-milling technology, it was observed that the specific energy increased from 61 $kWh.t^{-1}$ to 657 $kWh.t^{-1}$ when the sieve size was decreased from 8 mm to 0.25 mm (Deines and Pei, 2010). The maceration mill is one of the examples of such attempts, which has been repeatedly tested for the pretreatment of various kinds of phytomass (wheat, oats, rapeseed, rye straw, jatropha shells, sunflower, maize silage, etc.). Interactions between the hydromodules (5% to 40%), temperatures (20 °C to 95 °C), levels of disintegration (determined by physical and biological methods), hydraulic retention times (20 seconds up to 20 minutes) and energy demands have proved to be crucial for the overall financial performance. It was found that grinding, followed by maceration allows for a deeper penetration of the liquid into the seed mash and enhances the disintegration by pressure shockwaves (to be discussed later).

There are various traceable maceration methods in the literature (Krátký and Jirout, 2014). Much of the know-how is based on the energy efficient, "under-water" mode. This mode is especially useful where the pretreated phytomass will be subsequently subjected to steam explosion, acid hydrolysis or anaerobic fermentation (to be discussed later). In principle, the "under-water" mode removes the large quantities of microscopic bubbles that are formed due to the surface properties (surface wettability, surface tension, and the similar spectrum of physical phenomena), captured on the pretreated phytomass and inside of the fibers frayed by previous mechanical disintegration (harvest, mowing, etc.). Better permeation of the processing fluid (removal of the microscopic tiny bubbles) through the phytomass has several benefits. This pretreatment method increases the efficiency of the screw pump and significantly minimizes the pressure turbulences in the subsequent processing by the high-pressure fermenters (steam-explosion, acid hydrolysis, etc.). Alternatively, the removal of the air bubbles accelerates the formation of the anaerobic environment that increases the efficiency of anaerobic fermentation or enzymatic hydrolysis.

According to the review by Taherzadeh and Karimi (2008), mechanical pretreatment, such as maceration, is mandatory due to the high stability of the phytomass towards enzymatic or bacterial attacks. In addition, the manifestations of the hot maceration process were studied in an attempt to washout the labile pools of organic carbon while simultaneously increasing the phytomass reaction surface area (Maroušek et al., 2014). It was revealed that mass flow, operating temperature, retention time, hydromodule, and operating pressure play the key roles in relation to the micropore area, the formation of inhibitors, and especially the dynamics of the obtained biodegradability. The average maceration cost is most often in the range of 5–20 USD t^{-1} and, therefore, it is used quite often in turning phytowaste into other products (feed, soaps etc.). Nevertheless, the interactions between all these factors vary among various harvests of one individual plant genus. Therefore, the processing parameters must be precisely tailored to each phytowaste.

2.2 Acid hydrolysis

Acid hydrolysis is one of the most universal and widely-used methods for phytowaste pretreatment (most often rice husks, sunflower stalks, bagasse, and straw) (Sun and Cheng, 2002). This treatment can be classified as one of two types, namely dilute acid hydrolysis and concentrated acid hydrolysis. The concentrated acid hydrolysis method uses sulphuric acid or hydrochloric acid at the low temperature of around 100 °C and low pressure. Dilute acid pretreatment processes can be classified within high temperature (temperatures greater than 160 °C) and low temperature (temperatures less than 160 °C) treatments (Shahbazi and Zhang, 2010). The acid hydrolysis of the phytowaste is accelerated by a combination of high pressure and high temperature. The cost of hydrolysis is most often in the range of 20–60 USD t^{-1}, therefore, its commercial use towards phytowaste is limited to special applications (sucrose hydrolysates, feedstock for the chemical industry). The detailed analysis of the composition of organic matter, with an emphasis on glucose equivalents, is mandatory when assessing the efficiency of the process from a biochemical point of view (Lavarack et al., 2002). Inhibitory by-products produced under acidic conditions include common aliphatic carboxylic acids (such as acetic acid, formic acid, and levulinic acid), phenolic aromatic carboxylic acids (e.g., ferulic acid and 4–hydroxybenzoic acid), and non–phenolic aromatic carboxylic acids, such as cinnamic acid (Jönsson and Martín, 2016). This pretreatment method enables the conversion of cellulosic biomass into sugar, thus providing raw material for fermentation into ethanol. However, a neutralization of the pH is necessary for the downstream enzymatic hydrolysis or fermentation processes (Sun and Cheng, 2002).

Currently, the research is focused on disintegrating phytomass into its constituents in a market competitive and environmentally sustainable way through the minimal usage of energy and chemicals (Lavarack et al., 2002). Choosing the optimum pretreatment method of cellulose disintegration will alter or remove structural and compositional impediments to hydrolysis in order to improve the rate of enzyme hydrolysis and increase the yields of the fermentable sugars from cellulose or hemicellulose (Carvalheiro et al., 2008). The disintegration of the lignocellulose increases the efficiency of hydrolysis from 5% to 25% and reduces the residence time in the digester from 23% to 59% (Hendriks and Zeeman, 2009). An effective pretreatment is characterized by several criteria. Most authors analyze the effectiveness by using fiber analyzers, however, the acid hydrolysis (H_2SO_4) that determines the carbon proportions (Labile Pool I—obtained by hydrolysis with 5 N H_2SO_4 at 105 °C for 30 min, Labile Pool II—obtained by hydrolysis with 26 N H_2SO_4 at room temperature overnight and then with 2 N H_2SO_4 at 105 °C for 3 h, and Recalcitrant Pool—the unhydrolyzed residue) is considered to be a more practical approach. Waste phytomass based on starch does not usually require any pretreatment or can be hydrolyzed easily with weak acid hydrolysis or enzymatically by amylases or by using acidification bioreactors with mixed populations (Arapoglou et al., 2010).

2.3 Enzymatic hydrolysis

Enzymatic hydrolysis of lignocellulose phyto-wastes are conducted using cellulolytic enzymes. In this process, cellulose is degraded to reducing sugars by the cellulases. Enzymatic hydrolysis of the lignocellulose usually follows the steam-explosion pretreatment (to be discussed later), which disrupts the lignocellulosic structure (Singh et al., 2015). Pretreatment methods generally seek to remove lignin and hemicellulose from the biomass, leaving cellulose, which is more readily hydrolyzed when it is free of hemicellulose and lignin fractions. The process conditions for enzymatic hydrolysis are usually not demanding due to the natural origin of the cellulases. Cellulases are produced both by bacteria and fungi. Cellulase enzyme loadings in hydrolysis vary from a 7 to 33 FPU/g (filter paper unit) substrate, depending on the type and concentration of the substrates (Sun and Chen, 2002). A mixture of hemicellulases or pectinases with cellulases exhibited a significant increase in the extent of cellulose conversion (Beldman et al., 1984). The development of the hydrolyzing mixtures is based on the assumption that the plants are evolutionarily evolved to be resistant to the action of the digestive system of ruminants, fungi or bacteria (Himmel et al., 2007). The utilization of hemicellulase-rich enzyme extracts may be a potential solution to the use of pretreatments seeking to hydrolyze the hemicellulose fraction via thermochemical methods, in which enzymatic hydrolysis would result in significantly fewer inhibitors. The initial pretreatment as well as the subsequent processing and dosage of the hydrolyzing enzymes must, therefore, respect the physical and chemical nature of the processed phytomass (Jensen et al., 2010). An efficient hydrolysis of lignocellulose substrates to soluble sugars requires the interplay and synergistic interaction of multiple enzymes (Billard et al., 2012). Enzymatic hydrolysis is usually conducted under mild conditions (pH 4.8, temperature 45–50 ºC) and, thus, the utility cost is lower in comparison to acid or alkaline hydrolysis. It was repeatedly confirmed that each type of lignocellulose is unique and requires different process conditions (hydromodule; pH; hydraulic retention time; temperature etc.) (Himmel et al., 2007). Studies showed that additional attention should be paid to the specific surface, the average degree of cellulose polymerization, the amount of lignin and other parameters which, on the whole, can be called "hydrolyzability", within the meaning of compliance to the subsequent hydrolysis process (Singh et al., 2015). Lignin is one of the major inhibitors to enzymatic hydrolysis. The degradation of lignin is conducted using intense pretreatment methods, but its degradation products, such as phenolic acids, inhibit cellulase enzymes (Visser et al., 2015). Since the price of hydrolyzing enzymes has been significantly reduced over the past decade, the average cost of enzymatic hydrolysis was lowered to 5–20 USD t^{-1} and new profitable applications are defined, particularly in the feed industry (sugar syrups for beekeeping, artificial forage for cattle).

2.4 Biogas production

The most widely-used method of phytowaste processing is its anaerobic fermentation for biogas production. This is due to the relatively low purchase cost, proven technology and also often due to the subsidized prices for the produced electricity (Hall et al., 2012). Several methods have been designed to improve anaerobic fermenters and to optimize the quality and quantity of the biogas released (Kolář et al., 2011). A plethora of experiments on UASB (up flow anaerobic sludge blanket) and EGSB (expanded granular sludge blanket) reactors can be traced in the literature (Gebrezgabher et al., 2010). However, regarding phytowaste utilization and following the financial aspects, most investors prefer the less efficient, slowly agitated containers, in which the waste is gradually processed without any pretreatment. Absolutely primitive batch technologies are very popular (processing cost as low as 4 USD t^{-1}), especially in countries where labor is cheap (Schievano et al., 2009). There are indications that it is financially promising to separate the hardly biodegradable components of organic matter from feedstock. The removal of the lignocellulose ballast enables the minimization of the fermenter dimensions from thousands to tens of cubic meters and the reduction of the retention time from months to hours (Maroušek et al., 2018). Moreover, the waste energy obtained from the co-generation of methane can be used to run the overall technology, including the continuous charcoal kiln, to produce biochar from the fermented phytomass (Maroušek, 2014). The removal of the hardly fermentable ballast from the maize silage was investigated on a commercial scale in a bid to reduce the retention times, volume of fermenters and associated heating requirements. The technology consisted of the under-hot-water maceration followed by decantation and the double screw press to separate the most labile pools of carbon from the ballast organic matter. This procedure minimized the inhibitor formations in the subsequent steam explosion process that was followed by the enzymatic hydrolysis. The products of hydrolysis were squeezed out from the lignocellulose ballast by the rotary dewatering press, fused with liquids previously obtained and anaerobically fermented, providing higher methane yields in a shorter time period. The rigid briquettes from the ballast at the rotary dewatering press were charcoaled. As previously proposed, all the technology was designed to run on the waste heat from the flue gases originating in the co-generation unit (Maroušek, 2013b).

Experiments on the deeply pretreated phytomass showed that the relationship of carbon and nitrogen is ordinarily analyzed incorrectly (considering the total levels instead of the microbiologically available fractions). This mistake is endlessly repeated in literature and may be one of the reasons for the poor techno-economic performance of many biogas stations (Linton et al., 2007). Laboratory simulations as well as pilot scale experiments showed that there exists a lot of space for technical and economic optimization. Given that an increase in process efficiency reduces the operating volume, this finding represents various additional savings (Shehu et al., 2012). Kolář et al. (2008), confirmed that the sludge from anaerobic fermentation is of organic nature, therefore, it must first be mineralized by soil biota before the nutrients will become available to plant nutrition. Legislation ignores this and, thus, generates great losses for farmers (Arthurson, 2009).

2.5 Composting

Compost is a stabilised and sanitised product obtained from the composting of organic substances derived from urban and agro-industrial, biodegradable solid wastes with a pH value around 8 and is subjected to partial microbial fermentation (Sharma et al., 1997). Kolář et al. (2011) states that the main requirement, by the legal standard, is the prescribed nutrient content, minimum amount of dry matter, absence of hazardous elements and the complete decomposition of the original organic material, such that the origin of the material cannot be identified. Such 'pseudo' composts are often offered to farmers at a very low price since their production costs are usually paid by producers of biodegradable waste, who need to get rid of voluminous waste. However, these composts do not have any agronomical or financial advantage. According to Kolář et al. (2011), compost should contain partly humified organomineral material, in which a significant part of its organic component was stabilized by soil biota into the mineral colloid fraction. It should demonstrate high ion exchange capacity, a high buffering capacity and resistance to fast mineralisation. It is crucial for the compost to maintain nutrients in the soil by its ion exchange reactions and to protect them against elution from topsoil and subsoil layers to bottom soil or even to groundwater, no matter where these plant nutrients originate (Kolář et al., 2011).

Price et al. (2003) state that one of the widely-used methods to control the yield and stability of biotechnological processes is the analysis of the ratio of carbon, nitrogen and phosphorus present in the feedstock. Other studies also confirm that the management of nutrients is a crucial indicator, however, its bioavailability is not sufficiently considered on a commercial scale, which has significant negative economic impacts (Carmona et al., 2013). It is considered as proven (Andersen et al., 2011) that the process runs slowly in an abundance of carbon (which results in the need for bigger production facilities and other costs). On the other hand, increased amounts of nitrogen lead to the formation of ammonia (Troy et al., 2012). Although repeated attempts were made in order to accurately determine the species of microorganism and their enzymes, the result was never successful (Salinas-Martınez et al., 2008). Due to difficulties in the management of production parameters at a sufficient level of stability, the composting process is performed mainly by indirect biotechnological analyses, i.e., those which are examined rather by manifestations of various population dynamics or various life stages of consortia (Aleer et al., 2014). An example could be the metabolic factor, specific enzymatic activity, etc. (Jurado et al., 2014). Generally speaking, carbon, nitrogen and phosphorus are the most important elements. Nevertheless, the valuation of compost by the level of these elements does not provide much logic (Scharlemann et al., 2015). Composting cost significantly varies according to location (2–10 USD t^{-1} on average), therefore, it is profitable in the hobby segment rather than in the conventional agriculture.

2.6 Steam explosion

Steam explosion pretreatment represents a method of deep disintegration of phytomass cells that is caused by the rapid expansion of the highly-pressurized phytomass into the atmospheric pressure, resulting in disruption of the cell walls (Taherzadeh and Karimi, 2008). Steam explosion is routinely initiated at temperatures between 160–260 °C (Sun and Cheng, 2002), however, studies have showed that many of the other operating parameters might also play important roles. Firstly, it should be noted that every apparatus is partly unique, and the operating parameters should be tailored to the biochemical nature of the phytomass. Secondly, it should be understood that the hydraulic retention time, operating pressure and pressure gradient affect the level of the disintegration. In addition, it should be recognized that the combination of these operating parameters significantly affects the possible formation of inhibitors, like furan or hydroxymethylfurfural, that may slow down or even stop the subsequent biotechnological processing. Therefore, the optimal process conditions must be precisely tailor-made for each feedstock, considering the running cost (10–30 USD t^{-1} on average), product (most often feed) and the overall financial performance.

Once the steam-explosion reactors routinely operate at pressures up to 3 Mpa, it is mandatory for the phytomass to be in the form of a runny mash that is pumped in by a high-pressure screw pump. The high-pressure is formed by preheated steam that is pumped into the thick-walled, horizontally placed high-pressure reactor, equipped by horizontally-laid helix. The pressure is controlled by the temperature at the steam generator, while the overall volume of the superheated steam (routinely 120 °C to 240 °C) and the residence time (routinely 2 to 20 minutes) are controlled by the rotation of the helix. The high-pressure reactor is ended by the turn-slide tourniquet where the final cell wall disruption by cavitation is caused by the sudden change in pressure. Many physio-chemical, structural and compositional factors hamper the digestibility of the cellulose present in the lignocellulosic biomass. Obstacles in pretreatment processes are based on the connective ability of lignin, which is like the ballast binder of cellulose and hemicelluloses. Several studies have shown a good correlation between the pore volume or population (accessible surface area for cellulases) and the enzymatic digestibility of lignocellulosic materials (Himmel et al., 2007). The effect of these areas correlates with the degree of crystallinity, lignin protection, hemicellulose presentation and other factors (Taherzadeh and Karimi, 2008).

Neither of the two long-recognized parameters on steam explosion severity (Explosion Power Density and Severity factor) could fully characterize the complexity of the product's process suitability for further biochemical processing. In addition, a study on the mineralization kinetics showed that the operating conditions of the steam explosion treatment are significantly connected with the operating conditions and the nutrient management.

From a technical point of view, the specific surface obtained (microporosity) may be analyzed using the BET (Brunauer–Emmett–Teller) (Naderi, 2015) techniques, such as micropore area, external surface area, micropore volume, etc. However, these do not always correlate with the severity of the pretreatment parameters. The reason lies again in the manifestation of the natural resistance of the plant cell walls to microbial and enzymatic deconstruction (Himmel et al., 2007). Unfortunately, it showed that the recalcitrance lignocellulose ballast may also inhibit the digestion of readily fermentable compounds through their inclusion (Brodeur et al., 2011). Without well-chosen pretreatment methods and corresponding operating parameters, there are low methane yields, low conversion efficiencies and high retention times in fermenters (Herrmann et al. 2007).

*2.7 **Shockwave pretreatment***

Shockwave pretreatment is an emerging technique for phytomass disintegration based on intensive physical forces (Higa et al., 2014). The method is based on the formation of pressure shockwaves that occur after plasma expansion, which is caused by high-voltage, discharges into the water environment. The pressure shockwaves, following the water plasma expansion after the underwater high voltage discharge in reinforced vessels, are intensively studied as a new and promising phytomass disintegration method (Vochozka and Maroušková, 2017). As the shockwave comes in contact with any biomass, it results in destructive effects that are related to its intensity and biomass structure. Repeated efforts are made in order to transfer the technology into a continuous apparatus and its commercial application. It was observed that grinding, mixing, and hot maceration in an excess of methanol is advantageous due to methanol acting as both a liquid carrier for the pressure shockwaves and a solvent that increases the efficiency of oil extraction from the oilseeds while remaining usable for the subsequent esterification into biodiesel. However, findings regarding biodiesel are hardly financially viable nowadays, since the cost of the shockwave pretreatment has yet been cut below some 250 USD t^{-1}. The influence of the shockwave energy, interference of the shockwaves and the number of shockwaves together with the intensity of maceration was investigated in relation to the oil yields. It showed that the proportion of the energy input was inadequate compared to the oil yields additionally released. It was verified on a laboratory scale by the Soxhlet apparatus that almost 100% of oil extraction is achievable from most of the phytomass after shockwave treatment, without a temperature increase and the involvement of any solvents (Higa et al., 2014). However, following the price of the purchasing cost and the energy requirements, the economy seems to be viable only for essential oils, special food products, etc.

2.8 Pyrolysis

Pyrolysis involves the thermo–chemical conversion of biomass into renewable biofuels under oxygen deficient conditions (Lehto et al., 2014). Based on the process parameters, like temperature, level of oxygen, pressure, humidity, retention time and carbon content in biomass, pyrolysis results in three main by-products: Solid char (also known as biochar), liquid products (bio-oil) and non-condensable pyrolytic gas. There are indications that the pyrolysis of phytowaste has significance from agricultural, industrial, ecological and economical perspectives (Sohi et al., 2010). Biochar is appreciated for its soil-improving properties. In addition, extensive research on biochar has resulted in its multifaceted application, involving remediation, carbon sequestration, metallurgical applications, heat and power production, construction material, etc. (Sohi et al. 2010). Further, studies have suggested that the underlying mechanism of biochar efficacy can be related to the microporosity, which provides a high surface area to many ion-exchange reactions, binding important nutritive anions as well as some cations (Lehmann and Joseph, 2012). In addition, this fine structure acts as a good substrate for different soil micro flora and micro fauna. Also, biochar (priced around 250 USD t^{-1}) increases the pH and cation-exchange capacity, thereby contributing significantly to the overall soil fertility. Following the statements above, the average payback time of pyrolysis technology that is incorporated into farm systems is no more than 4 years. The production of biochar and its storage in soil have been suggested as a means of abating climate change by sequestering carbon, while simultaneously providing nutrients and increasing crop yields (Lehmann and Joseph, 2012).

The liquid pyrolytic product, bio-oil, can be used as fuel in industrial boilers/furnaces. It can also be upgraded into equivalents to fossil fuels or can also be used to produce value-added chemicals (Lehto et al., 2014). Furthermore, the energy-rich pyrolytic gas (a mixture of combustible constituents such as carbon monoxide, hydrogen, methane, carbon dioxide and a broad range of other volatile organic compounds) can be recycled back into the reactor to support the pyrolysis or can be refined into a plethora of value-added chemicals.

The pyrolysis of agricultural phytowaste (post–harvest residues, fermentation residues from biogas plant, etc.), is advantageous in comparison to wood biomass and other purpose-grown phytomass, as these are understood to be crop disease vectors, and brings low levels of energy that is available to soil biota.

Studies have shown that pyrolysis decomposition temperatures vary for different lignocellulose components. Hemicellulose decomposition occurs at temperature range of 220–315 °C, cellulose decomposition takes place in the temperature range of 314–400 °C, whereas the decomposition of lignin occurs on wide temperature range of 160–900 °C (Lehto et al., 2014). The heat requirement for the pyrolysis of several agricultural and wood-based biomasses can range from 207–434 kJ kg^{-1}. The pyrolysis of cellulose was observed to be endothermic, while that of hemicellulose and lignin was exothermic. Studies have reported that the pyrolysis of pure cellulose results in depolymerization to levoglucosan, with yields as high as 59%. In the case of lignin, 20% of lignin can be converted to phenolic monomers (can be upgraded to fuels) and 40% to char. Solid pyrolytic products are usually favored at low temperatures. The formation of condensable products is favored at higher temperatures and short residence times whereas non-condensable gaseous products are favored at higher temperatures and longer residence times due to the secondary reactions. The major concerns with the establishment of phytomass pyrolysis include the bulky agricultural residue limited to a local market due to transport costs, the inconsistent and variable feedstock composition, presence of moisture, in addition to the expensive process parameters (Lehto et al., 2014).

3. Summary

Choosing the corresponding processing techniques plays a key role in phytowaste utilization. It should be noted that most of these methods are energy intensive and, hence, the treatment results must be techno-economically balanced against their impact on the cost of the downstream processing steps and trade-off between acquisition and running costs towards revenues (Table 1). The best financial results can be achieved provided that (1) the processed phytowaste brings savings from waste management (avoided

Table 1. Simplified techno-economic analysis of presented phytowaste processing techniques best performance is achieved if phytowaste processing techniques are tailor-made to the feedstock, combined and incorporated into biorefinery complex (M = maceration, AH = acid hydrolysis, EH = enzymatic hydrolysis, BP = biogas production, C = composting, SE = steam explosion, SP = shockwave pretreatment, P = pyrolysis).

	M	**AH**	**EH**	**BP**	**C**	**SE**	**SP**	**P**
feedstock	moderately biodegradable	moderately biodegradable	easily and moderately biodegradable	easily and moderately degradable	any but disintegrated	moderately and hardly biodegradable	easily and moderately degradable	any but disintegrated
products	disintegrated and soaked phytomass	hydrolyzate	hydrolyzate	biogas, fermentation residues	compost	deeply disintegrated phytomass	deeply disintegrated phytomass	charcoal or biochar, pyrolytic gas, pyrolytic oil
running cost (USD t^{-1})	9	25	11	5	3	17	300	8
acquisition costs (k USD t^{-1} h^{-1})	32	110	45	80	10	75	190	170
payback period (year)	2	6	5	6	9	6	5	3

landfilling, field application etc.), (2) the treatment is carried out in a biorefining process (individual steps bring synergies from its combinations) and (3) the products obtained are of a high added value (cosmetics, functional feed etc.). It should be highlighted that many of the chosen techniques are environmentally favorable (nutrient recycling, carbon sequestration, soil improvement, etc.).

References

Aleer, S., Adetutu, E.M., Weber, J., Ball, A.S. and Juhasz, A.L. 2014. The potential impact of soil microbial heterogeneity on the persistence of hydrocarbons in contaminated subsurface soils. Journal of Environmental Management 136: 27–36.

Andersen, J.K., Boldrin, A., Christensen, T.H. and Scheutz, C. 2011. Mass balances and life cycle inventory of home composting of organic waste. Waste Management 31: 1934–1942.

Arapoglou, D., Varzakas, T., Vlyssides, A. and Israilides, C. 2010. Ethanol production from potato peel waste (PPW). Waste Management 30: 1898–1902.

Arthurson, V. 2009. Closing the global energy and nutrient cycles through application of biogas residue to agricultural land-potential benefits and drawback. Energies 2: 226–242.

Atkinson, C.J., Fitzgerald, J.D. and Hipps, N.A. 2010. Potential mechanisms for achieving agricultural benefits from biochar application to temperate soils: a review. Plant and Soil 337: 1–18.

Bajpai, P. 2016. Structure of lignocellulosic biomass. *In*: Pretreatment of Lignocellulosic Biomass for Biofuel Production. SpringerBriefs in Molecular Science. Springer, Singapore.

Beldman, G., Rombouts, F.M., Voragen, A.G.J. and Pilnik, W. 1984. Application of cellulase and pectinase from fungal origin for the liquefaction and saccharification of biomass. Enzyme and Microbial Technology 6: 503–507.

Billard, H., Faraj, A., Ferreira, N.L., Menir, S. and Heiss-Blanquet, S. 2012. Optimization of a synthetic mixture composed of major Trichoderma reesei enzymes for the hydrolysis of steam-exploded wheat straw. Biotechnology for Biofuels 5: 9.

Bovea, M.D., Ibáñez-Forés, V., Gallardo, A. and Colomer-Mendoza, F.J. 2010. Environmental assessment of alternative municipal solid waste management strategies. A Spanish case study. Waste Management 30: 2383–2395.

Brodeur, G., Yau, E., Badal, K., Collier, J., Ramachandran, K.B. and Ramakrishnan, S. 2011. Chemical and physicochemical pretreatment of lignocellulosic biomass: A review. Enzyme Research, 1–17.

Carmona, G., Varela-Ortega, C. and Bromley, J. 2013. Participatory modelling to support decision making in water management under uncertainty: Two comparative case studies in the Guadiana river basin, Spain. Journal of Environmental Management 128: 400–412.

Carvalheiro, F., Duarte, L.C. and Gírio, F.M. 2008. Hemicellulose biorefineries: A review on biomass pretreatments.

Coady, D., Parry, I., Sears, L. and Shang, B. 2017. How large are global fossil fuel subsidies? World Development 91: 11–27.

Deines, T.W. and Pei, Z.J. 2010. Power consumption study in knife milling of wheat straw. Trans. NAMRI/SME 38: 191–196.

Demirbaş, A. 2005. Potential applications of renewable energy sources, biomass combustion problems in boiler power systems and combustion related environmental issues. Progress in Energy and Combustion Science 31: 171–192.

Dhyani, V. and Bhaskar, T. 2018. A comprehensive review on the pyrolysis of lignocellulosic biomass. Renewable Energy 129: 695–716.

Gebrezgabher, S.A., Meuwissen, M.P., Prins, B.A. and Lansink, A.G.O. 2010. Economic analysis of anaerobic digestion—A case of green power biogas plant in the Netherlands. NJAS–Wageningen Journal of Life Sciences 57: 109–115.

Hall, J., Matos, S., Sheehan, L. and Silvestre, B. 2012. Entrepreneurship and innovation at the base of the pyramid: A recipe for inclusive growth or social exclusion? Journal of Management Studies 49: 785–812.

Hendriks, A.T.W.M. and Zeeman, G. 2009. Pretreatments to enhance the digestibility of lignocellulosic biomass. Bioresource Technology 100: 10–18.

Herrmann, C., Heiermann, M., Idler, C. and Scholz, V. 2007. Parameters influencing substrate quality and biogas yield. pp. 809–819. *In*: Proceedings of the 15th European Biomass Conference and Exhibition, Berlin, Germany, Florence, Italy: ETA-Renewable Energies.

Higa, K., Matsui, T., Hanashiro, S., Higa, O. and Itoh, S. 2014. Evaluation of the contact switch materials in high voltage power supply for generate of underwater shockwave by electrical discharge. The International Journal of Multiphysics 8: 359–366.

Himmel, M.E., Ding, S.Y., Johnson, D.K., Adney, W.S., Nimlos, M.R., Brady, J.W. and Foust, T.D. 2007. Biomass recalcitrance: Engineering plants and enzymes for biofuels production. Science 315: 804–807.

Homolka, J., Slaboch, J. and Švihlíková, A. 2014. Evaluation of effectiveness of investment projects of agricultural bio-gas stations. Agris online Papers in Economics and Informatics 6: 45–57.

Jensen, J.W., Felby, C., Jørgensen, H., Rønsch, G.Ø. and Nørholm, N.D. 2010. Enzymatic processing of municipal solid waste. Waste Management 30: 2497–2503.

Johnson, K.E. 2017. Pretreatment optimization methods for increased sugar yields from biomass pyrolysis.

Jönsson, L.J. and Martín, C. 2016. Pretreatment of lignocellulose: Formation of inhibitory by-products and strategies for minimizing their effects. Bioresource Technology 199: 103–12.

Jurado, M.M., Suárez-Estrella, F., Vargas-García, M.C., López, M.J., López-González, J.A. and Moreno, J. 2014. Evolution of enzymatic activities and carbon fractions throughout composting of plant waste. Journal of Environmental Management 133: 355–364.
Kolář, L., Kužel, S., Peterka, J., Štindl, P. and Plát, V. 2008. Agrochemical value of organic matter of fermenter wastes in biogas production. Plant, Soil and Environment 54: 321–328.
Kolář, L., Kužel, S., Peterka, J. and Borová-Batt, J. 2011. Utilisation of waste from digesters for biogas production. Biofuel's Engineering Process Technology. ISBN 978-953-307-480-1.
Krátký, L. and Jirout, T. 2014. Energy-efficient size reduction technology for wet fibrous biomass treatment in industrial biofuel technologies. Chemical Engineering & Technology 37: 1713–1720.
Lavarack, B.P., Griffin, G.J. and Rodman, D. 2002. The acid hydrolysis of sugarcane bagasse hemicellulose to produce xylose, arabinose, glucose and other products. Biomass and Bioenergy 23: 367–380.
Lehmann, J. and Joseph, S. (eds.). 2012. Biochar for environmental management: Science and technology. Routledge.
Lehto, J., Oasmaa, A., Solantausta, Y., Kytö, M. and Chiaramonti, D. 2014. Review of fuel oil quality and combustion of fast pyrolysis bio-oils from lignocellulosic biomass. Applied Energy 116: 178–90.
Linton, J.D., Klassen, R. and Jayaraman, V. 2007. Sustainable supply chains: An introduction. Journal of Operations Management 25: 1075–1082.
Mardoyan, A. and Braun, P. 2015. Analysis of czech subsidies for solid biofuels. International Journal of Green Energy 12: 405–408.
Maroušek, J., Itoh, S., Higa, O., Kondo, Y., Ueno, M., Suwa, R., Tominaga, J. and Kawamitsu, Y. 2013. Enzymatic hydrolysis enhanced by pressure shockwaves opening new possibilities in Jatropha Curcas L. processing. Journal of Chemical Technology & Biotechnology 88: 1650–1653.
Maroušek, J. 2014. Biotechnological partition of the grass silage to streamline its complex energy utilization. International Journal of Green Energy 11: 962–968.
Maroušek, J., Myšková, K. and Žák, J. 2015. Managing environmental innovation: Case study on biorefinery concept. Revista Técnica de la Facultad de Ingeniería Universidad del Zulia 38: 3.
Maroušek, J., Stehel, V., Vochozka, M., Maroušková, A. and Kolář, L. 2018. Postponing of the intracellular disintegration step improves efficiency of phytomass processing. Journal of Cleaner Production 199: 173–176.
Maroušek, J., Stehel, V., Vochozka, M., Kolář, L., Maroušková, A., Strunecký, O., Peterka, J., Kopecky, M. and Shreedhar, S. 2019. Ferrous sludge from water clarification: Changes in waste management practices advisable. Journal of Cleaner Production, 459–464.
Moore, J.M., Klose, S. and Tabatabai, M.A. 2000. Soil microbial biomass carbon and nitrogen as affected by cropping systems. Biology and Fertility of Soils 31: 200–210.
Naderi, M. 2015. Surface Area: Brunauer–Emmett–Teller (BET). In Progress in Filtration and Separation (pp. 585–608). Academic Press.
Piippo, S., Juntunen, A., Kurppa, S. and Pongrácz, E. 2014. The use of bio-waste to revegetate eroded land areas in Ylläs, Northern Finland: Toward a zero-waste perspective of tourism in the Finnish Lapland. Resources, Conservation and Recycling 93: 9–22.
Price, G.A., Barlaz, M.A. and Hater, G.R. 2003. Nitrogen management in bioreactor landfills. Waste Management 23: 675–688.
Pubule, J., Blumberga, A., Romagnoli, F. and Blumberga, D. 2015. Finding an optimal solution for biowaste management in the Baltic States. Journal of Cleaner Production 88: 214–223.
Qian, E.W. 2014. Chapter 7—Pretreatment and Saccharification of Lignocellulosic Biomass. Research Approaches to Sustainable Biomass Systems, pp. 181–204.
Rillig, M.C., Ingraffia, R. and de Souza Machado, A.A. 2017. Microplastic incorporation into soil in agroecosystems. Frontiers in Plant Science 8: 1805.
Salinas-Martínez, A., de los Santos-Córdova, M., Soto-Cruz, O., Delgado, E., Pérez-Andrade, H., Háuad-Marroquín, L.A. and Medrano-Roldán, H. 2008. Development of a bioremediation process by biostimulation of native microbial consortium through the heap leaching technique. Journal of Environmental Management 88: 115–119.
Sami, M., Annamalai, M. and Wooldridge, M. 2001. Co–firing of coal and biomass fuel blends. Progress in Energy and Combustion Science 27: 171–214.
Scarlat, N., Fahl, F. and Dallemand, J.F. 2018. Status and opportunities for energy recovery from municipal solid waste in Europe. Waste and Biomass Valorization 1–20.
Scharlemann, J.P.W., Tanner, E.V.J., Hiederer, R. and Kapos, V. 2015. Global soil carbon: Understanding and managing the largest terrestrial carbon pool. Carbon Management 5: 81–91.
Schievano, A., D'Imporzano, G. and Adani, F. 2009. Substituting energy crops with organic wastes and agro-industrial residues for biogas production. Journal of Environmental Management 90: 2537–2541.
Shahbazi, A. and Zhang, B. 2010. Dilute and concentrated acid hydrolysis of lignocellulosic biomass. pp. 143–158. *In*: Bioalcohol Production. Woodhead Publishing.
Sharma, V.K., Canditelli, M., Fortuna, F. and Cornacchia, G. 1997. Processing of urban and agro-industrial residues by aerobic composting: Review. Energy Conversion and Management 38: 453–478.

Shehu, M.S., Manan, Z.A. and Alwi, S.R.W. 2012. Optimization of thermo–alkaline disintegration of sewage sludge for enhanced biogas yield. Bioresource Technology 114: 69–74.

Singh, J., Suhag, M. and Dhaka, A. 2015. Augmented digestion of lignocellulose by steam explosion, acid and alkaline pretreatment methods: A review. Carbohydrate Polymers 117: 624–631.

Sohi, S.P., Krull, E., Lopez-Capel, E. and Bol, R. 2010. A review of biochar and its use and function in soil. pp. 47–82. *In*: Advances in Agronomy. Academic Press, Vol 105.

Sun, Y. and Cheng, J. 2002. Hydrolysis of lignocellulosic materials for ethanol production: A review. Bioresource Technology 83: 1–11.

Taherzadeh, M.J. and Karimi, K. 2008. Pretreatment of lignocellulosic wastes to improve ethanol and biogas production: A review. International Journal of Molecular Sciences 9: 1621–1651.

Troy, S.M., Nolan, T., Kwapinski, W., Leahy, J.J., Healy, M.G. and Lawlor, P.G. 2012. Effect of sawdust addition on composting of separated raw and anaerobically digested pig manure. Journal of Environmental Management 111: 70–77.

Vaneeckhaute, C., Michels, E., Tack, F. and Meers, E. 2012. Nutrient recycling from bio-waste as green fertilizers. In 17th Symposium on Applied Biological Sciences 77(1): 251–256.

Visser, E.M., Leal, T.F., de Almeida, M.N. and Guimarães, V.M. 2015. Increased enzymatic hydrolysis of sugarcane bagasse from enzyme recycling. Biotechnology for Biofuels 8: 5.

Vochozka, M. and Maroušková, A. 2017. Assessment of shockwave pretreatment in biomass processing. Energy Sources, Part A: Recovery, Utilization, and Environmental Effects 39(11): 1195–1199.

Wu, X., Yao, W., Zhu, J. and Miller, C. 2010. Biogas and CH_4 productivity by co-digesting swine manure with three crop residues as an external carbon source. Bioresource Technology 101: 4042–4047.

CHAPTER 8

Anaerobic Digestion for Energy Recovery and Carbon Management

Akihisa Kita[1] *Yutaka Nakashimada*[2] and *Shohei Riya*[3,*]

1. Introduction

In recent years, global environmental problems, such as global warming, have been highlighted. The importance of developing environmentally friendly science and technology centered on sustainable and renewable resources with low greenhouse gas (GHG) emissions is strongly recognized because of the impact of rising prices of fossil resources and carbon dioxide (CO_2) emission regulations. Anaerobic digestion technology has carbon (C) neutral properties, and it is possible to use biomass containing various organic substances (carbohydrate, protein, lipid, cellulose and others), including sewage sludge, garbage and livestock waste, as raw materials. Therefore, anaerobic digestion technology has an advantage over other processes for bioenergy production and low energy requirements for operation. Furthermore, the advantages of anaerobic digestion include the following: (1) Energy recovery from organic matter in biomass as methane (CH_4) results in reduced usage of fossil resources and enables the reduction of CO_2 emissions; (2) because most of the organic matter in the biomass is gasified, the volume of anaerobic digestion residue is reduced, and the facilities for post-treatment and disposal can be small; (3) Anaerobic digestion residues are easy to compost and increase the epidemiological safety against pathogenic bacteria and viruses; (4) Anaerobic digestion residues can also be used as liquid fertilizers; (5) The large capacity anaerobic digester has the function of a buffer tank, therefore, the latter process can be operated with ease. Anaerobic digestion has many excellent functions; it not only benefits from energy recovery, but also reduces environmental impact by resource recycling and can be a component in preventing global warming and encouraging carbon recycling.

2. Principle of Anaerobic Digestion

Anaerobic digestion reduces and decomposes organic matter in biomass by the action of anaerobic bacteria and archaea, and its products are CH_4, CO_2 and ammonia. The decomposition of organic matter

[1] Department of Applied Chemistry and Biotechnology, National Institute of Technology (KOSEN), NIIHAMA College.
[2] Graduate School of advanced Sciences of Matter, Hiroshima University.
[3] Graduate School of Engineering, Tokyo University of Agriculture and Technology.
* Corresponding author: sriya@cc.tuat.ac.jp

into gas by Anaerobic digestion is generally represented by Buswell's equation, as shown in the following (Buswell and Sollo, 1948):

$$C_nH_aO_b + \left(n - \frac{a}{4} - \frac{b}{2}\right)H_2O \longrightarrow \left(\frac{n}{2} + \frac{a}{8} + \frac{b}{4}\right)CH_4 + \left(\frac{n}{2} + \frac{a}{8} + \frac{b}{4}\right)CO_2 \tag{1}$$

However, it is a complex reaction that is mainly performed in four stages (Figure 1; Schink, 1997). In the first stage, polymer macromolecules in organic matter are hydrolyzed to lower molecular weight substances by the action of hydrolytic enzymes, and then mainly fermented to volatile fatty acids (VFAs), hydrogen and CO_2. VFAs, except for acetate, are further oxidized to acetate with the generation of hydrogen anaerobically by acetogenic bacteria. Finally, acetate, hydrogen and CO_2 are converted into CH_4 by methanogenic archaea. The multistep reaction for CH_4 formation is accompanied by co-operation of a huge variety of bacteria and archaea, which have different catabolic activities for fermentation intermediates.

Many bacteria have the ability to secrete extracellular enzymes and hydrolyze solid components, such as carbohydrates, proteins and fats, into their constituent monomers (monosaccharides, amino acids, glycerol and fatty acids). The hydrolysates are metabolized by hydrolyzing bacteria or other bacteria to become VFAs (such as acetate, formate, propionate, butyrate and valerate), hydrogen and CO_2 (Vavilin et al., 1996; Sanders et al., 2000). Under certain conditions, lactic acid and alcohol are produced from sugar. Thus, hydrolysis/acidogenesis is a very complicated reaction, and the bacterial groups involved in these reactions are collectively referred to as acidogenic bacteria. Briefly, genera *Clostridium*, *Bacteroides*, *Bacillus*, *Lactobacillus* and some others are known as carbohydrate-degrading bacteria. Among them, cellulolytic bacteria include genera *Thermoanerobacter, Caldicellulosiruptor, Cellulosilyticum, Clostridium, Cellulomonas, Streptomyces* and *Bacteroides* (Koeck et al., 2014). Genera *Bacteroides*, *Butyrivibrio*, *Clostridium*, *Staphylococcus*, *Eubacterium*, *Peptococcus* and *Bacillus* are known as proteolytic bacteria, and the amino acids are converted to VFAs, such as acetate, propionate and butyrate (Liu et al., 2012; Siebert and Toerien, 1969).

VFAs such as propionate, lactate and ethanol produced by acidogenesis (Figure 1) are converted to acetate and hydrogen by hydrogen-forming acetogenesis (Table 1). At this time, reactions [1, 2] and [4] are the endergonic reactions. For example, in the conversion of propionate to acetate and hydrogen, the free energy change (ΔG_0') is 76.1 kJ. Thus, symbiosis with hydrogen-utilizing methanogens or sulfate-reducing bacteria is required in order to make the reaction proceed, as in reactions [5] and [6] (Boone and Bryant, 1980; McInerney et al., 1981). The hydrogen-utilizing methanogens play a role in lowering the hydrogen partial pressure, which makes it possible for hydrogen-producing acetogens to degrade propionate and obtain chemical energy for growth (Schink, 1997).

The phenomenon, in which one bacterium produces hydrogen, and a series of reactions proceed while being consumed by another bacterium, is called inter-species H_2 transfer, and was first reported

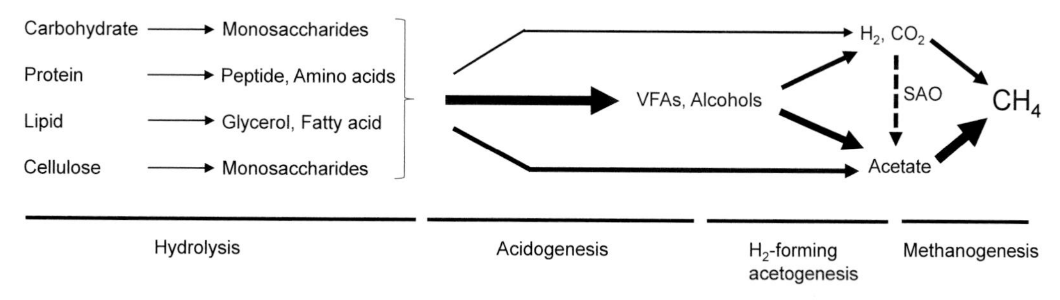

Figure 1. Methane production from organic matter by multistep reaction. VFAs, volatile fatty acids; SAO, syntrophic acetate oxidation. Modified from reference (Nakashimada and Nishio, 2017).

Table 1. Range of free energy change in methanogenesis reactions from lower fatty acids, lactic acid and ethanol.

Reaction	Reaction formula	Free energy change ($\Delta G^{0\prime}$ kJ/reaction)
[1]	$CH_3CH_2COO^- + 3H_2O \leftrightharpoons CH_3COO^- + HCO_3^- + H^+ + 3H_2$	76.1
[2]	$CH_3CH_2CH_2OO^- + 2H_2O \leftrightharpoons 2CH_3COO^- + HCO_3^- + H^+ + 2H_2$	48.1
[3]	$CH_3CHOHCOO^- \leftrightharpoons CH_3COO^- + HCO_3^- + H^+ + 2H_2$	–4.2
[4]	$C_2H_5OH + H_2O \leftrightharpoons CH_3COO^- + H^+ + 2H_2$	9.6
[5][a]	$4CH_3CH_2COO^- + 3H_2O \leftrightharpoons 4CH_3COO^- + HCO_3^- + H^+ + 3CH_4$	–102.4
[6][a]	$4CH_3CH_2COO^- + 3SO_4^{2-} \leftrightharpoons 4CH_3COO^- + 4HCO_3^- + H^+ + 3HS^-$	–151.3

[a] For the reaction to proceed as in reactions [5] and [6], symbiosis with hydrogen utilizing methanogenic bacteria or sulfate reducing bacteria is required (Boone and Bryant, 1980; McInerney et al., 1981).

by Bryant et al. (1967). Several hydrogen-forming acetogenic bacteria have been isolated, including *Synthrophomonas sapovorans* (C4–C18 fatty acids) (Roy et al., 1986), *Syntrophobacter wolinii* (propionate) (Boone and Bryant, 1980), *Syntrophomonas wolfei* (C4–C8 fatty acids) (McInerney et al., 1981), *Syntrophus buswellii* (benzoate) (Mountfort et al., 1984), *Syntrophococcus sucromutans* (fructose) (Krumholz and Bryant, 1986) and *Desulfovibrio vulgaris* (lactate) (Seitz et al., 1988).

The methanogens are responsible for the final step of the anaerobic digestion process. In methanogenesis, the production of CH_4 from H_2 and CO_2 is promoted by the activity of hydrogenotrophic methanogens, such as genera *Methanobacterium*, *Methanobrevibacter*, *Methanococcoides*, *Methanococcus*, *Methanocorpusculum*, *Methanoculleus*, *Methanogenium*, *Methanohalophilus*, *Methanolacinia*, *Methanospirillum* and *Methanothermus* (Demirel and Scherer, 2008).

Acetate degradation can proceed through two different pathways: Direct cleavage of acetate by acetoclastic methanogens or syntrophic acetate oxidation (SAO) (Hao et al., 2011). In the direct cleavage of acetate by acetoclastic methanogens, the methyl group of acetate is converted into CH_4 by the acetoclastic cleavage pathway, while the carboxyl group is converted into CO_2 (Yang et al., 2015). Only two genera, *Methanosarcina* and *Methanosaeta*, are known to be acetoclastic methanogens. Unlike *Methanosarcina*, which prefers methylated compounds, such as methanol and methylamines to acetate, *Methanosaeta* is a specialist that uses only acetate (Smith and Ingram-Smith, 2007).

SAO is a two-step reaction, in which acetate is oxidized to H_2 and CO_2 by SAO bacteria, followed by the subsequent reduction to CH_4 by hydrogenotrophic methanogens (SAO-HM pathway) (Hao et al., 2011). Recent studies have reported that the SAO-HM pathway contributes significantly to CH_4 production under inhibitory conditions related to the ammonium-nitrogen level, acetate concentration, synergetic stress of acids and ammonium, dilution rate, temperature and general methanogenic population structure (Hao et al., 2011; Mountfort and Asher, 1978; Yang et al., 2015). It is generally assumed that most of the CH_4 (about 65%–80%) is generated via direct cleavage of acetate by acetoclastic methanogens although the value depends on environmental conditions (Hao et al., 2011; Mackie and Bryant, 1981; Smith and Mah, 1966; Mountfort and Asher, 1978). Hence, acetoclastic methanogens play an important role in CH_4 production by anaerobic digestion.

Methanogens inhabit a reductive environment with an oxidation-reduction potential of −300 mV or less (Demirel and Scherer, 2008). The optimum pH of any methanogen is within a narrow neutral range (pH 6.0–8.0) while the optimum temperature is in a wide range of 20–98 °C depending on the species. They can be divided into four major groups: Low and ambient temperature (20–25 °C), middle temperature (mesophilic, 30–40 °C), high temperature (thermophilic, 50–65 °C) and hyperthermophilic (higher than 80 °C). Most methanogens prefer a salt concentration of about 0.1–0.7 mol/l, but halotolerant and halophilic *Methanosarcina* (Sowers, Baron and Ferry, 1984) and *Methanosaeta* (Kita et al., 2016a; Mori et al., 2012; Ma et al., 2006) have been discovered in the marine environment (see section 4–2 below for further information).

3. Various Processes for Anaerobic Digestion

3.1 Conventional process

In general, a continuously stirred tank reactor (CSTR) is used for treatment of organic waste, because of its simple structure and cost effectiveness. CSTR technology is suitable to digest waste with a percentage of total solid (TS) of 3%–8% and is applied to the treatment of various organic wastes such as kitchen waste, food waste, animal manure and sewage sludge (Gunaseelan, 1997). The contents in the CSTR can be thoroughly mixed by mechanical stirring (Figure 2A), gas blowing (Figure 2B), or a combination thereof. Continuous stirring ensures equalization of the temperature in the tank, sufficient mixing of the input substrate and the digested sludge, prevention of the occurrence of scum, and prevention of localized high concentration areas by the prevention of short circuit flow. The sludge retention time (SRT) is equal to the hydraulic retention time (HRT) because the contents in the reactor are thoroughly mixed. Although the HRT can vary depending on the properties of the organic waste and the operating conditions (such as operation temperature), it is typically in the range of 20–30 days with less than 3 kg-COD/m^3/d (the COD indicates Chemical Oxygen Demand) of organic loading rate.

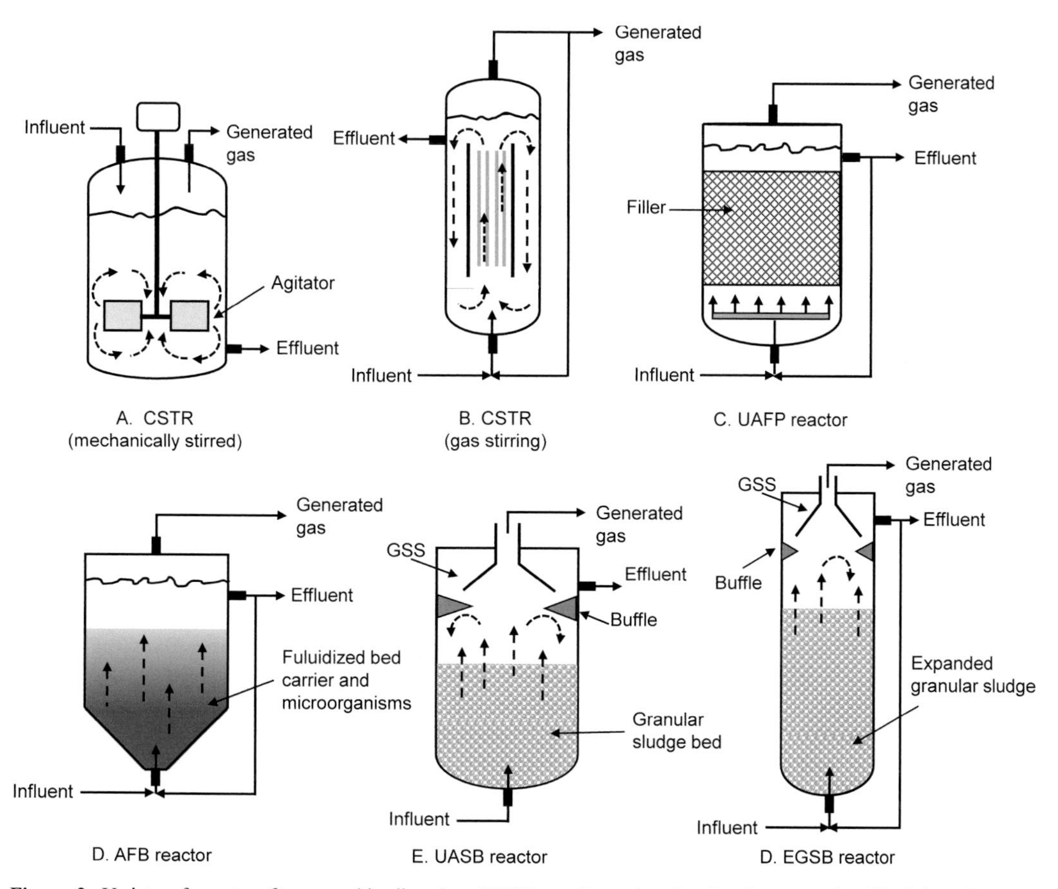

Figure 2. Variety of reactors for anaerobic digestion. CSTR, continuously stirred tank reactor (modified from De Mes et al., 2003); UAFP, up-flow anaerobic filter process (modified from Goswami et al., 2016); AFB, anaerobic fluidized bed (modified from Goswami et al., 2016); UASB, up-flow anaerobic sludge blanket (modified from Seghezzo et al., 1998); EGSB, expanded granular sludge bed (modified from Seghezzo et al., 1998).

3.2 High rate process

To ensure stable treatment performance and treatment under high-organic loading conditions, it is necessary to maintain methanogens with low specific growth rate in high concentration in the reactor. The organic loading rate of organic wastewater is lower than that of organic waste and higher in volume. Consequently, various kinds of treatment methods to increase the organic loading rate, in other words, to reduce the HRT, have been developed.

The up-flow anaerobic filter process (UAFP) is a process developed by Young and McCarty (1969). In UAFP, the wastewater introduced from the lower part of the filter bed rises slowly through the filter bed of plastic filler or crushed stone to which anaerobic bacteria and archaea are attached as a biofilm (Figure 2C). The thickness of the biofilm is about 1–4 mm, and the concentration of sludge differs in the vertical direction of the reactor. UAFP processing performance is strongly influenced by the type of filler. The bacteria and archaea in the sludge trapped by the filler greatly contribute to the anaerobic digestion. Because UAFP can be operated with high cell concentration, it can shorten the HRT. Furthermore, there are features such as easy start-up, no need to return sludge, and resistance to fluctuations in water quality and quantity. Therefore, the application range of UAFP is broad, and it can be applied to a wide range of concentrations in various types of drainage. Typical tank loads are 0.1–15 kg-COD/m^3/d although the HRT differs depending on the type of wastewater, suspended solid concentration, biodegradability of organic matter, water temperature and required treated water quality.

The anaerobic fluidized bed (AFB) is a method of suspending particulate adhesion carriers such as sand, anthracite, lightweight aggregate and the like with a particle size of about 0.2–1.0 mm in a fluid state, and attaching and propagating anaerobic bacteria and archaea on the surface (Figure 2D) (Stronach et al., 1986; Switzenbaum and Jewell, 1980). Fluidized bed technology was originally developed for aerobic biochemical oxygen demand removal, nitrification and denitrification at the beginning of the 1970s, and was later introduced to anaerobic treatment processes as AFB (Switzenbaum and Jewell, 1980; Stronach et al., 1986; Jeris, 1983; Jewell et al., 1981). AFB is promoted as a process that can treat low concentration of wastewater because of the high concentration of cells in the reactor, no clogging, high contact efficiency and diffusivity of the substrate into the biofilm because of the thinness of the biofilm (Heijnen et al., 1989). The potential disadvantages are that scale-up is relatively difficult, and there are technical issues such as power cost for carrier flow, peeling of the biofilm because of friction between particles and outflow of carrier particles.

The up-flow anaerobic sludge blanket (UASB) is a high efficiency anaerobic wastewater treatment process developed in the Netherlands in the 1970s (Lettinga et al., 1980). This technology does not use an adhesive carrier like UAFP and AFB, but uses the properties of aggregation and granulation of anaerobic bacteria to form and retain granular sludge with a particle size of about 0.5–2.0 mm with excellent sedimentation in the reactor. The structure of the UASB treatment tank is provided with an influent supply device at the lower part of the reaction tank and a gas-solid three-phase separator (GSS: gas-solid separator) at the upper part (Figure 2E). Waste water flows uniformly upward from the bottom of the reaction tank to produce granular sludge and is subjected to anaerobic biological treatment. In UASB, it is important to maintain and continuously grow granular sludge in the tank. For that purpose, it is necessary to create an environment where methanogens can preferentially grow in the UASB treatment tank, and in particular, methanogenesis of the genus *Methanosaeta*, which forms the skeleton of granular sludge and grows with acetate as the sole substrate. It is the most widely used system for medium- and high-concentration wastewater treatment in the food industry, because high organic loading treatment (greater than 10 kg COD/m^3/d) is possible, and the sludge layer does not block the outlet.

The expanded granular sludge bed (EGSB) is a high load modified version of the UASB method. In the EGSB method, the treated water is positively returned to the reactor in order to give a high upward flow rate so as to flow the granules and bring the water flow close to complete mixing, and allow control of the pH and accumulation of VFAs in the reactor (Figure 2F) (Kato et al., 1994). It enables high load processing that is two to three times higher than that of the UASB process.

3.3 Two-phase anaerobic digestion

In the acidification process of solid waste digestion, the low pH, high hydrogen partial pressure and high concentration of VFAs can cause inhibition of both acetogenic and methanogenic processes. Therefore, a two-phase anaerobic digestion method that divides anaerobic digestion into VFAs production and gas generation reaction was developed. In a two-phase system, the different microbial phases are separated. In such systems, the hydrolytic and acidification phases occur in the first reactor, while acetogenesis and methanogenesis occur in the second reactor. The concept of two-stage digestion is driven by separate optimization of each step, in which different types of bacteria and archaea with different physiology participate. Therefore, the two phase system has advantages over conventional one-phase processes, such as increased stability of the process by controlling the acidification phase to prevent overloading and the formation of inhibitors, and higher organic loading rates and shorter HRT, resulting in potentially higher yields of biogas in smaller digesters than in one-stage reactors (Nizami and Murphy, 2010; Demirer and Chen, 2005).

3.4 Dry anaerobic digestion

Anaerobic digestion is classified as either wet or dry anaerobic digestion by TS content in the digester. Wet anaerobic digestion is traditionally used for digestion at 1%–10% TS content (Ge et al., 2016), while dry anaerobic digestion is anaerobic digestion conducted at TS content higher than 15% (Andre et al., 2018) and is usually applied for high TS content waste, such as crop residue or livestock waste (Andre et al., 2018). Furthermore, 1–10 cm biomass pieces can be digestible during dry digestion (Iino et al., 2011; Riya et al., 2018), thus, there is no need for pulverization of biomass as a pretreatment. Due to the low water content in the reactor, dry anaerobic digesters have much smaller (about 1/10th) volume than wet anaerobic digesters. Therefore, higher volumetric CH_4 production and thermal efficiency can be achieved with dry digestion than with wet digestion (Guendouz et al., 2008). Another advantage of dry anaerobic digestion over wet anaerobic digestion is the production of solid digestate instead of liquid digestate from wet anaerobic digestion (Figure 3). Liquid digestate contains nutrients (nitrogen (N), phosphorus (P), potassium (K) and others) from the feedstock and should be treated before discharge, if no land is available for spreading. While the solid digestate also contains these nutrients, handling of the solid digestate is easier than that of the liquid. Therefore, liquid waste can be treated by mixing high-TS-content feedstocks (co-digestion) without liquid digestate production. It should be noted that, in dry anaerobic digestion, the ammonia concentration tends to be higher than that in wet digestion because of the lower water content (Wang et al., 2013). High ammonia concentration would lead to inhibition of anaerobic digestion, so it is necessary to monitor the ammonia concentration for stable operation.

Figure 3. Wet digestate (left) and dry digestate (right).

4. Case Studies for Carbon Recycling with Anaerobic Digestion

4.1 Carbon recycling with anaerobic digestion in land environments

4.1.1 Recycling and treatment system of pig manure with rice cultivation

Pig manure management is an ongoing environmental concern. Generally, pig manure is separated into solid and liquid fractions in Japan. The solid fraction is composted, while the liquid fraction is treated by an activated sludge system. In terms of treatment of the liquid fraction, it has been difficult to satisfy the N effluent standard, especially for small farmers who cannot manage the biological wastewater treatment facility well. In addition, 87% of concentrated feed, which is used as pig feed, is imported from foreign countries (MAFF, 2019), meaning that nutrients such as N are imported and accumulated in the environment.

Another environmental problem is CH_4 emission from rice fields in Japan and other countries where rice is produced. During rice cultivation, rice fields are flooded and organic matter is decomposed anaerobically, leading to CH_4 production and emission to the atmosphere (Le Mer and Roger, 2001). Generally, rice straw is left in the field after rice harvest. Watanabe et al. (1999) found that about half of the CH_4 emitted from the rice fields was produced from rice straw incorporated before rice cultivation. Therefore, improvements in rice straw C management would significantly affect CH_4 emission.

To solve these problems, we focused on C and nutrient management using dry anaerobic digestion and rice cultivation (Figure 4) with a high-yielding rice variety developed in Japan for livestock feeding (Kato, 2008). In this system, pig manure is co-digested with rice straw in a dry anaerobic digestion system (see below). The solid digestate produced is applied to rice cultivation as fertilizer. Harvested rice grain is fed to pigs, and straw is used as substrate for dry anaerobic digestion. Thus, a sustainable C and nutrient recycling system could be achieved using dry anaerobic digestion and rice cultivation.

Several factors need to be optimized in order to achieve this pig manure management system: First, the mixing ratio of pig manure and rice straw; second, the operation conditions of dry anaerobic digestion such as TS content, mixing and inoculation; and finally, the digestate application method for rice and/or other crop cultivation. Here, we summarize our recent results.

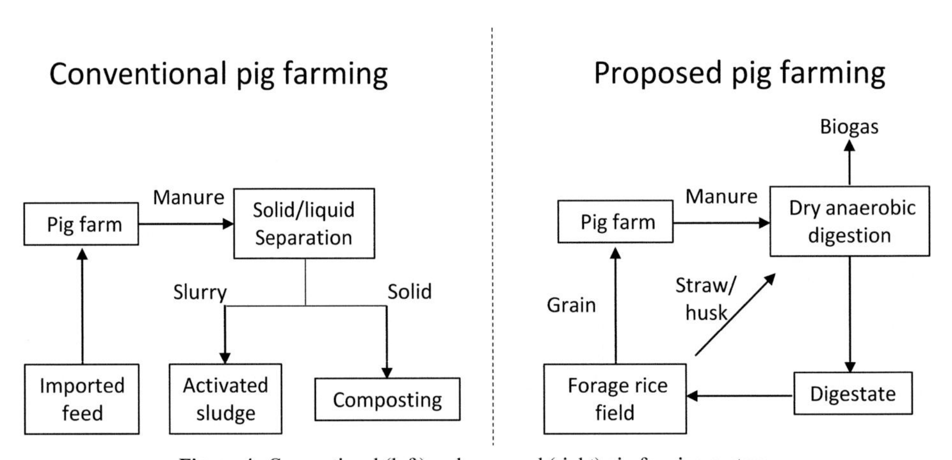

Figure 4. Conventional (left) and proposed (right) pig farming system.

4.1.1.1 Mixing ratio of pig manure and rice straw

Anaerobic digestion fed with pig manure alone inhibits digestion because of excess ammonia (free [unionized] ammonia N, FAN) (Rajagopal et al., 2013). At low concentrations, ammonia is important for bacterial growth, while high concentrations of ammonia can cause a severe drop in performance of anaerobic digestion. It has been thought that (i) excess ammonia inhibits the enzymes responsible for CH_4

production directly, or (ii) the ammonia molecule might diffuse passively into bacterial cells and cause proton imbalance or potassium ion (K^+) deficiency (Gallert et al., 1998). Since ammonia is equilibrated with ammonium (NH_4^+) and hydroxide (OH^-) ions, inhibition by excess ammonia depends on the NH_4^+ concentration, temperature and pH. Higher NH_4^+ concentration, higher temperature, and/or higher pH results in higher concentration of FAN (Fernandes et al., 2012).

Numerous strategies to overcome ammonia inhibition have been studied, including acclimation of microflora, pH or temperature control, ammonia stripping and others (Nakashimada et al., 2008; Rajagopal et al., 2013). Among many strategies, adjustment of the C/N ratio by co-digestion is a cost-effective method of reducing ammonia inhibition (Rajagopal et al., 2013). It has been reported that co-digestion of chicken manure with agricultural wastes (with low ammonia content) decreased the ammonia concentration, and increased biogas production when compared with digestion of chicken manure alone (Abouelenien et al., 2014). Therefore, pig manure should be co-digested with other substrates in order to dilute the FAN concentration in the reactor. Iino et al. (2011) found that both wet and dry thermophilic anaerobic digestion of rice straw resulted in CH_4 yield of about 300 ml/g VS, which was comparable to other biomass such as grass or wheat straw (Weiland, 2010). This suggests that rice straw can be used not only as an FAN diluent, but also as a C source in the anaerobic digestion process.

Our previous study showed that mixing rice straw with pig manure is a good strategy for stable biogas production (Zhou et al., 2016; Riya et al., 2016). Zhou et al. (2016) compared semi-batch dry-thermophilic anaerobic digestion of pig manure alone, and mixtures of pig manure and rice straw at different mixing ratios (95:5, 78:22 and 65:35 w/w resulting in C/N ratios of 10, 20 and 30). The average CH_4 yield of pig manure alone was 35 ml/g VS, while ca. 100, 244, and 258 ml/g VS was obtained for C/N ratios of 10, 20 and 30, respectively. Therefore, co-digestion of pig manure and rice straw can increase CH_4 production. This result was similar to that of Wang et al. (2012b), who observed an increase in biogas production with increasing C/N ratio in co-digestion of manure and wheat straw. Zhou et al. (2016) found that hydrogenotrophic methanogenesis was the major CH_4 production pathway because *Methanothermobacter wolfeii* was predominant in the thermophilic reactor fed with both pig manure alone, and the mixtures of pig manure and rice straw. It has been reported that hydrogenotrophic methanogens were dominant in thermophilic anaerobic digestion (50–55 °C) of organic solid waste, paper, or sludge (Goberna et al., 2009; Sasaki et al., 2011; Tang et al., 2011). In contrast, Riya et al. (2016) also operated a semi-batch dry thermophilic reactor fed with a mixture of pig manure and rice straw for a longer time period than that of Zhou et al. (2016). They found that the FAN concentration in the reactor fed with pig manure alone exceeded 2000 mg N/kg wet weight, which was higher than the threshold for inhibition (1100 mg N/l) (Yenigün and Demirel, 2013), while ca. 1100 and 800 mg N/kg wet weight of FAN were observed for C/N ratios of 20 and 30, respectively (Riya et al., 2016). They also found that hydrogenotrophic methanogens, such as *Methanothermobacter* and *Methanoculleus*, were predominant after 120 days of operation in manure only and for C/N ratios of 20 and 30, while acetoclastic methanogens (*Methanosarcina*) were dominant after 180 days in the semi-batch dry thermophilic reactor fed with a mixture of pig manure and rice straw (C/N ratios of 20 and 30; Riya et al., 2016). These results suggest that the acetoclastic pathway would be important for CH_4 production during the thermophilic co-digestion of pig manure and rice straw. *Methanosarcina* is known to be more resistant to FAN inhibition than other acetoclastic and hydrogenotrophic methanogens (Wiegant and Zeeman, 1986).

4.1.1.2 TS content

High TS content in the dry anaerobic reactor also affects its performance because water dissolves nutrients and substrates, and transports them to bacteria by diffusion (Lay et al., 1997; Mora-Naranjo et al., 2004; Pommier et al., 2007). For example, Abbassi-Guendouz et al. (2012) reported that the CH_4 yield of cardboard was about 200 ml/g VS at 10% TS content in a batch reactor, while it decreased to ca. 25 ml/g VS at 35% TS content. According to Le Hyaric et al. (2012), specific methanogenic activity increased with decreasing TS content for cellulose, acetate and propionate as substrate. In the dry anaerobically digested residue, 50% of the water was free water, and the remaining water was bound water (Garcia-Bernet et al., 2011). This might cause an increase in the rheological properties of sludge (Garcia-Bernet et al., 2011) and affect digestibility at high TS content. Thus, dry anaerobic digesters should be operated at an appropriate TS content in the reactor.

Riya et al. (2018) determined the optimal substrate TS content based on the response of performance and TS content in semi-batch thermophilic dry-anaerobic digestion of pig manure and rice straw. They fed substrate (rice straw and pig manure) at a constant organic loading rate with varying substrate TS contents (27%, 32%, 37%, and 42%) by adding different amounts of water (0%–36% of the total substrate). Stable and high biogas production (564 ± 13–580 ± 36 N m^3/t VS) was observed between 18% and 27% TS in the reactor. However, reductions in biogas production, and accumulation of VFAs occurred when the TS content of the reactor exceeded 28%. Furthermore, they constructed a simple TS balance model, and found that a substrate TS content of < 32% could maintain the TS content in the reactor at < 28% (Riya et al., 2018). These results suggest that rice straw with less than 67% of TS is acceptable to achieve a substrate TS content of < 32%. The TS content of rice straw at harvest is ca. 40%, and varies between 40%–80% depending on the weather conditions before collection (Otani 2012). Therefore, the substrate TS content might need to be adjusted according to the timing of collection and storage duration of the straw.

4.1.1.3 Mixing

Because of limited mass transfer in the medium of dry anaerobic digestion, contact of substrate with inoculum is an important factor. Several studies have attempted to determine diffusivity in the dry anaerobic digestion medium. These studies reported two orders of magnitude lower diffusivity at 15%–25% of TS than that of pure water (Bollon et al., 2013; Zhang et al., 2016). Therefore, mixing of the contents in the dry digester is important. According to a review by André (2018), four types of stirring mode are used in the continuous dry process: (i) recirculation of the digestate, (ii) recirculation of biogas, (iii) mechanical stirring, or (iv) no stirring; for batch processes, liquid recirculation is mainly used. In addition to liquid recirculation in batch dry anaerobic digestion, solid digestate recirculation was also reported in a full-scale reactor treating agricultural wastes (Chiumenti et al., 2018).

In terms of liquid recirculation during batch dry anaerobic digestion, most previous studies have focused on optimization of the leachate-to-substrate ratio, the recirculated leachate volume, and recirculation frequency (Benbelkacem et al., 2010; Degueurce et al., 2016). Meng et al. (2019b) tested two liquid circulation modes (percolation and immersion) during batch thermophilic dry-anaerobic digestion of rice straw using pig urine for liquid circulation. In the percolation mode, leachate was poured on the rice straw filled in the mesh bag, while liquid content was passed through the bag. For immersion, the rice straw-filled mesh bag was immersed in the leachate for the designated contact time. Leachate recirculation by percolation might cause non-uniform leachate flow because of the heterogeneous structure of the medium (André et al., 2015), while it is expected that most of the substrate in the bag could be in contact with the leachate by immersion. The CH_4 yield of the immersion mixture of rice straw and solid digestate into leachate was higher than that of percolation of leachate. Furthermore, the CH_4 yield increased from 1 to 24 h of the immersion period, while it decreased after longer than 24 h of immersion. Therefore, pig urine can be used as liquid recirculation medium under certain conditions.

For simpler and more economical mixing, Meng et al. (2019a) evaluated the mixing frequency in the contents of a dry anaerobic digester without leachate recirculation. In this approach, the reactor contents were mixed in the open (under atmosphere) during digestion, simulating turning the contents using heavy farm machinery, such as a loader. The CH_4 yields with mixing once per 3 or 7 days were comparable with no mixing. However, mixing once per day showed lower CH_4 yields and accumulation of acetate, suggesting that frequent mixing resulted in inhibition of acetoclastic methanogenesis by oxygen. Considering the net energy production, no mixing during digestion was more effective than any of the mixing options tested. However, large scale validation is needed.

4.1.1.4 Inoculation

In addition to liquid recirculation during dry anaerobic digestion, the source of microorganisms (the inoculum) also plays an important role. In batch anaerobic digestion, the inoculum is mixed with the substrate at the start of digestion. Generally, liquid digestate obtained from wet anaerobic digestion was used as inoculum in laboratory testing of dry anaerobic digestion (Li et al., 2011; Zhou et al., 2017). The ratio of substrate (feeding) and inoculum (F/I ratio, expressed as the ratio of VS amount) is an important parameter used to express the extent of inoculation. Generally, it showed strong and inverse correlation with CH_4 yield (Xu et al., 2014). Meng et al. (2018) found that a low F/I ratio (0.5), that is, a high inoculation rate of solid digestate, showed higher CH_4 yield than those of high F/I ratios (of 1, 2, and 3) for dry anaerobic digestion of pig manure and rice straw. Furthermore, the structures of the bacterial and archaeal communities of digestates were similar between F/I ratios of 0.5 and 3, suggesting that digestate can be used as the inoculum for subsequent digestion at F/I ratios of between 0.5 and 3 (Meng et al., 2018).

In contrast to studying the inoculation amount, there are few reports on the inoculation method (Ge et al., 2016). Zhu et al. (2015) compared complete and partial premixing of substrate and liquid digestate in batch dry digestion of corn stover. Partial premixing by placing the inoculum on the top and in the middle of the feedstock showed higher CH_4 yield than complete premixing at overall F/I ratios of 4 and 6, because a highly inoculated (low F/I ratio) layer in the partial premixing digester can reduce VFA accumulation (Zhu et al., 2015). In contrast, Meng et al. (2019b) combined partial premixing at the start of digestion and leachate recirculation during digestion. They placed a slurry of solid digestate and pig urine on the rice straw for premixing at an overall F/I ratio of 3, and conducted leachate recirculation during digestion. They found no significant difference in CH_4 yield when compared with complete premixing (Meng et al., 2019b). Therefore, partial premixing would be more applicable to solid digestion with liquid recirculation.

4.1.1.5 Digestate for fertilizer and its environmental impacts

In anaerobic digestion, plant nutrients, such as N, P and K, are not removed. Therefore, the digestate can be used as fertilizer. Dried digestate from a thermophilic dry anaerobic digester fed with pig manure and rice straw contained 30.3% and 1.69% of total C and total N (dry basis), respectively (Riya et al., 2014). Thus, the C/N ratio was close to 18, which was lower than 25 (Paul 2007), suggesting that N in the digestate would be released into the soil rather than immobilized by microbes. However, the above-ground biomass of rice grown with digestate application (7.9–10.7 t dry/ha) was lower than that grown with chemical fertilizer (15.2 t dry/ha). Therefore, the quality of digestate and its fertilization application method should be improved in the future.

The input of organic matter to the rice fields causes substantial increases in CH_4 emission because organic matter is anaerobically decomposed in the flooded rice soil. Riya et al. (2014) reported that the CH_4 emission from rice fields fertilized with dry digestate of pig manure and rice straw was 509 kg C/ha, which was 9.3% of the incorporated digestate C, and higher than that of rice fields fertilized with chemical fertilizer (105 kg C/ha). In general, CH_4 emission from rice fields fertilized with compost or fermented residue is lower than that treated with feedstock material because readily degradable material is already degraded (Yagi and Minami, 1990; Wassmann et al., 1993). For example, 0.4%–14% and 2.1%–3.1% of C were emitted as CH_4 by incorporating rice straw and compost, respectively (Yagi and Minami, 1990). In contrast, when considering both the dry anaerobic digestion process and rice cultivation, the CH_4 emitted during rice cultivation was 10% of the CH_4 produced in the digestion. Therefore, it suggested that most of the C was recovered as energy in dry anaerobic digestion.

4.1.1.6 System sustainability

Finally, we evaluated N loading to the environment, GHG emissions, and the economic balance for a conventional system, and the proposed system for 1000 pigs (Figure 4), based on the above results and literature values (Hosomi et al., 2014). In the proposed system, two scenarios (C/N ratios of 20 and 30 for thermophilic dry-anaerobic digestion) were evaluated because these C/N ratios showed stable biogas production, assuming that the distance between each facility (pig farm, dry digester, and rice field) was 10 km. In the calculation, we further assumed that 10% of the pig feed was substituted with harvested rice grain. For C/N ratios of 20 and 30, the area of rice field required to supply straw to meet the target C/N ratios were 34 and 84 ha, respectively, based on the mass balance.

The N loading to the environment by effluent from the wastewater treatment of the conventional system was 3.8 t N/year, while it was 0.3 and 0.5 t N/year by leaching from rice fields for C/N ratios of 20 and 30 in the proposed system, respectively. If normalized to the rice cultivation area, the N leaching values were 0.010 and 0.0065 t N/ha/year for C/N ratios of 20 and 30, respectively. The lower N leaching in C/N 30 than C/N 20 was because of the different N application rate (244 and 158 kg N/ha in C/N 20 and 30, respectively). These results suggest that dry-anaerobic digestion of manure and recycling digestate for rice production can significantly reduce N loading to the environment.

In the proposed system, overall GHG emissions are lower than those of conventional systems when taking into consideration the GHG sink (Figure 5). In the conventional system, pig farming, composting, and the wastewater treatment plant (WWTP) emitted 390 t CO_{2eq} (carbon dioxide equivalent) annually with no GHG sink. GHG emissions in the proposed system were higher than that of the conventional one because of CH_4 emissions from rice paddy fields. A C/N ratio of 30 exhibited higher GHG emissions than that of C/N 20, because C/N 30 required the cultivation of a larger area of paddy fields to produce rice straw for dry digestion, thus requiring more fossil fuel for operation of the digester, and cultivation of the rice paddy field. However, the proposed system has potential GHG sinks, such as power generation and heat recovery by co-generation of CH_4, and soil C storage by adding digestate to rice paddy fields. The total GHG reductions of these sinks in C/N 20 and 30 were 460 and 913 t CO_{2eq}/year, respectively. Thus, net GHG emissions of the proposed system were 105 and 99 t CO_{2eq}/year, which were 73% and 75% lower than those of the conventional system, respectively. Therefore, the proposed system is a low-C pig manure management system. It should be noted that using digestate as a fertilizer can replace chemical fertilizer, resulting in further decrease in GHG emission by fossil fuel use for chemical fertilizer production. However, this was not considered in the calculations above, because land management in the conventional system was not evaluated. Therefore, the influence of land use change on GHG emission from rice fields should be evaluated in the future.

The economic balance showed that the proposed system could have a net positive income, but not enough to support both pig and rice farmers. The proposed system is operated with the cooperation of both pig and rice farmers. Based on the statistics of the Ministry of Agriculture, Forestry, and Fisheries (MAFF), the sum of net income for both the pig farmer and the rice farmer needs to be 0.17 and 0.38 million US dollars for C/N 20 and 30, respectively (MAFF, 2013a; MAFF, 2013b). These values were

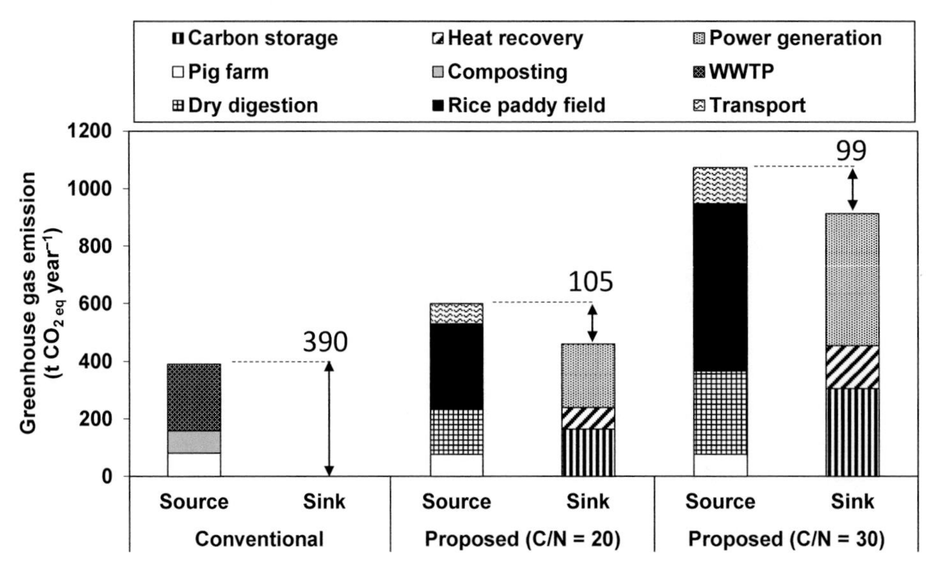

Figure 5. Greenhouse gas emission from conventional and proposed systems (1000 pigs). Values on the bar graph indicate net greenhouse gas emission calculated by subtracting sink from source. WWTP: wastewater treatment plant.

higher than the net income values in C/N 20 and 30 of 0.091 and 0.25 million US dollars, respectively (Figure 6). Therefore, the proposed system is not economically feasible at the 1000 pigs scale.

Based on the dependence of cost and benefit on the size of the pig farm, a farm of more than 7000 pigs was economically and environmentally sustainable (Hosomi et al., 2014). Table 2 shows the summary of GHG and economic balance at the 7000 pigs scale, with 10 km in distance between each facility. Reductions in GHG emissions of 81% and 89% were achieved for C/N 20 and 30, respectively. In addition, the net profit met the required income for both pig and rice farmers. For the proposed systems, GHG emissions and costs normalized to rice cultivation area (values in parenthesis in Table 2) were lower in C/N 30 than in C/N 20. This was because of the utilization of larger scale of dry anaerobic digester and rice field in C/N 30 than those of C/N 20, resulting in a more energetic and cost-effective operation in C/N 30. It should be noted that the benefit to the rice farmer was supplemented with a subsidy from the government in order to support farmers conducting rice cultivation, and in cooperation with the livestock farmer. Therefore, our case study can work in Japan or countries with a similar subsidy policy. Further research is required in order to develop systems that are more applicable to situations without additional subsidy.

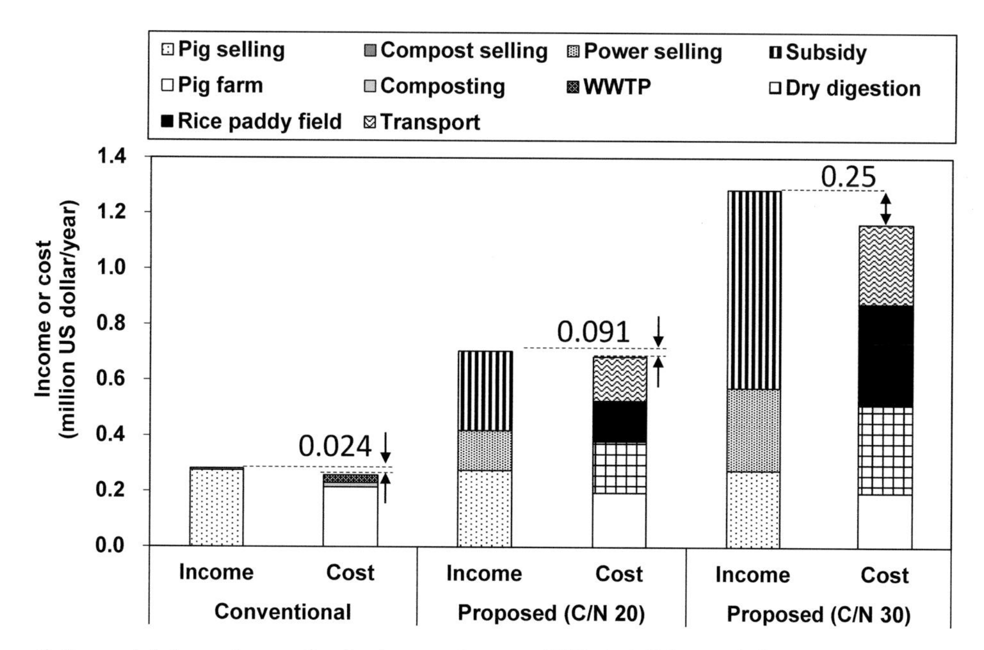

Figure 6. Economic balance of conventional and proposed systems (1000 pigs). Values on the bar graph indicate net income calculated by subtracting cost from income. Cost was the sum of capital and operational costs. A subsidy is provided from the government to support farmers that conduct rice cultivation, and cooperate with livestock farmers. WWTP: wastewater treatment plant.

Table 2. Summary of greenhouse gas emissions and economic balances for 7000 pig farming.

Scenario	Greenhouse gas emission ($tCO_{2\,eq}$/year)			Money (million US dollars/year)		
	Emission	**Sink**	**Net emission**	**Benefit**	**Cost**	**Net benefit**
Conventional	2731	0	2731	1.97	1.81	0.168
Proposed (C/N 20)	3731 (15.8)[a]	3220 (13.6)	511 (2.17)	4.93 (20.9)	3.70 (15.7)	1.23 (5.19)
Proposed (C/N 30)	6719 (11.4)	6392 (10.9)	326 (0.553)	9.00 (15.3)	6.30 (10.7)	2.70 (4.59)

[a] Greenhouse gas emission (tCO_{2eq}/ha/year) and money (1000 US dollars/ha/year) normalized to area of rice field (236 and 589 ha for C/N 20 and 30, respectively).

Other considerations include substrate availability for dry anaerobic digestion, and how to use the rest of the grain that was not fed to pigs. Pig manure is produced every day. In contrast, production of rice straw is seasonal, being produced once or twice a year, depending on the region. Therefore, storage of the straw or use of other biomasses have to be considered. The rice straw can be compactly collected and stored by baling (Otani, 2012), although this entails additional cost. However, it should be noted that, if the water content of the baled straw is higher than 25%, fermentation occurs (Kadam et al., 2000), which reduces VS content, and hence CH_4 yield is reduced. Therefore, these reductions should be taken into account or other storage methods should be considered. In terms of other biomasses, Riya et al. (2017) found that plants grown in riverbanks (seven kinds of grass biomasses, and three kinds of woody biomasses) were digestible using dry-thermophilic anaerobic digestion. Riverbank biomasses have been harvested at regular intervals in Japan. Therefore, the riverbank biomasses would be one of the options for biomass substitution, while energy consumption, GHG emission, and additional cost for collection and harvest should be considered. Rice grain can also be used as a substrate for dry anaerobic digestion (see below). Using biomass other than rice straw as substrate for dry anaerobic digestion can reduce the required area of rice fields producing straw for dry digestion, which allows farmers to treat pig manure at smaller scale.

Usage of the grain is another potential problem. Because a low (10%) substitution rate of grain feeding to pigs was assumed in this analysis, 80%–90% of the harvested grain was not used. One option to use the grain is feeding to other animals. It has been reported that the rice grain can be fed to other livestock (dairy cows, beef cattle, laying-hens or broiler chickens; Tsunekawa, 2016). A higher substitution rate than 10% might also be possible, as Katsumata et al. (2015) reported that 75% substitution of pig feed with grain had no influence on the pigs' growth. Rice grain can be added to the dry digester for energy production, because the CH_4 yield of the grain is higher than that of leaf or stem (Iino et al., 2011). The selection of the options to use grain would depend on the situation within which the proposed system is applied. Evaluation of these options (straw storage, using other biomass as substrate for dry digestion, excess grain usage) would be helpful for provision of a more balanced and sustainable system.

4.1.2 Future direction

Our studies suggest that dry anaerobic digestion can effectively produce CH_4 from pig manure and rice straw. The proposed system could significantly reduce GHG emissions when compared with conventional pig farming systems. However, more research is needed, especially for economic viability without subsidy. Our economic analysis suggested that the cost for dry anaerobic digestion and rice paddy fields accounted for about 50% of the potential benefit (Figure 6). Therefore, research into low cost dry anaerobic digestion and rice cultivation, as well as increasing benefit in the system, is required.

In addition, incorporation of dried digestate showed lower growth of rice than chemical fertilizer. Therefore, digestate treatment methods should be studied. Recently, biochar production has been gaining attention. Pyrolysis produces bio-oil, syngas, and biochar (Laird et al., 2009). Bio-oil and syngas can be used as energy sources, while biochar can be used as a soil amendment to improve nutrient and moisture retention (Laird et al., 2009) and provide minerals such as P (Wang et al., 2012a). Biochar amendment increased N use efficiency (He et al., 2018) and reduced the cost of fertilizer and pesticide applications (Wang et al., 2018). Additionally, biochar can potentially reduce GHG emission through mitigation of CH_4 and nitrous oxide emissions (Yang et al., 2019) and increase in C sequestration (Woolf et al., 2010). To maximize these potential benefits, optimization of biochar production conditions and system analysis are required.

4.2 C recycling with anaerobic digestion in marine environments

Although anaerobic digestion is applicable for the treatment of various kinds of organic matter, including manure, food, and municipal waste, its use is restricted to organic waste with relatively low salt concentration because methanogenesis is strongly inhibited by a sodium (Na) concentration of more than 10 g/l (Lefebvre and Moletta, 2006). However, there are some reports that high-salt-containing organic waste and wastewater could be treated by Anaerobic digestion (Chowdhury, 2010; Roesijadi et al., 2010; Shi et al., 2014).

The wastewater generated by a fish-processing industry is rich in salts because fish processing industries use a large amount of salt for fish preservation (Lefebvre and Moletta, 2006). Although marine macroalgae are sometimes considered as hazardous organic wastes (Nkemka and Murto, 2010), the ash content of the wet macroalgae can reach 3% (w/w) (Roesijadi et al., 2010). Although the application of Anaerobic digestion for such high-salt-containing organic matter requires dilution with fresh water, this operation decreases the economic efficiency of the process because of the requirement for increased reactor volume and aerobic treatment of the effluent subsequent to Anaerobic digestion (Shi et al., 2014). Therefore, it is necessary to develop Anaerobic digestion processes that are applicable to organic matter with high salinity.

Generally, acetoclastic methanogenesis shows low salt tolerance and is responsible for the high-salt intolerance of the Anaerobic digestion process (Rinzema et al., 1988; Oren, 1999). Therefore, it is important to isolate high-salt-tolerant methanogens and understand their physiology for the development of industrial Anaerobic digestion processes for high-salt-containing marine feedstock.

Some hydrogenotrophic methanogens are known to inhabit environments of extremely high salt concentration, such as salt lakes and deep-sea sediments. For instance, *Methanococcoides methylutens*, which was isolated from the Sumner Branch of Scripps Canyon located near La Jolla, California, exhibited growth under saline conditions, with optimal sodium chloride (NaCl) concentration of 0.24–0.64 M (corresponding to 1.4%–3.7% (w/v) of NaCl) (Sowers and Ferry, 1983). *Methanohalophilus mahii*, isolated from the Great Salt Lake, was another moderately halophilic methanogen, which exhibited optimum growth at an NaCl concentration of 12% (w/v) (Paterek and Smith, 1988; Ollivier et al., 1994). Furthermore, the extremely halophilic methanogen *Methanohalobium evestigatum*, isolated from a saline lagoon in Crimea, showed growth at 25% salinity (Ollivier et al., 1994; Zhilina and Zavarzin, 1990).

Two genera of acetoclastic methanogens are known (genus *Methanosarcina* and genus *Methanosaeta*; Smith and Ingram-Smith, 2007). Given that 70% of the CH_4 generated during Anaerobic digestion is derived from the cleavage of acetate (Smith and Mah, 1966), an important role is played by acetoclastic methanogens in this process. A few species of halophilic *Methanosarcina* have been isolated to date. The optimal range of NaCl for the growth of *Methanosarcina acetivorans* has been reported as 0.1–0.6 M, with growth not observed in the absence of NaCl (Sowers et al., 1984). *Methanosarcina mazei* Gö1 showed growth at an Na^+ concentration of 18,000 mg/l, corresponding to a 4.6% (w/v) solution of NaCl (De Vrieze et al., 2012; Spanheimer and Müller, 2008). The single report available on halophilic species of *Methanosaeta* showed that *Methanosaeta pelagica* 03d30qT, isolated from a tidal flat sediment, exhibited growth at Na^+ concentrations of 0.20–0.80 M (corresponding to 1.2%–4.7% of NaCl), with optimum growth at 0.28 M (1.6% of NaCl) (Mori et al., 2012). Kita et al. (2016a) reported that the acetoclastic methanogen *Methanosaeta* sp. H_A, enriched from marine sediment collected from Hiroshima Bay, exhibited growth at 2.06 M (corresponding to 12% NaCl; Figure 7).

Marine sediments can be used for the reactor development of ultra-high-speed production of CH_4 from acetate under saline conditions (Kita et al., 2016b). By UAFP using marine sediment as the immobilization carrier and a source of halophilic acetoclastic methanogens, the maximum CH_4 production efficiency of 16.8 l l^{-1} d^{-1} was obtained at a dilution rate of 16.2 d^{-1} and a salt concentration of 3%. Observation with a scanning electron microscope revealed the formation of granules and the presence of *Methanosaeta*-like filamentous structures and *Methanosarcina*-like cocci (Figure 8).

These results suggest that marine sediments are excellent not only as a microbial source of salt-tolerant Anaerobic digestion, but also in granule formation that enables ultra-fast CH_4 production from acetate. It also suggests that organic matter from both land and marine environments are major sources of CH_4 emission, although a portion of the CH_4 is oxidized by methane-oxidizing bacteria and archaea in the marine environment (Reeburgh, 2007). Thus, in order to avoid the unregulated emission of CH_4, which is a strong GHG from organic matter, it is necessary to manage both land and marine organic resources to ensure that GHG emissions are reduced. Anaerobic digestion is a promising candidate to achieve this outcome because it enables the recovery of CH_4 from organic resources, its use as renewable energy and finally its release as CO_2. Furthermore, by using marine sediment as a source of salt-tolerant bacteria and

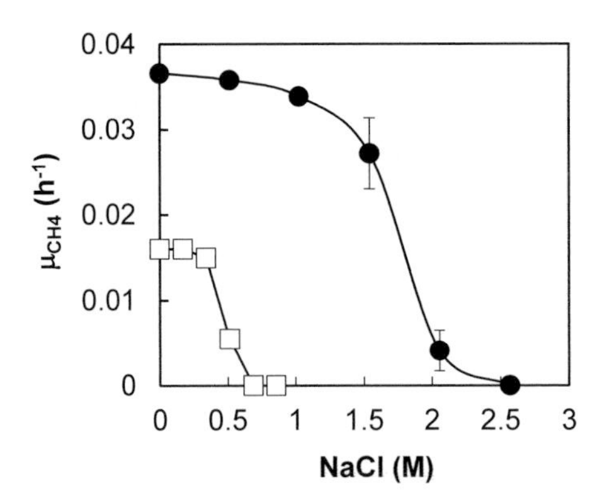

Figure 7. Specific growth rate of *Methanosaeta* sp. H_A and *M. concilii* at various concentrations of NaCl (Kita et al., 2016a). Symbols: Squares, *Methanosaeta concilii* DSM 6752; circles, enriched strain H_A. Cells were cultured at 37 °C in media of pH 7.0 containing 61 mM acetate and 0–2.57 M NaCl. Error bars represent standard deviation of the cultures in triplicate.

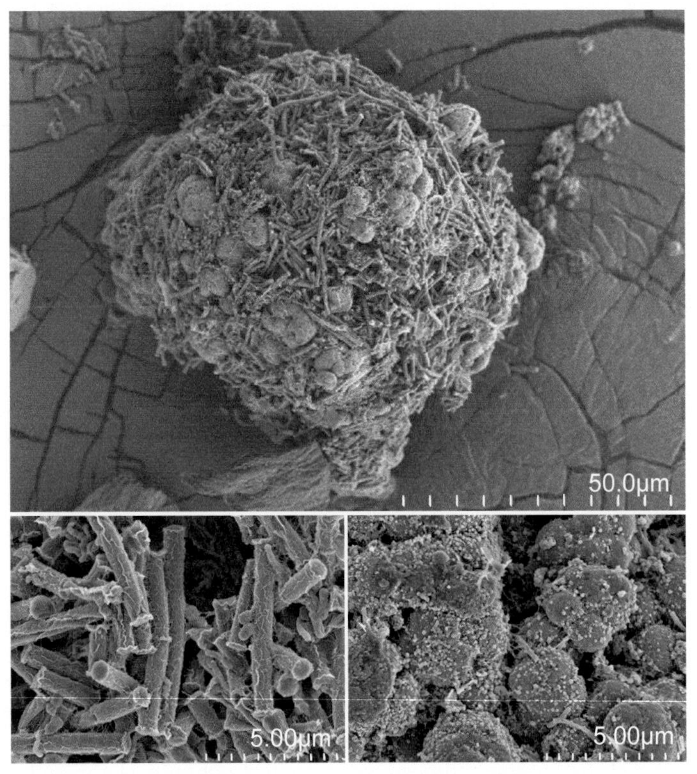

Figure 8. SEM photographs of a granule-like structure (A), filamentous *Methanosaeta*-like cells (B), and coccoid *Methanosarcina*-like cells (C) in the sediment from the fixed-bed reactor (Kita et al., 2016b).

archaea, it is possible to utilize high salt content renewable resources, such as seaweeds, that are difficult to process in conventional anaerobic digestion, and contribute to the reduction of fossil fuel utilization, such as natural gas.

4.2.1 Anaerobic digestion of marine bioresources seaweed and chitin

Abundant elements of marine ecosystems, such as seaweed, are hardly utilized as a food resource and possess low or no lignin content, making them a suitable alternative to land biomass for energy generation (John et al., 2011; Jung et al., 2013; Kita et al., 2016c). Macroalgae are photoauxotrophic and, thus, produce and store organic C by utilizing either CO_2 or HCO_3^- (Gao and McKinley, 1994; Jung et al., 2013). Therefore, seaweed has been receiving increasing attention as a third-generation biomass (Nigam and Singh, 2011). Furthermore, in recent years, studies of life cycle assessment have reported the superiority of seaweed utilization for biofuel production (Aitken et al., 2014; Ertem et al., 2017).

Miura et al. (2014) found that marine sediment has equal or higher activity not only for methanogenesis, but also for hydrolysis/acidogenesis and acetogenesis, when compared with freshwater methanogenic sludge (Table 3). In that study, several marine sediments were evaluated as microbial sources for Anaerobic digestion of *Saccharina japonica*, a brown alga, at seawater salinity. A marine sediment completely converted the produced VFAs to CH_4 within a short period, showing that marine sediments can be used as microbial sources for CH_4 production from algae under high-salt conditions without dilution.

Thus, a salt-tolerant methanogenic microbial community was developed from the marine sediment. Fed-batch cultivation was used to acclimate the salt-tolerant methanogenic microbial community (Miura et al., 2015). When the fed-batch cultivation was conducted by adding dry seaweed at 1 wt% TS per round of cultivation, the CH_4 production rate and salinity level increased 8-fold and 1.6-fold, respectively, at the tenth round of cultivation in which the salinity was equivalent to 5% NaCl.

Miura et al. (2016) also developed the semi-continuous methanogenesis ability of brown algae in halotolerant Anaerobic digestion because it is required in order to produce CH_4 efficiently from macroalgae. After the accumulation of a methanogenic microbial consortium, a semi-continuous culture was conducted. When the culture was started with an organic loading rate of 2.0 g VS (volatile solid)/kg/day (HRT of 39 days), stable CH_4 production was confirmed. When the organic loading rate was raised to 2.9 g VS/kg/day (HRT of 28 days), organic acids partially accumulated. At this time, the salt concentration of the culture solution was about 2%, which was equivalent to the calculated concentration when all ash contained in the substrate was dissolved. These results indicated that the acclimated marine sediment culture was able to produce CH_4 semi-continuously from raw brown algae without dilution and nutrient supply under steady state. The average CH_4 yield was 346 ml/g VS substrate, and the CH_4 production rate was 602 ml/l/day, respectively (Table 4).

In conclusion, it is possible to carry out continuous anaerobic digestion by bacterial consortia of marine salt-tolerant methanogens using non-hydrophobic brown algae, and it is considered that highly efficient anaerobic digestion is possible.

Table 3. Activity of marine sediments at each step of methane production at seawater salinity (Miura et al., 2014).

Microbial source	NaCl	Conversion activity (mg-COD/g-VS/d)				
		Hydrolysis/ acidogenesis	Acetogenesis		Methanogenesis	
			Propionate	Butyrate	H_2-CO_2	Acetate
Mesophilic granule	–	1735	17	184	3250	54
	+	608 (35.0)[a]	ND[b] (0)	27	2940 (90.5)	ND (0)
Marine sediment-1	+	405	79	121	1170	48
Marine sediment-2	+	613	ND	554	3480	102
Marine sediment-3	+	1304	ND	1476	5690	79
Marine sediment-4	+	617	ND	719	1130	45
Marine sediment-5	+	697	35	315	1040	30

[a] Percentage of remaining activity in the presence of 3% NaCl compared to that without NaCl addition.
[b] Not detected during 60 d cultivation.

Table 4. Comparison of medium temperature continuous anaerobic digestion performance of large brown algae.

Fermented materials	Organic load (g-VS/L/d)	HRT (days)	Addition of nutrition	CH_4 yield (ml/g-VS)	CH_4 production rate (ml/l/day)	References
Fresh water dilution conditions						
Laminaria hyperborean	1.65	24	–	280	462	(Hanssen et al., 1987)
Laminaria saccharina	1.65	24	–	230	380	(Hanssen et al., 1987)
Laminaria hyperborea	1.00	20	+	230	230	(Hinks et al., 2013)
Undiluted condition						
Saccharina latissimi	1.73	38	+	238	412	(Jard et al., 2012)
	2.17	33	+	137	297	(Jard et al., 2012)
Saccharina japonica	2.00	39	–	358	695	(Miura et al., 2016)
	2.87	28	–	335	961	(Miura et al., 2016)
	1.74	46	–	346	602	(Miura et al., 2016)

Modified from reference (Miura et al., 2016).

Besides seaweed, chitin is another target to be managed or used as an abundant marine biological resource. Chitin is the main component of crustacean shells, such as crab, shrimp, and crawfish; it is a target to be managed for linear polymer production and it is the second most abundant natural biopolymer on Earth after cellulose (Yan and Chen, 2015). It makes the shortlist of highly abundant biopolymers with enormous global production rates estimated at approximately 10^{10}–10^{11} tons per year (Gooday, 1990; Beier and Bertilsson, 2013). The large amount of chitin-containing crustacean waste generated by fishery processing of shrimp and crab is a promising renewable source estimated to be ~6–8 million tons annually worldwide (Yan and Chen, 2015). However, in developing countries, waste shells are often just dumped in landfill or the sea (Yan and Chen, 2015).

The CH_4 generated from the decomposition of the dumped chitinous waste can be released to the atmosphere. Therefore, management of treatment and effective use of chitinous waste is important. If microorganisms can be used to produce bio-fuels and valuable functional chemicals from chitinous waste, it might be possible to construct chitin biorefineries on a worldwide scale.

In recent years, the use of chitin decomposition products and chitin nanofiber derived from shrimp and crab shells in the health food and medical fields has expanded (Mao et al., 2017; Chen et al., 2016). However, it is extremely difficult to use all of the chitin-containing waste, because of the problems of demand and supply volume and the quality of the waste (impurities, corruption and other problems). Several studies on bio-fuel production from chitin-waste and chitin decomposition products such as N-acetylglucosamine (GlcNac) by microorganisms have also been reported (Table 4). Inokuma et al. (2013) reported that *Mucor circinelloides* NBRC 6746 produced 18.6 ± 0.6 g/l of ethanol from 50 g/l GlcNAc after 72 h and the maximum ethanol production rate was 0.75 ± 0.1 g/l/h. Furthermore, *M. circinelloides* NBRC 4572 produced 6.00 ± 0.22 and 0.46 ± 0.04 g/l of ethanol from 50 g/l of colloidal chitin and chitin powder after 16 and 12 days, respectively (Inokuma et al. 2013). Meanwhile, it was reported that yeast strain *Scheffersomyces stipitis* NBRC 10063 achieved the highest reported GlcNAc consumption (81.2 g/l) and ethanol titer (23.3 g/l) from 100 g/l GlcNAc after 240 h (Inokuma et al., 2016). Hydrogen and CH_4 production in a two-phase anaerobic bioprocess using raw chitin was investigated by Gorrasi et al. (2014). After a preliminary aerobic pre-hydrolysis of the raw chitin by the chitinolytic fungus *Lecanicillium muscarium* CCFEE 5003, hydrogen was obtained by dark fermentation and CH_4 was subsequently produced by further digestion using a bacterial consortium. The highest hydrogen (147 ml/l) and CH_4 (6585 ml/l) production were obtained after 24 days dark fermentation and 30 days of digestion, respectively. Evvyernie et al. (2001) reported that *Clostridium paraputrificum* M-21 could produce 1500 ml/l of hydrogen from chitin after 16 h. For raw lobster shell, the total amount of hydrogen evolved was 395 ml/l as a result of the consumption of 2.6 g of shrimp shell/l. However, hydrogen evolution from the lobster shells ceased after 10 h of cultivation. Thus, development of efficient and stable microbial conversion technology for untreated chitin and chitin containing waste is underway. The largest impediment is that the chitin polysaccharide is not easily decomposable and cannot be degraded by microorganisms generally used industrially, such as *Escherichia coli* and yeast.

In contrast, our previous study found that alginate, which is similarly difficult to decompose like chitin, was easily degraded by bacterial consortia in the marine environment (Kita et al., 2015; Kita et al., 2016c). Furthermore, since no significant accumulation of shrimp and crab shells was found in the marine environment, it was predicted that there would be a highly efficient chitin degradation mechanism in the marine environment by the microbial symbiosis system. In our study, a bacterial consortium capable of anaerobically degrading untreated chitin and chitinous wastes was obtained from marine sediments (unpublished data). The chitin degradation rate of this bacterial consortium increased with each repeated cultivation, and it showed a CH_4 production rate of up to about 90 ml-CH_4/g/d when 5 g/l untreated shrimp shell was used as a substrate. Stable and efficient microbial conversion of "untreated chitin powder and chitinous waste" has not previously been reported. Therefore, it is very interesting to elucidate the mechanism by which marine bacterial consortia can easily degrade the untreated chitin powder and chitin containing crustacean waste. This is because the physiological and ecological findings of the bacterial consortium involved in the degradation of chitin-containing substances, such as shrimp and club shells in the marine anaerobic environment, are obtained. Furthermore, the findings are very important for the development of an effective disposal method for raw chitinous waste and its application to material production.

The elucidation of the symbiotic system and chitin degradation mechanism is a very important subject for the development of an effective disposal method for raw chitinous waste and its application to material production.

6. Concluding Remarks

The management of various organic substances generated in nature and from industry is very important for environmental protection. Furthermore, depletion of fossil resources and concern over global warming from their use is a crucial issue, resulting in the demand for energy reduction and recovery. Many types of anaerobic digesters have been developed to treat many kinds of land and marine organic substances with high organic loading rate. Indeed, the number of industrial anaerobic digestion plants is increasing all over the world. In this context, anaerobic digestion is an attractive technology for C management of such organic substances with cost-effective energy recovery. However, anaerobic digestion is a complicated system, requiring the participation of many kinds of microorganisms with different roles. Thus, operational stability is significantly affected by operation parameters, such as pH, temperature, and concentration of inhibitory compounds. For stable operation of anaerobic digestion and improvement of the process performance, it is necessary to understand not only technologies for plant development and operation, but also microbial physiology and ecology of the process in more detail.

Many types of minerals remain in the residue slurry after anaerobic digestion. The main component of the residue is total N, of which ammonium is a major fraction. Ammonia must be treated prior to disposal, for example by nitrification and denitrification during wastewater treatment; otherwise, it would cause environmental pollution by eutrophication of rivers and lakes. However, this treatment is an energy and cost-intensive process. Because K and phosphate are also present in the residue slurry, the residue can be used as fertilizer as demonstrated in section 4.1.1. Moreover, recycling of ammonia from wastes would be helpful for protecting the environment through decreasing the demand for synthetic chemical fertilizer. Ammonia is mainly produced from fossil fuels by Haber–Bosch synthesis. The total annual production of ammonia was 142 million tons of N in 2017 (http://minerals.usgs.gov/minerals/pubs/commodity/-nitrogen/). Although synthetic chemical fertilizer might currently be cheaper than that recovered from organic wastes, the recycling of C and N in organic substances by means of anaerobic digestion and fertilization should be a promising methodology to protect the environment through management of both energy and pollutants such as C and N.

References

Abbassi-Guendouz, A., Brockmann, D., Trably, E. et al. 2012. Total solids content drives high solid anaerobic digestion via mass transfer limitation. Bioresour. Technol. 111: 55–61.

Abouelenien, F., Namba, Y., Kosseva, M.R., Nishio, N. and Nakashimada, Y. 2014. Enhancement of methane production from co-digestion of chicken manure with agricultural wastes. Bioresour. Technol. 159: 80–87.

Aitken, D., Bulboa, C., Godoy-Faundez, A. and Turrion-Gomez, J.L. 2014. Life cycle assessment of macroalgae cultivation and processing for biofuel production. J. Clean. Prod. 75: 45–56.

André, L., Durante, M., Pauss, A., Lespinard, O., Ribeiro, T. and Lamy, E. 2015. Quantifying physical structure changes and non-uniform water flow in cattle manure during dry anaerobic digestion process at lab scale: Implication for biogas production. Bioresour. Technol. 192: 660–669.

André, L., Pauss, A. and Ribeiro, T. 2018. Solid anaerobic digestion: State-of-art, scientific and technological hurdles. Bioresour. Technol. 247: 1027–1037.

Beier, S. and Bertilsson, S. 2013. Bacterial chitin degradation-mechanisms and ecophysiological strategies. Front. Microbiol. 4: 149.

Benbelkacem, H., Bayard, R., Abdelhay, A., Zhang, Y. and Gourdon, R. 2010. Effect of leachate injection modes on municipal solid waste degradation in anaerobic bioreactor. Bioresour. Technol. 101(14): 5206–5212.

Bollon, J., Benbelkacem, H., Gourdon, R. and Buffière, P. 2013. Measurement of diffusion coefficients in dry anaerobic digestion media. Chem. Eng. Sci. 89: 115–119.

Boone, D.R. and Bryant, M.P. 1980. Propionate-degrading bacterium, *Syntrophobacter wolinii* sp. nov. gen. nov., from methanogenic ecosystems. Appl. Environ. Microbiol. 40(3): 626–632.

Bryant, M.P., Wolin, E.A., Wolin, M.J. and Wolfe, R.S. 1967. Methanobacillus omelianskii, a symbiotic association of two species of bacteria. Archiv für Mikrobiologie 59(1): 20–31.

Buswell, A.M. and Sollo, F.W. 1948. The mechanism of the methane fermentation. J. Am. Chem. Soc. 70(5): 1778–1780.

Chen, X., Yang, H. and Yan, N. 2016. Shell biorefinery: Dream or reality? Chem. Eur. J. 22(38): 13402–13421.

Chiumenti, A., da Borso, F. and Limina, S. 2018. Dry anaerobic digestion of cow manure and agricultural products in a full-scale plant: Efficiency and comparison with wet fermentation. Waste Manage. 71: 704–710.

Chowdhury, P., Viraraghavan, T. and Srinivasan, A. 2010. Biological treatment processes for fish processing wastewater—A review. Bioresour. Technol. 101(2): 439–449.

De Mes, T.Z.D., Stams, A.J.M., Reith, J.H. and Zeeman, G. 2003. Methane production by anaerobic digestion of wastewater and solid waste. pp. 58–102. *In*: Reith, J.H., Wijffels, R.H. and Barten, H. (eds.). Bio-methane and Bio-hydrogen. Petten: Dutch Biological Hydrogen Foundation.

De Vrieze, J., Hennebel, T., Boon, N. and Verstraete, W. 2012. Methanosarcina: The rediscovered methanogen for heavy duty biomethanation. Bioresour. Technol. 112: 1–9.

Degueurce, A., Tomas, N., Le Roux, S., Martinez, J. and Peu, P. 2016. Biotic and abiotic roles of leachate recirculation in batch mode solid-state anaerobic digestion of cattle manure. Bioresour. Technol. 200: 388–395.

Demirel, B. and Scherer, P. 2008. The roles of acetotrophic and hydrogenotrophic methanogens during anaerobic conversion of biomass to methane: A review. Rev. Environ. Sci. Bio. 7(2): 173–190.

Demirer, G.N. and Chen, S. 2005. Two-phase anaerobic digestion of unscreened dairy manure. Process Biochem. 40(11): 3542–3549.

Ertem, F.C., Neubauer, P. and Junne, S. 2017. Environmental life cycle assessment of biogas production from marine macroalgal feedstock for the substitution of energy crops. J. Clean. Prod. 140: 977–985.

Evvyernie, D., Morimoto, K., Karita, S., Kimura, T., Sakka, K. and Ohmiya, K. 2001. Conversion of chitinous wastes to hydrogen gas by *Clostridium paraputrificum* M-21. J. Biosci. Bioeng. 91(4): 339–43.

Fernandes, T.V., Keesman, K.J., Zeeman, G. and van Lier, J.B. 2012. Effect of ammonia on the anaerobic hydrolysis of cellulose and tributyrin. Biomass Bioenerg. 47: 316–323.

Gallert, C., Bauer, S. and Winter, J. 1998. Effect of ammonia on the anaerobic degradation of protein by a mesophilic and thermophilic biowaste population. Appl. Microbiol. Biotechnol. 50(4): 495–501.

Gao, K. and McKinley, K. 1994. Use of macroalgae for marine biomass production and CO_2 remediation: A review. J. Appl. Phycol. 6: 45–60.

Garcia-Bernet, D., Loisel, D., Guizard, G., Buffiere, P., Steyer, J.P. and Escudie, R. 2011. Rapid measurement of the yield stress of anaerobically-digested solid waste using slump tests. Waste Manage. 31(4): 631–635.

Garcia-Bernet, D., Buffiere, P., Latrille, E., Steyer, J.P. and Escudie, R. 2011. Water distribution in biowastes and digestates of dry anaerobic digestion technology. Chem. Eng. J. 172(2-3): 924–928.

Ge, X., Xu, F. and Li, Y. 2016. Solid-state anaerobic digestion of lignocellulosic biomass: Recent progress and perspectives. Bioresour. Technol. 205: 239–49.

Goberna, M., Insam, H. and Franke-Whittle, I.H. 2009. Effect of biowaste sludge maturation on the diversity of thermophilic bacteria and archaea in an anaerobic reactor. Appl. Environ. Microbiol. 75(8): 2566–2572.

Gooday, G.W. 1990. The ecology of chitin degradation. *In*: Marshall, K.C. (ed.). Advances in Microbial Ecology. Boston, MA: Springer US.

Gorrasi, S., Izzo, G., Massini, G., Signorini, A., Barghini, P. and Fenice, M. 2014. From polluting seafood wastes to energy. Production of hydrogen and methane from raw chitin material by a two-phase process. J. Environ. Prot. Ecol. 15(2): 526–536.

Goswami, R., Chattopadhyay, P., Shome, A., Banerjee, S.N., Chakraborty, A.K. Mathew, A.K. and Chaudhury, S. 2016. An overview of physico-chemical mechanisms of biogas production by microbial communities: A step towards sustainable waste management. 3 Biotech. 6: 72.

Guendouz, J., Buffiere, P., Cacho, J., Carrere, M. and Delgenes, J.P. 2008. High-solids anaerobic digestion: Comparison of three pilot scales. Water Sci. Technol. 58(9): 1757–1763.

Gunaseelan, V.N. 1997. Anaerobic digestion of biomass for methane production: A review. Biomass Bioenergy 13: 83–114.

Hanssen, J.F., Indergaard, M., Østgaard, K., Bævre, O.A., Pedersen, T.A. and Jensen, A. 1987. Anaerobic digestion of *Laminaria* spp. and *Ascophyllum nodosum* and application of end products. Biomass 14(1): 1–13.

Hao, L.P., Lü, F., He, P.J., Li, L. and Shao, L.M. 2011. Predominant contribution of syntrophic acetate oxidation to thermophilic methane formation at high acetate concentrations. Environ. Sci. Technol. 45(2): 508–513.

He, T.H., Liu, D.Y., Yuan, J.J. et al. 2018. A two years study on the combined effects of biochar and inhibitors on ammonia volatilization in an intensively managed rice field. Agric. Ecosyst. Environ. 264: 44–53.

Heijnen, J.J., Mulder, A., Enger, W. and Hoeks, F. 1989. Review on the application of anaerobic fluidized bed reactors in waste-water treatment. Chem. Eng. J. 41(3): B37–B50.

Hinks, J., Edwards, S., Sallis, P.J. and Caldwell, G.S. 2013. The steady state anaerobic digestion of *Laminaria hyperborea*—effect of hydraulic residence on biogas production and bacterial community composition. Bioresour. Technol. 143: 221–30.

Hosomi, M., Riya, S., Toyoda, K. and Akisawa, A. 2014. Development of low-environmental-impact co-benefit system by piggery wastewater treatment and high-yielding forage rice production (1B-1103) (in Japanese). In: Environment Research and Technology Development Fund Report.

Iino, H., Zhou, S., Sagehashi, M. et al. 2011. Thermophilic anaerobic digestion characteristics of forage rice. Environ. Sci. (in Japanese) 24: 462–471.

Inokuma, K., Takano, M. and Hoshino, K. 2013. Direct ethanol production from N-acetylglucosamine and chitin substrates by *Mucor* species. Biochem. Eng. J. 72: 24–32.

Inokuma, K., Hasunuma, T. and Kondo, A. 2016. Ethanol production from N-acetyl-D-glucosamine by *Scheffersomyces stipitis* strains. AMB Express 6(1): 83.

Jard, G., Jackowiak, D., Carrere, H. et al. 2012. Batch and semi-continuous anaerobic digestion of Palmaria palmata: Comparison with Saccharina latissima and inhibition studies. Chem. Eng. J. 209: 513–519.

Jeris, J.S. 1983. Industrial wastewater treatment using anaerobic fluidized bed reactors. Water Sci. Technol. 15(8): 169–176.

Jewell, W.J., Switzenbaum, M.S. and Morris, J.W. 1981. Municipal wastewater-treatment with the anaerobic attached microbial film expanded bed process. J. Water Pollut. Control Fed. 53(4): 482–490.

John, R.P., Anisha, G.S., Nampoothiri, K.M. and Pandey, A. 2011. Micro and macroalgal biomass: A renewable source for bioethanol. Bioresour. Technol. 102: 186–193.

Jung, K.A., Lim, S.R., Kim, Y. and Park, J.M. 2013. Potentials of macroalgae as feedstocks for biorefinery. Bioresour. Technol. 135: 182–190.

Kadam, K.L., Forrest, L.H. and Jacobson, W.A. 2000. Rice straw as a lignocelulosic resource: Collection, processing, transportation, and environmental aspects. Biomass Bioen. 18: 369–389.

Kato, H. 2008. Development of rice varieties for Whole Crop Silage (WCS) in Japan. Jpn. Agric. Res. Q 42(4): 231–236.

Kato, M.T., Field, J.A., Versteeg, P. and Lettinga, G. 1994. Feasibility of expanded granular sludge bed reactors for the anaerobic treatment of low-strength soluble wastewaters. Biotechnol. Bioeng. 44 (4): 469–479.

Kita, A., Miura, T., Okamura, Y. et al. 2015. *Dysgonomonas alginatilytica* sp. nov., an alginate-degrading bacterium isolated from a microbial consortium. Int. J. Syst. Evol. Microbiol. 65: 3570–3575.

Kita, A., Suehira, K., Miura, T. et al. 2016a. Characterization of a halotolerant acetoclastic methanogen highly enriched from marine sediment and its application in removal of acetate. J. Biosci. Bioeng. 121(2): 196–202.

Kita, A., Kobayashi, K., Miura, T. et al. 2016b. High-rate fermentation of acetate to methane under saline condition by aceticlastic methanogens immobilized in marine sediment. J. Jpn. Petrol. Inst. 59(1): 9–15.

Kita, A., Miura, T., Kawata, S. et al. 2016c. Bacterial community structure and predicted alginate metabolic pathway in an alginate degrading bacterial consortium. J. Biosci. Bioeng. 121(3): 286–292.

Koeck, D.E., Pechtl, A., Zverolv, V.V. and Schwarz, W.H. 2014. Genomics of cellulolytic bacteria. Curr. Opin. Biotechnol. 29: 171–183.

Krumholz, L.R. and Bryant, M.P. 1986. *Syntrophococcus sucromutans* sp. nov. gen. nov. uses carbohydrates as electron donors and formate, methoxymonobenzenoids or *Methanobrevibacter* as electron acceptor systems. Arch. Microbiol. 143(4): 313–318.

Laird, D.A., Brown, R.C., Amonette, J.E. and Lehmann, J. 2009. Review of the pyrolysis platform for coproducing bio-oil and biochar. Biofuel. Bioprod. Bior. 3(5): 547–562.

Lay, J.J., Li, Y.Y., Noike, T., Endo, J. and Ishimoto, S. 1997. Analysis of environmental factors affecting methane production from high-solids organic waste. Water Sci. Technol. 36(6-7): 493–500.

Le Hyaric, R., Benbelkacem, H., Bollon, J., Bayard, R., Escudie, R. and Buffiere, P. 2012. Influence of moisture content on the specific methanogenic activity of dry mesophilic municipal solid waste digestate. J. Chem. Technol. Biotechnol. 87(7): 1032–1035.
Le Mer, J. and Roger, P. 2001. Production, oxidation, emission and consumption of methane by soils: A review. Eur. J. Soil Biol. 37(1): 25–50.
Lefebvre, O. and Moletta, R. 2006. Treatment of organic pollution in industrial saline wastewater: A literature review. Water Res. 40(20): 3671–3682.
Lettinga, G., van Velsen, A.F.M., Hobma, S.W., de Zeeuw, W. and Klapwijk, A. 1980. Use of the upflow sludge blanket (USB) reactor concept for biological wastewater treatment, especially for anaerobic treatment. Biotechnol. Bioeng. 22(4): 699–734.
Li, Y., Zhu, J., Wan, C. and Park, S.Y. 2011. Solid-state anaerobic digestion of corn stover for biogas production. Trans. Asabe 54(4): 1415–1421.
Liu, H., Wang, J., Liu, X., Fu, B., Chen, J. and Yu, H.-Q. 2012. Acidogenic fermentation of proteinaceous sewage sludge. Water Res. 46(3): 799–807.
Ma, K., Liu, X. and Dong, X. 2006. *Methanosaeta harundinacea* sp. nov., a novel acetate-scavenging methanogen isolated from a UASB reactor. Int. J. Syst. Evol. Microbiol. 56(1): 127–131.
Mackie, R.I. and Bryant, M.P. 1981. Metabolic activity of fatty acid-oxidizing bacteria and the contribution of acetate, propionate, butyrate, and CO_2 to methanogenesis in cattle waste at 40 and 60 °C. Appl. Environ. Microbiol. 41(6): 1363–1373.
Mao, X., Guo, N., Sun, J. and Xue, C. 2017. Comprehensive utilization of shrimp waste based on biotechnological methods: A review. J. Clean Prod. 143: 814–823.
Mclnerney, M.J., Bryant, M.P., Hespell, R.B. and Costerton, J.W. 1981. *Syntrophomonas wolfei* gen. nov. sp. nov., an anaerobic, syntrophic, fatty acid-oxidizing bacterium. Appl. Environ. Microbiol. 41(4): 1029–1039.
Meng, L., Xie, L., Riya, S., Terada, A. and Hosomi, M. 2019a. Impact of turning waste on performance and energy balance in thermophilic solid-state anaerobic digestion of agricultural waste. Waste Manage. 87: 183–191.
Meng, L., Maruo, K., Xie, L., Riya, S., Terada, A. and Hosomi, M. 2019b. Comparison of leachate percolation and immersion using different inoculation strategies in thermophilic solid-state anaerobic digestion of pig urine and rice straw. Bioresour. Technol. 277: 216–220.
Meng, L.Y., Xie, L., Kinh, C.T. et al. 2018. Influence of feedstock-to-inoculum ratio on performance and microbial community succession during solid-state thermophilic anaerobic co-digestion of pig urine and rice straw. Bioresour. Technol. 252: 127–133.
Ministry of Agriculture, Forestry, and Fisheries (MAFF). 2013a. Production cost of rice, Statistics of agricultural production cost (in Japanese). Available from http://www.maff.go.jp/j/tokei/kouhyou/noukei/seisanhi_nousan/index.html#r3.
MAFF. 2013b. Profitability of rowing-finishing pig farm (in Japanese), Statistics of livestock production cost (in Japanese). Available from http://www.maff.go.jp/j/tokei/kouhyou/noukei/seisanhi_tikusan/#r.
MAFF. 2019. Situation of feed (in Japanese). Available from http://www.maff.go.jp/j/chikusan/sinko/lin/l_siryo/.
Miura, T., Kita, A., Okamura, Y. et al. 2014. Evaluation of marine sediments as microbial sources for methane production from brown algae under high salinity. Bioresour. Technol. 169: 362–366.
Miura, T., Kita, A., Okamura, Y. et al. 2015. Improved methane production from brown algae under high salinity by fed-batch acclimation. Bioresour. Technol. 187: 275–281.
Miura, T., Kita, A., Okamura, Y. et al. 2016. Semi-continuous methane production from undiluted brown algae using a halophilic marine microbial community. Bioresour. Technol. 200: 616–23.
Mora-Naranjo, N., Meima, J.A., Haarstrick, A. and Hempel, D.C. 2004. Modelling and experimental investigation of environmental influences on the acetate and methane formation in solid waste. Waste Manage. 24(8): 763–773.
Mori, K., Iino, T., Suzuki, K., Yamaguchi, K. and Kamagata, Y. 2012. Aceticlastic and NaCl-requiring methanogen *Methanosaeta pelagica* sp. nov., isolated from marine tidal flat sediment. Appl. Environ. Microbiol. 78(9): 3416–23.
Mountfort, D.O. and Asher, R.A. 1978. Changes in proportions of acetate and carbon dioxide used as methane precursors during the anaerobic digestion of bovine waste. Appl. Environ. Microbiol. 35(4): 648–54.
Mountfort, D.O., Brulla, W.J., Krumholz, L.R. and Bryant, M.P. 1984. *Syntrophus buswellii* gen. nov., sp. nov.: A benzoate catabolizer from methanogenic ecosystems. Int. J. Syst. Evol. Microbiol. 34(2): 216–217.
Nakashimada, Y., Ohshima, Y., Minami, H., Yabu, H., Namba, Y. and Nishio, N. 2008. Ammonia-methane two-stage anaerobic digestion of dehydrated waste-activated sludge. Appl. Microbiol. Biotechnol. 79(6): 1061–1069.
Nakashimada, Y. and Nishio, N. 2017. Energy recovery from organic waste. pp. 545–572. *In*: Yoshida, T. (ed.). Applied Bioengineering: Innovations and Future Directions. Weinheim: Wiley-VCH Verlag GmbH & Co. KGaA.
Nizami, A.S. and Murphy, J.D. 2010. What type of digester configurations should be employed to produce biomethane from grass silage? Renew. Sust. Energ. Rev. 14(6): 1558–1568.
Nkemka, V.N. and Murto, M. 2010. Evaluation of biogas production from seaweed in batch tests and in UASB reactors combined with the removal of heavy metals. J. Environ. Manage. 91(7): 1573–1579.
Ollivier, B., Caumette, P., Garcia, J.L. and Mah, R.A. 1994. Anaerobic bacteria from hypersaline environments. Microbiol. Rev. 58(1): 27–38.

Oren, A. 1999. Bioenergetic aspects of halophilism. Microbiol. Mol. Biol. Rev. 63(2): 334–348.

Otani, R. 2012. Rice straw rapid drying harvesting system utilizing conventional combine (in Japanese). J. Jpn. Soc. Agric. Mach. 74: 348–352.

Paterek, J.R. and Smith, P.H. 1988. *Methanohalophilus mahii* gen. nov., sp. nov., a methylotrophic halophilic methanogen. Int. J. Syst. Evol. Microbiol. 38(1): 122–123.

Paul, E.A. 2007. Soil microbiology ecology and biochemistry in perspective. pp. 3–24. *In*: Paul, E.A. (ed.). Soil Microbiology, Ecology, and Biochemistry (3rd edition). Academic Press.

Pommier, S., Chenu, D., Quintard, M. and Lefebvre, X. 2007. A logistic model for the prediction of the influence of water on the solid waste methanization in landfills. Biotechnol. Bioeng. 97(3): 473–482.

Rajagopal, R., Masse, D.I. and Singh, G. 2013. A critical review on inhibition of anaerobic digestion process by excess ammonia. Bioresour. Technol. 143: 632–641.

Reeburgh, W.S. 2007. Oceanic methane biogeochemistry. Chem. Rev. 107: 486–513.

Rinzema, A., van Lier, J. and Lettinga, G. 1988. Sodium inhibition of acetoclastic methanogens in granular sludge from a UASB reactor. Enzyme Microb. Technol. 10(1): 24–32.

Riya, S., Katayama, M., Takahashi, E., Zhou, S., Terada, A. and Hosomi, M. 2014. Mitigation of greenhouse gas emissions by water management in a forage rice paddy field supplemented with dry-thermophilic anaerobic digestion residue. Water Air Soil Pollut. 225(9): 2118.

Riya, S., Suzuki, K., Terada, A. and Hosomi, M. 2016. Influence of C/N ratio on performance and microbial community structure of dry-thermophilic anaerobic co-digestion of swine manure and rice straw. J. Med. Bioeng. 5(1): 11–14.

Riya, S., Sawayanagi, K., Suzuki, K., Zhou, S., Terada, A. and Hosomi, M. 2017. Digestibility of riverbed plants by dry-thermophiilc anaerobic digestion (in Japanese). Kagaku Kogaku Ronbun 43: 224–230.

Riya, S., Suzuki, K., Meng, L., Zhou, S., Terada, A. and Hosomi, M. 2018. The influence of the total solid content on the stability of dry-thermophilic anaerobic digestion of rice straw and pig manure. Waste Manage. 76: 350–356.

Roesijadi, G., Jones, S.B., Snowden-Swan, L.J. and Zhu, Y. 2010. Macroalgae as a Biomass Feedstock: A Preliminary Analysis. United States: Department of Energy, USA.

Roy, F., Samain, E., Dubourguier, H.C. and Albagnac, G. 1986. *Synthrophomonas sapovorans* sp. nov., a new obligately proton reducing anaerobe oxidizing saturated and unsaturated long chain fatty acids. Arch. Microbiol. 145(2): 142–147.

Sanders, W.T.M., Geerink, M., Zeeman, G. and Lettinga, G. 2000. Anaerobic hydrolysis kinetics of particulate substrates. Water Sci. Technol. 41(3): 17–24.

Sasaki, D., Hori, T., Haruta, S., Ueno, Y., Ishii, M. and Igarashi, Y. 2011. Methanogenic pathway and community structure in a thermophilic anaerobic digestion process of organic solid waste. J. Biosci. Bioeng. 111(1): 41–46.

Schink, B. 1997. Energetics of syntrophic cooperation in methanogenic degradation. Microbiol. Mol. Biol. Rev. 61(2): 262–280.

Seitz, H.-J., Conrad, R. and Schink, B. 1988. Thermodynamics of hydrogen metabolism in methanogenic cocultures degrading ethanol or lactate. FEMS Microbiol. Lett. 55(2): 119–124.

Seghezzo, L., Zeeman, G., van Lier, J.B., Hamelers, H.V.M. and Lettinga, G. 1998. A review: The anaerobic treatment of sewage in UASB and EGSB reactors. Bioresour. Technol. 65: 175–190.

Shi, X., Lefebvre, O., Ng, K.K. and Ng, H.Y. 2014. Sequential anaerobic–aerobic treatment of pharmaceutical wastewater with high salinity. Bioresour. Technol. 153: 79–86.

Siebert, M.L. and Toerien, D.F. 1969. The proteolytic bacteria present in the anaerobic digestion of raw sewage sludge. Water Res. 3(4): 241–250.

Smith, K.S. and Ingram-Smith, C. 2007. Methanosaeta, the forgotten methanogen? Trends Microbiol. 15(4): 150–155.

Smith, P.H. and Mah, R.A. 1966. Kinetics of acetate metabolism during sludge digestion. Appl. Microbiol. 14(3): 368–371.

Sowers, K.R. and Ferry, J.G. 1983. Isolation and characterization of a methylotrophic marine methanogen, *Methanococcoides methylutens* gen. nov., sp. nov. Appl. Environ. Microbiol. 45(2): 684–90.

Sowers, K.R., Baron, S.F. and Ferry, J.G. 1984. *Methanosarcina acetivorans* sp. nov., an acetotrophic methane-producing bacterium isolated from marine sediments. Appl. Environ. Microbiol. 47(5): 971–978.

Spanheimer, R. and Müller, V. 2008. The molecular basis of salt adaptation in *Methanosarcina mazei* Gö1. Arch. Microbiol. 190(3): 271.

Stronach, S.M., Rudd, T. and Lester, J.N. 1986. Anaerobic Digestion Processes in Industrial Wastewater Treatment. Springer-Verlag.

Switzenbaum, M.S. and Jewell, W.J. 1980. Anaerobic attached-film expanded-bed reactor treatment. J. Water Pollut. Control. Fed. 52(7): 1953–1965.

Tang, Y.Q., Ji, P., Hayashi, J., Koike, Y., Wu, X.L. and Kida, K. 2011. Characteristic microbial community of a dry thermophilic methanogenic digester: Its long-term stability and change with feeding. Appl. Microbiol. Biotechnol. 91(5): 1447–1461.

Tsunekawa, I. 2016. Problems and possibilities for reducing expenses in the circulation and processing of rice for feed (in Japanese). Jpn. J. Farm Manage. 53: 6–16.

Vavilin, V.A., Rytov, S.V. and Ya Lokshina, L. 1996. A description of hydrolysis kinetics in anaerobic degradation of particulate organic matter. Bioresour. Technol. 56(2): 229–237.

Wang, L., Li, L.Q., Cheng, K. et al. 2018. An assessment of emergy, energy, and cost-benefits of grain production over 6 years following a biochar amendment in a rice paddy from China. Environ. Sci. Pollut. Res. 25(10): 9683–9696.
Wang, T., Camps-Arbestain, M., Hedley, M. and Bishop, P. 2012a. Predicting phosphorus bioavailability from high-ash biochars. Plant Soil 357(1-2): 173–187.
Wang, X., Yang, G., Feng, Y., Ren, G. and Han, X. 2012b. Optimizing feeding composition and carbon-nitrogen ratios for improved methane yield during anaerobic co-digestion of dairy, chicken manure and wheat straw. Bioresour. Technol. 120: 78–83.
Wang, Z., Xu, F. and Li, Y. 2013. Effects of total ammonia nitrogen concentration on solid-state anaerobic digestion of corn stover. Bioresour. Technol. 144: 281–287.
Wassmann, R., Wang, M.X., Shangguan, X.J. et al. 1993. 1st records of a field experiment on fertilizer effects on methane emission from rice fields in Hunan-province (peoples-republic-of-China). Geophys. Res. Lett. 20(19): 2071–2074.
Watanabe, A., Takeda, T. and Kimura, M. 1999. Evaluation of origins of CH_4 carbon emitted from rice paddies. J. Geophys. Res. Atmos. 104(D19): 23623–23629.
Weiland, P. 2010. Biogas production: Current state and perspectives. Appl. Microbiol. Biotechnol. 85(4): 849–60.
Wiegant, W.M. and Zeeman, G. 1986. The mechanism of ammonia inhibition in the thermophilic digestion of livestock wastes. Agric. Wastes 16(4): 243–253.
Woolf, D., Amonette, J.E., Street-Perrott, F.A., Lehmann, J. and Joseph, S. 2010. Sustainable biochar to mitigate global climate change. Nat. Commun. 1.
Xu, F., Wang, Z.W. and Li, Y. 2014. Predicting the methane yield of lignocellulosic biomass in mesophilic solid-state anaerobic digestion based on feedstock characteristics and process parameters. Bioresour. Technol. 173: 168–176.
Yagi, K. and Minami, K. 1990. Effect of organic-matter application on methane emission from some japanese paddy fields. Soil Sci. Plant Nutr. 36(4): 599–610.
Yan, N. and Chen, X. 2015. Sustainability: Don't waste seafood waste. Nature 524(7564): 155–7.
Yang, S.L., Tang, Y.Q., Gou, M. and Jiang, X. 2015. Effect of sulfate addition on methane production and sulfate reduction in a mesophilic acetate-fed anaerobic reactor. Appl. Microbiol. Biotechnol. 99(7): 3269–3277.
Yang, S., Xiao, Y., Sun, X., Ding, J., Jiang, Z. and Xu, J. 2019. Biochar improved rice yield and mitigated CH_4 and N_2O emissions from paddy field under controlled irrigation in the Taihu Lake Region of China. Atmos. Environ. 200: 69–77.
Yenigün, O. and Demirel, B. 2013. Ammonia inhibition in anaerobic digestion: A review. Process Biochem. 48: 901–911.
Young, J.C. and McCarty, P.L. 1969. The anaerobic filter for waste treatment. J. Water Pollut. Control Fed. 41(5): 160–173.
Zhang, Y., Li, H., Cheng, Y. and Liu, C. 2016. Influence of solids concentration on diffusion behavior in sewage sludge and its digestate. Chem. Eng. Sci. 152: 674–677.
Zhilina, T.N. and Zavarzin, G.A. 1990. Extremely halophilic, methylotrophic, anaerobic bacteria. FEMS Microbiol. Lett. 87(3): 315–321.
Zhou, S., Nikolausz, M., Zhang, J.N., Riya, S., Terada, A. and Hosomi, M. 2016. Variation of the microbial community in thermophilic anaerobic digestion of pig manure mixed with different ratios of rice straw. J. Biosci. Bioeng. 122(3): 334–340.
Zhou, Y., Li, C., Nges, I.A. and Liu, J. 2017. The effects of pre-aeration and inoculation on solid-state anaerobic digestion of rice straw. Bioresour. Technol. 224: 78–86.
Zhu, J., Yang, L. and Li, Y. 2015. Comparison of premixing methods for solid-state anaerobic digestion of corn stover. Bioresour. Technol. 175: 430–435.

CHAPTER 9

Critical Aspects in Developing Sustainable Biorefinery Systems Based on Bioelectrochemical Technology with Carbon Dioxide Capture

Jhuma Sadhukhan

1. Introduction

Biorefineries utilizing residues and waste materials that are inherently heterogeneous in characteristics to produce multiple added value products need to be actively sought for sustainable development to fulfil societal demands for goods and services. Such integrated biorefineries are yet to be realized. The main reason why a biorefinery was not successfully deployed is the lack of favorable economics that could be due to many reasons, in particular, small quantity of biomass availability, its heterogeneity and low energy density. Many start-up facilities have been abandoned due to lack of favorable finance (Lane, 2017). Fossil-based economy is still the key driver of macroeconomics (Net Zero 2050, 2019). Serious de-fossilization of carbon is, therefore, urgently sought by alternative sustainable biorefinery systems.

Biomass is considered as carbon neutral or to be a source of renewable carbon alternative to environmentally damaging finite fossil resources, coal, crude oil and natural gas. During biomass growth, carbon is sequestrated or captured and the embedded carbon is released during its use, forming a full carbon cycle. Cereals, crops, starch and sugars can be fermented to produce bioethanol. Oily crops, such as vegetable oils, can be transesterified to biodiesel. However, these feedstocks have indirect land use issues. Indirect land use change has been defined as the displaced land for food and feed production by cultivation of crops for "biofuels, bioliquids and biomass fuels" (Directive 2009/28/EC). These feedstocks are not sustainable in long term due to their indirect land use change due to potential competition with food and feed. Lignocelluloses (agricultural, forestry and garden wastes and residues) made up of lignin, cellulose and hemicellulose are present in plant cell walls available as residues of food or feed and, thus, more suitable for biorefinery utilization (Sadhukhan et al., 2014). Lignocelluloses can, after physical or mechanical operations, be thermochemically or biochemically processed if the moisture content is less than 10% or biochemically processed when the moisture content is less than 25% by mass to produce chemical or biofuel, along with combined heat and power (CHP) (Wan et al., 2016a; Wan et al., 2016b). Biomass feedstocks, after drying and pretreatment, undergo the main valorization step thermochemical conversion (combustion, pyrolysis, gasification, torrefaction, hydrothermal liquefaction,

University of Surrey, UK.

hydro-processing, etc.), or biochemical conversion (fermentation), followed by separation and water and energy recovery steps (Sadhukhan et al., 2018). The biorefinery process should be able to meet its energy and resource demands by systematic process integration (Sadhukhan et al., 2014). This is essential in order for a biorefinery to be economically viable. However, full application of process integration in a biorefinery for heat recovery and energy self-sufficiency is limited by the scale of operation of the biorefinery. The scale of the biorefinery operation is in turn constrained by the availability, quality and energy density of biomass and level of financing. Thus, these important aspects need to be holistically, rigorously and systematically analyzed in order to inform decision makers.

Some technologies, such as anaerobic digestion, have received widespread attention from the government and investors (Sadhukhan, 2014). Electricity produced from biogas from anaerobic digestion of organic residues and wastes receives feed-in-tariff. Bioethanol blending has become mandatory in many countries in order to ensure penetration of fermentation technologies (IEA Bioenergy, 2016). Policy can, therefore, help biorefinery development in two ways: Mandatory blending or energy mix and incentives or subsidies. Policy can also help by imposing a carbon tax on fossil processing technologies (Karstensen and Peters, 2018). However, these are not yet adequate to curb carbon dioxide from the atmosphere. Amongst all the greenhouse gases, carbon dioxide is responsible for 80% of climate change impact (IPCC, 2018). Thus, the recent thrust from the Intergovernmental Panel on Climate Change (IPCC) is on the technologies showing promises of capturing carbon dioxide directly from the atmosphere (IPCC, 2018). These are the emerging technologies actively sought in this chapter.

Earlier works have considered reuse of CO_2 rich streams in chemical reactions to produce syngas, hydrogen, formic acid, methane, ethylene, methanol, dimethyl ether, urea, Fischer-Tropsch liquid, succinic acid, etc., as well as other materials, such as epoxides acetals and orthoesters, as important precursors to many polymers and carbonates through mineralization reactions (Ng et al., 2013). CO_2 as a reactant in Sabatier's reaction for methane formation, in reactions to produce calcite and succinic acid and CO_2 reuse for the growth of algae to produce biofuels (e.g., bioethanol and biodiesel), have been analyzed for techno-economic viability (Sadhukhan et al., 2015). CO_2 reduction reaction examples using microbial electrosynthesis are the new concepts being developed in this area. Lower hydrocarbons, such as methane, acetic acid, methanol, formaldehyde and formic acid, are being investigated. The reactions in Equations (1–5) to produce these molecules show spontaneous reactions for methane, acetic acid and methanol production. $\Delta G^{o'}_{r,cathode}$ is the Gibbs free energy for the cathode side reaction under standard conditions (25 °C temperature and 1 atm pressure) and pH 7 (Sadhukhan et al., 2016).

$$HCO_3^- + H^+ + 4H_2 \rightarrow CH_4 + 3H_2O \qquad \Delta G^{o'}_{r,cathode} = -135.6\ kJ \tag{1}$$

$$2HCO_3^- + 2H^+ + 4H_2 \rightarrow CH_3COOH + 4H_2O \qquad \Delta G^{o'}_{r,cathode} = -64.6\ kJ \tag{2}$$

$$HCO_3^- + H^+ + 3H_2 \rightarrow CH_3OH + 2H_2O \qquad \Delta G^{o'}_{r,cathode} = -23\ kJ \tag{3}$$

$$HCO_3^- + H^+ + 2H_2 \rightarrow HCHO + 2H_2O \qquad \Delta G^{o'}_{r,cathode} = 21.8\ kJ \tag{4}$$

$$HCO_3^- + H^+ + H_2 \rightarrow HCOOH + H_2O \qquad \Delta G^{o'}_{r,cathode} = 38.5\ kJ \tag{5}$$

Medium chain fatty acids used as biofuels, such as caproate and caprylate, can also be produced from acetate at a biocathode using mixed microbial cultures (where *Clostridium kluyveri* was the predominant microorganism). The cathode reactions to caproate and butyrate are shown in Equations (6) and (7). Negative Gibbs free energies for the production of caproate and butyrate also imply spontaneous reactions.

$$3C_2H_3O_2^- + 2H^+ + 4H_2 \rightarrow Caproate(C_6H_{11}O_2^-) + 4H_2O \qquad \Delta G^{o'}_{r,cathode} = -96.6\ kJ \tag{6}$$

$$2C_2H_3O_2^- + H^+ + 2H_2 \rightarrow Butyrate(C_4H_7O_2^-) + 2H_2O \qquad \Delta G^{o'}_{r,cathode} = -48.3\ kJ \tag{7}$$

In principle, the recovery of aromatic compounds and cyclic alkanes from organic wastes is possible. Similar to the cathode chamber, the anode chamber can also generate a number of products, bioethanol or biofuel or bulk chemical, however, relying on the specific substrate, e.g., glucose, acetate, glycerol, etc. (Sadhukhan et al., 2016). Also, polymers can be decomposed into pyruvate, formate and fatty acids by microbial oxidation in the anode chamber. Microbial electrosynthesis has been investigated to produce formate by carbon dioxide utilization in the cathode by electron and proton harvesting through wastewater treatment in the anode (Shemfe et al., 2018).

In a microbial electrosynthesis process, electrons are harvested by substrate oxidation in anode and taken up after flowing through an external circuit in cathode, reducing carbon dioxide to an organic product. By the method of carbon dioxide reduction reaction and reuse, microbial electrosynthesis shows the promises of carbon dioxide capture and reuse from the atmosphere. Thus, microbial electrosynthesis can be an important technology for consideration to meet the carbon dioxide reduction target recommended by the IPCC (IPCC, 2018).

Carbon dioxide is a stable molecule. Its thermodynamic stability means significant energy input in reacting with hydrogen to make a product. The production of hydrogen is also energy intensive. All these indicate that the only way carbon dioxide capture and reuse from the atmosphere will be feasible is if the energy requirement is fulfilled by renewable energy. In order to make the finance work, the product should have a high market value so as to offset the cost of energy to produce hydrogen as the reducing agent for carbon dioxide. Moreover, microbial electrosynthesis technologies are at the nascent stage of development and this means that both the scale of operation and energy cost will not favor the finance for investment in the technology.

Microbial electrosynthesis needs to embody integrated preprocessing and post-processing product separation and purification unit operations as well as utility systems for heat recovery in order to demonstrate economic viability. Such a complete conceptual process synthesis should follow the twelve principles of green chemistry as well as process integration. Both the aspects are under-exploited or have not been actively evaluated because the main challenge has been to make the technology work as a proof of concept. Thus, this chapter focuses on fundamental systematic methodology that should be routinely applied in order to develop a nascent technology that shows promise for carbon dioxide capture and utilization.

2. Discussions on Essential Tools for Biorefinery Design and Development

The green chemistry principles are essential to apply at the laboratory scale and throughout the concept development and engineering design (Sadhukhan et al., 2019). Any toxic material input to the system should be avoided. Such material input should be avoided not only at the main processing stage, but also in the supply chains of the materials inputted to the main processing stage. Thus, life cycle assessment is essential and a life cycle assessment practitioner should be engaged throughout the development of a technology from the proof of concept through the pilot and the demonstration to the deployment stages (Sadhukhan et al., 2014). The practitioners' and specialists' inputs are imperative to scrutinize every material and its production pathway or supply chain. Using life cycle assessment, a material toxicity can be effectively checked as well as any impact on the environment from the supply chain of the material (Sadhukhan et al., 2019). This needs collaboration between a scientist and a life cycle assessment practitioner.

Life cycle assessment according to the International Organization for Standardization (ISO) 14040-44 series comprises interactive four stages: goal and scope definition, inventory analysis, impact assessment and interpretation (Sadhukhan et al., 2014). Each stage of a product or service life cycle should be considered for the assessment of impacts. The life cycle stages include feedstock acquisition, manufacturing, logistics, use and waste management. Following the waste disposal hierarchy, reduce, reuse, recycling and recovery should be applied in the order of decreasing priority throughout the development of the process concept. Geographic locations and impact categories beyond carbon management should be considered. Life cycle assessment should be done holistically, rigorously and systematically.

Process integration bridges the two disciplines of biotechnology and systems' life cycle assessment. Process integration helps to develop reaction, separation, heat recovery and utility systems using step by step systematic synthesis approach (Sadhukhan et al., 2014). The methodology includes but not limited to pinch analysis, total site utility system design, reaction and separation unit operations' synthesis, in-process recycling, process intensification, etc. A process integration study is done using computer-aided process design analysis (Ng et al., 2015). Aspen Plus and SuperPro Designer are amongst the popular tools for such analysis (Martinez-Hernandez et al., 2018; Martinez-Hernandez et al., 2017). A process systems engineer should be involved at an early design stage to bridge between the scientist and the life cycle assessment practitioner.

Techno-economic assessment should be applied to the integrated process for a robust evaluation of the entire process. Techno-economic assessment includes estimations of capital cost, operating cost, cost of production, value on processing and discounted cash flow (Martinez-Hernandez et al., 2014; Sadhukhan et al., 2008). Sensitivity analysis should be accompanied to present the effect due to a whole range of uncertain techno-economic and macroeconomic parameters. Although each of the above analyses should be rigorous in nature, the detailed report should give rise to an evidence based executive summary for decision makers to enable appropriate recommendations. It should also be emphasized that each of the above analyses should be rigorous and robust in nature and interactive to support objectively a truly sustainable system. The detailed report should be transparent and reproducible. Figure 1 shows the interactive methodological framework to enable informed decision making. The following sections discuss the challenges with new renewable technology deployment and the ways to overcome the challenges.

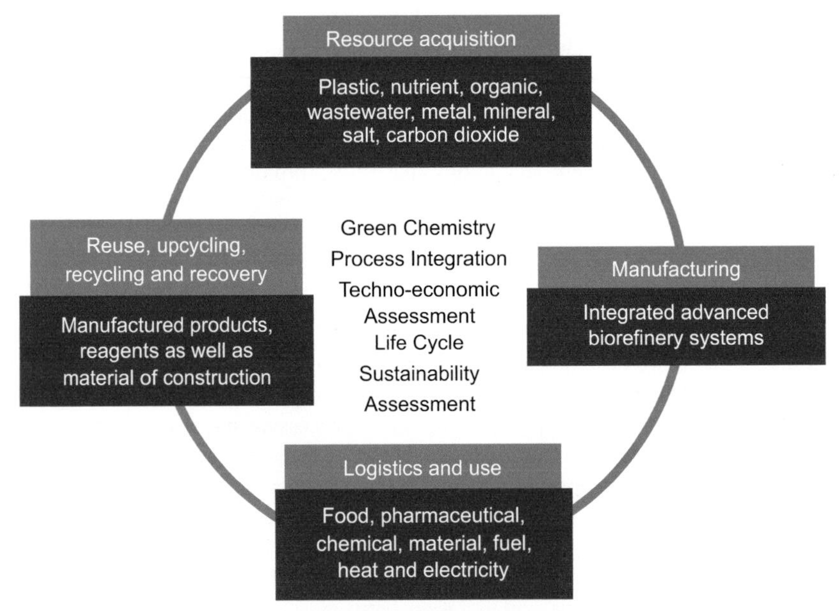

Figure 1. Interactive methodological framework to enable optimal design and decision making of integrated advanced biorefinery systems.

3. Biorefinery Configurations with Microbial Electrosynthesis for Carbon Capture and Utilization

Previous works have reported waste streams as the feedstocks to both the anode and the cathode chambers to feed the microbes as the sustainable way forward for microbial electrosynthesis of products by carbon capture and reuse (Shemfe et al., 2018). Any energy requirement should be fulfilled by renewable energy supply. Electrolytes should be environmentally friendly and supplied by sustainable supply chain. Electrodes and any membrane material should not be made of critical or rare earth materials, should be non-toxic and safe for the environment as well as stable over a long duration of operation. The fossil input to any material and energy needed for the microbial electrosynthesis should be minimal.

The main purpose of the anode in a microbial electrosynthesis process is to harvest electrons and provide the energy needed to drive the reduction reaction of carbon dioxide in the cathode. If electrons and energy could be harvested from a waste stream containing organics measured by chemical oxygen demand, then the anode process could be considered renewable (Sadhukhan et al., 2017).

The main purpose of the cathode in the microbial electrosynthesis process for desired organic production is to reduce carbon dioxide by the intake of protons transported through the electrolyte and a membrane and electrons transferred via an external circuit, both harvested by the oxidation of organics in the waste stream in the anode. The cathode process is considered renewable if it uses protons and electrons harvested from waste streams and carbon dioxide is utilized from the atmosphere.

All these above aspects are challenging that restrict the development of the technology through the scale. However, with industrial initiative, climbing up the technology readiness level is possible.

Industrial deployment for lactic acid production and succinic acid production can be illustrated in order to learn lessons on integrated separation technologies to adhere to the green chemistry principles (Sadhukhan et al., 2014). Separation unit operations should be designed in an integrated manner so as to enhance techno-economic feasibility of a biorefinery. The outlet stream from fermentation or microbial electrosynthesis contains a broth of desired product, however not at a marketable purity. The fermentation or microbial electrosynthesis process is operated in semi-batch or batch or continuous mode. While the continuous mode gives higher productivity but lower purity, the batch mode gives higher purity and lower productivity. A semi-batch or fed-batch operation can be optimized to offer the best trade-off between purity and productivity. A fed-batch biochemical reactor configuration may include continuous or semi-continuous feeding of substrates to the fermenter. The fermentation broth containing the desired product can be removed as an outlet stream when the substrate is depleted or product formation stops. A fed-batch configuration can be optimized to remove any inhibition reactions to product.

Integration between various separation technologies helps to overcome major barriers in downstream processing of fermentation or other biochemical process outlets. The outlet stream from a biochemical reaction process would contain bacterial cells and proteins or nutrients, etc. These need to be separated from the desired product and recycled back to the biochemical reaction process. The fermentation broth can first be clarified by microfiltration to remove bacterial cells, proteins, etc. Ultrafiltration can further be used to separate cells from the broth and allow recycling of cells back to the biochemical reaction process. For a biochemical or bioelectrochemical reaction broth after micro- and/or ultra-filtration, electrodialysis is a separation unit operation of choice for industrial deployment. Electrodialysis has been proven as an effective separation technique for acid recovery, learnt from successful industrial deployment (Sadhukhan et al., 2014).

Dialysis involves separating molecules or ions based on difference in sizes. It is mainly used to separate solutes with a large difference in their rates of diffusion through a semi-permeable membrane. Dialysis process is driven by a concentration gradient across the membrane. The flux in this process is generally lower than in other membrane processes. To enhance the separation process, the dialysis can be performed under the influence of an electrical field. This process is called electrodialysis. Electrodialysis is especially useful to separate charged particles, such as ions. Electrodialysis separates an aqueous solution into a dilute solution (called the diluent) and a concentrated solution (called the concentrate).

Electrodialysis is based on the principle that positively charged cations and negatively charged anions migrate towards the electrodes with opposite charges, cathode and anode, respectively. A simple electrodialysis cell will be formed by two membranes, i.e., a cation selective membrane and an anion selective membrane and two electrodes, i.e., cathode and anode, respectively. When an electrical current is applied between the electrodes, the anions migrate towards the positively charged electrode (anode). The anion-selective membrane allows only anions to pass. The cations move in the opposite direction towards the negatively charged electrode (cathode). The cation-selective membrane allows only cations to pass.

Advances in electrodialysis have enabled the development of an integrated separation process called the double electrodialysis process. This process combines electrodialysis followed by electrodialysis bipolar membrane (also called water-splitting electrodialysis). In practical applications, multi-compartment electrodialysis stacks are used. In electrodialysis bipolar membrane, feed enters the various compartments formed by the anion- and cation-selective membranes in alternate arrangement. The anions from a compartment move towards the anode through the anode-selective membrane and enter the adjacent compartment. However, they are not able to pass further because the next membrane is cation-selective. As a result, the anions are trapped between the two membranes of the compartment. A similar process occurs to the cations which move in the opposite direction and are also trapped in an adjacent compartment. Cations neutralize any excess charge due to accumulation of anions. As a result, the solution in the original compartment becomes diluted while the solution in the adjacent compartment becomes concentrated. By using this process, a highly concentrated solution of the salt (e.g., lactate) can be obtained in the concentrate stream. The electrochemical separation unit operations are discussed in great depth by Sadhukhan et al. (2014) with schematics and applications in biorefineries.

The electrodialysis process with bipolar membranes combines the conventional electrodialysis with biopolar membrane for the conversion of a salt into its corresponding acid and base. Bipolar membranes induce the splitting of water into hydrogen protons (H^+) and hydroxide ions (OH^-). Because of this, the process is also called water-splitting electrodialysis. Bipolar membranes consist of a cation exchange (i.e., cation selective) membrane, an anion exchange (i.e., anion-selective) membrane and a catalytic intermediate layer that promotes the splitting of the water. The anions are neutralized by the protons generated at the surface of the anion exchange membrane. The cations are neutralized by the hydroxide ions generated at the surface of the cation exchange membrane. By using this process an organic salt can be converted into the free acid form. The base used for pH adjustment may also be recovered.

The electrodialysis process with bipolar membranes requires a feed that has been treated to remove large molecules. Clarification of the fermentation broth is first performed by microfiltration. An important issue for the electrodialysis process with bipolar membranes operation is the presence of multivalent cations, such as Ca^{2+}, Mg^{2+} and Al^{3+}. These cations form insoluble hydroxides at the interface of the bipolar membrane, where the ions separate. Thus, their concentrations must be reduced to about 1 ppm. Bipolar membranes require integration with other separation processes. The organic acid (e.g., lactic acid) obtained after the electrodialysis process with bipolar membranes can be further purified by ion exchange resins. Then, the electrodialysis step removes the inorganic salts.

Despite a good separation by electrodialysis and electrodialysis with bipolar membranes, an intermediate chelation step is required for efficient operation of the arrangement electrodialysis + chelation + electrodialysis with bipolar membranes. Chelation refers to the separation of ions by forming coordinated complex arrangements with organic compounds, wherein the separated ions are the central atoms. After chelation, the electrodialysis with bipolar membranes step enables the conversion of salt into its acid. Then, polishing is carried out in ion-exchange columns. Finally, evaporation is used for further concentration and purification in order to produce technical grade acid product. To produce high purity product, the esterification-hydrolysis strategy used in the conventional process can be used.

The integrated biorefinery flowsheet is shown in Figure 2 (Sadhukhan et al., 2014).

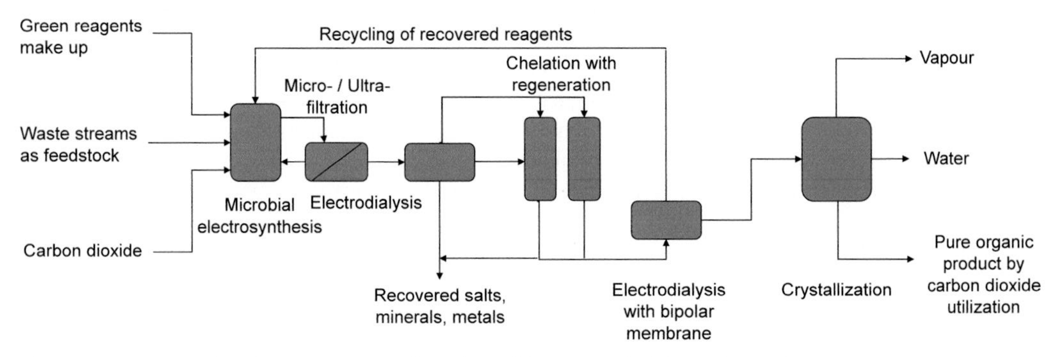

Figure 2. Integrated biorefinery system configuration with carbon dioxide utilization by microbial electrosynthesis (Sadhukhan et al., 2014).

Table 1. Separation unit operations for biorefinery systems (Sadhukhan et al., 2014).

Separation process	Driving force	Application example
Distillation, flash columns	Relative volatilities	Ethanol, biodiesel separation
Extraction, absorption	Solubility in liquid solvent	Levulinic acid extraction, CO_2 absorption
Adsorption	Solubility in solid sorbent	CO_2 adsorption
Membrane-based separations:		
Microfiltration	Pressure gradient	Yeast cell separation
Ultrafiltration	Pressure gradient	Bacteria cell separation
Nanofiltration	Pressure gradient	Proteins, enzymes, sugars, amino acids, colorants
Reverse osmosis	Pressure gradient	Organic acid concentration
Dialysis	Concentration gradient	Non-charged particles
Electrodialysis	Electrical field	Organic acid separation
Electrodialysis with bipolar membrane	Electrical field	Organic acid separation
Pervaporation	Pressure gradient	Ethanol dehydration
Crystallisation	Difference in solubility and supersaturation	Succinic acid production
Ion exchange	Electrostatic attraction	Organic acids separation
Centrifugation	Centrifugal force	Algae harvesting, solids separation
Sedimentation	Difference in density between solids and liquid	Algae harvesting
Coagulation-flocculation	Electrostatic attraction	Algae harvesting
Precipitation	Solubility	Organic acids separation

There are many choices of separation unit operations shown in Table 1, along with the driving forces for their operations and their applications in biorefinery systems (Sadhukhan et al., 2014). This makes the selection of separation unit operations challenging. All possible options should first be considered through sketching of various flowsheets with different separation unit operations and interconnections. Each of them will undergo rigorous and comprehensive process simulation (e.g., in Aspen Plus) in order to analyze the mass and energy balance and sizes of individual unit operations (Sadhukhan et al., 2014). Detailed techno-economic analysis can evaluate the economic margins, payback times and discounted cash flows, etc. (Sadhukhan et al., 2014). This exercise is essential in determining the most economically feasible flowsheet. Also, any potential tradeoffs must be considered through life cycle assessment and, when applicable, through life cycle sustainability assessment, so that all dimensions of sustainability are adequately taken into account for an overall optimal biorefinery flowsheet.

The scale of separation unit operations and, thus, their economics are an important factor that determines their choices. Electrochemical unit operations define the finer end of separation and also work at smaller scales compared to distillation and extraction/absorption/adsorption unit operations that tend to work at bulk or large scale separations. The learning curve effect can be applied to forecast the future scenarios with the electrochemical-based technologies and biorefinery systems (Shemfe et al., 2018). Also, the electrochemical-based biorefinery systems are an important avenue of research and development that should be considered for the recovery of resources from even trace quantities present in waste streams, where other processes fail to recover or valorize waste streams into resources (Sadhukhan et al., 2016). Thus, the electrochemical-based biorefinery systems can achieve a circular economy, leaving behind no waste to the environment. The present challenge for industrial deployment of the electrochemical technologies is the up-scaling (Srikanth et al., 2014; Kumar et al., 2019). Multiple unit operations, proven at pilot scale, can be deployed in order to attain a desired scale of operation, but are not effective in terms of utilizing capital investments and optimizing operating costs. The electrochemical process development from pilot through demonstration to deployment stage often gets stuck due to lack of industrial initiative and investment. An example of an exception is 500 L electrolyte capacity investigation based on the design basis as an upscaling exercise by the Indian Oil Corporation (Srikanth, 2019). Policies are not in place to overcome the barriers of economics for the bioelectrochemical technologies (Marshall et al. 2019). In this situation, industries have to come forward to make a feasible progress through the scale. A viable way forward is to develop an industrial cluster, whereby an industry takes waste streams of certain quantities from another in order to investigate the electrochemical technologies of a desired scale to turn into added value products. By profit as well as capital cost sharing and a long term sustainable business model, both types of industry, problem holder and solution provider, can benefit from a win-win scenario. However, such an initiative is on hold due to the lack of policy support and a strong commitment to de-fossilization of carbon in the value chains. Now that the IPCC emphasizes the critical role of carbon capture and utilization in product synthesis, policies to explicitly support optimal technologies and systems in order to curb carbon dioxide from the atmosphere should be developed.

Some process synthesis heuristics regarding choices of electrochemical reaction and separation technologies and their scales can be developed, learnt and applied from previous successful and failed industrial deployments. However, there is a limit to the application of heuristics in selecting the separation unit operations, instinctively. Optimization based on mathematical programming is an effective approach to ensure a thorough analysis and that no opportunities for better techno-economic performance are missed (Foong et al., 2019). Optimization can be mathematical programming based, where individual unit operations' models are developed and their interactions are considered. An objective function is set to maximize or minimize within defined process and market constraints. However, solution of computer aided mathematical programming becomes challenging for large scale process systems, especially when heat exchanger network and utility system synthesis are considered simultaneously with reaction and separation process synthesis.

A systematic synthesis analysis by applying the twelve principles of green chemistry and process integration gives a better control over the design and decision making. These principles deploy in common, benign solvents, reagents and chemicals, production of benign products, and energy efficiency and atom economy measures. They also emphasize green catalyst design and selection. Waste mitigation, least hazardous routes to synthesis, renewable feedstock selection, real time monitoring, control and analysis and safety are also the common ethos across the two disciplines. The two disciplines can, therefore, go hand-in-hand in order to create a holistic optimal system. By strictly adhering to these principles, the designer can have a full control over optimal decision making.

Process integration has provided effective powerful tools for conceptual process design and synthesis, pinch analysis being the most common one. Furthermore, pinch analysis applied to maximize heat recovery within a site and heat exchanger network synthesis has been adapted to optimize hydrogen network, water network, total site utility system, carbon dioxide recovery and utilization, and bioethanol recovery and utilization, etc., giving rise to the corresponding pinch analysis tools, such as hydrogen pinch, water pinch, utility pinch, carbon dioxide pinch and bioethanol pinch; an example is given by Martinez-Hernandez et al. (2013). These tools work inherently to minimize waste and emission and resource depletion and maximize efficiency. The tools also promote cost optimization by maximizing the use of capital and operating costs in economic margin generation. A robust value analysis tool has been developed for differential and graphical economic analysis of individual streams and pathways in a complex network, thus optimizing the network for the highest economic performance (Sadhukhan et al., 2003).

4. Techno-economic Analysis

A comprehensive techno-economic analysis methodology should be applied to estimate the capital and operating costs in order to estimate the overall economic performance in terms of (but not limited to) economic margin, payback time, cost of production, value on processing and discounted cash flow (Sadhukhan et al., 2004).

Total capital cost comprises direct cost, indirect cost and working capital, factored with respect to delivered cost of equipment. The individual factors for direct (delivered cost of equipment, installation, instrumentation and control, piping, electrical systems, buildings including services, yard improvements, service facilities), indirect (engineering and supervision, construction expenses, legal expenses, contractor fee, contingency) and working capital costs are called Lang factor (Sadhukhan et al., 2014). The summation of all the Lang factors makes 4.67, 5.03 and 5.93 times the delivered cost of equipment, for solid, solid-fluid and fluid processing systems, respectively, in order to calculate the total capital investment (Sadhukhan et al., 2014). Furthermore, the total capital investment can be annualized by applying an annual capital charge, for the estimation of (but not limited to) economic margin, payback time, cost of production, value on processing and discounted cash flow. The delivered cost of equipment at a base year is estimated for a given size from a base size by applying a scaling factor, usually 0.6–0.8 (Sadhukhan et al., 2014). The scaling factor less than 1 shows the benefit of economy of scale. Furthermore, the delivered cost of equipment at a given year should be updated from a base year by applying the Annual Chemical Engineering Plant Cost Index (Sadhukhan et al., 2014).

Total operating cost is 20–30% greater than the summation of fixed and variable operating costs, the latter includes cost of feedstock and any raw materials and utilities imported. Fixed operating costs include personnel, laboratory, supervision and overhead costs dependent on labor cost and maintenance, insurance, local tax and royalty, etc., dependent on annual indirect operating cost (Sadhukhan et al., 2014). A relevant set of techno-economic variables that require data collection for carbon dioxide capture and utilization in microbial electrosynthesis-based biorefinery systems is shown in Table 2 (Sadhukhan et al., 2014).

Table 2. A relevant set of techno-economic variables that need data collection for carbon dioxide capture and utilization in microbial electrosynthesis-based biorefinery systems.

Capital cost components	**Operating cost components**	**Product revenues**
Anode	Cost of waste streams as feedstock	Target chemical
Cathode	Energy cost	Salts, minerals, metals
Current collectors	Anolyte	Nutrients
Membranes	Catholyte	Water
Micro-/Ultra-filtration	Make up reagents	Electricity
Electrodialysis	Maintenance	Organic byproducts
Chelation	Personnel	
Electrodialysis with bipolar membrane	Laboratory, supervision and overhead costs	
Crystallization	Insurance	
Installation	Local tax	
Instrumentation and control	Royalty	
Piping	Annual capital charge	
Electrical systems	Miscellaneous	
Buildings including services		
Yard improvements		
Service facilities		
Engineering and supervision		
Construction expenses		
Legal expenses		
Contractor fee		
Contingency		

Chemical oxygen demand is an indicator of organic loading of wastewaters. The higher the chemical oxygen demand, the greater the organic loading of wastewaters. Organic in wastewaters is the substrate for bacterial or microbial growth in the anode of a microbial electrosynthesis system. Thus, the higher the chemical oxygen demand, the greater the microbial growth rate. This is proportional to the productivity by carbon dioxide capture and utilization in the cathode of a microbial electrosynthesis system. This is because greater number of electrons are harvested from greater strength wastewaters by microbes. Carbon dioxide participates in the reduction reaction in the cathode by accepting electrons and protons harvested from the anode in a microbial electrosynthesis system. Thus, the cost of production of cathodic product by carbon dioxide capture and utilization will decrease with increasing chemical oxygen demand in anodic wastewaters (Shemfe et al., 2018). Also, it can be observed that carbonic acid is more effective than dissolved carbon dioxide that, in turn, is more effective than atmospheric carbon dioxide in terms of cathodic productivity. Hence, there should be a mechanism to turn atmospheric carbon dioxide into carbonic acid before the carbon dioxide capture and utilization in the cathodic chamber of the microbial electrosynthesis system. Applied voltage is another critical variable, more critical than the chemical oxygen demand of the anodic substrate to influence cathodic productivity by carbon dioxide capture and utilization (Shemfe et al., 2018). With increasing applied potential across a microbial electrosynthesis cell, productivity by carbon dioxide capture and utilization in the cathode increases (Shemfe et al., 2018). The cost of production of the product by carbon dioxide capture and utilization in the cathode decreases with increasing applied potential, due to the fact that the rate of increase in productivity is much greater than the rate of increase in applied potential and the value of the product is much higher than the cost of applied potential. These trends have to be considered for techno-economic viability, scale optimization and learning curve generation for microbial electrosynthesis systems. Furthermore, maintenance is an important consideration from economic and environmental impact costs' perspectives (Shemfe et al., 2018). Efforts should be focused on enhancing the stability of microbial electrosynthesis systems and increasing the number of years before a cell needs to be replaced to more than 5 years.

The various economic indicators, such as economic margins, payback time, value on processing, cost of production and discounted cash flow analysis, are used to comprehensively present the techno-economic performance of microbial electrosynthesis systems and electrochemical-based biorefinery systems. The annual capital investment and total operating cost must be subtracted from the value of products in order to obtain the economic margin. A simple payback time can be estimated by the ratio between total capital investment and the difference between product value and total operating cost.

Any material stream in a process network can be evaluated in terms of its value on processing (VOP) and cost of production (COP). The difference between the two, VOP–COP provides the economic margin of the stream. The concept has been discussed by Martinez-Hernandez et al. (2014).

The evaluation of COP of a stream starts from known market price of feedstocks, added with the costs incurred from the production of the stream, thus proceeds in the forward direction until end products' evaluations. The COP of a stream is the summation of all associated cost components (i.e., the costs of feedstocks, utilities, operating and annualised capital costs) that have contributed to the production of that stream up to that point. This must mean inclusion of only those fractional costs involved with the stream's production. This must mean only those fractional costs involved with the stream's production.

The VOP evaluation proceeds in the backward direction. From known end products' market prices, the VOP of all intermediate streams and feedstocks to a process network is evaluated. The VOP of a stream at a point within the process is obtained from the prices of products that will ultimately be produced from it, minus the costs of auxiliary raw materials and utilities and the annualized capital cost of equipment that will contribute to its further processing into these final products. The cash flows can be discounted in order to take into account the time value of money by applying a discount factor. When the net present value, i.e., the cumulative discounted cash flow, becomes zero at the end of an estimated plant life time, the discount rate obtained iteratively is called internal rate of return. The higher the internal rate of return, the longer until the break-even point between the cumulative discounted cash flow or net cash flow and years.

Techno-economic analysis is, thus, a comprehensive bottom-up approach that links equipment sizing and operating inventories to the various economic performance indicators. A step by step systematic computer-aided process design can, therefore, lead conceptual process design through to techno-economic analysis (Sadhukhan et al., 2019). However, for overall sustainability, life cycle assessment approaches are needed for environmental impact assessment and mitigation and also to address social dimension of sustainability (Sadhukhan et al., 2019).

5. Life Cycle Assessment and Life Cycle Sustainability Assessment

Life cycle assessment is a rigorous, holistic and systematic methodology for the assessment of life cycle environmental impacts of a system from the ISO 14040-44 series. Life cycle assessment should be carried out beyond the process systems across the supply chains, so as to ensure that the entire system is sustainable. Life cycle assessment is, therefore, data intensive and can take three different ways of system consideration: Stand-alone system, attributional and consequential systems. In a stand-alone system case, only life cycle assessment of environmental impacts of the foreground system is carried out. This is useful to identify hot spots and ways to mitigate such hot spots. In an attributional system case, life cycle assessment of environmental impacts of a system is undertaken in order to analyze environmental impact saving by displacement of equivalent higher impactful system. An example is illustrated in Figure 3. It shows the environmental impact savings by the various services provided by the microbial electrosynthesis system, by the displacement of equivalent fossil-based services. In a consequential system case, life cycle assessment of environmental impacts of a system is carried out in order to analyze environmental impact saving in future scenarios. An example is that a microbial electrosynthesis system can utilize renewable electricity to make new products from carbon dioxide capture and utilization. The fact that the quality and quantity of wastewater as a substrate for the microbial electrosynthesis system would also change in future scenarios should also be taken into consideration for a holistic consequential life cycle assessment. All the three ways of carrying out a life cycle assessment should be investigated in order to ensure sustainability at the system level and in the current scenario, as well as in future scenarios.

Life cycle impact categories are an important consideration for informed decision making. The impact categories that are relevant and important for a given system and can make a difference between design choices should be carefully selected. There are many life cycle impact assessment methodologies suggesting different sets of environmental impact categories, such as the CML life cycle impact assessment methodology, Europe-based ILCD life cycle impact assessment methodology, USA-based Traci life

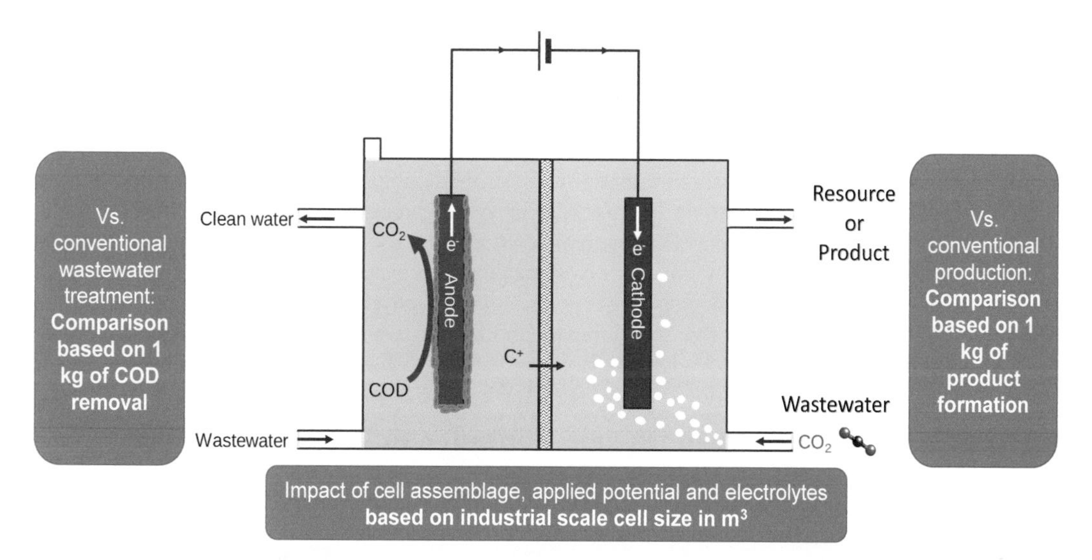

Figure 3. An example of microbial electrosynthesis system for attributional life cycle assessment of environmental impacts of the system to analyze environmental impact saving by displacement of equivalent higher impactful system (COD: chemical oxygen demand).

cycle impact assessment methodology and ReCiPe, etc., discussed in Sadhukhan et al. (2014). Different environmental impact categories from the different life cycle impact assessment methodologies can be chosen for the best assessment of a system. The selected list of environmental impact categories should not be exhaustive, but adequate to meaningfully help in design and decision making of the system under consideration. The environmental impact categories can correspond to emissions to the atmosphere, land and water. A recommended list of environmental impact categories includes global warming potential or climate change impact potential and air quality (especially particulate matters and urban smog) (atmospheric emission impact categories), water depletion and eutrophication (water emission impact categories) and terrestrial ecosystem toxicity potential, land use and land use change (land emission impact categories). In addition, a handful of human health-related impact categories are recommended in the various life cycle impact assessment methodologies.

Life cycle costing is required in order to assess economic viability beyond the process flowsheet across entire supply chains. The techno-economic analysis methodology discussed here can be extended to the various stages of the life cycle of a product or system, feedstock acquisition, manufacturing, logistics, use and waste management. Thus, the economic margin, payback time, cost of production, value on processing and discounted cash flow, etc., can be estimated across the supply chains or overall system. Like life cycle assessment, the ISO recommends the use of four stages, goal and scope definition, inventory analysis, impact assessment and interpretation for a comprehensive life cycle costing. The functional unit can be the same as the life cycle assessment.

In addition to life cycle environmental impact assessment and life cycle costing, a system should be socially acceptable, showing savings in various social impact categories. The United Nations Environmental Programme and the Society of Environmental Toxicology and Chemistry and the ISO 26000 set out five social themes and twenty two sub-themes regarding assessment for social sustainability (Sadhukhan et al., 2019). The social life cycle assessment methodology should follow the same guidelines as the life cycle assessment in terms of the four stages, goal and scope definition, inventory analysis, impact assessment and interpretation. Each stage of a product or service life cycle should be considered for the assessment of environmental as well as social impacts. The life cycle stages include feedstock acquisition, manufacturing, logistics, use and waste management. The same system boundary should be applied for both the environmental and social life cycle assessments. The functional unit is different in the two different methodologies. The functional unit for the life cycle assessment of a product or system should be related to its service or function. Similarly, for social life cycle assessment, economic value of the product or labor cost for the system is considered as the functional unit to estimate all themes and sub-themes of social life cycle assessment. Figure 4 shows life cycle assessment, life cycle costing and social life cycle assessment categories, relevant and important for biorefinery systems utilizing carbon dioxide in making products.

Life cycle costing is required in order to assess economic feasibility beyond the process flowsheet across entire supply chains. The techno-economic analysis methodology discussed here can be extended to the various stages of the life cycle of a product or system, feedstock acquisition, manufacturing, logistics, use and waste management. The economic margin, payback time, cost of production, value on processing and discounted cash flow, etc., can be estimated across the supply chains or overall system. In addition to life cycle environmental impact assessment and life cycle costing, a system should be socially acceptable by showing savings in five social themes and twenty-two sub-themes regarding assessment for social sustainability.

6. Biorefineries with Carbon Dioxide Utilization to Deliver the United Nations Sustainable Development Goals

Biorefineries with carbon dioxide utilization can deliver the United Nations Sustainable Development Goals (SDGs) (United Nations, 2015) because these systems are the only alternatives to fossil-based systems by de-fossilizing carbon. "The 2030 Agenda for Sustainable Development, adopted by all the United Nations Member States in 2015, provides a shared blueprint for peace and prosperity for people

Environmental impact potentials
- Atmospheric
  - Global warming (Global impact)
  - Ozone depletion (Global)
  - Acidification (Regional)
  - Urban smog (Local)
  - Particulate matter (Local)
- Aquatic (Global impact)
  - Eutrophication
- Health
  - Human health cancer effects
  - Human health non-cancer effects
  - Ionising radiation
  - Ecotoxicity (air)
  - Ecotoxicity (freshwater)
  - Ecotoxicity (marine water)
  - Ecotoxicity (soil or terrestrial)
- Resource depletion (Global)
  - Minerals
  - Elements
  - Fossils
  - Water
  - Land

Economic indicators
- Labour cost (gender dimension)
- Material cost
- Energy and utility cost
- Taxation
- Cost of Production (COP) or Minimum Selling Price (MSP)
- Value on Processing (VOP) or Netback
- Economic margin
- Return on Investment (ROI)
- Discounted cash flow analysis
- Net Present Value (NPV)

Social impact categories
- Labour rights and decent work
  - Child labour
  - Forced labour
  - Excessive working time
  - Wage assessment
  - Poverty
  - Migrant labour
  - Freedom of association
  - Unemployment
  - Labour laws
- Health and safety
  - Injuries and fatalities
  - Toxics and hazards
- Human rights
  - Indigenous rights
  - High conflicts
  - Gender equity
  - Human health issues
- Governance
  - Legal systems
  - Corruption
- Community infrastructure
  - Hospital beds
  - Drinking water
  - Sanitation
  - Children out of school
  - Smallholder vs commercial farms

Figure 4. Life cycle assessment, life cycle costing and social life cycle assessment categories, relevant and important for biorefinery systems with carbon dioxide utilization.

and the planet, now and into the future. At its heart are the 17 Sustainable Development Goals (SDGs), which are an urgent call for action by all countries—developed and developing—in a global partnership. They recognize that ending poverty and other deprivations must go hand-in-hand with strategies that improve health and education, reduce inequality, and spur economic growth—all while tackling climate change and working to preserve our oceans and forests." With these ethos, alternative biorefinery supply chain systems should be developed to preserve natural capital as well as end poverty and inequality. Biorefineries are believed to deliver the SDGs, in particular, 6, 7, 8, 9, 12 and 13, discussed below (Sadhukhan et al., 2019).

6.1 SDG 6—clean water and sanitation

Clean water and sanitation are foundational needs of the society. The microbial electrosynthesis systems intake waste streams, sludge, etc., as the feedstock or substrate for microbial growth and thereby decompose organic loading of the wastewaters and purify them. These systems operating at small scales can be integrated into individual sanitation systems for treatment to recycle water and at the same time harvest energy by electron and proton sourcing by decomposition of the organic load in wastewaters (Barışçı et al., 2016). It is believed that the system would have an increasingly important role to play in the development of sustainable toilets with multi-productions, recycled water, energy and compost, in order to deliver cost-effective clean water and sanitation systems.

6.2 SDG 7—affordable and clean energy

Like clean water and sanitation, affordable and clean energy constitutes a foundational need of society. The core of the innovation is to transform heterogeneous complex waste streams into clean energy generation. The microbial electrosynthesis systems are able to effectively harvest electrons and protons, the vital sources of energy from wastewaters (Seelam et al., 2018). Electrons can flow through the external circuit to generate a clean form of electricity. At a household scale or small distributed scale, the system can operate to utilize household wastewaters and transform into clean electricity by the way of recycling water, the system being called microbial fuel cells (Jadhav et al., 2018).

6.3 SDG 8—decent work and economic growth (by "economic productivity through diversification and technological upgrading and innovation") (Sadhukhan et al., 2019)

Integrated advanced biorefinery systems defined as a facility with integrated, efficient and flexible conversion of biomass feedstocks, through a combination of physical, chemical, biochemical and thermochemical processes, into multiple products (Sadhukhan et al., 2014) should be deployed to deliver the SDG 8. Biorefineries can create green jobs and offer decent work and clean economic growth.

6.4 SDG 9—industry, innovation and infrastructure (by "increased resource-use efficiency and greater adoption of clean and environmentally sound technologies and industrial processes") (Sadhukhan et al., 2019)

Multi-productions from multiple feedstocks by integrated multi-process biorefinery systems are the only effective way to confront the highly energy efficient but environmentally damaging fossil-based systems. Biorefineries with multi-productions from heterogeneous complex waste streams are extremely important for sustainable development, alternative to fossil based systems (Sadhukhan et al., 2014). Innovations, research and development should be focused on integrated biorefineries until these succeed in effectively replacing fossil-based systems.

6.5 SDG 12—responsible consumption and production

With alternative products and by maximizing resource efficiency and infrastructure sustainability, biorefineries are able to deliver SDG12.

6.6 SDG 13—climate action

By underwriting national policies, strategies and planning for eliminating climate change impact culprit, fossil derived products, by alternative integrated advanced biorefinery systems, SDG 13 can be delivered.

7. Conclusions

This chapter shows that integrated advanced biorefinery systems defined as a facility with integrated, efficient and flexible conversion of biomass feedstocks, through a combination of physical, chemical, biochemical and thermochemical processes, into multiple products can be industrially deployed. The biorefinery feedstocks should be waste materials, including lignocelluloses, and should not compete in terms of indirect land use for food or feed production. A successful industrial deployment depends on the scale of the process operations and biomass availability and how closely these two scales can be aligned to make economic sense. Integrated process systems, including hybrid and intensified unit operations, are able to process heterogeneous waste streams. In reality, heterogeneous waste streams are more usual than homogeneous. Thus, all technologies should be developed with an aim to process heterogeneous waste streams. Waste streams as potential feedstocks for biorefineries could be available in small quantities and dilute conditions, but may fail to comply with stricter environmental regulations. Thus, contaminants present in small quantities and in dilute conditions need to be removed by innovative technologies. Microbial synthesis is one such process that can remove and recover contaminants present in small quantities and in dilute conditions from waste streams as resources. Microbial electrosynthesis technologies are being developed at pilot and demonstration scales can thus be suitable to match the scale of feedstock availability. At the heart of its development is the ability to remove, capture and utilize carbon dioxide into added value products. By the method of carbon dioxide reduction reaction and reuse, microbial electrosynthesis shows the promise of carbon dioxide capture and reuse from the atmosphere. Microbial electrosynthesis can be an important technology for consideration to meet the carbon dioxide reduction target recommended by the IPCC. Microbial electrosynthesis needs to embody integrated preprocessing and post-processing product separation and purification unit operations as well as utility systems for heat recovery to demonstrate economic feasibility. Such a complete conceptual process synthesis should follow the twelve principles of green chemistry as well as process integration. Waste streams as the substrate to microbes or feedstocks to both the anode and the cathode chambers to feed the microbes is the sustainable way forward for microbial electrosynthesis of products by carbon capture and reuse. Any energy requirement should be fulfilled by renewable energy supply. Electrolytes should be environmentally friendly and supplied by sustainable supply chain. Electrodes and any membrane material should not be made of critical or rare earth materials, should be non-toxic and safe for the environment as well as stable over a long duration of operation. The outlet stream from a microbial electrosynthesis process would contain bacterial cells and proteins or nutrients, etc., which need to be recovered from the desired product by micro- and ultra-filtration, as appropriate, and recycled back to the biochemical reaction process. The broth without biomass, primarily containing salts, minerals and metals, in addition to a target desired product by carbon dioxide utilization, can be routed through electrodialysis, chelation and electrodialysis with bipolar membranes to separate salts, minerals and metals from the desired organic stream. Such a separation protocol can be developed bespoke for a desired product objective from the successful industrial deployment of succinic acid or lactic acid production processes from waste biomass. Furthermore, the desired product can be purified by the crystallization process. Organic products by carbon dioxide utilization can range from low to molecular chain volatile fatty acids with similar properties as transport or aviation fuels. A target organic product by carbon dioxide utilization can also be recovered as a chemical product with a higher market price than the price of biofuel. A systematic synthesis analysis, applying the twelve principles of green chemistry and process integration, gives an overall optimal flowsheet integrating reaction, separation, water and utility systems. These principles deploy in common, benign solvents, reagents and chemicals, production of benign products, and energy efficiency and atom economy measures. They also emphasize green catalyst design and selection. Waste mitigation, least hazardous routes to synthesis, renewable feedstock selection, real time monitoring, control and analysis and safety are also the common ethos across the two disciplines. The two disciplines can, therefore, go hand-in-hand in order to create a holistic optimal system. By strictly adhering to these principles, the designer can have a full control over optimal decision making. Life cycle sustainability assessment should be carried out beyond the process systems across the supply chains, so as to ensure that the entire system is sustainable. Life cycle sustainability assessment is a rigorous, holistic and systematic methodology regarding the assessment of life cycle environmental, economic and social impacts of a system. Life cycle impact categories are an important consideration for informed decision making. The impact categories that are relevant and important for a given system and are sensitive to make a difference between design configurations should be carefully selected. The environmental impact categories can correspond to emissions to the atmosphere, land and water. Furthermore, biorefineries utilizing microbial electrosynthesis and carbon dioxide are discussed to deliver the United Nations Sustainable Development Goals, in particular, 6, 7, 8, 9, 12 and 13.

References

Barışçı, S., Särkkä, H., Sillanpää, M. and Dimoglo, A. 2016. The treatment of greywater from a restaurant by electrosynthesized ferrate (VI) ion. Desalination and Water Treatment 57(24): 11375–11385.

Directive 2009/28/EC https://eur-lex.europa.eu/legal-content/EN/ALL/?uri=celex%3A32009L0028 (accessed 19 April 2019).

Foong, S.Z., Andiappan, V., Foo, D.C. and Ng, D.K. 2019. Flowsheet synthesis and optimisation of palm oil milling processes with maximum oil recovery. pp. 3–32. *In*: Green Technologies for the Oil Palm Industry. Springer, Singapore.

https://www.ieabioenergy.com/wp-content/uploads/2016/09/iea-bioenergy-countries-report-13-01-2017.pdf (accessed 19 April 2019).

https://www.ipcc.ch/2018/10/08/summary-for-policymakers-of-ipcc-special-report-on-global-warming-of-1-5c-approved-by-governments/ (accessed 19 April 2019).

Jadhav, D.A., Chendake, A.D., Schievano, A. and Pant, D. 2018. Suppressing methanogens and enriching electrogens in bioelectrochemical systems. Bioresource Technology 277: 148–156.

Karstensen, J. and Peters, G. 2018. Distributions of carbon pricing on extraction, combustion and consumption of fossil fuels in the global supply-chain. Environmental Research Letters 13(1): 014005.

Kumar, M., Sahoo, P.C., Srikanth, S., Bagai, R., Puri, S.K. and Ramakumar, S.S.V. 2019. Photosensitization of electro-active microbes for solar assisted carbon dioxide transformation. Bioresource Technology 272: 300–307.

Lane, J. 2017. DowDuPont to exit cellulosic biofuels business. http://www.biofuelsdigest.com/bdigest/2017/11/02/breaking-news-dowdupont-to-exit-cellulosic-ethanol-business/(accessed 19 April 2019).

Marshall, R., Sadhukhan, J., Macaskie, L.E., Semple, K.T., Velenturf, A.P.M. and Jopson, J. 2018. The organic waste gold rush: Optimizing resource recovery in the UK bioeconomy. https://resourcerecoveryfromwaste.files.wordpress.com/2018/10/rrfw-ppn-the-organic-waste-gold-rush-web.pdf (accessed 19 April 2019).

Martinez-Hernandez, E., Sadhukhan, J. and Campbell, G.M. 2013. Integration of bioethanol as an in-process material in biorefineries using mass pinch analysis. Applied Energy 104: 517–526.

Martinez-Hernandez, E., Campbell, G.M. and Sadhukhan, J. 2014. Economic and environmental impact marginal analysis of biorefinery products for policy targets. Journal of Cleaner Production 74: 74–85.

Martinez-Hernandez, E., Amezcua-Allieri, M.A., Sadhukhan, J. and Anell, J.A. 2017. Sugarcane bagasse valorization strategies for bioethanol and energy production. *In*: Sugarcane-Technology and Research. IntechOpen.

Martinez-Hernandez, E., Ng, K.S., Allieri, M.A.A., Anell, J.A.A. and Sadhukhan, J. 2018. Value-added products from wastes using extremophiles in biorefineries: Process modeling, simulation, and optimization tools. pp. 275–300. *In*: Extremophilic Microbial Processing of Lignocellulosic Feedstocks to Biofuels, Value-Added Products, and Usable Power. Springer, Cham.

Ng, D.K., Tan, R.R., Foo, D.C. and El-Halwagi, M.M. 2015. Process Design Strategies for Biomass Conversion Systems. John Wiley & Sons.

Ng, K.S., Zhang, N. and Sadhukhan, J. 2013. Techno-economic analysis of polygeneration systems with carbon capture and storage and CO_2 reuse. Chemical Engineering Journal 219: 96–108.

Sadhukhan, J., Zhang, N. and Zhu, X.X. 2003. Value analysis of complex systems and industrial application to refineries. Industrial & Engineering Chemistry Research 42(21): 5165–5181.

Sadhukhan, J., Zhang, N. and Zhu, X.X. 2004. Analytical optimisation of industrial systems and applications to refineries, petrochemicals. Chemical Engineering Science 59(20): 4169–4192.

Sadhukhan, J., Mustafa, M.A., Misailidis, N., Mateos-Salvador, F., Du, C. and Campbell, G.M. 2008. Value analysis tool for feasibility studies of biorefineries integrated with value added production. Chemical Engineering Science 63(2): 503–519.

Sadhukhan, J. 2014. Distributed and micro-generation from biogas and agricultural application of sewage sludge: Comparative environmental performance analysis using life cycle approaches. Applied Energy 122: 196–206.

Sadhukhan, J., Ng, K.S. and Martinez-Hernandez, E. 2014. Biorefineries and Chemical Processes: Design, Integration and Sustainability Analysis. Wiley, Chichester.

Sadhukhan, J., Ng, K.S. and Martinez-Hernandez, E. 2015. Process systems engineering tools for biomass polygeneration systems with carbon capture and reuse. Process Design Strategies for Biomass Conversion Systems, pp. 215–245.

Sadhukhan, J., Lloyd, J.R., Scott, K., Premier, G.C., Eileen, H.Y., Curtis, T. and Head, I.M. 2016. A critical review of integration analysis of microbial electrosynthesis (MES) systems with waste biorefineries for the production of biofuel and chemical from reuse of CO_2. Renewable and Sustainable Energy Reviews 56: 116–132.

Sadhukhan, J., Joshi, N., Shemfe, M. and Lloyd, J.R. 2017. Life cycle assessment of sustainable raw material acquisition for functional magnetite bionanoparticle production. Journal of Environmental Management 199: 116–125.

Sadhukhan, J., Martinez-Hernandez, E., Murphy, R.J., Ng, D.K., Hassim, M.H., Ng, K.S., Kin, W.Y., Jaye, I.F.M., Hang, M.Y.L.P. and Andiappan, V. 2018. Role of bioenergy, biorefinery and bioeconomy in sustainable development: Strategic pathways for Malaysia. Renewable and Sustainable Energy Reviews 81: 1966–1987.

Sadhukhan, J., Gadkari, S., Martinez-Hernandez, E., Ng, K.S., Shemfe, B., Garcia, E.T. and Lynch, J. 2019. Novel macroalgae (seaweed) biorefinery systems for integrated chemical, protein, salt, nutrient and mineral extractions and environmental protection by green synthesis and life cycle sustainability assessments. Green Chemistry, in Press.

Seelam, J.S., Rundel, C.T., Boghani, H.C. and Mohanakrishna, G. 2018. Scaling up of MFCs: challenges and case studies. pp. 459–481. *In*: Microbial Fuel Cell. Springer, Cham.

Shemfe, M., Gadkari, S., Yu, E., Rasul, S., Scott, K., Head, I.M., Gu, S. and Sadhukhan, J. 2018. Life cycle, techno-economic and dynamic simulation assessment of bioelectrochemical systems: A case of formic acid synthesis. Bioresource Technology 255: 39–49.

Srikanth, S., Email communication on 14.02.2019.

Srikanth, S., Maesen, M., Dominguez-Benetton, X., Vanbroekhoven, K. and Pant, D. 2014. Enzymatic electrosynthesis of formate through CO_2 sequestration/reduction in a bioelectrochemical system (BES). Bioresource Technology 165: 350–354.

The United Nations Sustainable Development Goals https://www.un.org/sustainabledevelopment/sustainable-development-goals/(accessed 19 April 2019).

Towards fossil-free energy in 2050, 2019. https://europeanclimate.org/wp-content/uploads/2019/03/Towards-Fossil-Free-Energy-in-2050.pdf (accessed 19 April 2019).

Wan, Y.K., Sadhukhan, J., Ng, K.S. and Ng, D.K. 2016a. Techno-economic evaluations for feasibility of sago-based biorefinery, Part 1: Alternative energy systems. Chemical Engineering Research and Design 107: 263–279.

Wan, Y.K., Sadhukhan, J. and Ng, D.K. 2016b. Techno-economic evaluations for feasibility of sago-based biorefinery, Part 2: Integrated bioethanol production and energy systems. Chemical Engineering Research and Design 107: 102–116.

CHAPTER 10

Synthesis of Regional Renewable Supply Networks

Žan Zore, *Lidija Čuček* and *Zdravko Kravanja**

1. Introduction

Traditionally, production systems follow linear supply chains, from extraction, through (pre-)processing, to use and final disposal (Korhonen et al., 2018). However, such linear patterns of production and consumption are not sustainable and lead to resource depletion and environmental degradation (Merli et al., 2018). Circular economy provides an alternative that is cyclical and offers the potential for progress towards sustainable development (Schroeder et al., 2019). Circular economy, with its cradle-to-cradle (also closed-loop or zero-waste) philosophy, focuses on optimizing environmental unburdening ("doing more good"), as compared to optimizing environmental burdening ("doing less harm") (Toxopeus et al., 2015). It offers an opportunity for sustainable production and consumption based on limitless resources and continuous growth (Govindan and Hasanagic, 2018) and is closely connected to resource sustainability, emission reduction and waste management (Koh et al., 2017).

Circular economy has recently gained considerable attention in industry (Niero et al., 2018), from several governments (Korhonen et al., 2018), researchers and others (Merli et al., 2018). It could be viewed as the integration of self-sustaining production supply networks (Genovese et al., 2017) and sustainability, with concomitant closing of material and energy loops. It is interlinked with the industrial ecology concept, which aims at a continuous exchange of energy and materials within production systems in a sustainable way (Arbolino et al., 2018), with the ultimate goal of achieving zero impact on the environment (Leigh and Li, 2015). A suitable means of designing closed-loop or circular production systems is the concept of sustainable supply chain networks (Winkler, 2011).

Sustainable supply chain networks need to balance economic, environmental and social performance. However, views on what constitutes "sustainable", differ at various levels, at the levels of individuals, companies, local communities, governments and globally, owing to differences in values, interests, contexts (Mascarenhas et al., 2014) and diversities. Sustainability is, thus, defined and assessed differently at the micro (smaller-scale) and macro (larger-scale) levels (Zore et al., 2016). It is important that a priority for supply network designs should be to create them to be truly sustainable and not just less unsustainable

Faculty of Chemistry and Chemical Engineering, University of Maribor, Slovenia.
Emails: zan.zore@um.si; lidija.cucek@um.si
* Corresponding author: zdravko.kravanja@um.si

(Pagell and Shevchenko, 2014). A truly sustainable supply chain network exhibits at least non-negative economic performance, while doing at least no harm to environment and society (Pagell and Shevchenko, 2014). Synthesis of more sustainable and especially truly sustainable systems is, thus, a complex multi-objective problem (Azapagic et al., 2016). It is also important, that sustainability of systems be evaluated from a longer term perspective, across an entire system's lifetime (Zore et al., 2018b).

In this Chapter, the methodology for the synthesis of truly sustainable system-wide supply networks is introduced by using a mathematical programming approach, with the objective of maximizing Sustainability Net Present Value. Furthermore, the proposed methodology is demonstrated on the synthesis of larger, subcontinental-scale supply networks of Central Europe, producing food, fuel and renewable electricity. The Central Europe-"subcontinental" scale case study demonstrates partial transition from the current, mainly fossil-based energy supply, to renewable energy supply over a time horizon of 20 years (between 2020 and 2040).

2. Concept and Approach for the Synthesis of Truly Sustainable Renewable Supply Networks

In this section, a methodology for truly sustainable synthesis of renewable-based supply networks for producing energy and bioproducts is described. It is based on (i) a concept of System-Wide Supply Networks (SWSN) and the related multi-layer superstructure, (ii) a concept of Sustainability Net Present Value, and (iii) the use of a mathematical programming approach.

2.1 Concept of system-wide supply networks and a multi-layer superstructure

A system-wide supply network is a network comprising multiple supply networks, e.g., agricultural, chemical, biochemical and energy supply networks for shared production of food, fuel and electricity. It can represent fossil- and/or renewable-based production spread locally, regionally or even globally; for example, Figure 1 illustrates a continental (European) SWSN. Each constituent supply network is further composed of separate elements. The synthesis of (bio)chemical supply networks deals with the integration of layers from molecules and process units, to production plants and their networks. In the case of food and biomass supply and demand networks, we deal with planting and harvesting, collecting and pre-processing, processing in biorefineries, and distributing products to end users. Similarly, the energy supply network is composed of a basic layer where energy is released by means of chemical, nuclear or other physical phenomena; the next layers deal with conversion and the last one with energy distribution and use. Each of these supply networks is accompanied by a suitable superstructure; for example, Figure 2 shows a four-layer superstructure for a larger-scale biomass supply and demand network where the arrows represent the flows q between and within the layers. For more details regarding the four-layer superstructure and its mathematical formulation the readers are referred to the paper by Čuček et al. (2014a). Note that in a SWSN, all these four layers belonging to different supply networks are merged into an overall network and constitute an overall superstructure.

2.2 Concept of sustainability net present value

Generally, the synthesis of sustainable systems is a complex, multi-criteria problem, since it should offer good solutions from all three sustainability pillars: Economic, environmental and societal. Research studies have shown that different criteria provide different "optimal" solutions (Novak Pintarič and Kravanja, 2015). Multi-criteria optimization is of special importance if we want to achieve the best sustainable solutions. Among the methods that enable simultaneous optimization of sustainability indicators (economic, environmental and social) are the methods called Sustainability Profit (SP) (Zore et al., 2017) and Sustainability Net Present Value (SNPV) (Zore et al., 2018b) from different micro- and macroeconomic perspectives (Zore et al., 2018a), which are defined on a monetary basis and,

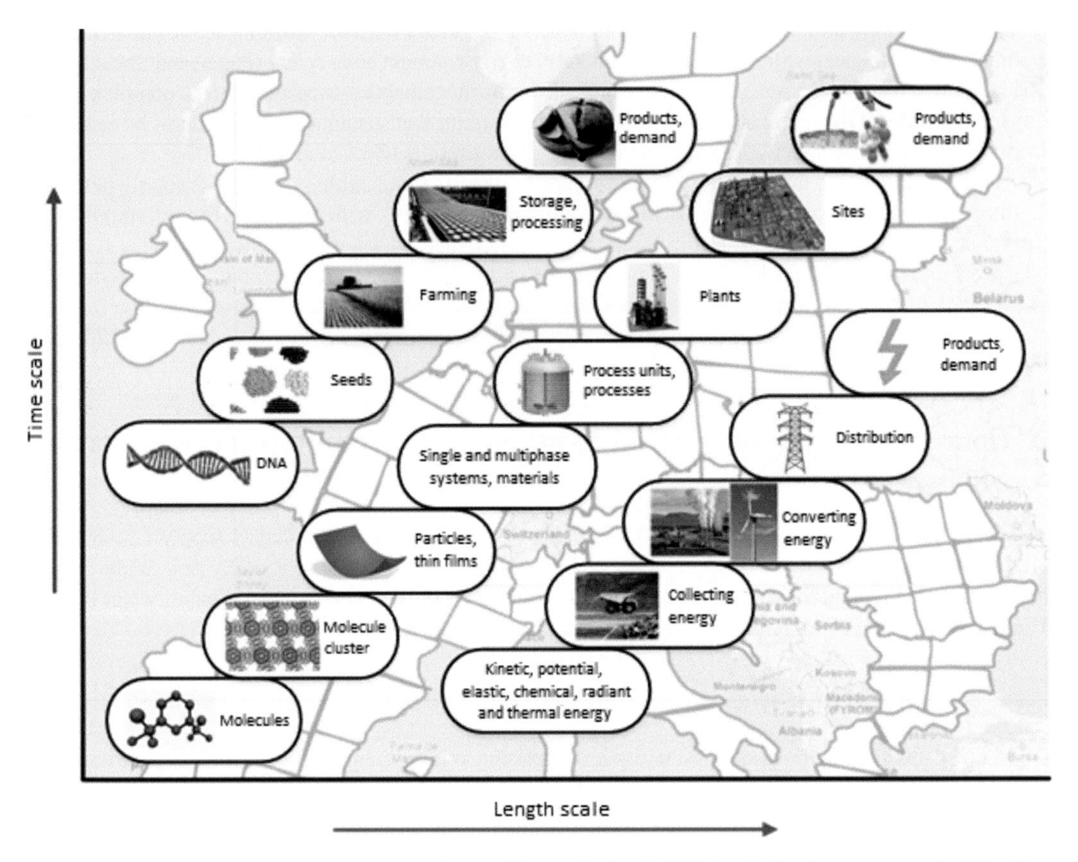

Figure 1. Continental system-wide supply network comprising agricultural, (bio)chemical and energy networks (after Zore et al., 2018b, based on representation of chemical supply chains by Marquardt et al., 1999).

thus, enable simple merging of all three indicators. In the case of SNPV, it is composed of Economic ($NPV^{Economic}$), Eco (NPV^{Eco}), and Social (NPV^{Social}) net present values, see equation (1):

$$\max SNPV = NPV^{\text{Economic}} + NPV^{\text{Eco}} + NPV^{\text{Social}} \quad (1)$$

This economic-based method helps to avoid the use of weighting, normalization and multi-dimensionality of the problem and is easy to interpret and understand. Since separate criteria are now merged, a multi-objective optimization problem is reduced to a single-objective one, which is particularly convenient for the synthesis of larger scale systems—and regional renewable supply networks are among such larger scale systems. Note that SNPV is an extension of SP, since the annual cash flows used to calculate NPV are almost identical to SP. Although SNPV is somewhat more difficult to calculate, it possesses two important advantages over SP; it considers the time value of money and is suitable for sustainability assessment over a system's entire lifetime, which, for strategic long-term investment planning, enables one to obtain appropriate trade-off solutions between economic efficiency, environmental (un)burdening and social responsibility. Note that, besides SNPV, which is located at the intersection of all three sustainability pillars, there are three different binary combinations: Viability NPV as Economic plus Eco NPVs, Equitability NPV as Economic plus Social NPVs, and Bearability NPV as Eco plus Social NPVs—see Figure 3. We can calculate SNPV from different viewpoints: Micro- (industrial), macro- (industrial plus governmental), and wider macro-economic (industrial, plus governmental, plus employees). Since sustainable development should be directed towards improving general welfare, a wider macro-economic view has been considered for SNPV calculation.

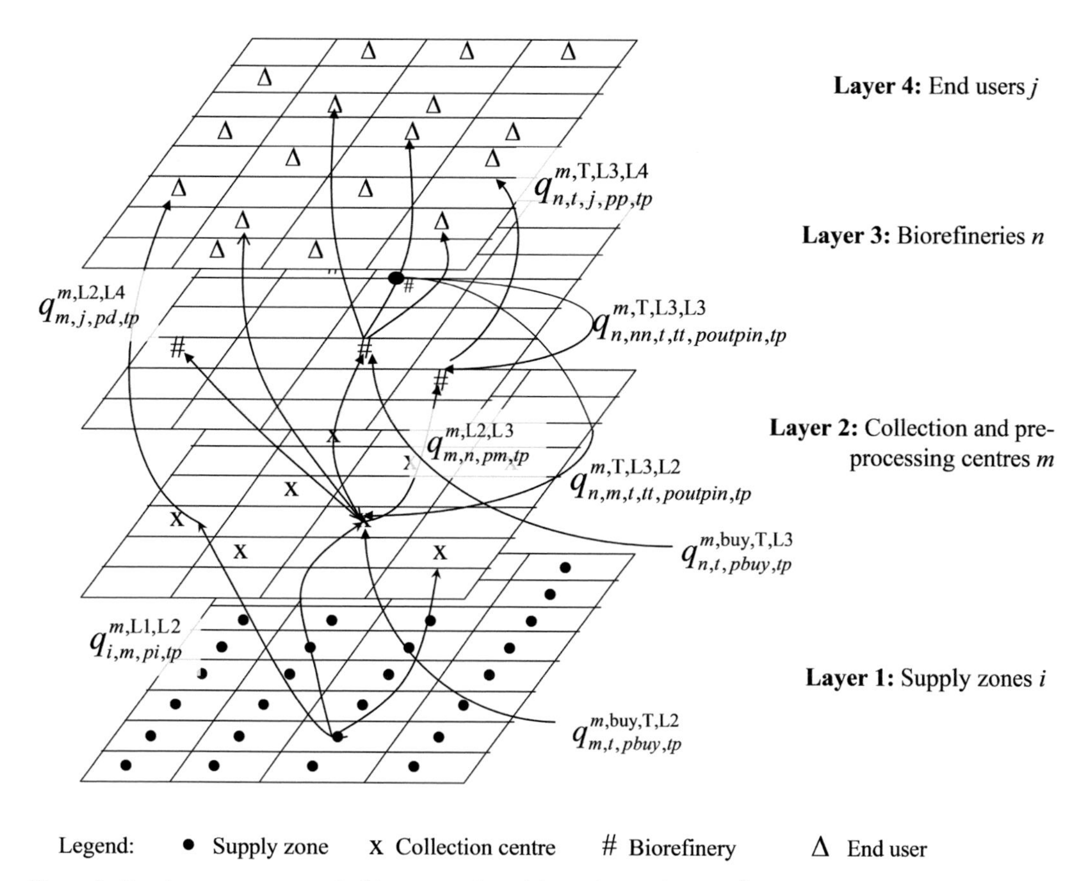

Figure 2. Four-layer superstructure for biomass supply and demand networks (after Čuček et al., 2014a, adapted from Lam et al., 2011).

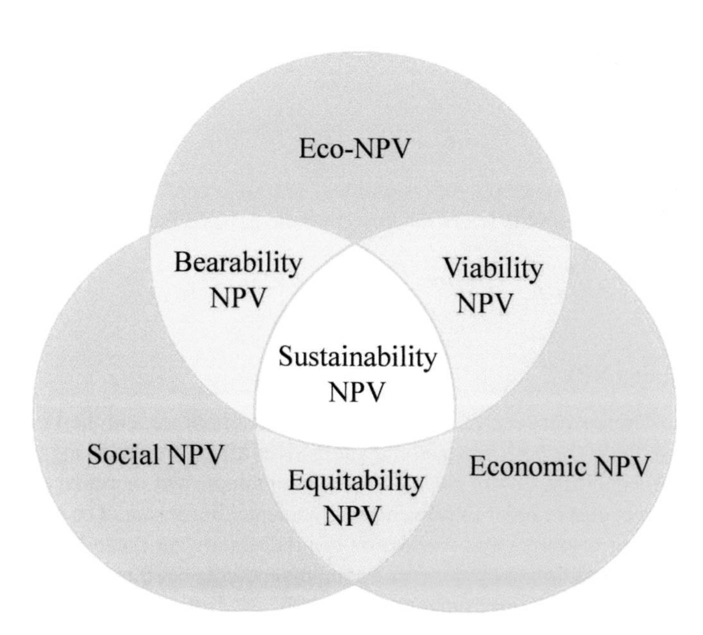

Figure 3. Sustainability net present value is located at the intersection of all three sustainability pillars (from Zore et al., 2018b, based on a representation of sustainability by Dréo, 2016).

2.2.1 *Derivation of sustainability net present value*

Note that SNPV is defined as incremental SNPV (ΔSNPV), i.e., as the difference between the new and old system's SNPVs, where each NPV is calculated over lifetime T as a sum of the corresponding incremental net annual cash flows (ΔFC_t), each one being discounted from its time t to the present time ($t = 0$) with a discount rate of r_d.

$$\begin{aligned}\Delta SNPV &= SNPV^{\text{New}} - SNPV^{\text{Old}} \\ &= \Delta NPV^{\text{Economic}} + \Delta NPV^{\text{Eco}} + \Delta NPV^{\text{Social}} \\ &= \sum_{t=0}^{T} \frac{\Delta FC_t^{\text{Economic}}}{(1+r_d)^t} + \sum_{t=0}^{T} \frac{\Delta FC_t^{\text{Eco}}}{(1+r_d)^t} + \sum_{t=0}^{T} \frac{\Delta FC_t^{\text{Social}}}{(1+r_d)^t}\end{aligned} \tag{2}$$

2.2.1.1 Incremental Economic NPV

This is defined as the difference between the Economic NPV of new alternatives and the Economic NPV of old alternatives. For the purpose of simplicity, we assume that old alternatives operate close to zero NPV. This assumption allows us to define incremental economic net annual cash flow only in terms of new alternatives. From micro-economic or industrial perspective, it is defined as the after-tax surplus of revenue (R_t) and additional revenue from governmental subsidies or higher redemption prices (AR_t) over expenditures (E_t), eco tax ($E_t^{\text{eco tax}}$) and annual depreciation (D_t^{Economic}); plus D_t^{Economic}, minus investments (I_t^{Economic}) plus possible salvage value (SV_t^{Economic}), all defined at year t.

$$FC_t^{\text{Company}} = (1-r_t)(R_t + AR_t - E_t - E_t^{\text{eco tax}} - D_t^{\text{Economic}}) + D_t^{\text{Economic}} - I_t^{\text{Economic}} + SV_t^{\text{Economic}} \tag{3}$$

Incremental Economic NPV from an industrial perspective is then defined:

$$\Delta NPV_{\text{Industry}}^{\text{Economic}} = \sum_{t=0}^{T} \frac{(1-r_t)(R_t + AR_t - E_t - E_t^{\text{eco tax}} - D_t^{\text{Economic}}) + D_t^{\text{Economic}} - I_t^{\text{Economic}} + SV_t^{\text{Economic}}}{(1+r_d)^t} \tag{4}$$

From the wider macro-economic perspective (industrial, plus governmental, plus employees), the definition of net annual cash flow becomes much simpler, because taxes paid by companies and governmental incomes from these taxes are cancelled out.

$$FC_t^{\text{Economic}} = FC_t^{\text{Company}} + FC_t^{\text{Goverment}} = R_t + AR_t - E_t - I_t^{\text{Economic}} + SV_t^{\text{Economic}} \tag{5}$$

Note that net annual cash flow for employees is omitted, since it is considered in Social NPV. Incremental Economic NPV from the wider macro-economic perspective is finally defined:

$$\Delta NPV^{\text{Economic}} = \sum_{t=0}^{T} \left(R_t + AR_t - E_t - I_t^{\text{Economic}} + SV_t^{\text{Economic}}\right) / (1+r_d)^t \tag{6}$$

2.2.1.2 Incremental Eco NPV

It is defined as the difference between the Eco NPV of new alternatives and the Eco NPV of old ones, where eco annual cash flows for both new and old alternatives are calculated using eco-cost coefficients (Delft University of Technology, 2019). Eco-cost coefficients represent marginal costs that one would have to spend per product unit in order to prevent environmental burdening. The eco-cost of a company (EC_t), related to the use of resources and production of products at year t, can be calculated as the sum of mass and energy flows multiplied by the corresponding eco-cost coefficients. Note that net eco-cost related to the transition from old, environmentally harmful production alternatives to greener ones is usually negative because the eco-cost related to new alternatives (EC_t^{new}) is usually smaller compared to that for old alternatives (EC_t^{old}). By shutting down old alternatives, charges related to their eco-cost are released and saved. Their EC_t^{old} thus represents the eco benefit (EB_t). The difference between EB_t and the eco-cost related to new alternatives (EC_t^{new}) is the eco-profit (EP_t) which, in the definition of Eco NPV, is equal to its incremental eco net annual cash flowΔFC_t^{Eco}. If production alternatives need subsidies or higher redemption product prices (AR_t) in order to operate with a positive economic NPV, this extra money resembles additional eco-cost charged in order to avoid burdening the environment. Incremental eco net annual cash flow is then defined as:

$$\Delta FC_t^{\text{Eco}} = -(EC_t^{\text{new}} - EC_t^{\text{old}}) - AR_t = EB_t - EC_t - AR_t \tag{7}$$

and incremental Eco NPV as:

$$\Delta NPV^{\text{Eco}} = \sum_{t=0}^{T} \left(EB_t - EC_t - AR_t \right) / (1 + r_{\text{d}})^t \tag{8}$$

2.2.1.3 Incremental Social NPV

Similarly, this is the surplus of the Social NPV of new alternatives over the Social NPV of old ones. Its incremental net annual cash flow is defined as a function of a net number of jobs (new jobs created, minus old jobs lost), ΔN_t^{jobs}, at year t:

$$\Delta FC_t = \Delta SS_t + \Delta SU_t - \Delta SC_t + \Delta SB_t^{\text{Employees}} - \Delta SC_t^{\text{Employees}} = \Delta GS \tag{9}$$

where terms represent:

- ΔSS_t: Social security contributions to the state paid by employees and employers defined as the difference between average gross and net salaries in the production sector, multiplied by ΔN_t^{jobs},
- ΔSU_t: Social unburdening of the state budget because of new job creation defined as the product of ΔN_t^{jobs} and the average state social transfer for unemployed people,
- ΔSC_t: Social cost of the state and the company for employees defined as the product of ΔN_t^{jobs} and the sum of an average state social transfer and an average company's social charge per employee,
- $\Delta SB_t^{\text{Employees}}$: Social benefit from the employees prospective corresponding to net salary plus social benefits provided to a net number (ΔN_t^{jobs}) of employees by the state and companies, and
- $\Delta SC_t^{\text{Employees}}$: From the newly employed employees prospective, this term represents the loss of their unemployment support from the government defined as the product of ΔN_t^{jobs} and the average state social transfer for unemployed people.

Note that, because from the wider macro-economic perspective most of the parts in the above terms cancel each other out, the incremental net annual cash flow simply becomes equal to the sum of gross salaries, ΔGS, defined as the product of ΔN_t^{jobs} and the average employee's gross salary.

Incremental Social NPV is then defined as:

$$\Delta NPV^{\text{Social}} = \sum_{t=0}^{T} \frac{\Delta GS}{(1 + r_{\text{d}})^t} \tag{10}$$

2.2.1.4 Incremental Sustainability NPV

SNPV comprising Economic, Eco and Social NPVs is finally defined:

$$\begin{aligned} \Delta SNPV &= \sum_{t=0}^{T} \frac{\Delta FC_t^{\text{Sustainability}}}{(1 + r_{\text{d}})^t} \\ &= \sum_{t=0}^{T} \left(R_t - E_t - I_t^{\text{Economic}} + SV_t^{\text{Economic}} + EB_t - EC_t + \Delta GS \right) / (1 + r_{\text{d}})^t \end{aligned} \tag{11}$$

Note that term AR_t is cancelled out. For more details, readers are referred to the papers by Zore et al. (2017, 2018a, b). For simplicity, the symbol "Δ" will be omitted in the continuation.

2.2.2 *Truly sustainable solutions*

As a very rough condition, it can be stated that a solution is sustainable if it has positive or at least nonnegative SNPV. However, it may happen that even when a solution possesses very high value of SNPV, one of its NPVs can still be negative. For a system to be truly sustainable, a necessary condition should, therefore, be consideration of all its constituent parts. Any serious deficiency in any sustainability pillar would cause such a system to not sustain current and future challenges and, hence, fail to endure. Therefore, a genuinely sustainable solution should stay within the intersection where all three sustainability pillars are at least nonnegative:

$$\begin{aligned} &NPV^{\text{Economic}} \geq 0 \\ &NPV^{\text{Eco}} \geq 0 \\ &NPV^{\text{Social}} \geq 0 \end{aligned} \tag{12}$$

An even stricter condition is that any solution obtained must also be economically feasible from the industrial point of view. A nonnegative constraint for economic NPV from the industrial perspective, see equation (13), should be imposed on a group of selected production plants as a whole:

$$NPV_{\text{Industry}}^{\text{Economic}} \geq 0 \tag{13}$$

Note that the above constraints form a feasible region for a family of genuinely sustainable solutions where the most sustainable is, in principle, the one that is obtained by maximizing SNPV subjected to the above nonnegativity constraints:

$$\begin{aligned} &\max SNPV = NPV^{\text{Economic}} + NPV^{\text{Eco}} + NPV^{\text{Social}} \\ &\text{s.t.} \quad NPV^{\text{Economic}} \geq 0 \\ &\qquad NPV^{\text{Eco}} \geq 0 \qquad\qquad \text{(SNPV)} \\ &\qquad NPV^{\text{Social}} \geq 0 \\ &\qquad NPV_{\text{Industry}}^{\text{Economic}} \geq 0 \end{aligned} \tag{14}$$

Note also that once maximal SNPV was obtained with its resulting NPVs being maximal, $NPV^{\text{Economic, Max}}$, $NPV^{\text{Eco, Max}}$ and $NPV^{\text{Social, Max}}$, it is possible to obtain any neighboring solution by imposing the following additional constraints:

$$\begin{aligned} &NPV^{\text{Economic}} \leq NPV^{\text{Economic, Max}} - d^{\text{a}} \\ &NPV^{\text{Eco}} \leq NPV^{\text{Eco, Max}} - d^{\text{b}} \\ &NPV^{\text{Social}} \leq NPV^{\text{Social, Max}} - d^{\text{c}} \end{aligned} \tag{15}$$

where d^a, b^b and d^c are the corresponding distances, in either positive or negative directions.

2.3 *General mathematical programming formulation*

Based on the above concept of Sustainability Net Present Value, a general mixed-integer (non)linear programming (MI(N)LP) formulation for a System-Wide Supply Network is posed and the (SWSN-MI(N)LP) problem is presented below:

$$
\begin{array}{lll}
\text{a)} & \multicolumn{2}{l}{\max SNPV = w^{\mathrm{a}} \cdot NPV^{\mathrm{Economic}}(y,x) + w^{\mathrm{b}} \cdot NPV^{\mathrm{Eco}}(y,x) + w^{\mathrm{c}} \cdot NPV^{\mathrm{Social}}(y,x)} \\
 & \text{s.t.} & \\
\text{b)} & h_{l,p,s,t}(x,y) = 0 & \\
\text{c)} & g_{l,p,s,t}(x,y) \le 0 & \forall l \in L,\ \forall p \in P,\ \forall s \in S,\ \forall t \in T \\
\text{d)} & B_{l,p,s,t}\, y + C_{l,p,s,t}\, x \le b_{l,p,s,t} & \\
\text{e)} & NPV^{\mathrm{Economic}}(y,x) \ge 0 & \text{(SWSN-MI(N)LP)} \qquad (16) \\
\text{f)} & NPV^{\mathrm{Eco}}(y,x) \ge 0 & \\
\text{g)} & NPV^{Social}(y,x) \ge 0 & \\
\text{h)} & NPV^{\mathrm{Economic}}_{\mathrm{Industry}}(y,x) \ge 0 & \\
 & x \in X = \left\{ x \in R^{n} : x^{LO} \le x \le x^{UP} \right\} & \\
 & y \in Y = \{0,1\}^{m} &
\end{array}
$$

Different objectives can be composed by setting suitable 0–1 values to the weightings w^a, w^b and w^c. When all values are 1, the objective function (a) would resemble the SNPV to be maximized. Equality constraints (b) usually represent mass and energy balances, and design equations. Inequality constraints (c) denote product quantity and quality specifications, operational, environmental, and feasibility constraints. Logical constraints (d) are used for the selection/rejection of sustainable alternatives, e.g., routes, zones, processes, etc. The remaining inequality constraints (e-h) represent nonnegativity sustainability constraints. Note that constraints (b-d) are defined at different levels $l \in L$, for different production technologies $p \in P$, across several supply networks, $s \in S$, at time periods $t \in T$ over the whole system's lifetime. Variables x in equation (16) represent continuous variables and y binary variables/decisions related to topology of supply network.

The following section presents an application of the above methodology to the synthesis of a supply network producing food, fuel and electricity for targeting the partial transition from fossil-based to renewable resources on a subcontinental scale of Central Europe over a time horizon of 20 years.

3. Case Study

The SWSN superstructure combines a heat-integrated biorefinery supply network, a renewable electricity supply network and a food supply network—see Figure 1. A case study of supply networks spread over Central Europe is considered, the chosen countries being Austria, Czech Republic, Germany, Hungary, Poland, Slovakia and Slovenia. Corn and wheat are feedstocks for food supply; solar photovoltaics, wind turbines and geothermal plants are used for electricity generation; and corn stover, wheat straw, miscanthus, algae oil, waste cooking oil and forest residue are feedstocks for biofuel supply. The main products are food, bioethanol and green gasoline as gasoline substitutes, biodiesel and Fischer-Tropsch (FT) diesel as diesel substitutes, as well as hydrogen and electricity. Possible technology routes are presented in Figure 4. The superstructure, which is an extension of the one shown in Figure 2, is based on four layers, L1–L4, including harvesting sites at L1, storage, pre-processing of raw materials to intermediate products and production of electricity at L2, bio-refineries at L3, and demand locations at L4 (Zore et al., 2018b). Each layer is divided into 33 zones across Central Europe. It is assumed that locations of sites are at zone centers. Transportation is calculated within and between zones. To reduce the size of the model, it is assumed that the distances for the transportation of biomass and waste, energy and products are limited (Lam et al., 2011). For solar, wind and geothermal energy, it is assumed that they cannot be transported until they have been transformed into electrical energy. Up to 10% of the total area of each zone is assumed to be devoted to satisfying food and biofuel demand, and up to 1% to producing

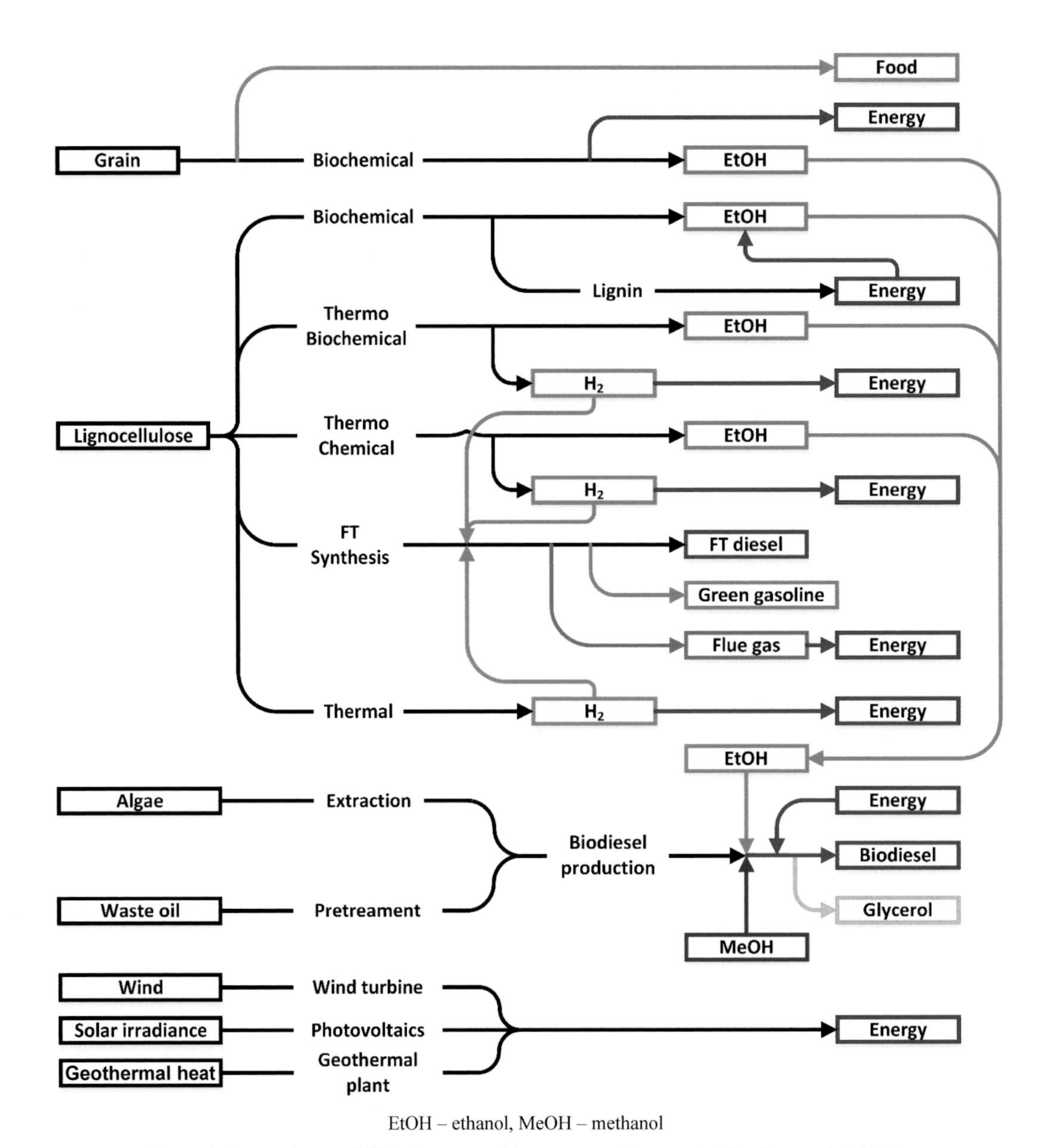

EtOH – ethanol, MeOH – methanol

Figure 4. Integration possibilities in a renewable energy supply network (after Zore et al., 2018b).

electricity from renewable sources, and when all areas are combined, a total of up to 11% of the area is assumed. Note that almost 8% of the land in Europe is already used for corn and wheat production (FAOSTAT, 2018). For the additional land area (2%), an eco-cost of 0.5 €/m^2 (Hendriks and Vogtländer, 2004) is assumed, except for afforestation; in this case, an eco-benefit of 0.5 €/m^2 is considered.

The supply network model is a multi-period, mixed-integer linear programming (MILP) model, which considers monthly time periods for biomass, food and biofuels, and hourly time periods for solar, wind and geothermal energy and electricity (Čuček et al., 2016). In addition, 20 yearly periods are defined over the 20-year time horizon for replacing fossil-based fuels with biofuels, and electricity with renewable electricity. Owing to the large computational times, the number of periods has been reduced to 6 periods per month and 4 periods per day (Zore et al., 2018b). All the patterns of demand and yields are the same as in the model used for the continental-size supply network by Zore et al. (2018b).

The demand for food is assumed to remain constant throughout the year, while the demand for fuel and electricity is assumed to increase over time. It is also assumed that all the zones in a specific country have the same demand for food, biofuel and electricity. The demand for food (corn and wheat) should be satisfied in each country, and the share of demand for biofuels and renewable electricity is increased over the years, while total demand for liquid fuels is decreased (by up to 50% of current consumption in 20 years) because of the transport sector switching from liquid fuel to plug-in vehicles, and so increasing electricity consumption. Note that, on the 11% total of land in Central Europe, it is not possible to achieve entire replacement of liquid fuel and electricity by biofuel and renewable electricity (100% of demand coming from renewable sources); thus, lower substitution levels were chosen (up to 30% for gasoline, 20% for diesel and 50% for electricity in the final year).

Data for the biorefinery and food supply network are taken from Čuček et al. (2014a, b), while for the electricity supply network, they are obtained from the U.S. Energy Information Administration (2016), and for the sake of price change projections, also from the U.S. Energy Information Administration (2013). The data related to sustainability NPV calculation are obtained from Zore et al. (2018b). All the results are based on a 3% interest rate, a 7% discount rate and a 20-year lifetime. The selected time horizon of 20 years is due to typical relatively shorter-term planning and assumed plant life of 20 years (Caputo et al., 2005). All the prices except those for fuel and electricity are kept constant, while higher redemption product prices (or subsidies) for fuels and electricity are needed in order to get some extra money, AR_t in equation (3), so as to achieve positive economic viability ($NPV^{Economic}$) and sustainability (*SNPV*) of such large-scale production of renewably-sourced electricity and biofuel. Therefore, a 2% price rise in fuel and electricity per year is introduced.

Table 1 shows the basic economic and environmental data for raw materials and products. Environmental data is represented by eco-cost coefficients as obtained from Delft University of Technology (2019). More detailed data considered in the case study are available in the literature (Čuček et al., 2014a, b), and the data related to social sustainability in Zore et al. (2017).

Table 1. Prices and eco-cost coefficients for raw materials and products.

Raw materials	**Price at year 1 (\$/kg)**	**Eco-cost coefficient (\$/kg)**	**Products**	**Price at year 1 (\$/kg)**	**Eco-cost coefficient (\$/kg)**
Corn grain	0.250	0.290	Corn grain	0.300	0.290
Wheat	0.300	0.100	Wheat	0.330	0.100
Corn stover	0.060	0.205	Ethanol	1.115* | 1.745**	0.378
Wheat straw	0.060	0.100	Green gasoline	1.775* | 2.778**	0.379
Miscanthus	0.030	0.050	MeOH-diesel	1.169* | 1.758**	0.430
Forest residue	0.040	0.041	EtOH-diesel	1.169* | 1.758**	0.430
Algae oil	0	0.000	FT-diesel	1.329* | 1.758**	0.437
Cooking oil	0.200	0.282	Hydrogen	1.580	1.054
Water	$2.5 \cdot 10^{-3}$	0.378	**Products**	**Price at year 1 (\$/kWh)**	**Eco-cost coefficient (\$/kWh)**
Ethanol	1.920***	1.989			
Hydrogen	2.200***	0.150	Electricity	0.100	
Methanol	0.350	0.290	- from wind turbines		0.010
			- from solar photovoltaics		0.028
			- from geothermal plants		0.002
			- from coal power plants		0.165

* market price.

** subsidized price.

*** assumed prices of products are slightly higher because of transport, additional storage and other cost.

The data for calculating the number of employees for the construction, operation and maintenance of processes are taken from the JEDI models (NREL, 2015), data on earnings from Wikipedia (2016), summarized from the national statistical offices, and the number of unemployed from Eurostat (2016).

The MILP models applied to a case study of Central Europe contain about 1,200,000 constraints, 8,000,000 continuous and 10,600 binary variables. The solutions presented in the following are obtained with an optimality gap of 5% in about 72 hours of CPU time, using a GAMS/CPLEX solver on an HPC server DL580 G9 CTO with 4 processors (32 core) Intel® Xeon® CPU E5-4627 v2 @ 3.30 GHz and 768 GB RAM.

Table 2 shows the main results when maximizing Economic NPV and Sustainability NPV from a wider macroeconomic perspective. Note that both objective values are significant, 194,517 M€ is maximal Economic NPV, and 378,438 M€ is maximal SNPV, of which Economic NPV is 130,589 M€, Eco NPV 106,359 M€ and Social NPV 141,491 M€. Note also that, in the calculation of Eco NPV, see equation (8), values of NPV of eco-benefit (EB), NPV of eco-cost (EC) and that from higher redemption prices for biofuels and electricity from alternative resources (AR) are 501,295 M€, 286,462 M€ and 108,474 M€. Rather than spending 501,295 M€ of eco-cost to avoid burdening the environment caused by fossil-based production plants (which corresponds to EB), new renewable-based production plants are proposed to be installed with eco-cost being significantly lower, 286,462 M€. The difference of the two corresponds to gross Eco NPV of 214,883 M€, from which 108,474 M€ are deducted, which are earned by higher redemption prices, thus obtaining the final value of Eco NPV (106,359 M€). Note that, by charging 286,462 M€ of eco-cost to new renewable alternatives, an assumption of zero waste operation is imposed (considering solely the waste inside the company gate; Zore et al., 2017). Without that, the final value of SNPV would be even higher. This assumption is important because it enables through maximization of SNPV identifying those unburdening alternatives that at the same time burden the environment the least. The final interpretation of the maximal SNPV value of 378,438 M€ is that in order to make modifications of this supply network producing food, fuel and electricity sustainable, instead of spending 501,295 M€ to avoid burdening the environment, owing to current, mainly fossil-based energy supply, 286,462 M€ would be spend for zero waste operation of renewable-based supply networks and 108,474 M€ would be charged through higher redemption prices (or subsidies) in order to make new production plants economically viable, while still achieving Economic NPV of 130,589 M€ and Social NPV of 141,491 M€, both calculated from a wider-macroeconomic perspective.

Different trade-offs between Economic NPV and eco-NPVs are obtained when optimization is performed with economic and sustainability criteria. When maximizing the economic criterion, higher NPV^{Economic}, but significantly smaller NPV^{Eco} and NPV^{Social} are obtained than in the case of maximizing the sustainability criterion. When maximizing SNPV, significant increases in NPV^{Eco} (for 89,848 M€) and NPV^{Social} (for 123,788 M€) are obtained, which together more than triple the increase, compared to the decrease in NPV^{Economic} (decrease of 63,928 M€), giving rise to a significant increase in SNPV—from 228,731 M€ to 378,439 M€.

When maximizing *SNPV*, the results show that even with their lower economic *NPV* value, renewable technologies are still the more sustainable solution when producing biofuel and electricity. The main differences between the most economic and the most sustainable solutions for satisfying market demand occur in the selection of raw materials, the type of product and the locations selected. Afforestation instead of miscanthus cultivation is the main reason for higher NPV^{Eco}. Maximizing *SNPV* largely promotes a

Table 2. Results when maximising economic and sustainability net present values.

Economic items	Maximization criteria	
	$NPV^{Economic}$	*SNPV*
NPV^{Economic} (M€)	**194,517**	130,589
NPV^{Eco} (M€)	16,511	106,359
NPV^{Social} (M€)	17,703	141,491
***SNPV* (M€)**	228,731	**378,439**

circular economy, renewable electricity production and afforestation, and thus solutions that allow for greater unburdening, and create less of a burden on the environment.

Table 3 shows the main results from the optimization of Economic and Sustainability NPVs in terms of area used, food supply, demand for fuel substitutes and electricity, the raw materials and technology used to produce renewable fuel and electricity and the number of employees in renewable energy production.

All the available area (10%) is used for food and biofuel production or for additional afforestation. The sustainability perspective would suggest that up to about 1.5% of the area be afforested, given the eco-benefits related to afforestation, as explained above. However, in year 20, the percentage of afforested area is lower, by about 0.01%, owing to the higher demands. On the contrary, from an economic perspective, the afforested area suggested is negligible.

With both criteria, the demand assumed for food, fuel and electricity is satisfied in all the years. When maximizing $NPV^{Economic}$, gasoline substitute requirements in the first year are satisfied by ethanol alone, while when *SNPV* is maximized, green gasoline is also produced. The capacity of both ethanol and green gasoline mainly increases over the years. For the substitution of diesel fuel, both maximization criteria suggest producing FT-diesel in larger quantities than biodiesel from waste cooking oil using methanol as a catalyst.

The capacity of both waste cooking oil and biodiesel production remains constant over the years. The biggest difference in raw material use is obtained for miscanthus and forest residue. $NPV^{Economic}$ prefers miscanthus because of its higher yield and lower price, and thus good economic performance, while SNPV prefers forest residue (afforestation) because of its sustainability. In the final year, when the share of fuel substitutes is the highest, both optimization criteria mainly suggest miscanthus cultivation and use.

Various technologies have been selected to produce biofuel and electricity. Among the technologies producing biofuel are gasification of lignocellulosic biomass and further syngas fermentation, FT synthesis for production of bioethanol, hydrogen, green gasoline and FT diesel, and transesterification of waste cooking oil with methanol as a catalyst. To produce renewable electricity, wind turbines are preferred over solar photovoltaics and geothermal power plants because of their better economic performance.

Table 3 shows that when maximizing *SNPV*, almost twice the number of employees is required in the final year as compared to the case when maximizing $NPV^{Economic}$. The requirements to satisfy demand over the years are met with about 295,000 employees when maximizing $NPV^{Economic}$ and with about 429,000 employees when maximizing *SNPV*. Most of the workers are required in the construction and manufacturing sectors for electricity production, specifically in building and operating wind turbines and photovoltaic panels. A higher number of employees is required for constructing and operating photovoltaic panels than for wind turbines and geothermal plants. From an economic perspective, only a small percentage of electricity demand is, thus, satisfied with solar photovoltaics.

Figure 5 shows suggested optimal production of gasoline substitutes over the next 20 years across Central Europe when maximizing *SNPV*. In year 1, production takes place mainly in Germany, western Austria and central Hungary. In year 10, when the demand for gasoline substitutes increases to 20%, higher amounts of biofuel are proposed in Germany and western Austria and in 2 new locations in Poland, but no longer in central Hungary. In the final year, with a further increase in production to 30%, an increase in the production capacity of already selected locations is suggested.

Figure 6 shows the distribution of diesel substitutes across Central Europe. At the start of the project, when 10% of diesel fuel should be replaced, the selected biorefineries for biodiesel production are allocated in the same locations as for gasoline substitutes. This is due to the lower transport and investment cost if biorefineries producing renewable ethanol and diesel are at the same location. In year 10, the increase in capacity is mainly observed in parts of Germany, western Austria and in 2 new locations in Poland. In the final year, when 20% of current diesel consumption should be replaced by renewable diesel fuel, the same locations are suggested but with increased capacity.

Figure 7 shows the distribution of hydrogen production over the years. Hydrogen is produced as a main product by gasification and hydrogen production from lignocellulosic biomass (Martín and Grossmann, 2011b), and as a by-product by gasification and syngas fermentation, gasification and catalytic

Table 3. Main results of economic and sustainability optimisations.

Supply network items	Maximization criteria					
	NPV^Economic^			*SNPV*		
	Year 1	*Year 10*	*Year 20*	*Year 1*	*Year 10*	*Year 20*
Area used (%)	11.00	11.00	11.00	11.00	11.00	11.00
afforested (%)	0.003	0.003	0.003	1.53	1.53	0.013
Food (kt/y)						
corn grain	24,769	24,769	24,769	24,769	24,769	24,769
wheat	37,226	37,226	37,226	37,226	37,226	37,226
Food demand satisfied (%)	100.00	100.00	100.00	100.00	100.00	100.00
Fuel demand satisfied (%)						
gasoline	10.00	20.00	30.00	10.00	20.00	30.00
diesel	10.00	15.00	20.00	10.00	15.00	20.00
Electricity demand satisfied (%)						
produced total	10.00	30.00	50.00	10.00	30.00	50.00
from wind	9.22	26.82	30.16	5.52	15.67	24.70
from solar	0.78	3.18	19.84	4.48	14.33	25.30
from geothermal	0	0	0	0	0	0
Raw materials (kt/y)						
corn stover	13,235	14,405	14,862	14,862	14,034	13,164
wheat straw	37,970	37,970	37,970	37,970	37,970	37,970
miscanthus	27,520	30,163	33,663	0	0	32,556
forest residue	0	0	0.369	18.365	26.512	1.441
algae	0	0	0	0	0	0
cooking oil	1,301	1,301	1,301	1,556	1,556	1,556
Technologies:*						
hydrogen[1]	•	•	•	•	•	•
dry-grind process[2]						
syngas fermentation[3]	•	•	•	•	•	•
catalytic synthesis[4]						
FT synthesis[5]	•	•	•	•	•	•
WCO methanol[6]	•	•	•	•	•	•
WCO ethanol[7]						
algae methanol[8]						
algae ethanol[9]						
photovoltaics (km^2)	25	102	576	136	445	820
wind turbines	8,714	35,066	44,169	4,757	16,637	31,802
binary cycle geothermal plants	0	0	0	0	0	0
Biofuels (kt/y)						
ethanol	4,469	4,680	9,575	2,917	6,332	9,747
green gasoline		2,673	2,406	974	1,636	2,297
et-diesel**						
me-diesel***	1,248	1,248	1,248	1,493	1,493	1,493
FT-diesel	3,880	6,369	8,858	3,665	6,154	8,643
hydrogen	4,197	4,140	1,844	1,599	1,374	1,348
Number of employees	23,192	51,086	295,145	79,265	237,785	428,730

* technologies:
[1] gasification and lignocellulosic hydrogen production (Martín and Grossmann, 2011a, b),
[2] dry-grind process (Karuppiah et al., 2008),
[3] gasification and syngas fermentation and [4] gasification and catalytic synthesis of lignocellulosic biomass (Martín and Grossmann, 2011a, b),
[5] gasification, FT synthesis and hydrocracking (Martín and Grossmann, 2011c),
[6] biodiesel production from waste cooking oil with methanol, and [7] ethanol, and from [8] algal oil with methanol (Martín and Grossmann, 2012), and [9] ethanol (Severson et al., 2013);
** biodiesel produced using ethanol and *** methanol as alcohol.

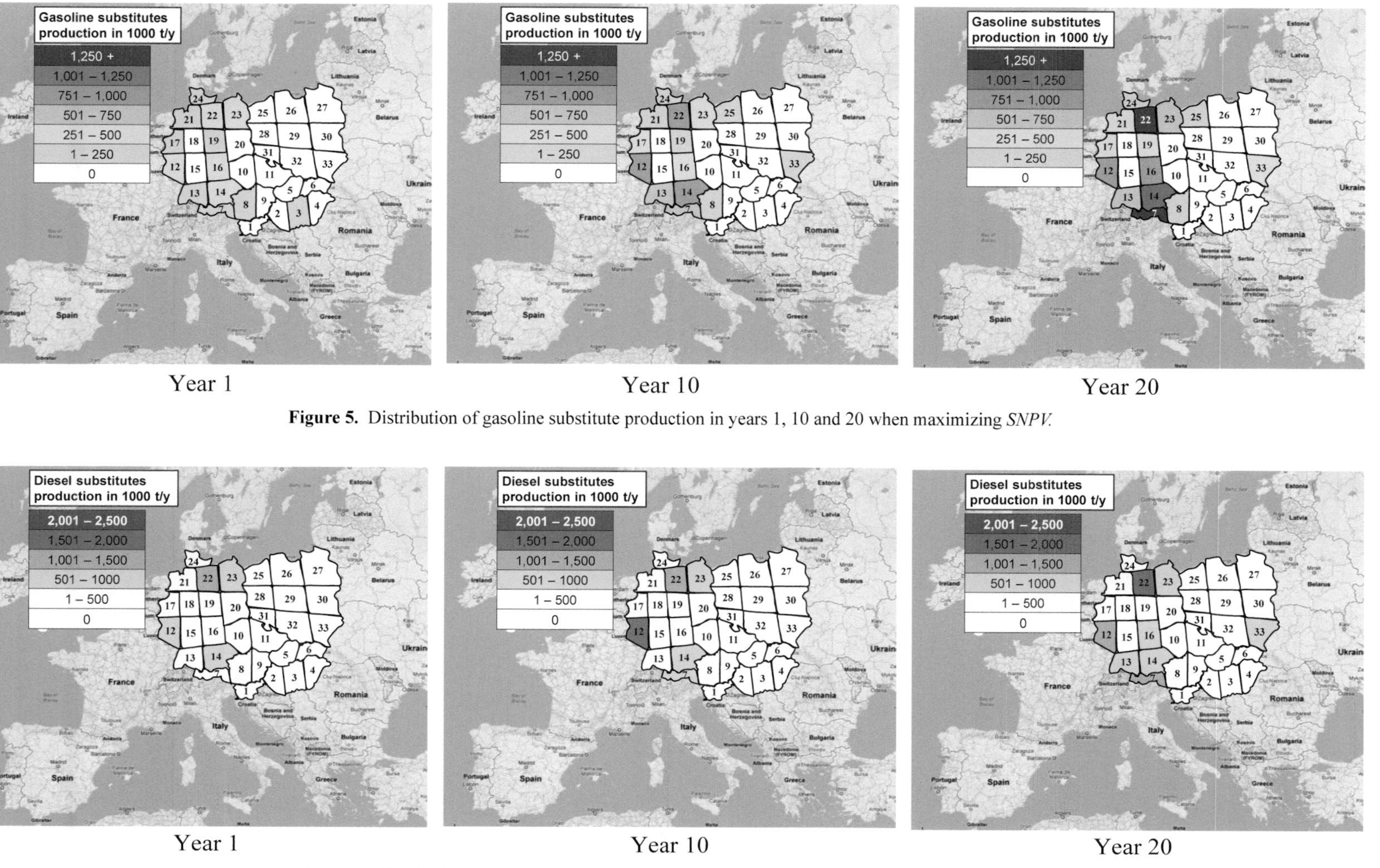

Figure 5. Distribution of gasoline substitute production in years 1, 10 and 20 when maximizing *SNPV*.

Figure 6. Distribution of diesel substitute production in years 1, 10 and 20 when maximizing *SNPV*.

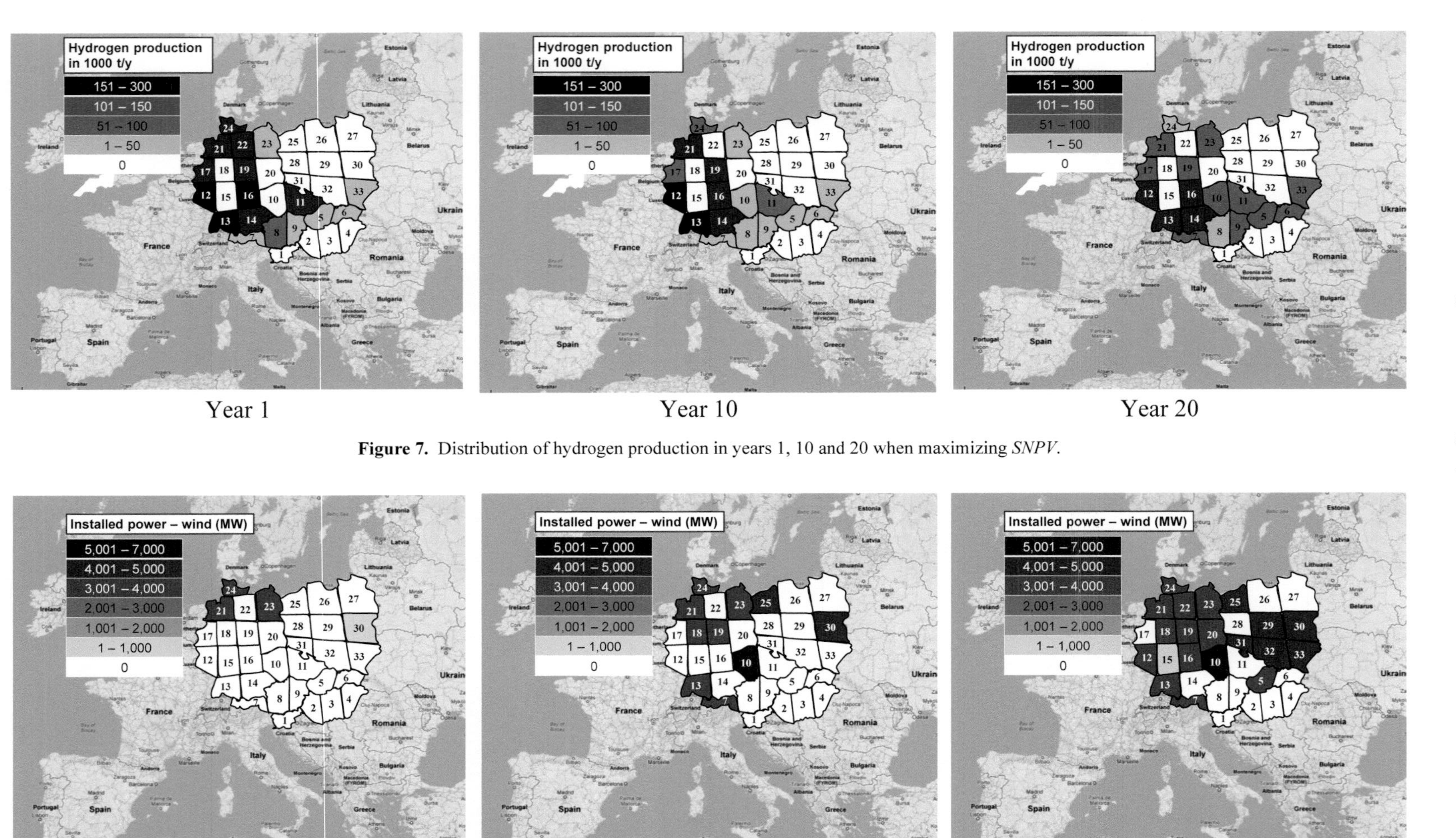

Figure 7. Distribution of hydrogen production in years 1, 10 and 20 when maximizing *SNPV*.

Figure 8. Distribution of installed wind turbines in years 1, 10 and 20 when maximizing *SNPV*.

synthesis (Martín and Grossmann, 2011a) and by gasification, FT synthesis and hydrocracking (Martín and Grossmann, 2011c) of lignocellulosic biomass. There were no requirements set for hydrogen itself; however, hydrogen could potentially play an important role in various sectors: electricity, heat, industry, transport and energy storage (Staffell et al., 2019). The results in terms of hydrogen production show that its production is reduced when the production of biofuel increases, owing to the lower production of hydrogen as a main product. The locations for hydrogen production are, in principle, the same as those for the production of biofuel.

Figure 8 shows the locations of wind turbines in years 1, 10 and 20 when maximizing *SNPV*. The wind, as a renewable source of electricity, takes precedence over photovoltaics, since it enables electricity production over nights and in winters. In Central Europe, there are only moderate solar resources and low ambient temperatures, especially in winter, therefore, electricity potential from photovoltaics is limited (Šúri et al., 2007).

In year 1, when 10% of electricity must be provided from renewable sources, 4,757 wind turbines are proposed, located in northern Germany and in continental locations with higher wind speed (eastern Poland). When demand for electricity rises to 30% of the final demand (including increases due to energy shift from liquid fuel to electricity), several new locations in central Germany, western Slovakia, western Austria and Poland are selected. At the same time, the number of wind turbines at previously selected locations, is increased, mainly in eastern Poland. In year 20, when the demand for renewable electricity rises to 50% (including transportation fuel shift), a total of 31,802 wind turbines are installed. Besides all the previous locations, in this scenario, wind turbines are selected in most of the zones in Germany, in several zones in Poland and in the western Czech Republic.

The locations of photovoltaic panel installations in years 1, 10 and 20 when maximizing *SNPV* are shown in Figure 9.

In year 1, when 10% of renewable electricity should be supplied, the selected locations for panels are mainly in southern Germany, with a few in northern Germany and eastern Poland, for an area of 136 km^2 in total. In year 10, with the increase in demand to 30%, the suggested locations for panels are in southern, central and eastern Germany, in two zones in Poland, western Austria and the western Czech Republic. To achieve a 50% share of renewable electricity (including additional demand due to the move towards electric vehicles) in year 20, it is suggested that 820 km^2 of photovoltaic panels be installed. Photovoltaic installations are proposed for various zones in Germany, Poland, western Czech Republic, western Austria and western Slovakia. It was demonstrated that, in order to meet the high demand for renewable energy in the future, a combination of technologies should be used.

4. Conclusions

This chapter has presented a methodology for synthesizing truly sustainable supply networks for producing food, fuels, electricity and other products from renewable resources. It is based on the concept of maximizing Sustainability Net Present Value to obtain single trade-off solutions between economic efficiency, environmental (un)burdening and social responsibility, over the whole lifetime of a System-Wide Supply Network. A general MI(N)LP problem has been proposed, which comprises Sustainability Net Present Value as an objective, a multi-period MI(N)LP model for a System-Wide Supply Network, and additional sustainability constraints to direct the optimization procedure towards truly sustainable solutions. The latter includes nonnegative constraints for Economic NPV, Eco NPV and Social NPV because truly sustainable solutions should be positive, or at least nonnegative, with respect to all three sustainability pillars. Since these solutions should be oriented towards improving general welfare, they are proposed to be evaluated and determined from the joint perspective of individuals, companies and governments. Finally, an additional nonnegative constraint for Economic NPV from the industrial perspective has been included in order to achieve economic feasibility of production plants.

The methodology has been applied to a case study of Central Europe supply network over the 20-year transition period from fossil- to renewable-based production of fuel and electricity (30% for gasoline, 20% for diesel and 50% for electricity in the final year) at a fixed demand for food. By

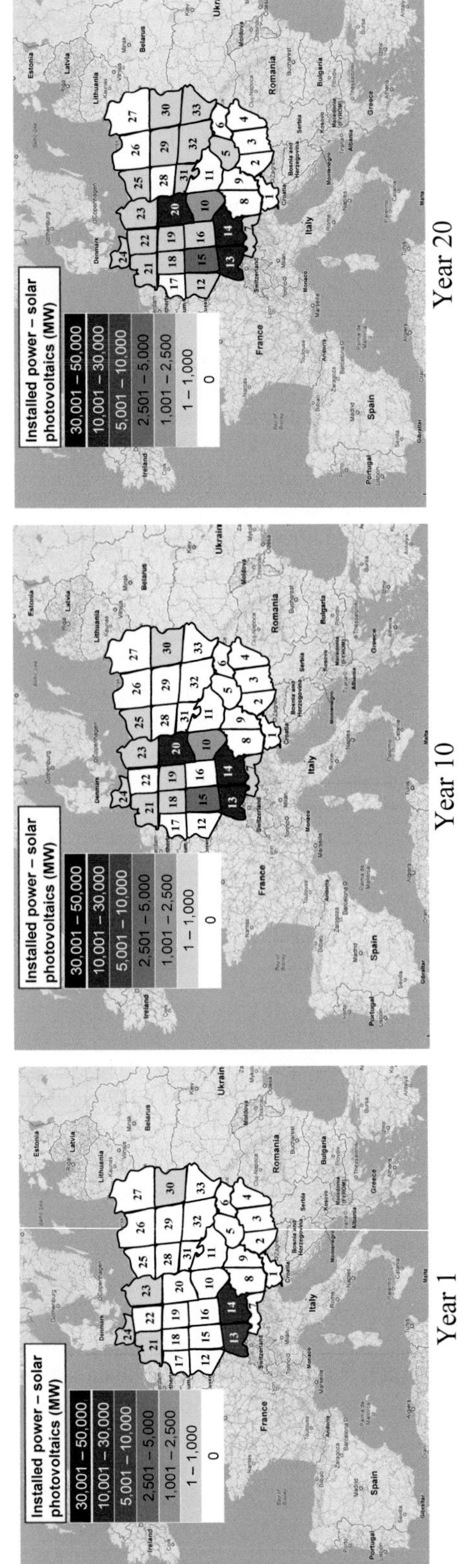

Figure 9. Distribution of installed photovoltaic panels in years 1, 10 and 20 when maximizing *SNPV*.

maximization of SNPV, a significantly positive objective value of 378,438 M€ is obtained. It was shown that around 501,295 M$ would be needed for this transition, according to a rough assessment based on eco-cost coefficients. The results indicate that significantly smaller amounts of money would be needed, with savings of 106,359 M$, which corresponds to Eco NPV, and that an additional 108,473 M$ would be realized as Economic NPV from self-sufficient production plants and 141,491 M$ as Social NPV from the surplus of new jobs created over those lost. Note that, from the industrial perspective, Economic NPV is also positive; however, somewhat higher redemption prices for renewable-based products would be needed (a 2% annual rise for fuels and electricity). This money would represent an additional eco-cost in order to make possible the transition from fossil- to renewable-based production. Note also that this additional eco-cost has been deducted from the Eco NPV. All these positive economic figures, obtained when maximizing SNPV, could be achieved through significantly better allocation of resources, selection of technologies that are suitable based on geographic variability, better production planning and more efficient distribution of products.

It should also be noted, that the production of food, biofuel and electricity, allocation of area for cultivating crops and for wind and solar electricity installation, transportation distances, plant locations, and other factors have been optimized; however, in practice, such solutions might not be attained. There are also significant uncertainty in yields, metheorological conditions, prices, area availability of other items. In some zones, alotting 11% of the area to food and energy production might be too high, while in some zones, the potential area could be more than 11%. Given these factors and other uncertainties, a significantly different optimal solution might be obtained in practice.

In future work, similar studies should be performed; however, these should be on a larger scale, such as the European Union, Europe and/or other continents, and for complete transition from fossil- to renewable-based fuel and electricity within a range of planning time horizons. Even better trade-off solutions between economic efficiency, environmental (un)burdening and social responsibility would be expected at larger scales because more synergy could be exploited than in the case of densely-populated Central Europe. Finally, it would be interesting to compare the results with the other methods of sustainability assessment.

Acknowledgement

The authors are grateful for the financial support from the Slovenian Research Agency (PhD research fellowship contract No. 1000-14-0552, activity code 37498 and core research funding No. P2 0032 and P2 0412).

References

Arbolino, R., De Simone, L., Carlucci, F., Yigitcanlar, T. and Ioppolo, G. 2018. Towards a sustainable industrial ecology: Implementation of a novel approach in the performance evaluation of Italian regions. Journal of Cleaner Production 178: 220–236.

Azapagic, A., Stamford, L., Youds, L. and Barteczko-Hibbert, C. 2016. Towards sustainable production and consumption: A novel DEcision-Support Framework IntegRating Economic, Environmental and Social Sustainability (DESIRES). Computers & Chemical Engineering 91: 93–103.

Caputo, A.C., Palumbo, M., Pelagagge, P.M. and Scacchia, F. 2005. Economics of biomass energy utilization in combustion and gasification plants: Effects of logistic variables. Biomass and Bioenergy 28: 35–51.

Čuček, L., Martín, M., Grossmann, I.E. and Kravanja, Z. 2014a. Multi-period synthesis of optimally integrated biomass and bioenergy supply network. Computers & Chemical Engineering 66: 57–70.

Čuček, L., Martín, M., Grossmann, I.E. and Kravanja, Z. 2014b. Large-scale biorefinery supply network—case study of the European Union. pp. 319–324. *In*: Klemeš, J.J., Varbanov, P.S. and Liew, P.Y. (eds.). Computer Aided Chemical Engineering (Vol. 33): Elsevier.

Čuček, L., Zore, Ž., Krajačić, G., Martín, M., Grossmann, I.E., Boldyryev, S. and Kravanja, Z. 2016. Synthesis of renewable energy supply networks considering different frequencies of fluctuations in supply and demand. *In*: 2016 AIChE Annual Meeting. San Francisco, CA, USA.

Delft University of Technology. 2019. The Model of the Eco-costs / Value Ratio (EVR). <www.ecocostsvalue.com/>. Last accessed: 27.4.2019.

Dréo, J. 2016. Sustainable development. <en.wikipedia.org/wiki/File:Sustainable_development.svg>. Last accessed: 27.4.2019.

Eurostat. 2016. Unemployment statistics. <ec.europa.eu/eurostat/statistics-explained/index.php/Unemployment_statistics>. Last accessed: 15.5.2016.

FAOSTAT. 2018. Crops. <faostat.fao.org/site/567/DesktopDefault.aspx?PageID=567#ancor>. Last accessed: 29.4.2019.

Genovese, A., Acquaye, A.A., Figueroa, A. and Koh, S.C.L. 2017. Sustainable supply chain management and the transition towards a circular economy: Evidence and some applications. Omega 66: 344–357.

Govindan, K. and Hasanagic, M. 2018. A systematic review on drivers, barriers, and practices towards circular economy: A supply chain perspective. International Journal of Production Research 56: 278–311.

Karuppiah, R., Peschel, A., Grossmann, I.E., Martín, M., Martinson, W. and Zullo, L. 2008. Energy optimization for the design of corn-based ethanol plants. AIChE Journal 54: 1499–1525.

Koh, S.L., Gunasekaran, A., Morris, J., Obayi, R. and Ebrahimi, S.M. 2017. Conceptualizing a circular framework of supply chain resource sustainability. International Journal of Operations & Production Management 37: 1520–1540.

Korhonen, J., Honkasalo, A. and Seppälä, J. 2018. Circular economy: the concept and its limitations. Ecological Economics 143: 37–46.

Lam, H.L., Klemeš, J.J. and Kravanja, Z. 2011. Model-size reduction techniques for large-scale biomass production and supply networks. Energy 36: 4599–4608.

Leigh, M. and Li, X. 2015. Industrial ecology, industrial symbiosis and supply chain environmental sustainability: A case study of a large UK distributor. Journal of Cleaner Production 106: 632–643.

Marquardt, W., von Wedel, L. and Bayer, B. 1999. Perspectives on lifecycle process modeling. pp. 192–214. *In*: Malone, M.F. and Trainham, J.A. (eds.). Proceedings of the 5th International Conference Foundations of Computer-aided Process Design, FOCAPD 1999. Breckenridge, Colorado: American Institute of Chemical Engineers.

Martín, M. and Grossmann, I.E. 2011a. Energy optimization of bioethanol production via gasification of switchgrass. AIChE Journal 57: 3408–3428.

Martín, M. and Grossmann, I.E. 2011b. Energy optimization of hydrogen production from lignocellulosic biomass. Computers & Chemical Engineering 35: 1798–1806.

Martín, M. and Grossmann, I.E. 2011c. Process optimization of FT-diesel production from lignocellulosic switchgrass. Industrial & Engineering Chemistry Research 50: 13485–13499.

Martín, M. and Grossmann, I.E. 2012. Simultaneous optimization and heat integration for biodiesel production from cooking oil and algae. Industrial & Engineering Chemistry Research 51: 7998–8014.

Mascarenhas, A., Nunes, L.M. and Ramos, T.B. 2014. Exploring the self-assessment of sustainability indicators by different stakeholders. Ecological Indicators 39: 75–83.

Merli, R., Preziosi, M. and Acampora, A. 2018. How do scholars approach the circular economy? A systematic literature review. Journal of Cleaner Production 178: 703–722.

Niero, M., Olsen, S.I. and Laurent, A. 2018. Renewable energy and carbon management in the Cradle-to-Cradle certification: Limitations and opportunities. Journal of Industrial Ecology 22: 760–772.

Novak Pintarič, Z. and Kravanja, Z. 2015. The importance of proper economic criteria and process modeling for single- and multi-objective optimizations. Computers & Chemical Engineering 83: 35–47.

NREL. 2015. JEDI—Jobs and Economic Development Impact Models. <www.nrel.gov/analysis/jedi/>. Last accessed: 17.5.2016.

Pagell, M. and Shevchenko, A. 2014. Why research in sustainable supply chain management should have no future. Journal of Supply Chain Management 50: 44–55.

Schroeder, P., Anggraeni, K. and Weber, U. 2019. The relevance of circular economy practices to the sustainable development goals. Journal of Industrial Ecology 23: 77–95.

Severson, K., Martín, M. and Grossmann, I.E. 2013. Optimal integration for biodiesel production using bioethanol. AIChE Journal 59: 834–844.

Staffell, I., Scamman, D., Abad, A.V., Balcombe, P., Dodds, P.E., Ekins, P., Shah, N. and Ward, K.R. 2019. The role of hydrogen and fuel cells in the global energy system. Energy & Environmental Science 12: 463–491.

Šúri, M., Huld, T.A., Dunlop, E.D. and Ossenbrink, H.A. 2007. Potential of solar electricity generation in the European Union member states and candidate countries. Solar Energy 81: 1295–1305.

Toxopeus, M.E., de Koeijer, B.L.A. and Meij, A.G.G.H. 2015. Cradle to cradle: Effective vision vs. efficient practice? Procedia CIRP 29: 384–389.

U.S. Energy Information Administration. 2013. Updated Capital Cost Estimates for Utility Scale Electricity Generating Plants. U.S. Department of Energy, Washington, USA <www.eia.gov/outlooks/capitalcost/pdf/updated_capcost.pdf>. Last accessed: 16.2.2017.

U.S. Energy Information Administration. 2016. Capital Cost Estimates for Utility Scale Electricity Generating Plants. U.S. Department of Energy, Washington, USA <www.eia.gov/analysis/studies/powerplants/capitalcost/pdf/capcost_assumption.pdf>. Last accessed: 30.12.2017.

Wikipedia. 2016. List of European countries by average wage. <en.wikipedia.org/wiki/List_of_European_countries_by_average_wage#cite_note-99>. Last accessed: 20.5.2016.

Winkler, H. 2011. Closed-loop production systems—A sustainable supply chain approach. CIRP Journal of Manufacturing Science and Technology 4: 243–246.
Zore, Ž., Čuček, L. and Kravanja, Z. 2016. Macro- and micro-economic perspectives regarding the syntheses of sustainable bio-fuels supply networks. pp. 2253–2258. *In*: Kravanja, Z. and Bogataj, M. (eds.). Computer Aided Chemical Engineering (Vol. 38): Elsevier.
Zore, Ž., Čuček, L. and Kravanja, Z. 2017. Syntheses of sustainable supply networks with a new composite criterion—Sustainability profit. Computers & Chemical Engineering 102: 139–155.
Zore, Ž., Čuček, L. and Kravanja, Z. 2018a. Synthesis of sustainable production systems using an upgraded concept of sustainability profit and circularity. Journal of Cleaner Production 201: 1138–1154.
Zore, Ž., Čuček, L., Širovnik, D., Novak Pintarič, Z. and Kravanja, Z. 2018b. Maximizing the sustainability net present value of renewable energy supply networks. Chemical Engineering Research and Design 131: 245–265.

CHAPTER 11

A Logistics Analysis for Advancing Carbon and Nutrient Recovery from Organic Waste

Edgar Martín-Hernández,[1,2] *Apoorva M Sampat,*[3] *Mariano Martin,*[2] *Victor M Zavala*[3] and *Gerardo J Ruiz-Mercado*[4,*]

1. Introduction

Modern societies are characterized by the generation of large amounts of waste, arising from the production of goods and services to satisfy social demands. The traditional manufacturing pattern is one-way linear, following the raw material extraction from the environment, manufacturing, usage, disposal of goods, energy consumption, and discarding the generated residues along the linear path. Hence, when considering the definition of sustainable development formulated by the World Commission on Environment and Development (WCED) in 1987, as "a development that meets the needs of the present without compromising the ability of future generations to meet their own needs", the linear production is an unsustainable production model, depleting the natural resources and degrading the environment.

Therefore, one of the main challenges of the current societies is to find processes and technologies that can transform the traditional linear production paradigm into a circular structure, where the materials are recovered to be reused, remanufactured, repaired, and finally closing the loop when these generated residues can be reused as raw material again. These circular production systems are aligned with the definition of a circular economy, which is formally defined by Korhonen et al. (2018) as follows:

"Circular economy is an economy constructed from societal production-consumption systems that maximizes the service produced from the linear nature-society-nature material and energy throughput flow. This is done by using cyclical materials flows, renewable energy sources, and cascade-type energy flows. Successful circular economy contributes to all the three dimensions of sustainable development. Circular economy limits the throughput flow to a level that nature tolerates and utilizes ecosystem cycles in economic cycles by respecting their natural reproduction rates".

[1] Oak Ridge Institute for Science & Education, hosted by U.S. Environmental Protection Agency, Office of Research & Development, Cincinnati, Ohio 45268, U.S.

[2] Department of Chemical Engineering, University of Salamanca. Pza. de los Caídos 1-5, 37008 Salamanca, Spain.

[3] Department of Chemical and Biological Engineering, University of Wisconsin-Madison, 1415 Engineering Dr., Madison, Wisconsin 53706, U.S.

[4] Office of Research and Development, Center for Environmental Solutions and Emergency Response, U.S. Environmental Protection Agency, 26 West Martin Luther King Drive, Cincinnati, Ohio 45268, U.S.

* Corresponding author: ruiz-mercado.gerardo@epa.gov

Therefore, the circular economy pursues both environmental and economic objectives. The environmental objectives are to maximize the generation of valuable products, minimize the needs of raw materials extracted from nature, and reduce releases to the environment by maximizing the reuse and recycling of materials. Regarding the economic goals, these consist in the reduction of raw materials and energy costs by applying material cycles and mitigating the costs associated with waste and emission management.

Within the purview of circular economy, the management of carbon and nutrients from organic waste can have a huge contribution due to the large amounts of organic residues generated from agricultural activities (e.g., livestock), food waste, biosolids, etc. In addition, the concept of waste to energy and value-added chemicals, which pursues the development of processes for the recovery of energy and production of high added-value chemicals from all kinds of residues, can be applied to organic waste in order to achieve a circular economy framework. To illustrate this, a diagram of a circular economy system built around organic waste is shown in Figure 1. Particularly, an anaerobic digestion process can be used to decompose the organic matter and produce biogas, constituted mainly of methane (CH_4) and carbon dioxide (CO_2). With proper treatment to recover the CO_2, the resulting high CH_4 purity gas can be used as a renewable alternative to the fossil-based natural gas (NG) in the production of power and heat. In addition, the sub-product from the anaerobic digestion, so-called digestate, is rich in nutrients (particularly nitrogen and phosphorous) that can be recovered and used as fertilizer for crops, thus closing the nutrient cycle loop.

The livestock agricultural sector is one of the largest economic activities in the world, as it must meet the needs of a demanding society. Additionally, the increasing income-spending potential of the middle class in developing countries has increased the demand for dairy and beef products. This results in some areas where livestock activities generate more GHGs than the transportation sector,[1] nutrient pollution, and other environmental challenges.

In developed societies, waste valorization to energy and material recovery represents a business opportunity for circular economy. Furthermore, organic waste management provides a great opportunity towards the production of sustainable resources and energy (WEC, 2016) and the capability of replacing fossil-based fuels. For instance, capturing carbon through the production of bio-based chemicals derived from the biogas generated after the anaerobic digestion of the organic wastes offers an alternative for replacing the generation of electricity and heat from fossil fuels with renewable sources throughout the production of bio-methane. It is reported that about 250 operational projects in the U.S. (Nov. 2017)[2] have avoided the emissions of 3.2×10^6 tons CO_2e (equivalent to annual emissions from 680,000 cars)[3] and generated 1.03×10^6 MWh of energy (enough to power 96,000 U.S. homes/yr).[4] Other studies have evaluated the potential of some regions to meet all their NG needs by using biogas instead (Taifouris and Martín, 2018). Therefore, waste-to-energy initiatives have gained support in the context of a circular economy philosophy to enhance the development of sustainable process alternatives (Korhonen et al., 2018). Among the organic waste treatment technologies, anaerobic digestion is deemed as an interesting and promising alternative that serves a double objective when processing residues, mitigating the potential environmental and human health issues and producing valuable products that are incorporated into the economic cycle in the form of energy and chemicals.

[1] http://www.fao.org/newsroom/en/news/2006/1000448/index.html.

[2] https://www.epa.gov/agstar/agstar-data-and-trends.

[3] A typical passenger vehicle emits about 4.7 metric tons of carbon dioxide per year, https://www.epa.gov/greenvehicles/greenhouse-gas-emissions-typical-passenger-vehicle.

[4] In 2016, the average annual electricity consumption for a U.S. residential utility customer was 10,766 kWh, https://www.eia.gov/tools/faqs/faq.php?id=97&t=3.

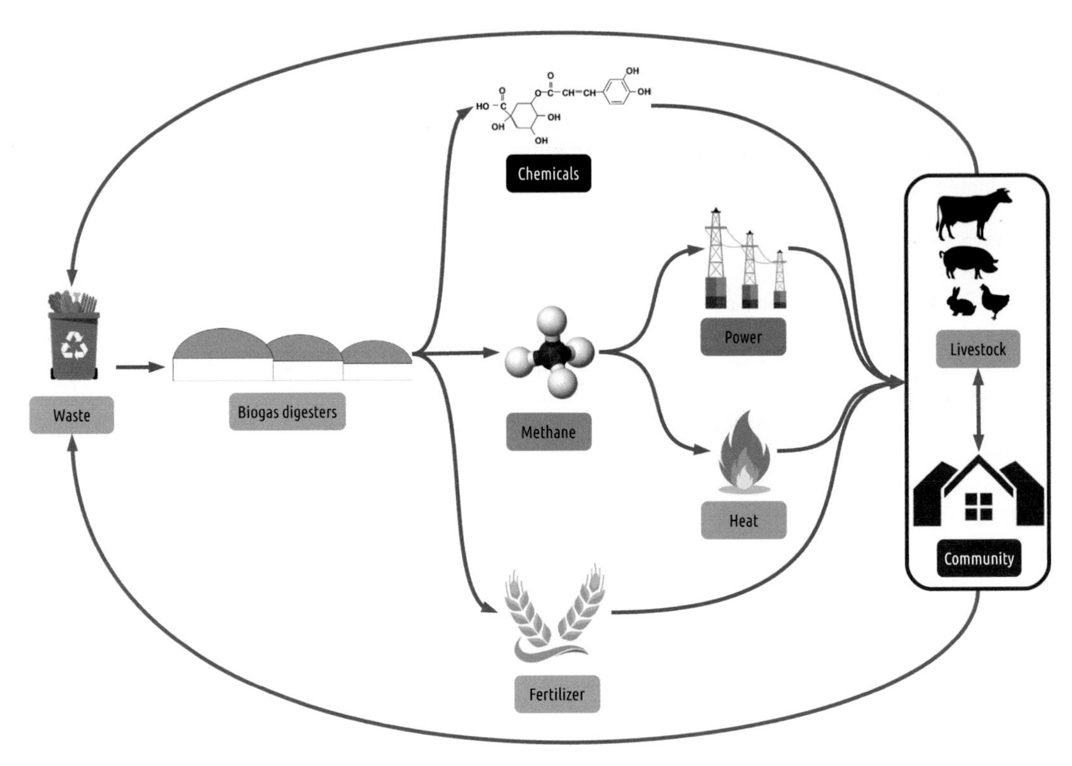

Figure 1. Circular economy generated around organic waste.

2. Biogas Upgrading and Carbon Capture

2.1 Bio-methane as a renewable source of energy

The biogas generated through anaerobic digestion of waste is mainly composed of CH_4, CO_2, small amounts of impurities (ammonia and hydrogen sulfide), and moisture. After removing these impurities, biogas represents a renewable source of CH_4 and CO_2. Biogas can be employed either to produce energy, as it is currently implemented in many industrial facilities, or as raw material to produce chemicals through a syngas/dry reforming path.

As a source of energy, the usage of biogas can be classified according to its quality. Raw biogas can be used for the co-generation of heat and electricity in combined heat and power (CHP) units for onsite consumption, mainly to supply heat to the digestor unit and produce hot water. In addition, it is possible to use biogas in gas turbines and power generators if upgraded by reducing the concentration of CO_2 and obtaining high purity methane (Somehsaraei et al., 2014; Reddy et al., 2016; León and Martín, 2016). Additionally, the large infrastructure available for the transportation and distribution of NG in Europe (Entsog, 2015) or the USA (EIA, 2018), can support the use of biomethane as a renewable alternative to NG. Likewise, high-quality biomethane can be compressed or liquefied to be transported using trucks or railways if a pipeline grid is not available. Finally, a remarkable feature of compressed or liquefied biomethane is its use as transportation fuel, replacing compressed-NG (CNG) and liquified-NG (LNG).

There are several pathways for upgrading biogas. For example, it is possible to hydrogenate the CO_2 content of the biogas into CH_4. This process is widely used as a purification step for ammonia production and has been evaluated for the production of biomethane from biogas at experimental level (Stangeland et al., 2017; Schidhauer and Biollaz, 2015) and lately full process engineering, including techno-economic analysis (Curto and Martín, 2019), have been reported. The main issue associated with the hydrogenation of CO_2 to obtain renewable biomethane is the high need for hydrogen from renewable energy sources instead of using fossil fuel sources since its production is energy intensive (Davis and Martin, 2014a, b). The high investment costs for renewable energy production using either photovoltaic panels and/or wind turbines limit the placement of these facilities to favorable locations where solar fields and wind farms can achieve competitive production costs of CH_4 from CO_2 (de la Cruz and Martín, 2016). Additionally, the recovered CO_2 has enough purity to be used as a raw material for other chemical processes.

Regarding the use of biogas as raw material to produce chemicals, some recent studies evaluate different routes to produce bio-synthesis gas (Hernández et al., 2017), methanol for biodiesel, and as a feedstock for algae-based oil production (Hernández and Martín, 2017). However, the investment and production costs of these chemicals through these novel routes are high, resulting in using biogas as a primary energy source alone.

2.2 Technical alternatives for biogas-carbon management

The common CO_2 capture techniques are based on separation processes, such as gas-liquid chemical absorption, gas-solid physical adsorption, and permeation processes using membranes. Currently, the most mature technologies for CO_2 capture, based on the studies available in the literature, are amine scrubbing, pressure swing adsorption (PSA), and membrane separation systems. Water scrubbing is another technology that is widely used for CO_2 capture, but it has some major disadvantages. Particularly, the air used for the degassing of water (Budzianowski et al., 2017) exits the unit with a high concentration of CO_2 that should be treated before being released to the atmosphere. Furthermore, this CO_2 can be used after being recaptured through additional processing steps.

2.2.1 Chemical absorption: Amines

For the CO_2 absorption using amines solution, the gas stream is passed counter-currently to an aqueous alkali solution containing the amine in order to promote a chemical absorption of the CO_2. As described in Figure 2, this process is carried out in a two-exit absorption column. The gas phase output stream is composed of the purified gas stream, while the liquid phase output stream consists of the aqueous solution with the absorbed CO_2. The aqueous stream is sent to a regeneration column in order to perform the desorption process by heating the solution. Even though the absorption process does not require external energy at ambient temperature, the regeneration process is energy intensive (Ünveren et al., 2017).

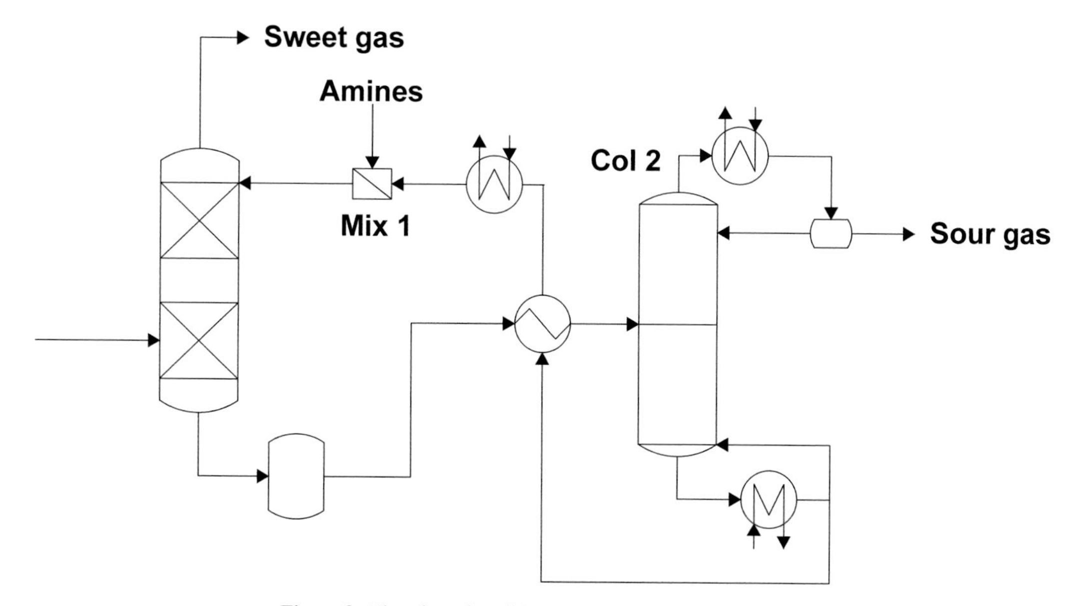

Figure 2. Flowsheet describing the amine absorption process.

The CO_2 absorption systems using amines typically operate at a temperature range of 25–30 ºC, partial pressures above 0.05 bar, and removal yields of 90%–95% (Zhang and Chen, 2013). However, the high concentration of CO_2 in the biogas, compared to post-combustion gases, results in lower operating pressures, reducing the compression costs.

Different amines, like monoethanolamine (MEA), diethanolamine (DEA), and methyl diethanolamine (MDEA) are among the common amines used for CO_2 capture, because of their high CO_2 affinity. The selection of the optimal amine to minimize the overall operating costs should be made considering several important aspects, such as the CO_2 to amine molar ratio, the heat of reaction (associated to the reactor refrigeration cost), and the absorption solution concentration and its price (Nuchitprasittichai and Cremaschi, 2013).

Amine absorption is a relatively simple CO_2 capture technology, which does not require the development of complex or novel equipment. The requirement of energy involved in the regeneration column is the largest share of the operating cost.

2.2.2 Pressure swing adsorption

A pressure swing adsorption (PSA) operation is based on physical adsorption at moderate pressure and low-pressure regeneration at a constant temperature. As shown in Figure 3, in order to perform the CO_2 adsorption, the raw gas is compressed prior to the adsorption stage. Then, the compressed gas is introduced into a column filled with the solid adsorbent material, forming a fixed bed. The gas is flowing across the bed, interacting with the adsorbent, and carrying out the CO_2 adsorption on the fixed bed through reaching a certain partial pressure for CO_2. After the bed reaches the saturation point, the operation stops, and the unit undergoes a desorption step. The desorption step consists of a concurrent flow of a gas stream with low CO_2 partial pressure which drags the CO_2. Commonly, PSA systems are composed of twice the amount of design-required adsorption columns, so while one-half of the units are in use, the other half are being regenerated.

Some of the most common adsorbent materials include zeolites, such as zeolites 13X and 4A. The CO_2 removal before the breakpoint using zeolites 13X and 4A is almost total, leaving only traces of CO_2 in the exit gas (Hauchhum and Mahanta, 2014). Additionally, some experimental results show that CO_2 adsorption onto zeolites can be modeled through theoretical isotherms, like the Langmuir isotherm (Hauchhum and Mahanta, 2014).

The adsorption capacity of the zeolites is directly related to the CO_2 partial pressure. Therefore, a system of compressors with an intermediate cooling system should be implemented. However, more detailed studies should be carried out in order to determine the optimal trade-off between compression cost and carbon capture efficiency.

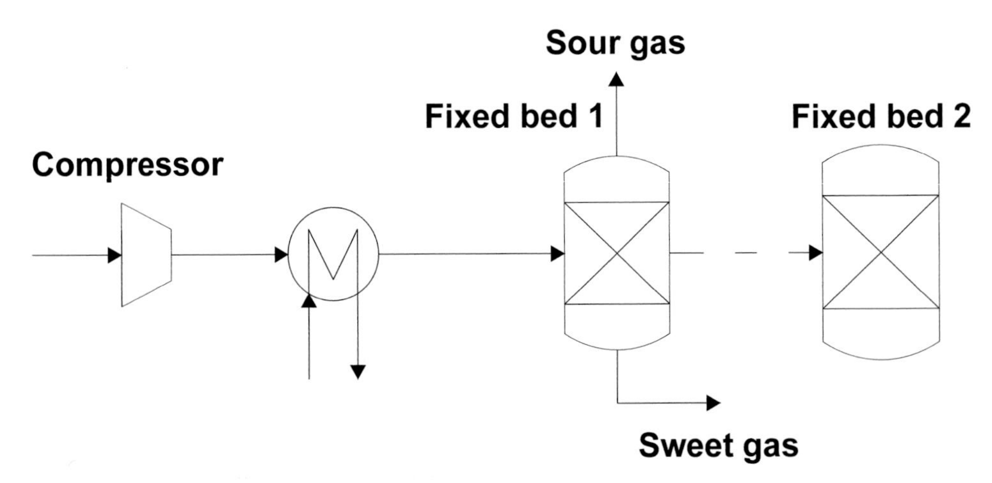

Figure 3. Flowsheet describing the pressure swing adsorption process.

2.2.3 Membrane separation systems

Membrane separation technologies are based on the permeability of different gases through a membrane. The species with higher permeability can cross the membrane, forming an outlet stream, i.e., permeated, while the species with lower permeability cannot go through the membrane barrier, being evacuated from the unit through an outlet stream, i.e., retentate.

As the driving force of the process is the gradient in the species concentration on both sides of the membrane, certain partial pressure values for the permeated compounds should be achieved before introducing the gas stream in the membrane unit. As shown in Figure 4, membrane systems consist of compressor units followed by a set of membrane modules. For the design of membrane systems, two variables should be considered: The membrane material and the arrangement of the units, since the membrane units can be arranged in different configurations such as single-stage or multi-stage

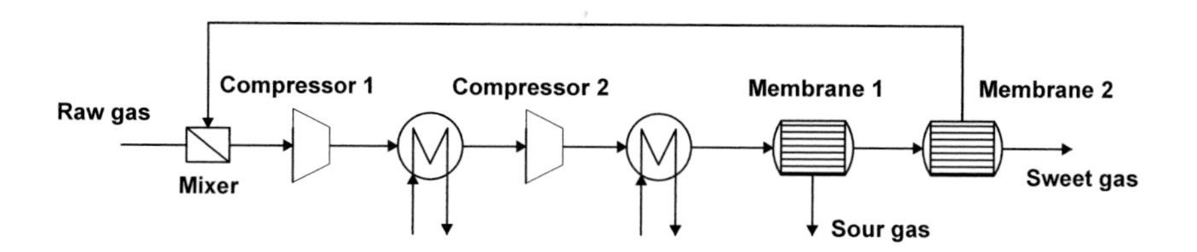

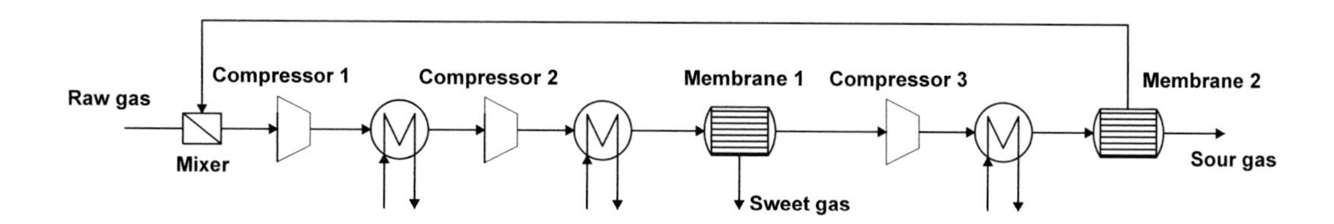

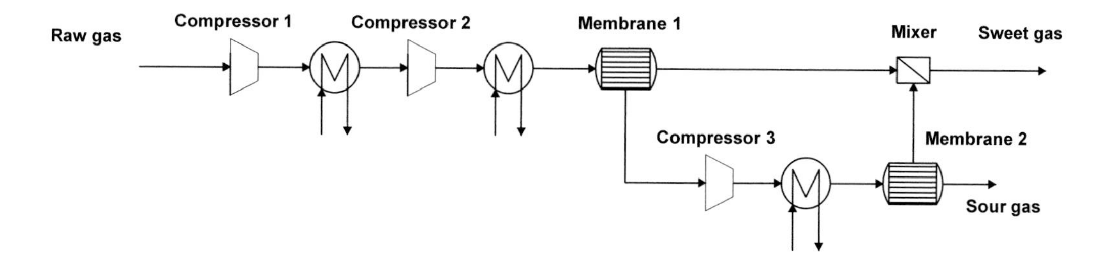

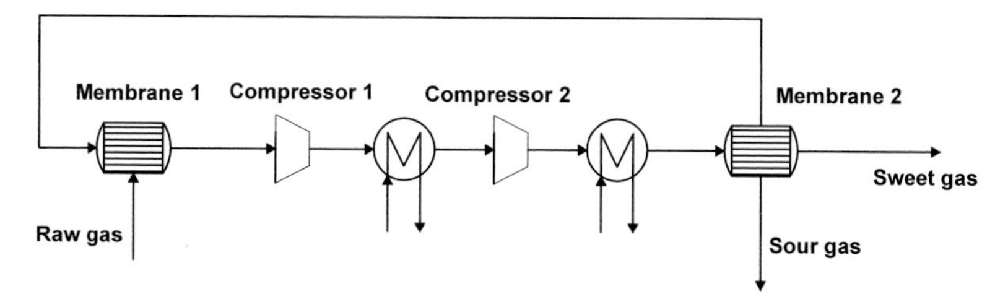

Figure 4. Multi-stage membrane system arrangements.

arrangements. A multi-stage configuration results in larger methane recovery and lower operational costs (Scholz et al., 2013). Among the multiple membrane stage systems, one can find configurations with one compression stage (Makaruk et al., 2010; Scholz et al., 2013) or multiple compression stages (Molino et al., 2013; Scholz et al., 2013). Some studies suggest that dual-stage membrane systems with a single compression stage can be the most flexible option, obtaining between 97% and 80% CO_2 recovery yields for CO_2 contents in the feed of 10–40% (mass) (Kim et al., 2017).

Since membrane materials are characterized by their gas permeability, many of these materials have been evaluated for CO_2 recovery, from traditional materials, including polymers, to the latest developed membrane materials, such as carbon molecular sieves. Among the common materials with larger CO_2 selectivity values, one can identify cellulose acetate, polyimide, and polycarbonate. Other membrane materials for the separation of CO_2 and CH_4 can be found in several reviews (Zhang et al., 2013; Chen et al., 2015; Vrbová and Ciahotny 2017).

2.3 Carbon recovery by chemical production

Several chemicals can be produced from the gases generated in the anaerobic digestion process. In particular, Hernández et al. (2017) developed a platform approach to produce dimethyl ether (DME), methanol, ethanol, and synthetic fuels based on the Fischer-Tropsch process. The production of hydrocarbons is based on the dry reforming of biogas to produce renewable syngas, which is their precursor. Due to the paramount importance of methanol, it was the first to be addressed (Hernández and Martín, 2016). In addition, DME, ethanol, and synthetic fuels can also be produced using the biosyngas as raw material by controlling the H_2 to CO ratio. Dry reforming is based on the equilibrium of the set of reactions shown in equation (1), which is dependent on the operating pressure and temperature. Dry reforming is an endothermic process, which can be heated by combusting a fraction of the generated raw biogas.

$$
\begin{aligned}
&CH_4 + CO_2 \leftrightarrow 2CO + 2H_2 \\
&CH_4 + H_2O \rightarrow CO + 3H_2 \\
&CO_{(g)} + H_2O_{(g)} \rightarrow CO_{2(g)} + H_{2(g)}
\end{aligned}
\tag{1}
$$

After the syngas production, methanol is produced following the set of reactions described in equation (2), catalyzed by a solid catalyst formed by a mixture of CuO–ZnO–AlO.

$$
\begin{aligned}
&CO + 2H_2 \leftrightarrow CH_3OH \\
&CO_2 + H_2 \leftrightarrow CO + H_2O \\
&CO_2 + 3H_2 \leftrightarrow CH_3OH + H_2O
\end{aligned}
\tag{2}
$$

The production of methanol is favored at high H_2/CO ratios, between 1.75 and 3, and it is considered that a CO_2 concentration between 2% and 8% is necessary due to the reaction mechanisms, see equation (2). Since the reaction equilibrium process is exothermic, and the concentration of species is lower at the right side of the reactions, the optimal operating conditions are defined by low temperatures and high pressures (Hernández and Martín, 2016). However, high pressure implies higher compression costs. Therefore, the trade-off between yield and operation cost is sought. The current industrial processes operate at pressures of 50–100 bar and temperatures of 473–573 K to activate the catalyst and boost the reaction (Hernández and Martín, 2016).

Furthermore, the production of ethanol and DME from biosyngas requires H_2/CO ratios of 1. Regarding the DME production, this is produced via methanol as an intermediate product using copper catalysts, followed by a dehydration process catalyzed by acidic catalysts. As it is shown in the set of equations described by equation (3), together with the methanol and DME production, the water gas shift reaction takes place as well.

$$CO_2 + 3H_2 \leftrightarrow CH_3OH + H_2O$$
$$CO + 2H_2 \leftrightarrow CH_3OH \qquad (3)$$
$$CO_2 + H_2 \leftrightarrow CO + H_2O$$
$$2CH_3OH \leftrightarrow CH_3OCH_3 + H_2O$$

Similar to methanol production, DME production is governed by chemical equilibrium and is usually carried out at temperatures between 508 K and 553 K. A more detailed description of DME production from biosyngas can be found on Peral and Martin (2015).

The production of ethanol from biosyngas can be achieved following two paths. One is the chemical synthesis of alcohols using catalysts, producing ethanol and other alcohols as byproducts. This is an energy intensive process since the biosyngas must be heated up to 573 K and it operates at high pressure. As several alcohols are synthesized, a further purification process must be implemented. Another alternative for ethanol production is the fermentation of the biosyngas in a bioreactor. This biological process operates at low temperature, 311 K, and low pressure. However, the water-ethanol separation, which is energy intensive, must be carried out. Additionally, to ensure the survival of the biota of the bioreactor, the ethanol concentration must be kept below 5% by feeding water in order to maintain the ethanol concentration under this threshold, thus limiting the yield of the process. More details of these processes can be found in Martin and Grossmann (2011).

Finally, synthetic fuels can be obtained from biosyngas based on Fischer-Tropsch (FT) processes. The FT processes are based on reactions equivalent to polymerization reactions where the chains of hydrocarbons grow due to the dissociation of CO over a catalyst, as shown in equation (4), where *nc* denotes the number of carbon atoms and *nh* the number of hydrogen atoms. Usually, the catalysts used in FT processes are based on cobalt or iron.

$$ncCO + (nc + \frac{nh}{2})\, H_2 \rightarrow C_{nc}H_{nh} + ncH_2O \qquad (4)$$

Other reactions that take place in the process are the methanation of the CO_2 and the water gas shift reaction, as described in equation (5).

$$CO + 3H_2 \rightarrow CH_4 + H_2$$
$$CO + H_2O \rightarrow CO_2 + H_2 \qquad (5)$$

FT synthesis can be carried out following two paths: High and low temperature processes. Both processes require H_2/CO ratios between 1 and 2. Each path is associated with a different distribution of products. On the FT high-temperature process, the fuel production is achieved at temperatures of 590–630 K and pressures between 10 and 40 bar, using iron-based catalysts. In this process, only traces of heavy fractions are produced, making further separation processes easier (Hernandez and Martin, 2018). This separation process consists of a first separation of the light hydrocarbon compounds as gases, the removal of water generated in the process, and a distillation column to separate the different liquid alkanes. The low-temperature FT process is performed at 440–530 K with cobalt or iron-based catalysts. This process produces larger amounts of heavy fractions than the high-temperature FT process. Therefore, a hydrocracking step must be included in order to break the long chain alkanes and enhance the conversion to gasoline and diesel. The rest of the separation step is similar to separation steps in the high-temperature FT process.

3. Carbon Footprint Mitigation by Nutrient Recovery

Anaerobic digestion processes generate a nutrient-rich residual stream, called digestate, which must be further processed in order to prevent soil and water contamination by nutrient pollution. Nutrient management is needed in order to prevent excessive releases of phosphorous and nitrogen to surface and underground water bodies (eutrophication) and can lead to the exponential growth of algae, cyanobacteria, and the occurrence of harmful algal blooms (HABs) (Sampat et al., 2018; García-Serrano et al., 2009).

Some of the harmful effects of HABs over the environment are reflected in both water and air. Moreover, the overgrowth of algae and cyanobacteria shifts the distribution of aquatic species, leads the depletion of oxygen due to algal decay, and produces cyanotoxins that are harmful to aquatic life, wildlife, and humans. In addition, the development of HABs and eutrophication increases the emissions of CH_4 and N_2O, which are strong greenhouse gases (GHGs). On a 100-yr time horizon, N_2O and CH_4 have 265 and 28 times the effect of CO_2 on atmospheric warming, respectively (DelSontro et al., 2018). Currently, GHG emissions from lakes and impoundments are around 20% of global CO_2-equivalent attributed to fossil fuel emissions, with 75% of the impact due to CH_4. In addition, global increases in eutrophication could cause a 5–40% increase in the GHG effects in the atmosphere. This is the equivalent effect of a 13% global increase in fossil fuel combustion emissions (DelSontro et al., 2018). Beaulieu et al. (2019) predict the total phosphorus concentrations in lakes and impoundments to increase by up to 3 times by the year 2050, leading the development of eutrophication processes with an associated increase of CH_4 emissions of 90%. Therefore, proper management of the nutrients contained in organic waste must be carried out in order to prevent the occurrence of eutrophication, HABs, and the subsequent damages to aquatic ecosystems and GHG emissions to the atmosphere.

However, when effective waste management practices are adopted, the nutrients in the digestate can be recovered as fertilizer, reducing the reliance on fossil-fuel based chemicals and the extraction of virgin feedstocks. Several processes can be used to recover the digestate nutrients. These range from simple mechanical separation, such as filters (Gustafsson et al., 2008), aerobic decomposition (U.S. Department of Agriculture (USDA), 2009), or chemical processing, such as struvite precipitation (Bhuiyan et al., 2008). The nutrient-rich material recovered in each case presents different characteristics affecting its final use, from a mixture of nutrients and organic matter to precipitates with a well-defined composition, as is the case of struvite production. The use of rich nutrient content products has been studied in the literature, showing the fertilizing efficiency of different materials (Hylander et al., 2006; Doyle and Parsons, 2002). Finally, it should be noted that quality variability in the recovered product, selling price, and production cost present complex trade-offs which should be considered at the decision-making step in order to select the most appropriate technology for each case (Martín-Hernández et al., 2018). There are a number of technologies that enable the recovery of nutrients from waste, among them we highlight the following:

3.1 Composting

Composting is one of the simplest ways to manage solid organic waste, obtaining a valuable product can be used as a soil conditioner, and improving the soil fertility (USDA, 2009).

In short, composting is the aerobic biological decomposition of organic matter. For the decomposition of organic waste, particularly when the source of the residue is from livestock facilities, a pretreatment stage has to be implemented in order to reduce the moisture of the waste, since the raw waste has a high-water content which should be eliminated in order to enhance the microbiological growth. Therefore, a liquid-solid pretreatment must be implemented. Nevertheless, a significant amount of inorganic fraction of nutrients, which are water soluble, will be eliminated in the liquid fraction and will not take part in the composting process, posing an environmental threat due to nutrient pollution. On the other hand, composting converts nitrogen in ammonia form, which is volatile and can be eventually released to the atmosphere. In addition, composting must be complemented with external nutrient sources as it contains low amounts of nutrients in relation to its total mass. Other advantages of composting are the reduction of organic matter, greater ease of handling over raw waste, and pathogen elimination.

Composting can be carried out by implementing three methods. Windrow composting consists of making large windrows of organic matter to enable natural biological decomposition. The piles should be turned periodically in order to expose the material to the air. However, the exposition of the waste to air leads to odor issues. A second method is to make a pile of organic matter over a structure where the air is forced or drawn to aerate the organic matter. To enhance the decomposition and improve odor control, the piles are usually covered. Finally, advanced compost systems run the decomposition in vessels where temperature, aeration, and moisture are controlled. However, the capital cost of these units is high and requires advanced management skills from the operators for its satisfactory performance.

The greatest disadvantage of composting methods for nutrient management is the variability of the final product. Since the ambient temperature, moisture and aeration conditions are not completely controlled, except in the case of high-cost sophisticated vessel composting systems, the recovered product characteristics are unpredictable (USDA, 2009).

3.2 Filtration

Filtration is a low-cost technology which can be suitable for facilities where the recovery of a high-value product is not necessary. This technology can employ a filter containing a reactive medium that enhances phosphorus removal. Reactive filtration occurs through various mechanisms, depending on the characteristics of the filter media. For instance, filter media made of compounds rich in cations under basic conditions, usually containing calcium silicates at pH values above 9, promote the formation of orthophosphate precipitates in the form of calcium phosphates, mainly as hydroxyapatite (Pratt et al., 2012). Metallurgical slag as the filtration medium captures phosphorous by adsorption over metal at pH values close to 7 (Pratt et al., 2012). Package filters are considered for small facilities, while gravity filtration systems are more appropriate for large installations, according to recommendations reported by the U.S. Environmental Agency (EPA) (US EPA, 1979). These filters are used for phosphorous recovery in wastewater treatment facilities (Gustafsson et al., 2008) and further analysis can be found in Shilton et al. (2006), reaching phosphorous removal efficiencies up to 90%. Note that the recovered product is not a high purity chemical compound (unlike struvite), since the recovered solid compound is a mixture of nutrients and organic matter.

3.3 Coagulation

Coagulation is based on the destabilization of colloidal suspensions, achieved by reducing the attraction forces between phases. This is followed by a flocculation process, where floccules from the previously destabilized colloids are formed, raising their density, and resulting in a precipitation step which separates the precipitated solid phase from the fluid phase. Therefore, the nutrients, along with other sedimented solids, are recovered by clarification. Regarding the mechanisms followed by nitrogen and phosphorous in the coagulation-flocculation process, phosphorus is removed primarily in the form of complexes with metal hydroxides, which is the dominant process operating at typical pH values between 5.5 to 7 (Szabó et al., 2008). On the other hand, nitrogen recovery is related to the removal of the colloidal phase (Aguilar et al., 2002). Different coagulation agents used for the recovery of nutrients are $FeCl_3$, $Fe_2(SO_4)_3$, $Al_2(SO_4)_3$, and $AlCl_3$.

The removal efficiency achieved is similar for different coagulant agents, with values up to 99% for phosphorus and 57% for nitrogen (Aguilar et al., 2002). The coagulation-flocculation process performance is strongly influenced by the initial ratio of metal to phosphorus, pH, and chemical oxygen demand (COD). In particular, some previous studies report that the initial metal-phosphorus molar ratio must be between 1.5 and 2.0 and a recommended pH range of 5.5 to 7. However, a COD increment has a negative impact on the removal efficiency (Szabó et al., 2008). As in the previous method, the recovered product is a mixture of nutrients in an organic phase.

3.4 Struvite production

Phosphorous and nitrogen can be recovered from the digestate through the formation of struvite, a mineral containing phosphate, ammonium, magnesium, and forming crystals with a chemical formula of $MgNH_4PO_4{\cdot}6H_2O$. The advantage of this technology is that the recovered struvite is a solid with well-known nutrient content, easy to transport, and can be used as slow-release fertilizer without any post-processing step (Doyle and Parsons, 2002). The removal of nutrients via struvite production follows the reaction equation described in equation (6), by adding magnesium chloride ($MgCl_2$) and resulting in the production of struvite crystals that can be recovered as solids. Due to the presence of potassium in the digestate, together with struvite, another product called potassium struvite or K-Struvite, can be produced, as described in equation (7), where the ammonia cation is substituted by the potassium cation (Wilsenach et al., 2007).

$$Mg^{2+} + NH_4^+ + H_nPO_4^{n-3} + 6H_2O \leftrightarrow MgNH_4PO_4 \cdot 6H_2O + nH^+ \quad (6)$$

$$Mg^{2+} + K^+ + H_nPO_4^{n-3} + 6H_2O \leftrightarrow KMgPO_4 \cdot 6H_2O + nH^+ \quad (7)$$

Some characteristics of livestock organic waste must be considered when designing a struvite production facility. Particularly, the high concentration of calcium ions that compete for phosphate ions and the presence of large amounts of suspended solids in the medium, which can interfere in the struvite formation and crystal growth step. However, some studies report successful struvite formation from livestock wastes (Schuiling and Andrade, 1999; Qureshi et al., 2008; Zhang et al., 2014; Zeng and Li, 2006).

At commercial scale, the advancement of the company Ostara©, which has developed commercial fluidized bed reactors for struvite formation in wastewater treatment plants (Ostara Nutrient Recovery Technologies Inc., 2019), is remarkable. This has opened a promising path to achieve technical and economic feasibility for struvite formation from agricultural residues, particularly regarding the processing of livestock waste.

4. Renewable Energy Incentives as Carbon Reduction Initiatives

In the context of carbon reduction initiatives, incentives to promote the installation of technologies to produce renewable energy are considered. Some studies describing the effects of incentives in the economic performance of these processes show how appropriate policies (in the form of incentives) can influence the decision-making procedure for accepting or rejecting the construction of facilities for bio-methane production, carbon reduction, and nutrient recovery.

4.1 European Union and United States

The European Commission established renewable energy targets of 20% final energy consumption from renewable sources by 2020 (Renewables Directive 2009/28/EC). This target enforces the European member states to reach their renewable energy objectives, promoting the development and installation of renewable energy production facilities. In this context, the production of bio-methane plays an important role due to its flexibility in electricity and heat production, as a transportation biofuel, and in its storage. Additionally, given the decentralized nature of the farms, the installation of biogas facilities can contribute to rural community development. In addition, at the mid-term time scale, there are additional targets that must be matched by 2030, such as a 32% of energy demand covered by renewable energy. To achieve these goals, the European Union provides funding on a sub-national level (i.e., regions or municipalities) through instruments as the European Agricultural Fund for Rural Development (EAFRD) (Regulation (EU) No 1305/2013 of the European Parliament and of the Council, of 17 December 2013) and the European Regional Development Fund (ERDF) (Regulation (EU) No 1301/2013 of the European Parliament and of the Council of 17 December). Additionally, tax exemptions are applied in order to promote biofuel production under Directive 2003/96/EC (Council Directive 2003/96/EC of 27 October 2003). Beyond the European legislation, each member state can promote its regulations to promote renewable energy production (European Commission, 2013).

In this context, there are four specific support schemes adopted in the European Union to promote the installation of organic waste treatment and the generation of renewable energy:

- The Feed-in tariff (FiT) scheme provides a remuneration per a certain amount of renewable energy produced. The value range of FiT varies across countries and the installed power capacity of the facility from 70 to 390 EUR/MWh (Deremince et al., 2017).
- Feed-in premium (FiP) are bonuses paid above the benchmark market price to subsidize the generation price of renewable energy. The value range of FiP varies across countries and the installed power capacity of the facility from 100 to 330 EUR/MWh (Deremince et al., 2017).
- Green Certificates (GC) are generated by the production of a unit of renewable energy and can be sold separately from the energy on the green certificates market, similarly to the Renewable Energy Credits implemented in the USA.
- Fiscal incentives are tax exemptions to make renewable energy more competitive in the market, as previously discussed (Deremince et al., 2017).

Additionally, another measure that would promote the reduction of carbon emissions from organic waste is the application of carbon credits to the emissions from organic waste. Carbon credits are a permit which allows the emission of a certain amount of GHGs. The amount of carbon emissions allowed for a certain sector is given by the number of permits emitted, and the GHG emissions allowed by each permit. These permits are tradable, as in the case of Green Credits, in a carbon market where companies can sell and buy the rights to emit GHGs. The goal of carbon credits and the carbon market is to control the GHG emissions through the purchase of a permit, which amount is limited. Therefore, following the market rules, their price will increase or decrease as a function of the GHG emission demand, and subsequently, the larger the emissions are the more amount of money the companies would have to pay. Other mechanisms for distributing carbon credits may be purchasing at auctions or awarding permits as offset for the implementation of clean technologies. Additionally, governments may distribute a certain number of permits at no charge, setting time-based emissions limits. As a result, only the facilities that exceed the limit set for a certain period of time must pay for additional permits, promoting compliance with the established standard.

Regarding the incentives available in the U.S., there are two main schemes: Renewable Energy Credits (RECs), and the Renewable Identification Numbers (RINs):

- Renewable Energy Credits is a mechanism created in the USA which guarantees that energy is generated from renewable sources, providing a system for trading produced renewable electricity. Each renewable megawatt-hour produced generates one REC that can be sold separately from the electricity commodity itself while representing the features of the generated renewable energy. RECs can be used to meet regulatory requirements by generators, trades, or end-users. The idea behind the implementation of RECs is to promote the use of renewable energy providing an incentive for its production by purchasing RECs similarly to the Green Certificates system developed in the European Union. The value of RECs fluctuates between 0.5 and 60 USD/MWh (Sampat et al., 2018).
- RIN is an identification number assigned to a batch of biofuel, allowing the tracking of its production, purchase, and final usage. Regarding the treatment of organic waste, the bio-methane produced through anaerobic digestion processes and its further upgrading can be eligible for the incentives associated with RINs in the form of liquified bio-methane under the Cellulosic RINs category (Sampat et al., 2018).

Another initiative that is available to promote the advancement and implementation of technologies for the management of carbon and nutrient from organic wastes is the Environmental Quality Incentives Program (EQIP) (USDA, 2019), managed by the Natural Resources Conservation Service (NRCS) of the U.S. Department of Agriculture. This program provides financial and technical support to agricultural producers so as to enable them to address improvements in their organic waste management installations and mitigate the environmental impact of their agricultural practices. The Rural Energy for America Program (REAP) (U.S. Department of Energy, 2019) provides loan financing and grant funding to agricultural producers so that they can install renewable energy systems. Anaerobic digestion facilities are eligible facilities among these programs (U.S. Department of Energy, 2019). Another incentive-based program for public institutions was the (now repealed) Clean Renewable Energy Bonds (CREBs) for financing anaerobic digestion projects, among other technologies, giving federal tax credits, replacing a fraction of the usual bond interest, resulting in a lower final interest rate. Finally, the Interconnection Standards for Small Generators set the interconnection standards for generators under 20 MW to be connected with the utility lines (Federal Energy Regulatory Commission, Docket No. RM16-8-000; Order No. 828). These power generators, when installed in biogas facilities, qualify for these incentives.

It is challenging to compare the equivalent target for U.S. renewable energy production with the Horizon 2020 goals set in the European Union since the renewable energy standards in the USA are set at the state level. There are 29 U.S. states with established targets, 8 states with voluntary goals, and 13 states not having any renewable energy standards or targets, as summarized in Table 1 (National Conference of State Legislatures, 2019). These renewable portfolio standards (RPS) require that a specified percentage of the electricity that utility companies sell comes from renewable resources.

In order to discuss the U.S. economic incentives associated with renewable energy and fuel recovery from livestock waste, three representative cases have been selected: a state with the most ambitious

Table 1. U.S. state renewable portfolio standards and goals (National Conference of State Legislatures, 2019).

State	Type of requirement	Requirement year	Renewable energy requirement (% or MW)	State	Type of requirement	Requirement year	Renewable energy requirement (% or MW)
AL	N/A	N/A	N/A	MT	Required	2015	15
AK	N/A	N/A	N/A	NE	N/A	N/A	N/A
AZ	Required	2025	15	NV	Required	2025	25
AR	N/A	N/A	N/A	NH	Required	2025	25.2
CA	Required	2045	100	NJ	Required	2030	50
CO	Required	2020	30	NM	Required	2020	20
CT	Required	2030	44	NY	Required	2030	50
DE	Required	2026	25	NC	Required	2025	12.5
FL	N/A	N/A	N/A	ND	Voluntary	2015	10
GA	N/A	N/A	N/A	OH	Required	2026	12.5
HI	Required	2045	100	OK	Voluntary	2015	15
ID	N/A	N/A	N/A	OR	Required	2050	50
IL	Required	2026	25	PA	Required	2021	18
IN	Voluntary	2025	15	RI	Required	2025	38.5
IA	Required	N/A	105 MW	SC	Voluntary	2021	2
KS	Voluntary	2020	20	SD	Voluntary	2015	10
KY	N/A	N/A	N/A	TN	N/A	N/A	N/A
LA	N/A	N/A	N/A	TX	Required	2025	10,000 MW
ME	Required	2017	40	UT	Voluntary	2025	20
MD	Required	2020	25	VT	Required	2032	75
MA	Required	2030	35	VA	Voluntary	2025	15
MI	Required	2025	35	WA	Required	2020	15
MN	Required	2025	26.5	WV	N/A	N/A	N/A
MS	N/A	N/A	N/A	WI	Required	2015	10
MO	Required	2021	15	WY	N/A	N/A	N/A

objective for renewable energy consumption, a state with intermediate goals, and a state with no established renewable energy requirements. As a measure of the determination to meet the different requirements, the fraction of energy that should be fulfilled from renewable sources and the type of obligation has been considered. Table 1 shows that the states of California and Hawaii present the most ambitious programs. For both states, their targets are to consume 100% renewable energy by 2045. However, California is selected as a more representative state because of its population, gross domestic product, and dimension. As an intermediate case, the average value of the renewable energy fraction covered from renewable sources was computed, obtaining a value of 29.7. The state with the value closest to this ratio is Colorado, where 30% of its energy demand should come from renewable sources. Finally, Alaska will be considered as representative of the states with no renewable energy targets (U.S. Department of Energy, 2019).

As a representative case of a determined renewable energy program, California has several incentive programs for the installation of anaerobic digestion technologies. Some of them consist of voluntary programs where the customers can support the development of renewable energy projects, such as the Pacific Power—Blue Sky Community Project Funds (Pacific Power, 2019). Furthermore, public funds are administered by the California Energy Commission (CEC), supporting renewable energy and research and development initiatives. Also, there are some feed-in tariff programs, such as the Marin Clean Energy—Feed-In Tariff, targeted towards small facilities (under 1 MW), or the Renewable Market Adjusting Tariff (ReMAT), available until the state cumulative capacity for the bioenergy ReMAT program reaches 250 MW. In addition, California offers a sales and use tax manufacturing exemption for electric power generation from sources other than a conventional power source and storage equipment under the "California Global Warming Solutions Act of 2006: Market-based compliance mechanisms: Fire prevention fees: Sales and use tax manufacturing exemption" (California Legislative Information, Assembly Bill 398 of 2017).

Regarding states with average renewable energy programs, Colorado offers incentives in the form of sales and use tax exemptions for renewable energy equipment across the state, as well as special tax incentives for economically distressed areas under the Colorado's Enterprise Zone (EZ) program. For both programs, anaerobic digestion systems are eligible as renewable energy equipment. There are two additional incentives in the form of a regulatory policy which enforces municipal electric utilities serving more than 40,000 customers to offer an optional green-power program that allows retail customers the choice of supporting emerging renewable technologies, and through an interconnection standard for generators under 10 MW to be interconnected to intra-state utility lines.

Finally, the state of Alaska, which currently has no official renewable portfolio standards and goals, enacted the House Bill 306 in the 2009–2010 legislature, stating that 50% of the electricity produced should come from renewable energy sources by 2025. However, this House Bill was not enacted as a codified statute (U.S. Department of Energy, 2019).

None of these incentives consider nutrient pollution, neither in Europe nor in the USA. However, it is expected that the implementation of incentives towards avoiding nutrient pollution, in the form of phosphorous credits (P credits), can provide a mechanism to mitigate many environmental problems. P credits would work in a similar way to carbon credits implemented in the European Union, developing a waste management market around phosphorus. P credits would be related to the emission of a certain amount of P to the environment, penalizing the emissions, or in the opposite way, being an incentive for the recovery of phosphorous instead of being released to the environment. In both ways, P credits will be an incentive to promote the installation of organic waste management and nutrient recovery technologies.

A study about the effect of Renewable Energy Credits (REC), Renewable Identification Number (RIN), and P credits can be found in Sampat et al. (2018). The conclusions obtained in this work show that with the current system of incentives, the optimal strategy to manage waste is to deploy technologies that conduct simultaneous recovery of liquefied biomethane and nutrients in order to take advantage of RIN and P credits incentives simultaneously, obtaining a payback period of 4.8 years in the best case. Additionally, a report about the incentives and the current development of biogas facilities in the European Union can be found in Deremince et al. (2017).

4.2 Other countries

Carbon policies adopted by other countries are also based on carbon taxes, by setting a price for the carbon emissions, but without defining a cap for the emissions. Another scheme is the emissions trading systems (ETS) based on carbon credits, where, as in the case of European Green Certificates or the RECs implemented in the USA, there are permits allowing the emission of a certain volume of GHGs. As detailed before, there is a market based on these permits, where companies can sell and buy the rights to emit GHGs. The following is a brief review of some national carbon pricing policies.

4.2.1 North America

- Canada: By 2019, Canada had adopted a carbon pricing system that introduced a federal regulation, allowing provinces and territories to adopt more restrictive policies or implement the federal administered system. This federal system, called output-based pricing system (OBPS), establishes a limit for emissions from each type of industrial facility based on their annual economic production. Those facilities whose emissions exceed their annual limit must pay for additional emission permits. The rate for the additional permits is 10 CAD per equivalent ton of CO_2 emitted in 2018, rising 10 CAD per year up to 50 CAD per equivalent ton of CO_2 emitted in 2022. Additionally, charges for GHG emitting fuels are considered. Four provinces, Quebec, Ontario, British Columbia and Alberta have their own carbon pricing system. Quebec and Ontario operate an integrated emissions trading system together with California for facilities emitting more than 25 kilotonnes of CO_2eq per year, releasing auctions for the purchase of emission permits. British Columbia developed its own OBPS system in 2012, establishing an initial rate of 30 CAD per additional ton of CO_2eq emitted. Since 2018, this rate will be raised by 5 CAD per year for four years. The agricultural sector is excluded from the OBPS system, while the liquefied natural gas sector is regulated under a specific OBPS system. Also, Alberta is regulated under its own OBPS system, charging 30 CAD per additional ton of CO_2eq emitted. Only the facilities emitting more than 100 kilotonnes of CO_2eq per year are subjected to the Alberta OBPS system. Saskatchewan, Manitoba, New Brunswick and Nova Scotia have announced their intention to implement their own carbon pricing system (Good, 2018).
- Mexico: In 2012, Mexico approved the Climate Change Law, taking the first steps towards the implantation of a voluntary emissions trading system. However, an amendment to this law approved in 2018 established that the Mexican ETS must be mandatory and adopted by 2022, after a pilot stage starting in 2019. The Ministry for the Environment and Natural Resources (SEMARNAT) is carrying out studies to set up the market rules for the pilot Mexican ETS. After the pilot-phase, the rules will be updated for the starting of the final emissions trading system (SEMARNAT, GIZ and German Federal Ministry for the Environment, Nature Conservation and Nuclear Safety, 2018).

4.2.2 Asia

- China: Pilot experiences for the implementation of emissions trading systems were implemented in the municipalities of Beijing, Chongqing, Shanghai, Tianjin and Shenzhen, and in the provinces of Guangdong and Hubei between 2013 and 2014, and in the province of Fujian in 2016. As a result of these experiences, in 2017, the National Development and Reform Commission (NDRC) of China announced the development of a national emissions trading system to be implemented by 2020. This system will cover the power generation industry at the beginning. The inclusion of other sectors in the ETS system has yet to be defined (Slater et al., 2018).
- Japan: Since 2010, the Tokyo Metropolitan Government launched a mandatory emissions trading system, with the objective of reducing CO_2eq emissions by 25% to 27% from 2020 to 2024 and 30% by 2030. The estimated set for carbon price by 2018 is 5.89 USD per ton of CO_2eq emitted (ICAP, 2019).
- Korea: In 2015, the Republic of Korea launched an emissions trading system called K-ETS. The Korean ETS system manages 69% of the national GHG emissions, establishing a GHG emission cap

of 538.5 megatonnes of CO_2eq for 2018. The exchange of permits is implemented in the Emission Permits Exchange in the Korea Exchange (KRX) through auctioning of the allowances (The Government of the Republic of Korea, 2017; World Bank and Ecofys, 2018).

4.2.3 Oceania

- Australia: Under the Emissions Reduction Fund and the Climate Solutions Fund, the Australian Government, through the Department of the Environment and Energy, promotes and supports the implementation of actions to improve the environmental sustainability of the stakeholders with the provision of 4.55 billion AUD. Among the eligible activities, there are actions to reduce the GHG emissions through the implementation of energy efficiency actions and waste management systems. The target is to reduce the emissions by 26%–28% below the 2005 emissions benchmark by 2030. Together with the government funding for the implementation of clean technologies and the deployment of measures to reduce the GHG emissions, Australia implemented an emissions trading system through the Australian carbon credit units (ACCUs). These permits can be sold to the Australian Government through the implementation of the mentioned actions to reduce the environmental impact of the stakeholders, or to other emitters who have to offset their emissions (Australian Government, Department of the Environment and Energy, 2019).

4.2.4 South America

- Argentina: Pursuing the discouragement of the consumption and production of GHGs emitting fuels, Argentina adopted a carbon tax for liquid fuels and coal in 2017, taxing 10 USD per ton of CO_2eq emitted. It should be noted that natural gas is exempted from this carbon tax. This tax will be applied gradually from 2019 at 10% of the total rate, increasing 10% each year until reaching a 10 USD full rate by 2028 (Government of Argentina, 2018; FARN, 2017).
- Chile: Since 2017, a carbon tax has been implemented in Chile, taxing the CO_2 emissions at a rate of 5 USD per ton of CO_2eq emitted. Only the facilities with a thermal power over 50 MW are subjected to the carbon tax (Government of Chile, 2019).
- Colombia: In 2017, a carbon tax was adopted by the Colombian government, including liquid and gas fossil fuels. The Ministry of Environment and Sustainable Development estimates that this tax will cover 27% of the total national emissions. The rate imposed is 15,000 COP (approx. 5 USD) per ton of CO_2eq emitted. However, the system contemplates exemptions from this tax for those stakeholders that implement measures to reduce GHG emissions or improve the energetic efficiency (MINAMBIENTE, 2017).

5. Effect of a Carbon Credit Policy Implementation in the USA: A Wisconsin Case Study

We illustrate how carbon credits can potentially influence the management of the organic waste generated by concentrated animal feeding operations (CAFOs). We developed a case study for the treatment of organic waste produced at the 100 largest dairy CAFOs in the state of Wisconsin. This case of study was originally proposed by Sampat et al. (2018) to demonstrate the results of a proposed supply chain framework to evaluate the simultaneous effects of economic and environmental analysis of livestock waste. The treatment facilities considered include the installation of anaerobic digestion facilities, coupled with the production of energy from generators powered by the produced biogas, and the further upgrading of biogas to obtain biomethane or liquefied methane. The digestate produced in the anaerobic digestion process can be further treated for nutrient recovery in order to mitigate the eutrophication of water bodies and avoid the generation of GHGs associated with the presence of HABs. More details of the case study, including waste composition, the recovered product market values, and a list of candidate technologies with corresponding investment and operating costs can be found in Sampat et al. (2018).

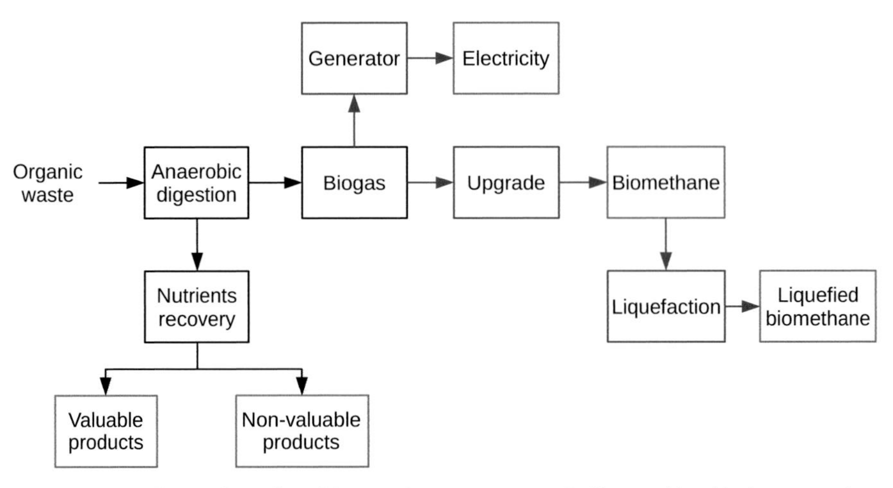

Figure 5. Possible configuration of the organic waste treatment facility considered in the case study.

The model determines the optimal configuration for each treatment facility that maximizes the total profit in the supply chain management network for livestock waste. Figure 5 shows the possible configuration of the waste treatment facility. The blocks outlined in black represent the common steps, a blue outline represents the different upgradation and refinement alternatives, and the red outlines denote the final products of the treatment process.

Sampat et al. (2018) show that the current product revenues obtained from the livestock waste treatment for nutrient and carbon recovery are not sufficient to make the system profitable. Therefore, in the base case scenario, they also consider the impact of economic benefits from the current U.S. incentives potentially available for these facilities, particularly RECs, RINs, and P credits. The values considered for these incentives are the following (Sampat et al., 2018):

- REC: 2 USD/MWh
- RIN: 2 USD/gal
- P credit: 22.04 USD/kg P

An alternative scenario proposed here would include carbon credits, assuming their current value in an existing reference carbon market, such as the European Union Emissions Trading System (European Commission, 2019b). This value is equal to 26.92 EUR per ton of CO_2 equivalent (Sandbag, 2019), which is equal to 0.11 USD per kg of C. The results from the base case scenario (considering REC, RIN, and P credits) are shown in Figure 6a, while the results from the alternative scenario when C credits are also included are shown in Figure 6b. The red dots indicate the location of CAFOs, yellow and green rings denote the placement of waste treatment facilities, blue arrows denote the transportation flows of the organic waste between the treatment facilities. Table 2 shows the results of both case studies. As a result of the implementation of the carbon credit policy, one can notice that, although the number of facilities for nutrient recovery facilities is the same (31), there is a change in the routes for the transport of waste reducing the transportation costs from 1.84 M USD/year to 1.80 M USD/year. Additionally, the location of the facilities installed is slightly different because of applying the carbon credit policy. Furthermore, additional revenue of 1.77 M USD per year is achieved, resulting in a slightly shorter payback period of 4.7 years in comparison with the payback period of 4.8 years if the carbon credits policy is not applied. The return of investment is larger with the application of the policy, 21.4% against 20.9% if the policy is not applied. It should be noticed that the application of the carbon credit incentives does not affect the type of technologies selected; for both scenarios, the production of liquefied methane and the recuperation of nutrients is selected because of RIN and P credit incentives. For both cases, RECs do not make electricity production economically sustainable.

Figure 6a. Base case scenario without considering carbon credits.

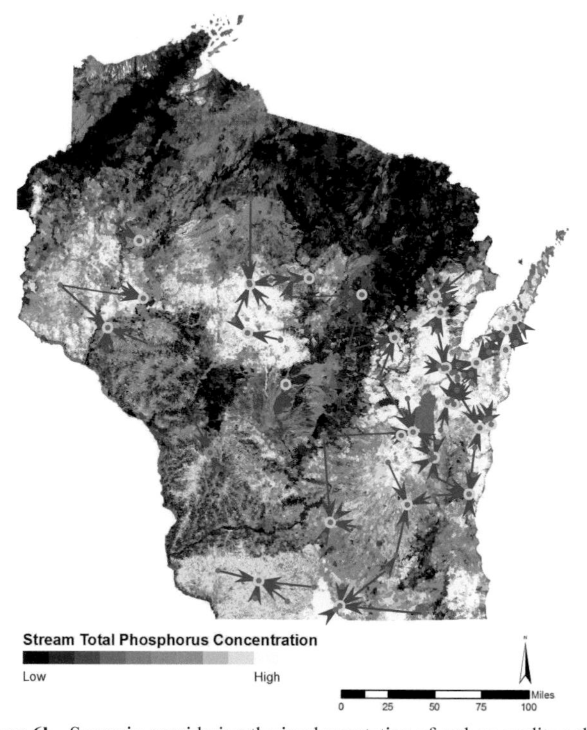

Figure 6b. Scenario considering the implementation of carbon credits policy.

Figure 6. Effects of the carbon credits policy implementation in the transport of manure in the state of Wisconsin (maps of the state of Wisconsin are adapted from U.S. Environmental Protection, 2014).

Table 2. Results of the case study considered with and without application of carbon credits policy.

	Without considering carbon credits	Considering carbon credits
Investment cost (USD)	4.36E+08	4.36E+08
Transportation cost (USD/year)	1.84E+07	1.80E+07
Operating cost (USD/year)	7.17E+07	7.17E+07
Revenue from products (USD/year)	3.78E+07	3.77E+07
Revenue from P credits (USD/year)	1.18E+08	1.18E+08
Revenue from C credits (USD/year)	0	1.77E+06
Revenue from RECs (USD/year)	0	0
Revenue from RINs (USD/year)	4.75E+07	4.74E+07
Return on investment (%)	20.9	21.4
Payback period (years)	4.8	4.7
Nutrient recovery facilities installed	31	31

6. Conclusions

The release of carbon and nutrients from organic waste to the environment without proper management can result in serious threats to both the environment and human health. These threats are mainly related to the pollution of water bodies, and are associated with the expensive remediation treatments required in order to mitigate the adverse effects of pollution, and the release of GHGs to the atmosphere from the decomposition of the organic waste and algal blooms because of eutrophication processes. Therefore, adequate management of organic residues must be developed in order to make the production processes, particularly the livestock agricultural sector, one of the largest economic activities over the world, more sustainable, preserving the ecosystems and avoiding expensive remediation processes. In this chapter, several options for the management of organic waste, focusing specifically on carbon and nutrient recovery, have been described, with the aim of promoting their development and implementation in the current and future facilities in order to ensure the long-term economic and environmental sustainability of livestock operations. Additionally, the current U.S. and European Union incentives are described in order to give an overview about how governmental institutions around the world are promoting the installation of waste treatment technologies for energy and material recovery and developing the necessary infrastructure to reach a sustainable growth based around a circular economy. In this context, the development of circular economy waste management systems has a vital role in achieving sustainable use of raw materials and reducing the generation of residues. As it has been presented and following the premises of the waste to energy and chemicals concepts, the management and repurposing organic waste is a promising field for developing a circular economy system where valuable products can be recovered and reincorporated into the current productive loop.

Finally, this work discusses potential opportunities for the process design and logistics management of organic waste material under various types of government incentives. The consideration of these aspects would allow communities, investors, policymakers, and other stakeholders to conduct studies on the effect of economic incentives, environmental impacts, and technical feasibility of technologies for developing more sustainable carbon and nutrient recovery from organic waste.

Acknowledgments

This work was supported by an appointment for E. Martín Hernández to the Research Participation Program for the Office of Research and Development, U.S. EPA, administered by the Oak Ridge Institute for Science and Education through Interagency Agreement No. DW-89-92433001. PSEM3 research group acknowledges funding from the Junta de Castilla y León, Spain, under grant SA026G18 and grant

EDU/556/2019. V.M. Zavala acknowledges funding from the U.S. National Science Foundation under award CBET-1604374.

Disclaimer

The views expressed in this article are those of the authors and do not necessarily reflect the views or policies of the U.S. EPA. Mention of trade names, products, or services does not convey, and should not be interpreted as conveying, official U.S. EPA approval, endorsement, or recommendation.

References

Aguilar, M.I., Sáez, J., Lloréns, M., Soler, A. and Ortuño, J.F. 2002. Nutrient removal and sludge production in the coagulation–flocculation process. Water Res. 36: 2910–2919.

Australian Government, Department of the Environment and Energy. 2019. About the Climate Solutions Fund—Emissions Reduction Fund.

Azapagic, A., Howard, A., Parfitt, A., Tallis, B., Duff, C., Hadfield, C., Pritchard, C., Gillett, J., Hackitt, J., Seaman, M., Darton, R., Rathbone, R., Clift, R., Watson, S. and Elliot, S. 2002. The sustainability metrics, sustainable development progress metrics recommended for use in the process industries. IChemE.

Beaulieu, J.J., DelSontro, T. and Downing, J.A. 2019. Eutrophication will increase methane emissions from lakes and impoundments during the 21st century. Nature Communications 10: 1375.

Bhuiyan, M.I.H., Mavinic, D.S. and Koch, F.A. 2008. Phosphorus recovery from wastewater through struvite formation in fluidized bed reactors: a sustainable approach. Water Sci. Technol. 57: 175–181.

Budzianowski, W.M., Wylock, C.E. and Marciniak, P.A. 2017. Power requirements of biogas upgrading by water scrubbing and biomethane compression: Comparative analysis of various plant configurations. Energy Conversion and Management 141: 2–19.

California Legislative Information. 2017. AB-398 California Global Warming Solutions Act of 2006: market-based compliance mechanisms: fire prevention fees: sales and use tax manufacturing exemption.

Chen, X.Y., Thang, H.V., Ramirez, A.A., Rodrigue, D. and Kallaguine, S. 2015. Membrane gas separation technologies for biogas upgrading. RSC Adv. 5: 24399–24448.

Council Directive 2003/96/EC of 27 October 2003 restructuring the Community framework for the taxation of energy products and electricity. Official Journal of the European Union L 283/51, 2003.

Curto, D. and Martín, M. 2019. Renewable based biogas upgrading. Journal of Cleaner Production 224: 50–59.

DelSontro, T., Beaulieu, J.J. and Downing, J.A. 2018. Greenhouse gas emissions from lakes and impoundments: Upscaling in the face of global change. Limnol. Oceanogr. Lett. 3: 64–75.

Deremince, B., Scheidl, S., Stambasky, J. and Wellinger, A. 2017. Data bank with existing incentives and subsequent development of biogas plants compared to national targets. EBA European Biogas Association.

Directive 2009/28/EC of the European Parliament and of the Council of 23 April 2009 on the promotion of the use of energy from renewable sources and amending and subsequently repealing Directives 2001/77/EC and 2003/30/EC. Official Journal of the European Union L 140/16, 2009.

Doyle, D.J. and Parsons, S.A. 2002. Struvite formation, control and recovery. Water Res. 36: 3925–3940.

[EIA] U.S. Energy Information Administration 2018. Natural gas pipelines. https://www.eia.gov/energyexplained/index.php?page=natural_gas_pipelines (accessed April 20, 2019).

[ENTSOG] European Network of Transmission System Operators for Gas 2015. The European Natural Gas Network.

European Commission. 2013. European Commission Guidance for the Design of Renewables Support Schemes.

European Commission. 2019b. EU Emissions Trading System (EU ETS). https://ec.europa.eu/clima/policies/ets_en (accessed April 20, 2019).

[FARN] Fundación Ambiente y Recursos Naturales. 2017. La reforma Tributaria y el Impuesto al Carbono. https://farn.org.ar/archives/23151 (accessed July 20, 2019).

García-Serrano, P., Ruano, S., Lucena, J.J. and Nogales, M. 2009. Guia práctica de la fertilización racional de los cultivos en España. MMAMRM. Spanish Government, ISBN 978-84-491-0997-3.

Good, J. 2018. Carbon Pricing Policy in Canada. Library of Parliament—Bibliothèque du Parlement, Publication 2018-07-E. Ottawa, Canada.

Government of Argentina. 2018. Argentina participó del diálogo sobre instrumentos de precio al carbono en las Américas.

Government of Chile. 2019. Precio al Carbono Chile. http://www.precioalcarbonochile.cl/preguntas-frecuentes (accessed July 24, 2019).

Gustafsson, J.P., Renman, A., Renman, G. and Poll, K. 2008. Phosphate removal by mineral-based sorbents used in filters for small-scale wastewater treatment. Water Res. 42: 189–197.

Hauchhum, L. and Mahanta, P. 2014. Carbon dioxide adsorption on zeolites and activated carbon by pressure swing adsorption in a fixed bed. Int. J. Energy Environ. Eng. 5: 349–356.

Hernández, B., León, E. and Martín, M. 2017. Bio-waste selection and blending for the optimal production of power and fuels via anaerobic digestion. Chem. Eng. Res. Des. 121: 163–172.

Hernández, B. and Martín, M. 2016. Optimal composition of the biogas for dry reforming in the production of methanol. Ind. Eng. Chem. Res. 55(23): 6677–6685.

Hernández, B. and Martin, M. 2017. Optimal integrated plant for production of biodiesel from waste. ACS Sust. Chem. Eng. 5(8): 6756–6767.

Hernández, B. and Martin, M. 2018. Optimization for biogas to chemicals via tri-reforming. Analysis of Fischer-Tropsch fuels from biogas. Energy Convers. Manag. 174: 998–1013.

http://www.environment.gov.au/climate-change/government/emissions-reduction-fund/about (accessed July 20, 2019).

https://www.entsog.eu/sites/default/files/2018-10/ENTSOG_CAP_MAY2015_A0FORMAT.pdf (accessed April 20, 2019).

Hylander, L.D., Kietlinska, A., Renman, G. and Simán, G. 2006. Phosphorus retention in filter materials for wastewater treatment and its subsequent suitability for plant production. Biores. Technol. 97: 914–921.

[ICAP] International Carbon Action Partnership. 2019. Japan-Tokyo Cap-and-Trade Program.

Kim, M., Kim, S. and Kim, J. 2017. Optimization-based approach for design and integration of carbon dioxide separation processes using membrane technology. Energy Procedia 136: 336–341.

Korhonen, J., Honkasalo, A. and Seppala J. 2018. Circular economy: The concept and its limitations. Ecol. Econ. 143: 37–46.

Makaruk, A., Miltner, M. and Harasek, M. 2010. Membrane biogas upgrading processes for the production of natural gas substitute. Separation and Purification Technology 74: 83–92.

Martín, M. and Grossmann, I.E. 2011. Energy optimization of bioethanol production via gasification of switchgrass. AIChE J. 57(12): 3408–3428.

Martín-Hernandez, E., Sampat, A., Martín, M. and Zavala. V.M. 2018. Optimal integrated facility for waste processing. Chem. Eng. Res. Des. 131: 160–182.

[MINAMBIENTE] Ministerio de Ambiente y Desarrollo Sostenible. 2017. Principales preguntas frente al impuesto nacional al carbono y la solicitud de no causación por carbono neutralidad. Government of Colombia.

Molino, A., Migliori, M., Ding, Y., Bikson B., Giordano, G. and Braccio, G. 2013. Biogas upgrading via membrane process: Modelling of pilot plant scale and the end uses for the grid injection. Fuel 107: 585–592.

[NCSL] National Conference of State Legislatures. 2019. State Renewable Portfolio Standards and Goals.

Nuchitprasittichai, A. and Cremaschi, S. 2013. Optimization of CO_2 capture process with aqueous amines—a comparison of two simulation–optimization approaches. Ind. Eng. Chem. Res. 52(30): 10236–10243.

Ostara Nutrient Recovery Technologies Inc. 2019. https://ostara.com/ (accessed April 20, 2019).

Pacific Power. 2019. Blue Sky Community Project Funds. https://www.pacificpower.net/blueskyfunds (accessed April 20, 2019).

Peral, E. and Martín, M. 2015. Optimal production of DME from switchgrass-based syngas via direct synthesis. Ind. Eng. Chem. Res. 54: 7465–7475.

Pratt, C., Parsons, S.A., Soares, A. and Martin, B.D. 2012. Biologically and chemically mediated adsorption and precipitation of phosphorus from wastewater. Curr. Opin. Biotechnol. 23: 890–896.

Qureshi, A., Lo, K.V. and Liao, P.H. 2008. Microwave treatment and struvite recovery potential of dairy manure. J. Environ. Sci. Health B 43: 350–357.

Reddy, K.S., Aravindhan, S. and Mallick, T.K. 2016. Investigation of performance and emission characteristics of a biogas fueled electric generator integrated with solar concentrated photovoltaic system. Renew. Energ. 92: 233–243.

Regulation (Eu) No. 1301/2013 of the European Parliament and of the Council of 17 December 2013 on the European Regional Development Fund and on specific provisions concerning the Investment for growth and jobs goal and repealing Regulation (EC) No. 1080/2006. Official Journal of the European Union L 347/289, 2013.

Regulation (Eu) No. 1305/2013 of the European Parliament and of the Council, of 17 December 2013, on support for rural development by the European Agricultural Fund for Rural Development (EAFRD) and repealing Council Regulation (EC) No. 1698/2005. Official Journal of the European Union L 347/487, 2013.

Sampat, A.M., Ruiz-Mercado, G.J. and Zavala, V.M. 2018. Economic and environmental analysis for advancing sustainable management of livestock waste: A wisconsin case study. ACS Sustainable Chem. Eng. 6(5): 6018–6031.

Sampat, A.S., Martín, E., Martín, M. and Zavala, V.M. 2017. Optimization formulations for multi-product supply chain networks. Comput. Chem. Eng. 104: 296–310.

Sandbag. 2019. Carbon price viewer. https://sandbag.org.uk/carbon-price-viewer/.

Schidhauer, T.J. and Biollaz, S.M.A. 2015. Reactors for catalytic methanation in the conversion of biomass to Synthetic Natural Gas (SNG). CHIMIA International Journal for Chemistry 69(10): 603–607.

Scholz, M., Melin, T. and Wessling, M. 2013. Transforming biogas into biomethane using membrane technology. Renewable and Sustainable Energy Reviews 17: 199–212.

Schuiling, R.D. and Andrade, A. 1999. Recovery of struvite from calf manure. Environmental Technology 20: 765–768.

SEMARNAT, GIZ and German Federal Ministry for the Environment, Nature Conservation and Nuclear Safety. 2018. Designing an Emissions Trading System in Mexico: Options for Setting an Emissions Cap. Deutsche Gesellschaft für Internationale Zusammenarbeit (GIZ) GmbH. Bonn, Germany.

Shilton, A.N., Elmetri, I., Drizo, A., Pratt, S., Haverkamp, R.G. and Bilby, S.C. 2006. Phosphorous removal by an 'active' slag filter—a decade of full-scale experience. Water Res. 40: 113–118.

Slater, H., de Boer, D., Shu, W. and Guoqiang, Q. 2018. 2018 China Carbon Pricing Survey. China Carbon Forum, Beijing.

Someharaeri, H.N., Majumerd, M.M., Breuhaus, P. and Assadi, M. 2014. Performance analysis of a biogas-fueled micro gas turbine using a validated thermodynamic model. Appl. Therm. Eng. 66: 181–190.

Stangeland, K., Kalai, D., Li, H. and Yu, Z. 2017. CO_2 methanation: The effect of catalyst and reaction conditions. Energy Procedia 105: 2022–2027.

Szabó, A., Takács, I., Murthy, S., Daigger, G.T., Licskó, I. and Smith, S. 2008. Significance of design and operational variables in chemical phosphorus removal. Water Environ. Res. 80: 407–4016.

Taifouris, M.R. and Martín, M. 2018. Multiscale scheme for the optimal use of residues for the production of biogas across Castile and Leon. J. Clean. Prod. 185: 239–251.

The Government of the Republic of Korea. 2017. Second Biennial Update Report of the Republic of Korea. Under the United Nations Framework Convention on Climate Change. Government Publications Registration Number 11-1092000-000046-11. Seoul, Republic of Korea.

United States of America Federal Energy Regulatory Commission, Docket No. RM16-8-000; Order No. 828. Requirements for Frequency and Voltage Ride Through Capability of Small Generating Facilities, 2006.

Ünveren, E.E., Monkul, B.O., Sarıoğlan, S., Karademir, N. and Alper, E. 2017. Solid amine sorbents for CO_2 capture by chemical adsorption: A review. Petroleum 3(1): 37–50.

[US DOE] U.S. Department of Energy. 2019. Database of State Incentives for Renewables & Efficiency (DSIRE).

[US EPA] U.S. Environmental Protection Agency. 1979a. Estimating Water Treatment Costs. Volume 2, Costs Curves Applicable to 1 to 200 mgd Treatment Plants.

[US EPA] U.S. Environmental Protection Agency. 1979b. Estimating Water Treatment Costs. Volume 3, Costs Curves Applicable to 2,500 gpd to 1 mgd Treatment Plants.

[US EPA] U.S. Environmental Protection Agency. 2014. Wisconsin Integrated Assessment of Watershed Health.

[USDA] United States Department of Agriculture. 2009. Agricultural Water Management Field Handbook.

[USDA] US Department of Agriculture. Environmental Quality Incentives Program (EQIP). https://www.nrcs.usda.gov/wps/portal/nrcs/main/national/programs/financial/eqip/# (accessed April 20, 2019).

[USDA] US Department of Agriculture. Rural Energy for America Program (REAP) Grants. https://www.energy.gov/savings/usda-rural-energy-america-program-reap-grants (accessed April 20, 2019).

Vrbová, V. and Ciahotný, K. 2017. Upgrading biogas to biomethane using membrane separation. Energy Fuel 31(9): 9393–9401.

[WCED] World Commission on Environment and Development. 1987. Our Common Future. Oxford University Press. New York.

[WEC] World Energy Council. 2016. World Energy Resources Waste to Energy.

Wilsenach, J.A., Schuurbiers, C.A.H. and van Loosdrecht, M.C.M. 2007. Phosphate and potassium recovery from source separated urine through struvite precipitation. Water Res. 41: 458–466.

World Bank and Ecofys. 2018. State and Trends of Carbon Pricing 2018. World Bank, Washington DC, US.

Zeng, L. and Li, X. 2006. Nutrient removal from anaerobically digested cattle manure by struvite precipitation. J. Environ. Eng. Sci. 5: 285–294.

Zhang, T., Li, P., Fang, C. and Jiang, R. 2014. Phosphate recovery from animal manure wastewater by struvite crystallization and CO_2 degasification reactor. Ecol. Chem. Eng. 21: 89–99.

Zhang, Y. and Chen, C.C. 2013. Modelling CO_2 absorption and desorption by aqueous monoethanolamine solution with Aspen rate-based models. Energ. Proc. 37: 1584–1586.

CHAPTER 12

Efficient and Low-Carbon Energy Solution through Polygeneration with Biomass

Kuntal Jana[1] and *Sudipta De*[2,*]

1. Introduction

To maintain the economic growth and to improve the living standard, secondary energy demand is always increasing for both developed and developing countries (IEA, 2015). Besides electricity, demand of several other energy utilities is also increasing. One such demand, environment-friendly fuel (gaseous and liquid) for transportation, is increasing over time. On the other hand, distribution of reserved fossil fuels (e.g., coal and petroleum) is not uniform. For example, availability of petroleum is concentrated within a few countries. Other countries depend on import of petroleum (BP, 2018). It affects the energy security of many developed and developing countries. Apart from economic growth, energy access is another critical challenge for developing and under-developed countries. About 1.1 billion people (14% of global population) has no access to electricity. Many people depend on biomass as fuel for cooking, having no access to any other clean fuel (IEA, 2017).

Apart from energy access and energy security, climate change is another critical challenge. The main contributor to climate change is anthropogenic CO_2 emission from fossil fuels and other industrial processes, as shown in Figure 2(a). Forestry and other land use change are also responsible for greenhouse gas (GHG) emission variations. Sector-wise, electricity and heat production, agriculture, forestry and other land use are the main sources of GHG emission (IPCC, 2014). Hence, in order to mitigate climate change, use of non-fossil fuels is a promising option for electricity and heat production as well as for other energy services.

Biomass is emerging as a promising source of secondary energy services, from electricity to transportation fuel, etc. (McKendry, 2002). Before the pre-industrial age, biomass was mainly used as the primary source of energy for heating and cooking. Discovery of fossil fuels (with high energy content) triggered rapid industrialization and replaced the use of biomass. However, in this present context of the search for sustainable energy solutions, biomass is regaining its significance. It is a widely available source of energy in both developing and developed countries (Bhattacharya, 2002). However, in many developing and under developed countries, it is under-utilized (say, as cattle fodder) and even treated as 'waste' to be disposed of (say, burned in agriculture field), with negative environmental impacts. Hence, an energy system design using biomass is imperative for a sustainable energy solution. The logistics of the conversion chain of biomass to the utility production of the energy system should be efficient from the viewpoint of economic, environmental and social performance.

[1] Department of Mechanical Engineering, National Institute of Technology Durgapur, Durgapur-713209, India.
Email: kuntaljana@gmail.com, kuntal.jana@me.nitdgp.ac.in

[2] Department of Mechanical Engineering, Jadavpur University, Kolkata 700032, India.

* Corresponding author: de_sudipta@rediffmail.com, sudipta.de@jadavpuruniversity.in

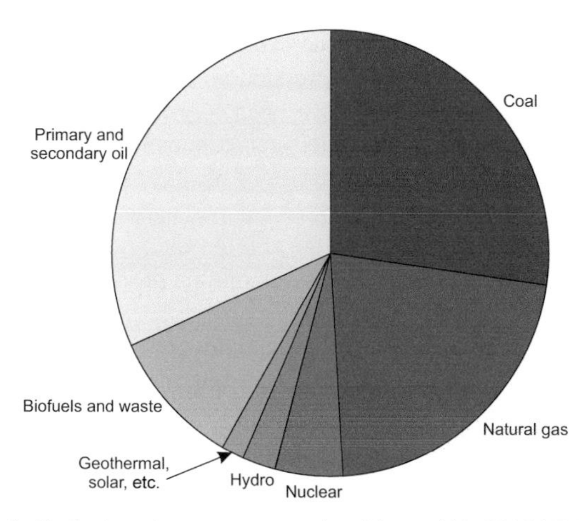

Figure 1. Fuel-wise primary energy supply of the world in 2016 (IEA, 2016).

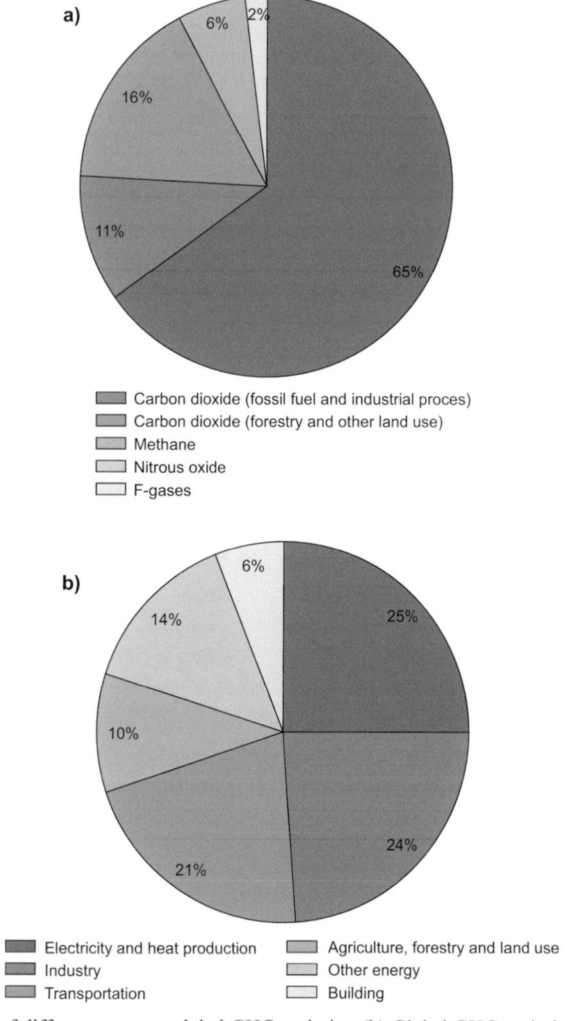

Figure 2. (a) Contribution of different gases to global GHG emission (b) Global GHG emissions from different economic sectors (IPCC, 2014).

Sources of biomass may vary widely. In some cases, it affects the food security, specifically when foodstuffs (e.g., sugarcane) are used for biofuel production. Unsustainable production of biomass from forest, or use of agricultural land for bioenergy harvesting may damage the biodiversity and food supply. Hence, proper planning is required for uninterrupted and sustainable supply of biomass without affecting the food security and environment. However, the use of residues from forest and agriculture can be a sustainable solution (Bentsen et al., 2014). Use of abandoned agricultural land may be another suitable source of biomass (Campbell et al., 2008). Sustainable production and use of energy crops may be another possible option. For third and fourth generation biofuel production, algae is an emerging option, as discussed later.

Polygeneration is an energy system for producing multiple utilities in a single plant through proper system integration. Apart from electricity, biofuels, chemicals, heating, cooling, etc., there are some other utilities produced in a polygeneration to cater to various energy service needs. With more social acceptance and better environmental and economic performance, biomass based polygeneration is a possible sustainable energy solution. In this chapter, availability and logistics of biomass are discussed. Then, several conversion technologies for biomass to bioenergy and its products through polygeneration are discussed. Different polygeneration system designs and their analysis are also studied in this chapter.

2. Availability of Biomass

For long term sustainable operation of a biomass-based energy system, the supply of biomass should be uninterrupted. However, the supply chain of biomass is dependent on availability. Various sources of biomass used as primary energy are given in Figure 3. From this figure, it is noted that mostly fuelwood is used. However, charcoal and agricultural biomass have good potential (Bauen et al., 2004). Wood from different residues are also a good source of energy. There is also the possibility of mixing biomass, though its usefulness is technology-dependent and sensitive to various factors, such as fermentation, pyrolysis, etc. Mauser et al. (2015) showed that, without changing the cropland area, biomass can meet the future demand, and that improved farm management and efficient spatial allocation will help to achieve this.

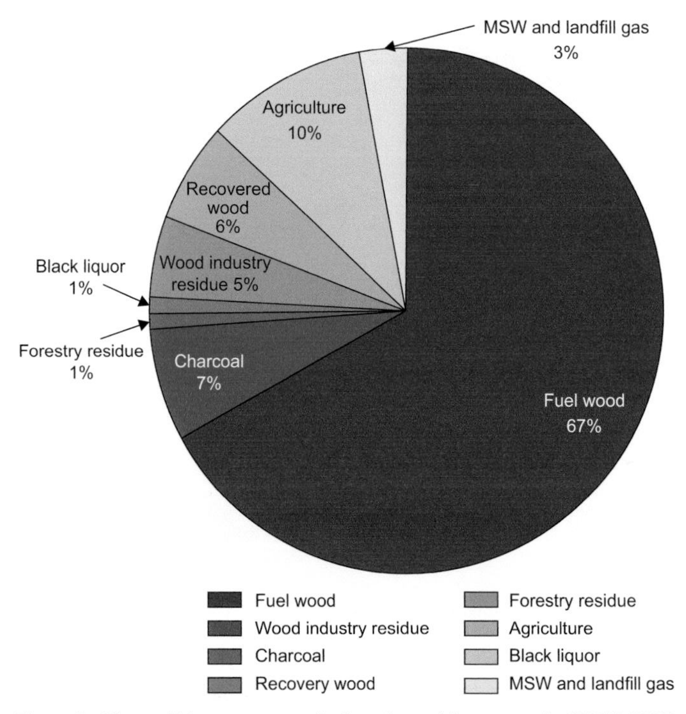

Figure 3. Share of biomass sources in the primary bioenergy mix (IPCC, 2007).

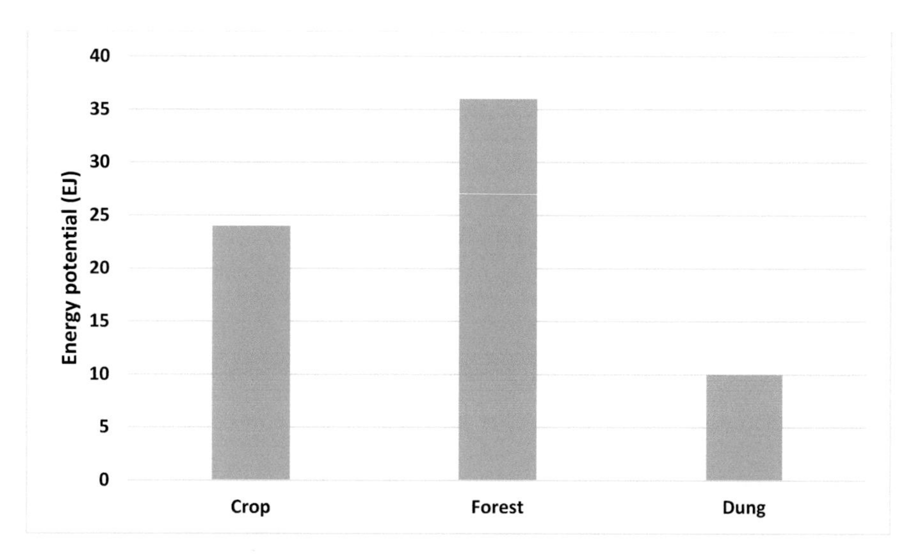

Figure 4. Worldwide energy potential (EJ) from residues (Bauen et al., 2004).

There is a significant scope of residue-based bioenergy, as shown in Figure 4. Forest residue, crop residue and animal waste have good potential in this respect. Depending on the availability of biomass supply, logistics will be designed along with the energy system design. The details of logistics are discussed in the subsequent section.

Determination of the size of a biomass-based plant is important, however, it should be noted that the availability of biomass (yearly/seasonally) has a significant effect on plant size. Optimization of the plant size may be done based on biomass distribution density, which affects the net GHG emission and economic performance of the plant due to preprocessing, transportation and handling of biomass. Depending on the supply area, the plant may be distributed or centralized. In case of distributed electricity generation, there is no high voltage transmission. However, a centralized plant supplies electricity through grid connection with high voltage transmission. Other utilities, e.g., biofuel, are supplied by long distance transportation. However, depending on the size of the plant, selection of technology and cost of utilities are also decided.

3. Logistics of Biomass

Logistics of biomass is very important for the planning and operation of biomass-based polygeneration, as it involves significant energy consumption and economic investment. There are several stages of logistics, as shown in Figure 5. From Figure 5, it is noted that biomass is collected from different sources. Then, according to the type of biomass, it is preprocessed for loading. There are different types of preprocessing, such as drying, baling, torrefaction and crushing. Reducing the volume and increasing the energy density are the main reasons behind the preprocessing before loading. After that, biomass is transported by small and medium trucks for short distances and by rail or large trucks for long distances, though long-distance transport of biomass is not generally recommended. Then, it is unloaded and handled for storage. Handling is done by screening and crushing. Then, biomass is fed to the suitable bioenergy conversion system through different kinds of feeders, like belts, conveyors, stokers and screws. Depending on the type of biomass and type of energy conversion plant and capacity of the plant, the feeder is selected.

4. Biomass Conversion Systems

Biomass available in nature is not in a suitable form to use directly for secondary energy or utility production. Hence, proper conversion of biomass to secondary energy or other utilities is essential. It may

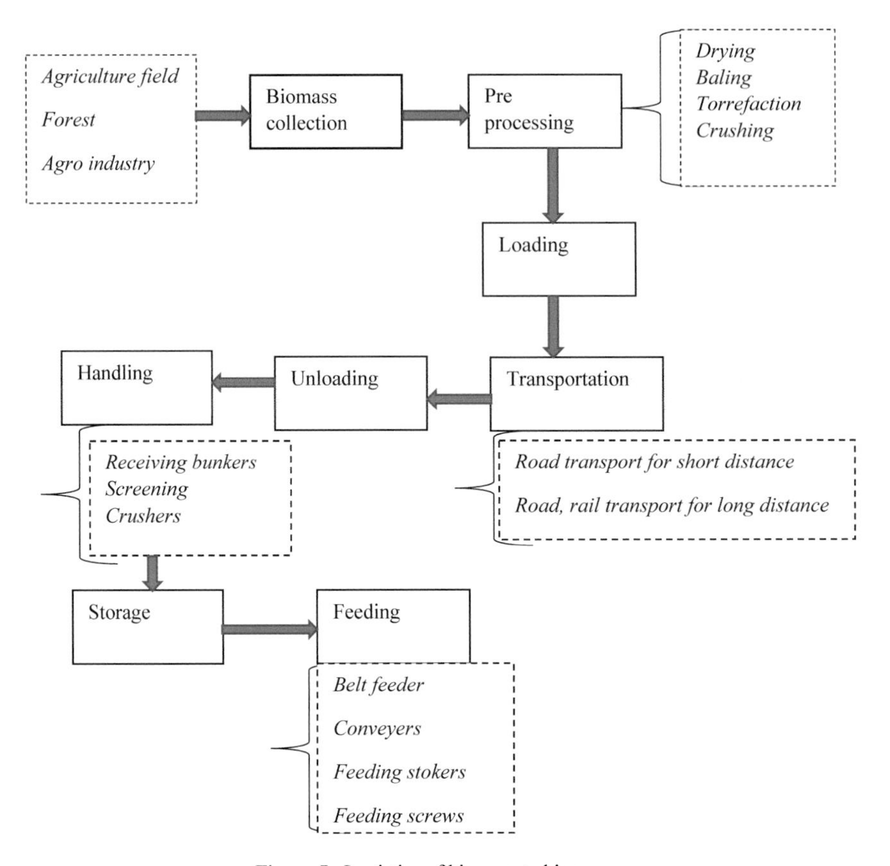

Figure 5. Logistics of biomass to bioenergy.

be converted through two main routes, as shown in Figure 6(a), i.e., biochemical and thermochemical. In the biochemical process, suitable biomass (e.g., cow dung or kitchen waste) is converted to liquid or gaseous products in the presence of air (aerobic digestion) or in the absence of air (anaerobic digestion). Biomethane is the most desired product of this process and may be used as an alternative for natural gas. Alcoholic fermentation is another biochemical process which is used to produce ethanol and mixed alcohol from suitable biomass (e.g., sugarcane). Combustion, gasification, pyrolysis and torrefaction are different thermochemical processes for biomass conversion. Biomass may also be directly used in a combustion process in order to produce heat in further downstream processes or it may be used with coal as cofiring. Gasification is another route of conversion of solid biomass to producer or syngas with good calorific value. Biomass is heated in the absence of any oxidizer to produce different liquids and gaseous products in a pyrolysis process.

During the gasification and pyrolysis process, several impurities are produced. This affects the performance of downstream processes. Tar, alkali metals, HCl, HCN, H_2S, COS, etc., are the impurities. However, tar is considered as the main impurity which creates problem during operation. It is deposited inside the pipeline in a downstream process when the temperature is below the tar condensation temperature. For long-term operation, tar in the product gas should be within a specified limit, e.g., for IC engine < 100 mg/Nm^3, for gas turbine < 5 mg/Nm^3, for methanol synthesis < 0.1 mg/Nm^3, for fuel cell < 1.0 mg/Nm^3, for FT synthesis < 1.0 μL/L (Rios et al., 2018). Tar concentration in the product gas may be reduced by an *in situ* tar reduction process. It avoids or depletes tar during its formation process. It can be done through adequate control of the process during operation by using a catalyst. Tar may be reduced in the post gasification process by physical removal or by a catalytic or thermal cracking process.

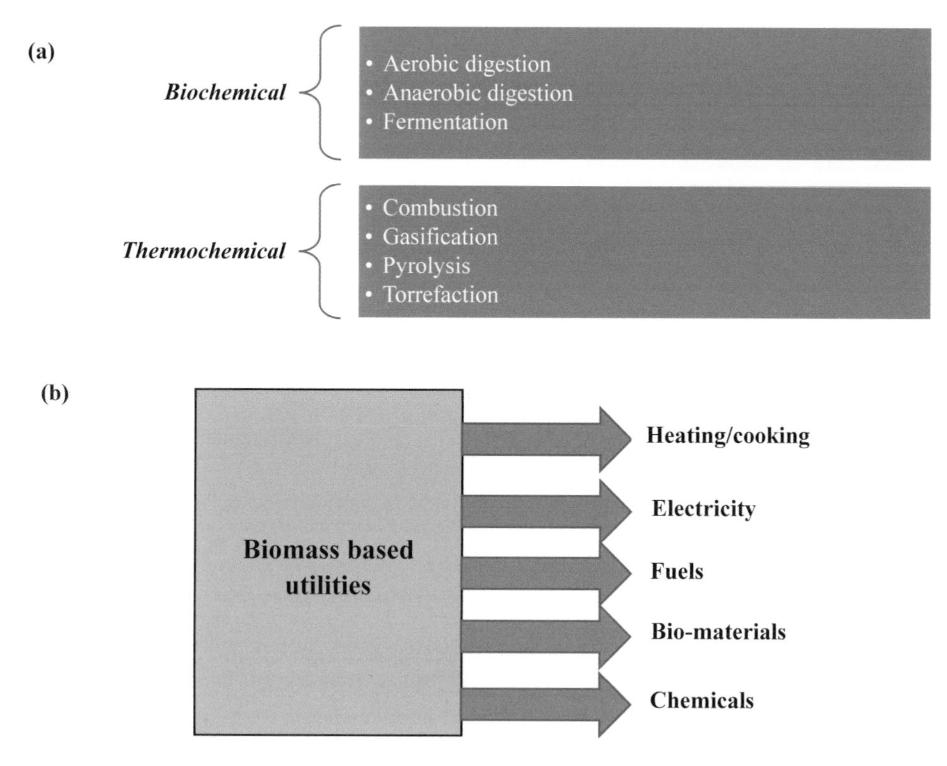

Figure 6. (a) Biomass conversion and (b) products.

In a torrefaction process, the moisture and volatile content of the biomass is removed in order to increase its energy density. Besides these two processes, in an agrochemical process, liquid biofuels are produced by using fuel extraction and esterification (Twidell and Weir, 2010). A brief discussion on biofuel production is given in subsequent section. Proper conversion of biomass and downstream utilization of the converted biomass make it an attractive option for different utility production. Biomass has been used primarily for heating and cooking since the pre-industrial age. Presently, biomass is a promising source of energy to produce electricity, fuels for transportation, chemicals and different bio-materials.

Liquid or gaseous biofuels are important utility outputs of polygeneration. Biofuels can be classified as primary or secondary (Alalwana et al., 2019). In the case of primary biofuels, biomass is used for heating or cooking directly without any conversion. Whereas, for production of secondary biofuels, biomass is converted through suitable conversion technology, as discussed earlier. Based on biomass types, there are four types of secondary biofuels, as shown in Figure 7. In first-generation technology, biofuels are produced from foodstuffs, e.g., sugarcane, rape seed, palm oil, sunflower, etc., ethanol and biodiesel being the outputs. One limitation of these biofuels is that their use affects food security and land use. Hence, a conflict between energy and food security arises. However, for second-generation biofuels, lingo-cellulosic biomass is used. This type of biomass contains lignin, cellulose and hemicellulose (e.g., agriculture waste, wood, etc.), and it does not affect food security. Algae-based biofuels belong to the third generation. However, in order to improve the production rate, genetically modified algae are used for biofuel production. During the production of such biomass, power and heat may either be produced or consumed, as discussed later. Hence, a suitable integration technique will improve the performance of the polygeneration plant. In Figure 8, biochemical conversion and esterification are shown. Sugar can be directly converted to ethanol by fermentation. However, hydrolysis is required in order to convert starch to ethanol. On the other hand, for second-generation biomass, pretreatment and hydrolysis are required

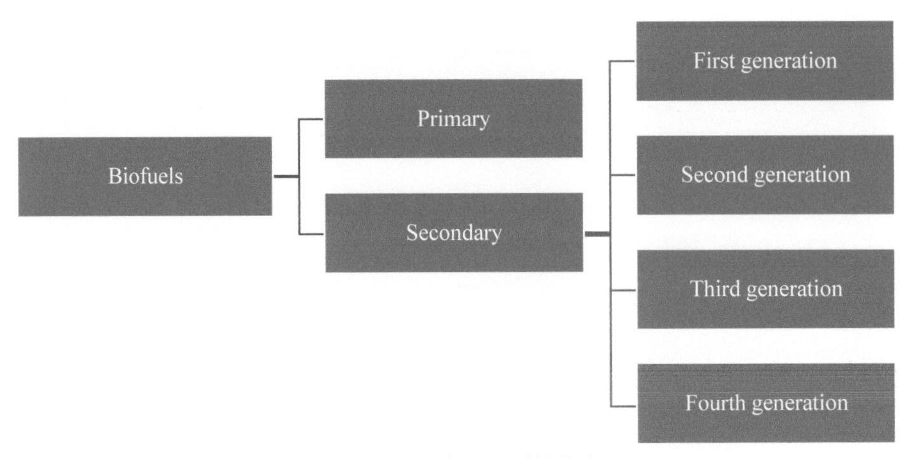

Figure 7. Classification of biofuels.

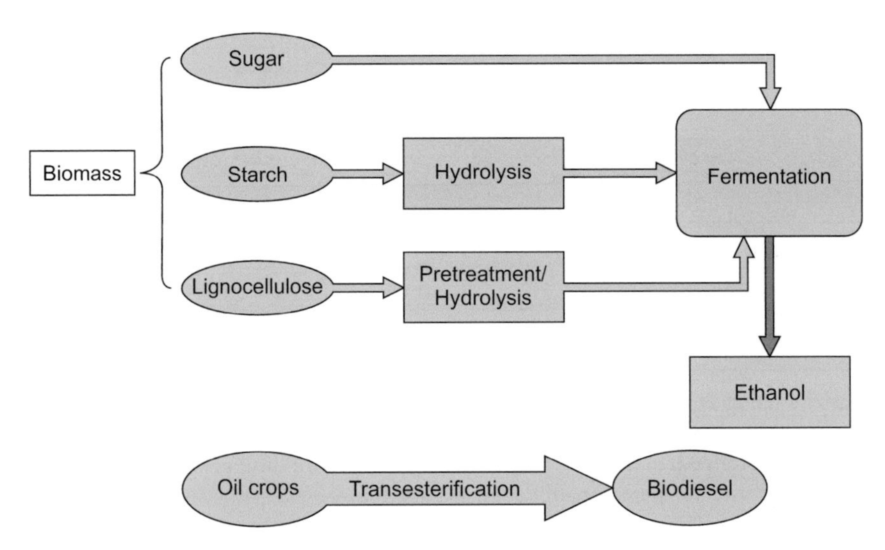

Figure 8. Biochemical and esterification conversion of biomass to biofuel (Jana et al., 2018).

in order to produce ethanol through fermentation. Oil crops (e.g., Jatropha, palm oil, rapeseed) can be converted to biodiesel by transesterification method. Gasification is a thermo-chemical way of producing biofuels, as shown in Figure 9. In this process, biomass is initially converted to syngas through a partial oxidation process by oxygen from air separation unit. Then, this gas is upgraded and biofuel synthesis is done. After biofuel synthesis, it is upgraded and separated to different mixed alcohols through distillation method. However, off-gas may be used for power and heat production in steam or gas turbine (GT) cycles, as shown in Figure 9.

During the conversion of biomass to biofuels, CO_2 is emitted and absorbed in several stages, as shown in Figure 10 as a representative case. In this figure, different stages of biomass to biofuels (biochemical) and associated CO_2 emissions are shown. From the figure, it is noted that CO_2 is absorbed during the growth of biomass. For harvesting, several machineries (e.g., pump, tractor, etc.) and fertilizers are used and cause CO_2 emission. For transportation (e.g. diesel use in trucks) and preprocessing of biomass (e.g., baling) significant amount of CO_2 is emitted. Hence, biomass is generally considered as carbon-neutral fuel but its utilization is not carbon neutral.

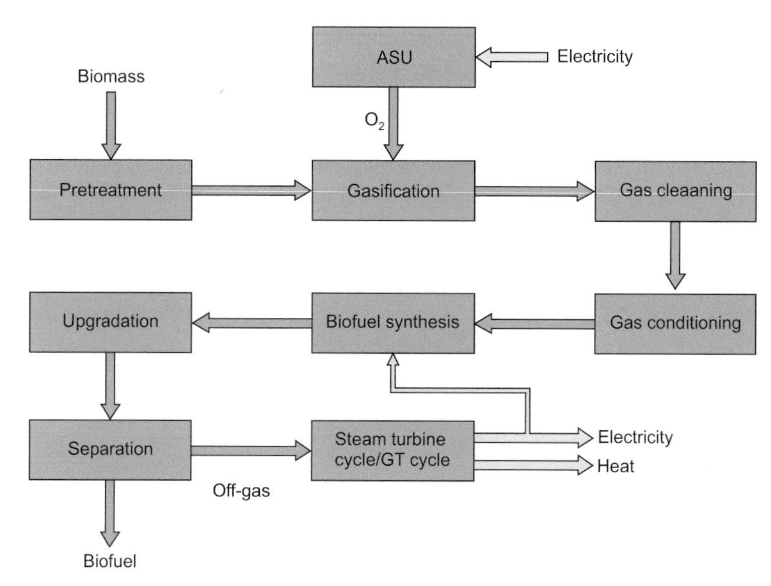

Figure 9. Thermochemical conversion of biomass to biofuel.

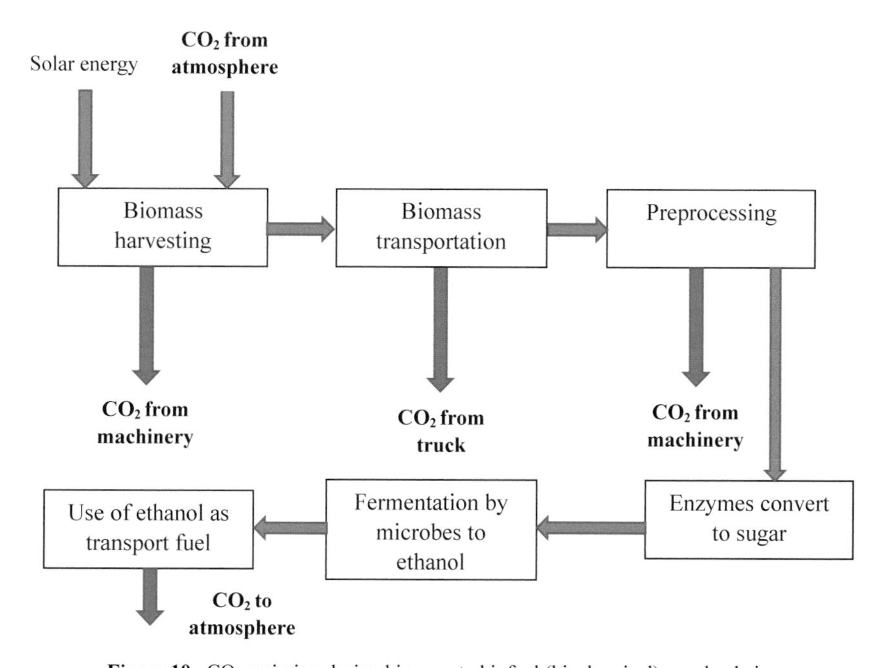

Figure 10. CO_2 emission during biomass to biofuel (biochemical) supply chain.

5. Polygeneration

In Figure 11, the concept of polygeneration is shown. From Figure 11, it is noted that power is the only utility output from a power plant. However, power and heat are the outputs from a cogeneration plant. In a trigeneration plant, output utilities are power, heat and cooling. In the case of a polygeneration plant, output utilities may be numerous, e.g., fuels and chemicals. Hence, polygeneration is a process integration for producing multiple utilities from a single plant. In the subsequent section, different types of polygeneration plants are shown to produce multiple utilities.

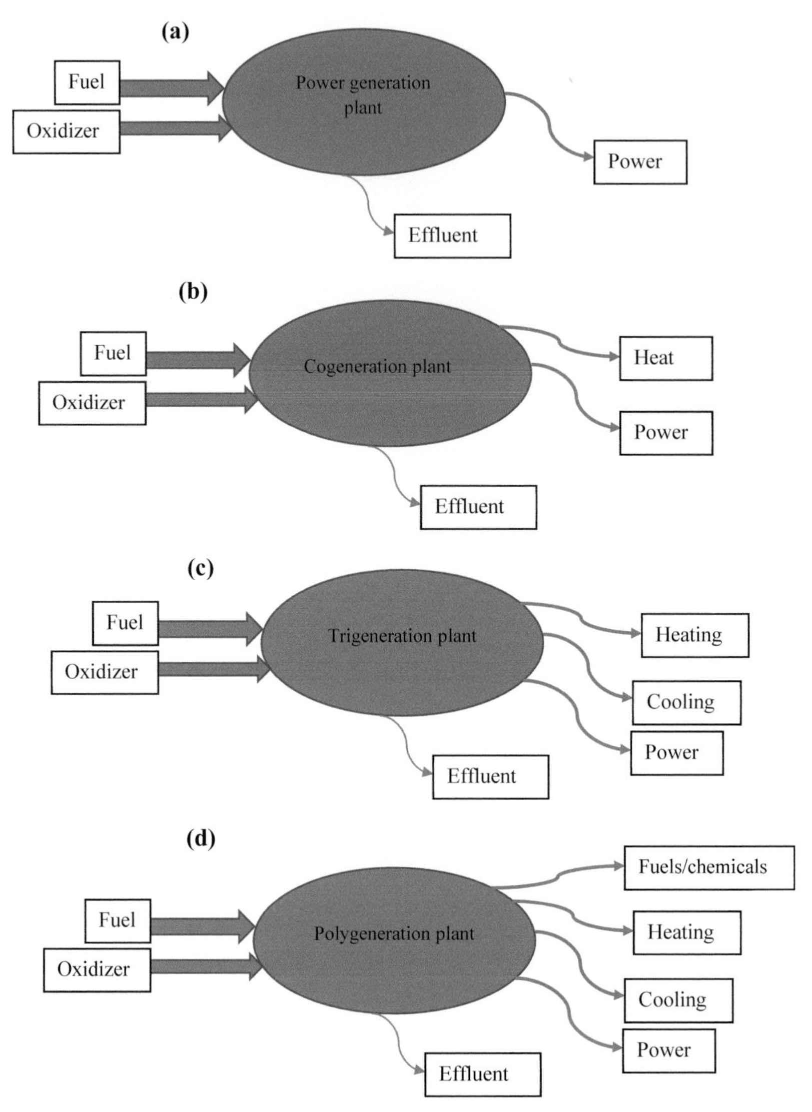

Figure 11. Concept of (a) power generation, (b) cogeneration, (c) trigeneration, (d) polygeneration.

5.1 Polygeneration using biomass and hybrid-biomass

In Table 1, the recent literature on biomass and hybrid biomass is shown. For a biomass-based polygeneration system, the input energy may be biomass only or it may be a hybrid energy system. In a hybrid biomass-based energy system, input energy of biomass is hybridized with another source of energy (e.g., solar energy) (Bajpai and Dash, 2012). Proper hybridization of biomass gives better supply and demand matching. It is noted that power, liquid and gaseous hydrocarbons, FT liquids, hydrogen, chemicals, char, tar, heating, cooling, etc., are different utility outputs. Different types of biomass, e.g., cotton stalk, black liquor, wood, algae, straw, bagasse, etc., are the input biomass. In some cases, biomass is used with fossil fuels (e.g., coal or natural gas). Pyrolysis, torrefaction and gasification are used in the conversion of biomass. Pyrolysis is done at different temperatures. However, gasification is done using different oxidizing media and in different types of gasifiers. Many authors have investigated this theoretically and a few papers on experimental study, specifically for pyrolysis, are available.

Table 1. Summary of recent literature (2013–2018) on biomass and hybrid biomass-based polygeneration.

Author (Year)	Products	Input fuel	Conversion technology	Remarks
Atsonios et al. (2017)	Liquid hydrocarbons, gaseous hydrocarbon	Biomass, coal	Pyrolysis	Simulation by using Aspen Plus®
Brau et al. (2013)	Hydrogen, high pressure steam	Butane, biomass	Indirect gasification	Simulation
Chen et al. (2014)	Char, liquid oil, biogas	Cotton stalk	Torrefaction, pyrolysis	Experimental study
Chen et al. (2017)	Bio-oil, char, gas	Cotton stalk	Torrefaction, pyrolysis	Experimental study
Dahlquist et al. (2017)	Syngas, hydrogen, power, heat	Black liquor, wood pellet	Gasification	Experimental and numerical study
Gadsboll et al. (2019)	SNG, power, heat	Wood, straw	Two-stage gasification	Experimental investigation of gasification
Graciano et al. (2018)	FT liquids, electricity, hydrogen, thermal energy	Algal biomass	Thermochemical conversion	Environmental and economic assessment
Ilic et al. (2014)	Biofuel, heat, electricity, biogas	Biomass	Gasification, anaerobic digestion	Theoretical assessment
Jana and De (2015a)	Power, desalinated water, heating, cooling	Coconut fiber	Gasification	Thermodynamic, economic and environmental study
Jana and De (2015b)	Power, ethanol, heating, cooling	Rice straw, sugarcane bagasse, coconut fiber	Gasification	Thermodynamic and economic study
Jana and De (2017)	Power, ethanol, heating, cooling	Rice straw	Gasification	Life cycle analysis
Kabalina et al. (2017)	SNG, syngas, char, heating, cooling, electricity	Refuse derived fuel	Gasification	Energy and economic assessment
Khanmohammadi and Atashkari (2018)	Electricity, hot water, fresh water	Biomass	Gasification	Modelling and optimization
Kraussler et al. (2018)	Hydrogen, electricity, heat	Biomass	Steam gasification	Experimental investigation on fluidized bed gasification
Kravanja et al. (2013)	Ethanol, heat	Straw	Biochemical	Heat integration technique (pinch analysis)
Liu et al. (2017)	Ethanol, xylose, heat and power	Corn cob	Saccharification and fermentation	Simulation and exergy analysis
Lythcke-Jorgensen et al. (2017)	Power, liquid fuel, natural gas, biogas/ syngas, district energy	Wood chip	Gasification	Optimization
Magdeldin et al. (2017)	Liquid fuels	Scots pine residue	Hydrothermal liquefaction	Techno-economic assessment
Naqvi et al. (2018)	SNG, electricity	Wheat straw, black liquor	Gasification	Theoretical study
Nguyen and Clausen (2018)	Bio-oil, SNG, hydrogen, methanol	Biomass	Catalytic hydropyrolysis	Thermodynamic analysis

Table 1 contd. ...

... *Table 1 contd.*

Author (Year)	Products	Input fuel	Conversion technology	Remarks
Sahoo et al. (2018)	Power, desalinated water, cooling	Biomass, solar	Combustion	Energy, exergy and economic analysis
Salkuyeh et al. (2016)	Methanol, ethanol, DME, olefins, FT liquids, electricity	Biomass, petcoke, natural gas	Gasification	Techno-economic analysis
Salman et al. (2018)	Dimethyl ether, methanol, heating, power	Refuse derived fuel	Gasification	Techno-economic analysis
Sigurjonsson and Clausen (2018)	Bio SNG, electricity and heat	Wood chips	Gasification	Techno-economic analysis
Truong and Gustavsson (2013)	Liquid fuels, electricity, heat, pellets	Woody biomass	Gasification	Techno-economic analysis
Vidal and Martin (2015)	Hydrogen, electricity, heat	Biomass, solar	Gasification	Optimization study
Wagner et al. (2015)	SNG, heat, power	Wood chips, corn silage	Gasification, anaerobic digestion	Thermodynamic and economic study
Wang et al. (2018)	Biogas, bio-oil, char	Corn stalk	Torrefaction, pyrolysis	Experimental investigation
Xin et al. (2013)	Anhydro-saccharides, light oxygenates and phenols	Cellulose, hemicellulose, lignin	Pyrolysis	Experimental investigation
Yang et al. (2016)	Charcol, biogas, woody tar, woody vinegar	Corn stalk, rice husk	Pyrolysis	Experimental investigation
Yang et al. (2018)	Syngas, charcoal, woody tar, electricity	Agriculture residues	Pyrolysis	Life cycle analysis
Zhang et al. (2018a)	Bio-oil, value added chemicals, syngas, bio-char	Corncob	Two-step pyrolysis	Experimental investigation
Zhang et al. (2018b)	Biogas, tar, vinegar, biochar	Biomass	Pyrolysis	Experimental investigation

6. Polygeneration System Design for Different Products

In this section, system layouts for producing different products/utilities are shown. In Figure 12, a polygeneration system for hydrogen production is shown. Biomass is gasified in a suitable gasifier. During the gasification process, heat is generated by exothermic reactions. This heat is used for generating steam, which is used in steam power cycle. Produced syngas is cooled, cleaned and conditioned. Recovered heat during cooling may be used in downstream processes. Tar and other acid gases are removed during cleaning of the syngas. Then, the water gas shift reaction occurs, increasing the concentration of H_2 and decreasing the concentration of CO. Then, H_2 is separated by a suitable method (e.g., adsorption, membrane). The remaining off-gas is used for power generation in a steam power cycle.

In Figure 13, the schematic of a pyrolysis-based polygeneration is shown. In this process, washing and torrefaction is done before the pyrolysis process (Chen et al., 2017). Bio-oil, bio-char and gas are the products of the pyrolysis process. After subsequent processing, fuels, char, chemicals, fertilizers, syngas, etc., may be produced. In Figure 14, the process for dimethyl ether (DME) production is shown. It is also a gasification-based process. After separation of DME, a fraction of CO_2 may be used in a gasification

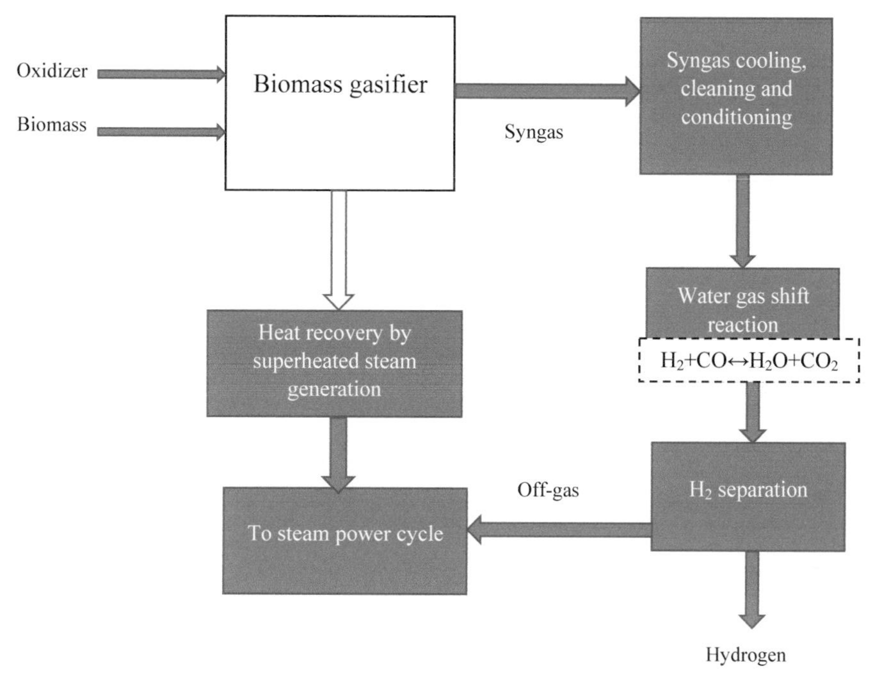

Figure 12. Polygeneration for hydrogen production.

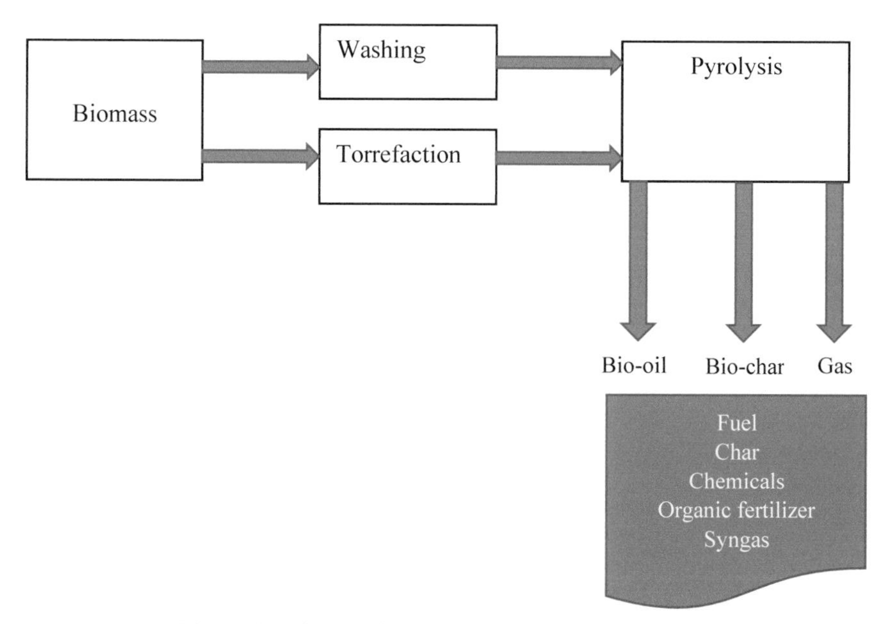

Figure 13. Polygeneration from cotton stalk through pyrolysis.

process. Generated steam using recovered heat may be used as the gasification medium. In Figure 15, a microalgae-based polygeneration process is shown (Graciano et al., 2018). For this, microalgae are harvested. Then, dewatering and drying processes are carried out using surplus energy. FT liquids, electricity, hydrogen and heat may be the utility outputs of this process. This process is a thermochemical conversion.

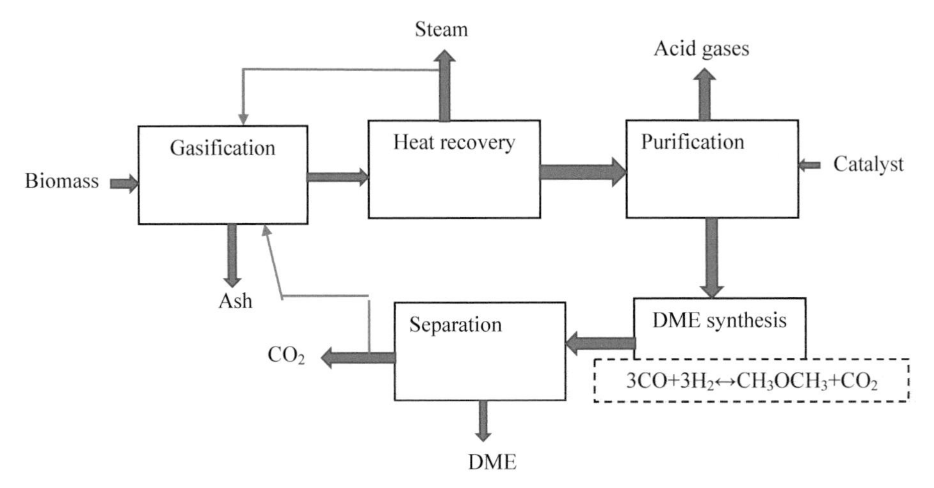

Figure 14. Thermo-chemical biomass-based polygeneration for chemical (DME) production.

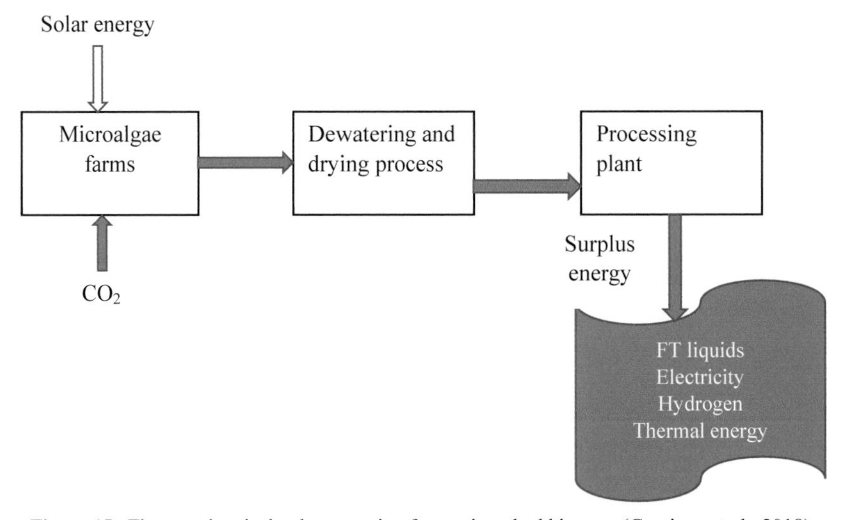

Figure 15. Thermo-chemical polygeneration from microalgal biomass (Graciano et al., 2018).

In Figure 16, rice straw-based polygeneration for producing ethanol, electricity, heating and cooling are shown (Jana and De, 2015a). It is a gasification-based conversion of biomass. In this process, rice straw is obtained as a by-product/waste of rice cultivation. Baling of rice straw is done for transportation. Then, it is fed to the gasifier in order to produce syngas/producer gas. This gas is utilized for ethanol production and for power generation through a combined cycle. This percentage of syngas may vary according to the demand of utilities. Otherwise, ethanol may be stored as energy storage when the power demand is less. During gasification and ethanol synthesis processes, heat is generated. This heat may be used for vapor absorption cooling and utility heat purposes. In Figure 17, production of power, desalinated water, cooling and utility heat is shown (Jana and De, 2015b). This polygeneration system is useful for coastal areas, where plenty of saline water is available. Coconut fiber is an input biomass which is available in coastal areas. Biomass-based polygeneration has the potential to capture CO_2, as shown in Figure 18. In this process, after gasification and gas cleaning, the water gas shift reaction is done in order to increase the concentration of CO_2 in the syngas. This CO_2 may be separated by suitable chemical or physical absorption and/or adsorption process. However, the produced H_2 may be utilized

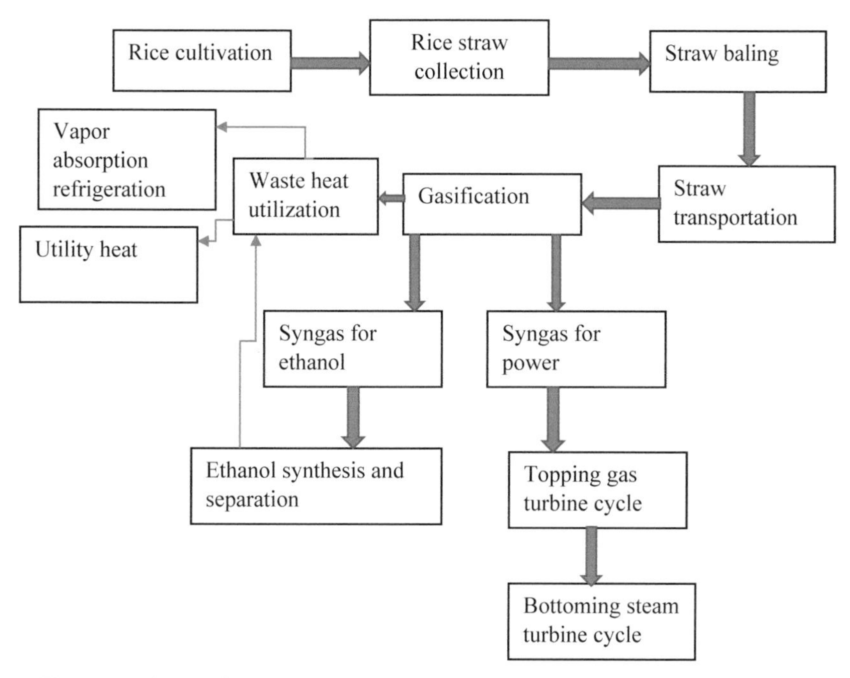

Figure 16. Thermo-chemical conversion of biomass for power, ethanol, heating and cooling.

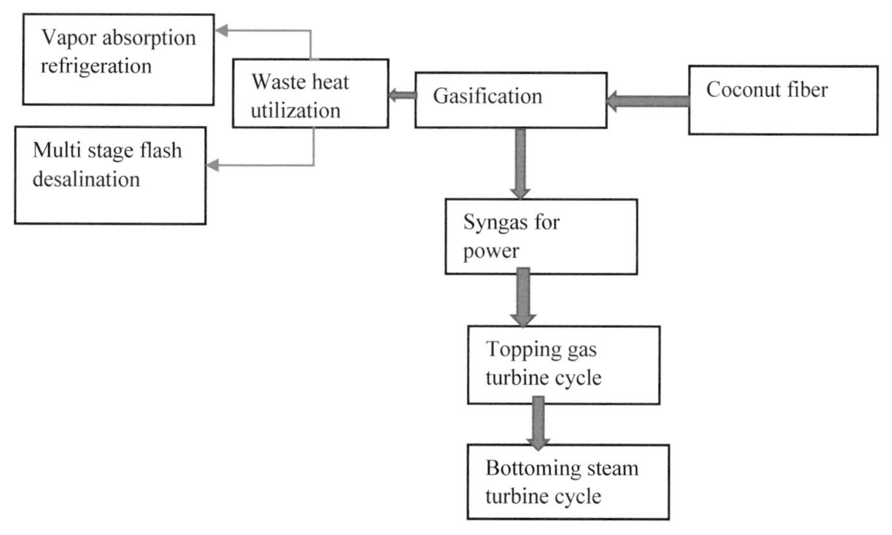

Figure 17. Thermo-chemical conversion of biomass for power, water, heating and cooling for coastal areas.

for power generation, fuel synthesis or chemical production. The captured CO_2 may be transported and stored or it may be utilized for value added products. In Figure 19, a summary of biomass to bioenergy is given. Different feedstocks, i.e., first-generation to fourth-generation biomass, and suitable conversion technology and potential products of this conversion route are shown.

In Table 2, the present developmental status of biomass upgradation and its conversion technology is shown. For biomass upgradation, palletization is commercially available technology. However, torrefaction and pyrolysis are at R&D and demonstration stages, respectively. Gasification technology

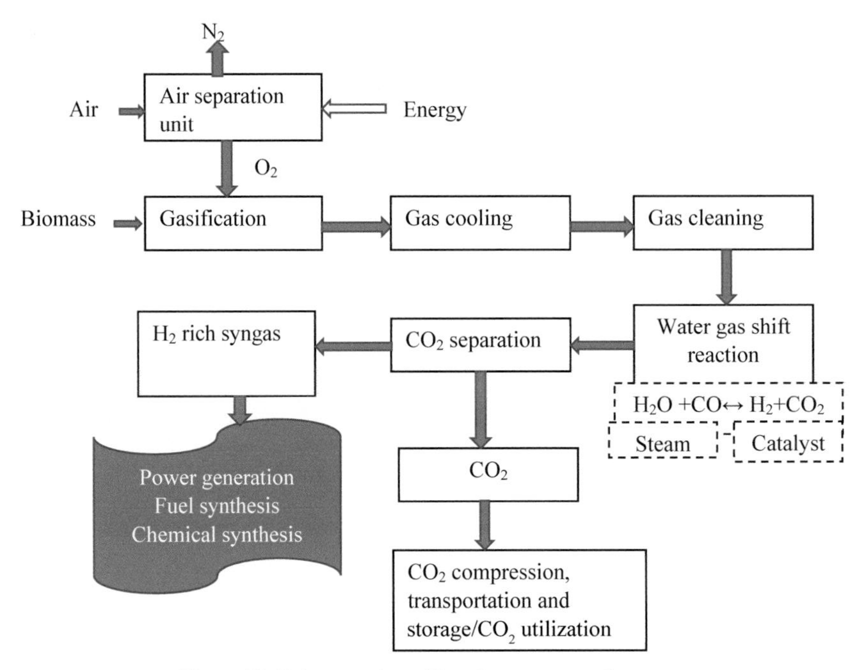

Figure 18. Polygeneration with carbon capture and storage.

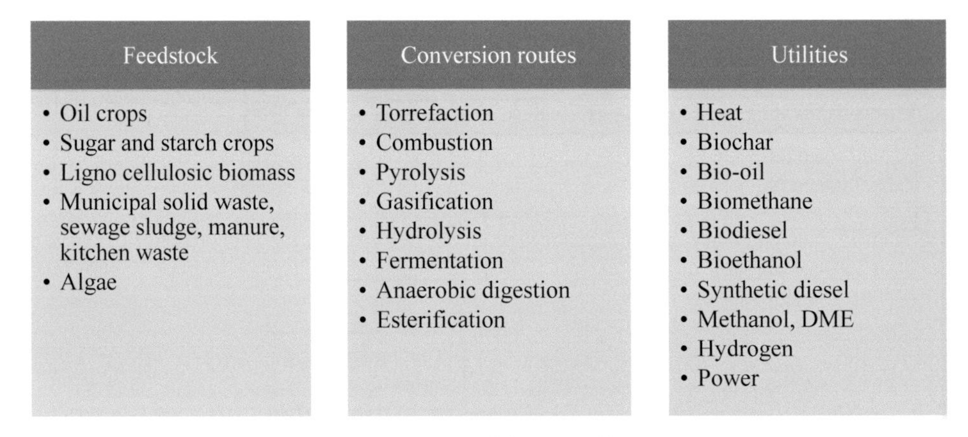

Figure 19. Summary of biomass to bioenergy.

Table 2. Development status of biomass upgradation and conversion (IEA, 2009).

Status/Technology	Research and development	Demonstration	Early commercial	Commercial
Biomass upgradation	Torrefaction	Pyrolysis		Pelletization
Biomass to gas			Gasification	Combustion
Thermodynamic cycle	Integrated gasification fuel cell	Organic rankine cycle, Stirling engine, Integrated gasification combined cycle (IGCC)		
Cofiring		Indirect cofiring	Parallel cofiring	Direct cofiring
Anaerobic digestion (AD)	Microbial fuel cell		Biogas upgrading, 2-stage AD	1-stage AD

is in its early commercial stage and combustion technology is a commercially available technology. To utilize the gasification process, integrated gasification fuel cells are at the R&D stage. Organic Rankine Cycle (ORC), Stirling engine and IGCC are in the demonstration phase. Indirect cofiring, parallel cofiring and direct cofiring of biomass with coal are at the stage of demonstration, early commercial and commercial use, respectively. Aerobic Digestion (AD) technology is commercially available. However, microbial fuel cells are at the R&D stage.

7. Sustainability Assessment of Biomass-based Polygeneration

The sustainability of a biomass-based polygeneration system is measured by economic, environmental and social parameters. Techno-economic analysis shows the economic viability of a polygeneration system, while socio-economic analysis shows the social acceptability of the polygeneration system. On the other hand, life cycle analysis (LCA) is a useful tool to study the life cycle environmental impacts of polygeneration system. There are three steps for LCA, i.e., goal and scope definition, inventory analysis, and impact assessment, and all the stages are interpreted simultaneously. System boundary is important to define the goal and scope. A representative system boundary is shown in Figure 20. In the case of waste biomass (agriculture waste), the growth phase is not included in the system boundary. However, for energy crops, the environmental impacts of biomass growth should be included in the system boundary, as shown in Figure 3. For polygeneration, allocation of environmental impacts for each utility is important. For a polygeneration system, an exergy-based allocation approach is useful (Jana and De, 2017). After allocation, environmental impacts are expressed for each utility. It may be compared with the other method of utility production.

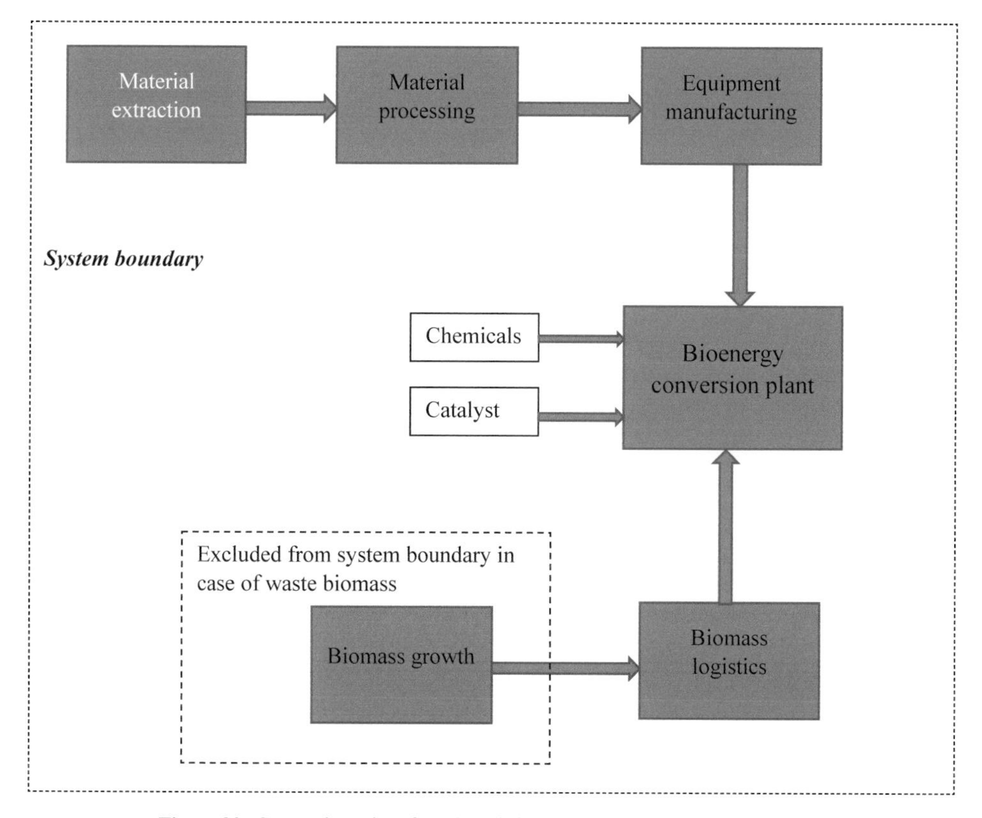

Figure 20. System boundary for LCA of biomass-based polygeneration system.

8. Conclusion

In the present context of energy security, energy access and climate change, biomass-based polygeneration may play an important role. It may increase the energy security by producing biofuels (liquid/gaseous) for many countries where liquid fossil fuel is not available to or sufficient for the transportation sector. It increases the energy access to rural areas where grid-based electricity is not always available. The carbon footprint of biomass-based polygeneration is significantly smaller than fossil fuel-based plants. Polygeneration has the ability to cater to the needs of local people by utilizing local resources. Thus, it increases social acceptability. Through polygeneration, various utilities can be produced, e.g., power, heating, cooling, liquid/gaseous fuels, chemicals, bio-materials, etc. Different kinds of biomass, e.g., edible oil crops, wood, agriculture waste, municipal solid waste, microalgae, etc., are the inputs to polygeneration. Thermo-chemical, bio-chemical and agro-chemical are the conversion technologies available for biomass to bioenergy conversion. Various biomass conversion technologies are commercially available and many technologies are at the R&D stage. This chapter shows that biomass-based polygeneration is a promising option as a sustainable energy solution, specifically for developing countries with abundantly available biomass.

References

Alalwana, H.A., Alminshid, A.H. and Aljaafari, H.A.S. 2019. Promising evolution of biofuel generation. Renewable Energy Focus 28: 127–139.

Atsonios, K., Kougioumtzis, M.A., Grammelis, P. and Kakaras, E. 2017. Process integration of a polygeneration plant with biomass/coal copyrolysis. Energy Fuels 31: 14408−14422.

Bajpai, P. and Dash, V. 2012. Hybrid renewable energy systems for power generation in stand-alone applications: A review. Renewable and Sustainable Energy Reviews 16: 2926–2939.

Bauen, A., Woods, J. and Hailes, R. 2004. Bioelectricity vision: Achieving 15% of electricity from biomass in OECD countries by 2020. WWF, Brussels, Belgium. Available at: www.panda.org/downloads/europe/biomassreportfinal.pdf. 05.07.2017.

Bentsen, N.S., Felby, C. and Thorsen, B.J. 2014. Agricultural residue production and potentials for energy and materials services. Progress in Energy and Combustion Science 40: 59–73.

BP. 2018. BP Statistical Review of World Energy, 2018. UK.

Bhattacharya, S.C. 2002. Biomass energy in Asia: A review of status, technologies and policies in Asia. Energy for Sustainable Development, VI, 3.

Brau, J.-F., Morandin, M. and Berntsson, T. 2013. Hydrogen for oil refining via biomass indirect steam gasification: Energy and environmental targets. Clean Technology and Environmental Policy 15: 501–512.

Campbell, J.E., Lobell, D.B., Genova, R.C. and Field, C.B. 2008. The global potential of bioenergy on abandoned agriculture lands. Environmental Science and Technology 42: 5791–5794.

Chen, Y., Yang, H., Yang, Q., Hao, H., Zhu, B. and Chen, H. 2014. Torrefaction of agriculture straws and its application on biomass pyrolysis poly-generation. Bioresource Technology 156: 70–77.

Chen, D., Cen, K., Jing, X., Gao, J., Li, C. and Ma, Z. 2017. An approach for upgrading biomass and pyrolysis product quality using a combination of aqueous phase bio-oil washing and torrefaction pretreatment. Bioresource Technology 233: 150–158.

Dahlquist, E., Naqvi, M., Thorin, E., Yan, J. and Kyprianidis, K. 2017. Experimental and numerical investigation of pellet and black liquor gasification for polygeneration plant. Applied Energy 204: 1055–1064.

Gadsboll, R.O., Clausen, L.R., Thomsen, T.P., Ahrenfeldt, J. and Henriksen, U.B. 2019. Flexible two stage biomass gasifier design for polygeneration. Energy 166: 939–950.

Graciano, J.E.A., Chachuat, B. and Alves, R.M.V. 2018. Enviro-economic assessment of thermochemical polygeneration from microalgal biomass. Journal of Cleaner Production 203: 1132–1142.

IEA. 2009. Bioenergy—A Sustainable and Reliable Energy Source. IEA Bioenergy: ExCo: 2009:06. Paris.

IEA. 2016. IEA World Energy Balances 2018 (https://webstore.iea.org/world-energy-balances-2018) Last accessed on 23.03.2019.

IEA. 2017. Energy Access Outlook 2017, International Energy Agency. Paris.

Ilic, D.D., Dotzauer, E., Trygg, L. and Broman, G. 2014. Integration of biofuel production into district heating part I: An evaluation of biofuel production costs using four types of biofuel production plants as case studies. Journal of Cleaner Production 69: 176–187.

IPCC. 2007. Mitigation of Climate Change, Working Group III, Chapter 4. 4th Assessment Report. Cambridge University Press. USA.

IPCC. 2014. Climate Change 2014: Synthesis Report. Contribution of Working Groups I, II and III to the Fifth Assessment Report of the Intergovernmental Panel on Climate Change [Core Writing Team, R.K. Pachauri and L.A. Meyer (eds.)]. IPCC, Geneva. Switzerland.

Jana, K. and De, S. 2015a. Polygeneration using agricultural waste: Thermodynamic and economic feasibility study. Renewable Energy 74: 648–660.

Jana, K. and De, S. 2015b. Sustainable polygeneration design and assessment through combined thermodynamic, economic and environmental analysis. Energy 91: 540–555.

Jana, K. and De, S. 2017. Environmental impact of biomass based polygeneration—A case study through life cycle assessment. Bioresource Technology 227: 256–265.

Jana, K., Mahanta, P. and De, S. 2018. Role of biomass for sustainable energy solution in India. *In*: Gautam, A. et al. (eds.). Sustainable Energy and Transportation, Springer, Singapore.

Kabalina, N., Costa, M., Yang, W. and Martin, A. 2017. Energy and economic assessment of a polygeneration district heating and cooling system based on gasification of refused derived fuels. Energy 137: 996–705.

Khanmohammadi, S. and Atashkari, K. 2018. Modelling and multiobjective optimization of a novel biomass feed polygeneration system integrated with multi-effect desalination unit. Thermal Science and Engineering Progress 8: 269–284.

Koh, L.P. and Ghazoul, J. 2008. Biofuels, biodiversity, and people: Understanding the conflicts and finding opportunities. Biological Conservation 2450–2460.

Kraussler, M., Binder M., Schindler, P. and Hofbauer, H. 2018. Hydrogen production within a polygeneration concept based on dual fluidized bed biomass steam gasification. Biomass and Bioenergy 111: 320–329.

Kravanja, P., Modarresi, A. and Friedl, A. 2013. Heat integration of biochemical ethanol production from straw—A case study. Applied Energy 102: 32–43.

Liu, F., Chen, G., Yan, B., Ma, W., Cheng, Z. and Hou, L. 2017. Exergy analysis of a new ligno-cellulosic biomass-based polygeneration system. Energy 140: 1087–1095.

Lythcke-Jorgensen, C., Clausen, L.R., Algren, L., Hansen, A.B., Munster, M., Gadsboll, R.O. and Haglind, F. 2017. Optimization of a flexible multi-generation system based on wood chip gasification and methanol production. Applied Energy 192: 337–359.

Magdeldin, M., Kohl, T. and Jarvinen, M. 2017. Techno-economic assessment of the by-products contribution from non-catalytic hydrothermal liquefaction of ligno-cellulose residues. Energy 137: 679–695.

Mauser, W., Klepper, G., Zabel, F., Delzeit, R., Hank, T., Putzenlechner, B. and Calzadilla, A. 2015. Global biomass production potentials exceed expected future demand without the need for cropland expansion. Nature Communications 6: 8946.

McKendry. 2002. Energy production from biomass (part 1): Overview of biomass. Bioresource Technology 83: 37–44.

Naqvia, M., Dahlquist, E., Yan, J., Naqvi, S.R., Nizami, A.S., Salman, C.A., Danishe, M., Farooq, U., Rehan, M., Khan, Z. and Qureshi A.S. 2018. Polygeneration system integrated with small non-wood pulp mills for substitute natural gas production. Applied Energy 224: 336–346.

Nguyen, T.-V. and Clausen, L.R. 2018. Thermodynamic analysis of a polygeneration system based on catalytic hydropyrolysis for the production of bio-oil and fuels. Energy Conversion and Management 171: 1617–1638.

Rios, M.L.V., González, A.M., Lora, E.E.S. and Olmo, O.A.A. 2018. Reduction of tar generated during biomass gasification: A review. Biomass and Bioenergy 108: 345–370.

Sahoo, U., Kumar, R., Singh, S.K. and Tripathi, A.K. 2018. Energy, exergy, economic analysis and optimization of polygeneration hybrid solar-biomass system. Applied Thermal Engineering 145: 685–692.

Salkuyeh, Y.K., Elkamel, A., The, J. and Fowler, M. 2016. Development and techno-economic analysis of an integrated petroleum coke, biomass, and natural gas polygeneration process. Energy 113: 861–874.

Salman, C.A., Naqvi, M., Thorin, E. and Yan, J. 2018. Gasification process integrated with existing combined heat and power plants for polygeneration of dimethyl ether and methanol: A detailed profitability analysis. Applied Energy 226: 116–128.

Sigurjonsson, H.A. and Clausen, L.R. 2018. Solution for the future smart energy system: A polygeneration plant based on reversible solid oxide cells and biomass gasification producing either electrofuel or power. Applied Energy 216: 323–337.

Truong, N.L. and Gustavsson, L. 2013. Integrated biomass-based production of district heat, electricity, motor fuels and pellets of different scales. Applied Energy 104: 623–632.

Twidell, J. and Weir, T. 2010. Renewable Energy Resources. Second edition. Taylor and Francis, USA.

Vidal, M. and Martín, M. 2015. Optimal coupling of a biomass based polygeneration system with a concentrated solar power facility for the constant production of electricity over a year. Computers and Chemical Engineering 72: 273–283.

Wagner, H., Wulf, C. and Kaltschmitt, M. 2015. Polygeneration of SNG, heat and power based on biomass gasification and water electrolysis—concepts and their assessment. Biomass Conversion and Biorefinery 5: 103–114.

Wang, X., Wu, J., Chen, Y., Pattiya, A., Yang, H. and Chen, H. 2018. Comparative study of wet and dry torrefaction of corn stalk and the effect on biomass pyrolysis polygeneration. Bioresource Technology 258: 88–97.

Xin, S., Yang, H., Chen, Y., Wang, X. and Chen, H. 2013. Assessment of pyrolysis polygeneration of biomass based on major components: Product characterization and elucidation of degradation pathways. Fuel 113: 266–273.

Yang, H., Liu, B., Chen, Y., Chen, W., Yang, Q. and Chen, H. 2016. Application of biomass pyrolytic polygeneration technology using retort reactors. Bioresource Technology 200: 64–71.

Yang, Q., Liang, J., Li, J., Yang, H. and Chen, H. 2018. Life cycle water use of a biomass-based pyrolysis polygeneration system in China. Applied Energy 224: 469–480.

Zhang, L., Li, S., Li, K. and Zhu, X. 2018. Two step pyrolysis of corncob for value-added chemicals and high quality bio-oil: Effects of pyrolysis temperature and residence time. Energy Conversion and Management 166: 260–267.

Zhang, X., Che, Q., Cui, X., Wei, Z., Zhang, X., Chen, Y., Wang, X. and Chen, H. 2018. Application of biomass pyrolytic polygeneration by a moving bed: Characteristic of products and energy efficiency analysis. Bioresource Technology 254: 130–138.

Section 2

Manufacturing and Construction (Batteries, Built Environment, Automotive, and other Industries)

CHAPTER 13

Urban Carbon Management Strategies

Joe F Bozeman III,[1,*] *John Mulrow,*[2]
Sybil Derrible[3] and *Thomas L Theis*[4]

1. Introduction

Urbanization is the shift in residence of the human population from rural to urban areas. It has been identified as one of the most important social transformations in the history of civilization (Ochoa et al., 2018), and has facilitated the emergence of societal benefits, such as improved economic access, the advent of public transportation, artistic expression, and other kinds of social innovations (Birch and Wachter, 2011; Guinard and Margier, 2018). Although urbanization has facilitated many social innovations, it has also led to adverse impacts to vital ecosystem services, including degradation of soil, air, and water quality (Ananda, 2018; Li et al., 2018). In this chapter, we explain how ecosystem impairment can be remediated following sustainable development—which is development that meets the needs of the present without compromising the ability of future generations to meet their own needs (Brundtland, 1987)—with a particular focus on carbon management strategies in urban ecosystems. Figure 1 shows the conceptual flow of this chapter. It is important to note that chapter references to 'carbon' or 'carbon emissions' may encompass other greenhouse gas (GHG) emissions which is sometimes expressed in the form of a carbon dioxide equivalency.

1.1 Global and regional urbanization trends

The world population has reached nearly 7.6 billion (UN DESA, 2017; US CB, 2019). Of this population, about 55% live in urban areas. By year 2050, the world population is expected to be about 9.8 billion, with the percentage of people that are expected to live in urban areas projected to increase to 68% (UN DESA, 2017; UN DESA, 2018a). This means approximately 2.5 billion people could be added to urban areas within the next few decades (UN DESA, 2018a).

[1] Institute for Environmental Science and Policy, University of Illinois at Chicago, 2121 West Taylor Street, MC 673, Chicago, IL 60607.
[2] Civil Engineering, University of Illinois at Chicago, 842 West Taylor Street, MC 246, Chicago, IL 60607.
Email: jmulrow@uic.edu
[3] Associate Professor, Civil and Materials Engineering Department, Institute for Environmental Science and Policy, University of Illinois at Chicago, 842 West Taylor Street, MC 246, Chicago, IL 60607.
Email: derrible@uic.edu
[4] Professor, Civil and Materials Engineering Department, Director, Institute for Environmental Science and Policy, 2121 West Taylor Street, MC 673, Chicago, IL 60607.
Email: theist@uic.edu
* Corresponding author: jbozem2@uic.edu

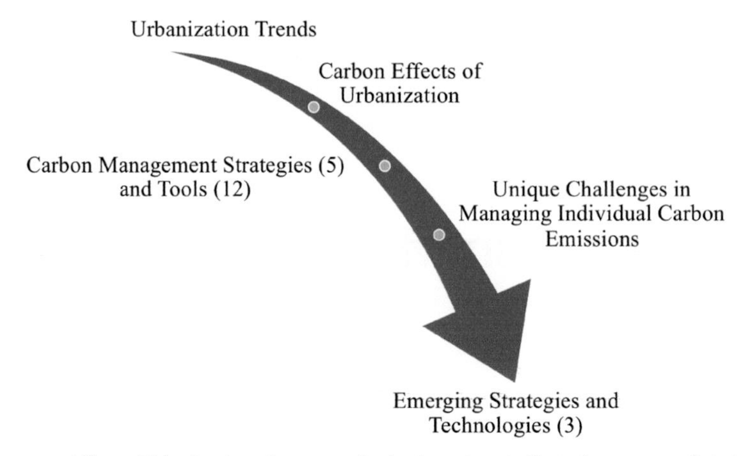

Figure 1. The conceptual flow of this chapter, where parenthesized numbers indicate the amount of strategies, tools and/or technologies.

Figure 2 shows population trends in the major regions of the world. North America has the greatest urbanized population. Much of this urbanization can be attributed to robust economic and public transportation developments in the United States (U.S.) over several decades (Auch et al., 2004; World Atlas, 2019).

Trends for all regions suggest that urbanization will continue through 2050, however, some European countries—such as Bulgaria, Croatia, Latvia, Lithuania, Poland, Moldova, Romania, Serbia, and Ukraine—are projected to experience population declines due to the persistence of low fertility rates

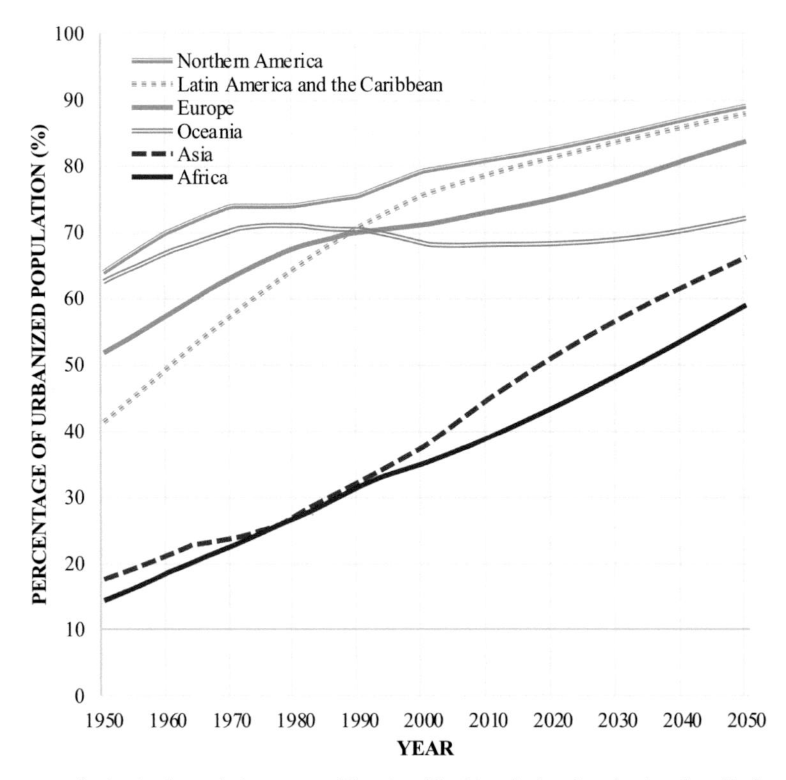

Figure 2. Percentage of urbanized population per world region. The historical and projection data displayed comes from UN DESA (2018b).

(UN DESA, 2017). Asia and Africa are expected to have sizeable increases in their urbanized populations through 2050 despite having relatively low current urbanization percentages. For Asia, an increasing urbanized population is projected despite Japan's ageing population challenges (Tamiya et al., 2011; UN DESA, 2018a). Year 2050 projections suggest that much of the world's urbanized population will be concentrated in the countries India and China of the Asian region and Nigeria of the African region (UN DESA, 2018a).

2. Carbon Effects of Urbanization

2.1 Urban classification

There are universal classifications for least developed and more developed countries. The intensity and types of emission effects vary based on the urbanization class. Least developed countries (LDCs) are defined as lower-income countries that have low levels of human assets (e.g., insufficient access to and/or ineffective health, nutrition and educational programming), are vulnerable to economic and environmental shocks, and have structural impediments to sustainable development (UN EAPD, 2019). The United Nations recognizes 47 LDCs: 33 in Africa, nine in Asia, four in Oceania, and one in Latin America and the Caribbean (UN CTAD, 2017). The North American and European regions have no identified LDCs. The remaining 148 countries of the world are classified as more developed countries (MDCs), which are higher-income countries with relatively high human assets that are capable of withstanding economic and environmental shocks. Urbanized areas can be further distinguished based on economic and population factors (UN WESP, 2014; UN SD, 2017). Urban classification is limited to LDC and MDC for the purposes of linking general carbon management strategies.

2.1.1 Carbon effects of MDCs

MDCs show trends of intensifying carbon effects. Developing countries or nations that fall into the MDC class tend to have high growth rates of carbon emission. Emission growth rates are found to be highest in places that have fast growing economies (Raupach et al., 2007). China and India have been two of the fastest growing economies in the world (Alam et al., 2016; Hewko, 2017). In comparison, developed countries, like the U.S., tend to have lower growth rates of carbon emissions over the last decade but more cumulative emissions over several decades (Hoesly et al., 2018; US EPA, 2017). Therefore, China, India, and the U.S. are used as exemplars of MDC carbon patterns.

Developing nations, such as China and India, have experienced rapid growth in carbon emissions and urbanization. China has the most carbon emissions of any country in the world (US EPA, 2017). These high emission rates can be attributed to energy-intensive industries and associated economic activities (Wang and Zhao, 2015). India, which is the third largest carbon emitting country in the world (US EPA, 2017), shares many of the same emission characteristics.

As a developed nation, the U.S. currently has lower carbon emission and urbanization growth rates than China yet is the second largest carbon emitter in the world (US EPA, 2017). The U.S. also has the highest cumulative emissions of any country in the world due to early economic development, which has resulted in the longest period of relatively high annual carbon emission rates, and the U.S. has had a disproportionately high carbon emission total over the last several decades, in comparison to China and India (Hoesly et al., 2018; Quere et al., 2016). Other developed MDCs, like those in Europe, also have relatively high carbon emission rates, urbanized population percentages, and per-capita emission rates. The U.S., however, remains the leader in cumulative carbon emissions and per-capita emission rates to date.

2.1.2 Carbon effects of LDCs

There are some common factors that tend to drive carbon effects in LDCs. One of the major factors is country-level attempts to attain economic stability. However, economic stability for LDCs tends to be linked to increased energy consumption. For instance, the causal links between carbon emissions and economic growth in Bangladesh, which is one of the nine LDCs in Asia, show that further economic growth is uni-directionally linked to increased energy consumption (Alam et al., 2012). This suggests that economic growth in Bangladesh also means growing energy use. Lower-income countries that are in an economic growth phase are much more likely to use traditional energy sources, such as coal, that emit high concentrations of carbon (Steckel et al., 2015). Taken together, this suggests that LDCs that are actively progressing towards economic stability are contributing to global carbon emissions in systemic ways. It has, therefore, been suggested that countries like these use renewable, clean energy technologies to provide the energy support needed for their continued economic growth (Kaygusuz, 2012). Facilitating economic growth using renewable and clean energy technologies, however, comes with unique political and economic challenges (Ito, 2017).

2.1.3 Other major carbon effects

Carbon effects associated with tree cover and soil are introduced and descriptions are applicable to urban areas in both MDCs and LDCs.

2.1.3.1 Soil organic carbon (SOC) effects

The soil can release carbon into the atmosphere. Soil organic matter (SOM) is the organic component of soil that includes plant residue, carbon, decomposing organic matter, and stable organic matter (Bot and Benites, 2005). SOC is vital to sustainable soil fertility (Wang et al., 2016). However, increased SOM decomposition from a warming climate could release increased carbon emissions, considering SOM contains almost three times as much carbon as the atmosphere (Cheng et al., 2017; Knorr et al., 2005).

2.1.3.2 Urban tree carbon effects

Urbanization activities, including deforestation and land-use changes, can impact carbon concentrations in the atmosphere. Trees sequester carbon through the stomata during photosynthesis by storing carbon as a biomass. Stomata are pores of plant leaves, stems, and other organs that facilitate gas exchange and are vital to the photosynthesis process (Lal and Augustin, 2012). For instance, U.S. estimates show that urban trees store carbon at a rate of 7.69 kilograms of carbon per square meter of tree cover annually (Nowak et al., 2013). Street and park trees have been shown to benefit urban areas through carbon sequestration and shading effects while also providing other benefits, such as improved air quality and stormwater runoff reductions (McPherson et al., 2005; Nowak et al., 2006). The removal of urban trees for land use changes—such as the removal of a forested, urban park to build an apartment complex—not only allows carbon that could have otherwise been sequestered to enter the atmosphere, it also facilitates further carbon emissions, since trees can emit stored SOC after tree death.

3. Carbon Management Strategies and Tools

Urban carbon management strategies and tools are expansive and have mutually-exclusive and synergistic qualities (i.e., they can be used singularly or combined to develop hybrid approaches that fit the needs and environmental profile of a given area). Major strategies and tools are described in this section.

3.1 Strategies

Five major strategies are highlighted for urban carbon management: (1) decentralized plant growth, (2) densification, (3) public transportation, (4) renewable energy and efficiency, and (5) urban agriculture.

3.1.1 Decentralized plant growth

Anthropogenic activities can significantly affect urban vegetation. These effects have come in the form of deforestation and similar land-use changes which have reduced carbon sequestration capacity. Urban planners have lobbied for the placement of tree and plant species throughout urban environments to, in part, address these reductions of carbon sinks (Ignatieva et al., 2011). Planners and scientists have since come to realize that there may be differences in the sustainable quality of vegetation that is strategically placed versus spontaneous vegetation. Thus, the net amount of carbon captured by urban vegetation, given photosynthetic and decomposition processes, varies based on how plant growth is managed.

Does interfering with existing vegetation and spontaneous plant growth in urban areas inhibit effective carbon management? In a recent Canadian study, existing forests within urban environments were shown to have four times more SOC stored compared to spontaneous vegetation and managed lawns, and had the largest leaf area index for carbon sequestration (Robinson and Lundholm, 2012). This suggests that any new urbanization activities should avoid the removal of existing forest to minimize the amount of carbon content that could be released into the atmosphere through SOC disturbance and decomposition (Jandl et al., 2007), and to maximize the amount of carbon that could be captured by the larger forest leaf area.

In circumstances where urbanization activities have already removed a considerable amount of forest cover, afforestation, the process of planting trees in an area of land without forest cover, may be suitable. However, urban practitioners and scientists should choose afforestation strategies carefully since many factors affect the amount of carbon that is sequestered and stored by trees, including tree size, tree condition, and environmental factors, such as moisture content. These factors vary across tree species and even amongst trees of the same species (Lal and Augustin, 2012). The type of tree species and tree spatiality used in afforestation can mean the difference in not only carbon management efficiencies but also human health. Some of urban trees produce pollen that is dispersed by wind and emit gaseous substances which can negatively affect air quality through photochemical reactions. Therefore, urban planners should not only consider the proximity of trees to pollutant sources when considering afforestation but also the allergenicity of certain tree pollen grains. The allergenicity of tree pollen can be modified by atmospheric pollutants and is dependent upon environmental factors, such as temperature and wind. For instance, pollen dispersion from the Cypress tree species (of the genus Cupressus) has been identified as a leading cause of respiratory allergies in southern Europe (Grote et al., 2016).

3.1.2 Densification

Regions around the world are projected to experience increasing urbanization in the coming years (refer to Figure 2). This means that urban areas may become denser, i.e., a city's population may increase without expansion of the city boundaries or urban land area. This phenomenon is called urban densification. As a strategy, densification is the implementation of design elements and infrastructure to address human population increases to an already urbanized area.

It is important to distinguish urban densification from urban sprawl. While densification generally occurs when a bounded urban area experiences, or has already experienced, a sizeable population increase, urban sprawl is the expansion of urban areas either through the development of new cities or expanding the bounds of existing ones. Urban sprawl can impact urban carbon management, particularly when low-building-density, detached houses are developed or utilized in these spaces rather than dense-buildings, such as apartments (Mohareb et al., 2016; Navamuel et al., 2018). Urban sprawl has carbon emission implications associated with increased land and resource use.

Densification presents unique challenges. New space utilization and logistical challenges emerge as more people dwell in bounded urban spaces (Wang et al., 2019). For instance, an urban area that has already incorporated carbon reduction measures into its infrastructure may be forced to seek further design and economic solutions in order to address the implications of increased population density in buildings, public transport, and the like (Derrible, 2018). This could manifest in existing building renovations, designing urban spaces to be more compact, and other costly measures (Conticelli et al., 2017). Another challenge of employing densification strategy comes with policy development and implementation since urban densification tends to be promoted as a way of limiting urban sprawl through policy (Tillie et al., 2018; Vergnes et al., 2014).

On the other hand, densification has benefits. For urban areas that are already equipped to withstand densification, they may experience an overall reduction in energy used per unit area compared to lesser-dense urban areas. There are also other benefits that could be realized, such as improvements in micro-climate and macro-climate conditions when cities integrate green space with water resources (Hong et al., 2017). Densification strategy, therefore, has significant implications for urban carbon management.

3.1.3 Public transportation

Public transportation has been associated with urbanization, carbon emissions, and air pollution. It correlates with economic activity and can be an indicator of future economic development (Duranton and Turner, 2012). When attempting to manage net carbon emissions in urban areas or cities while maintaining other co-benefits (e.g., reductions in traffic injuries, noise, congestion, and improvements in human respiratory outcomes), it is important to have the requisite share of private and public transportation modes accessible (Derrible et al., 2010; Kwan and Hashim, 2016).

Modal options include transportation by air, road, rail, and ferry; and these modes can be used for passenger commuting (e.g., traveling to work) and industrial purposes (e.g., shipping materials and equipment for urban construction). There are many transport options available within each mode that have varying carbon management and social benefits. For example, large cities stand to benefit from increased access to public transport because of the possibility of higher modal shifts (Kwan and Hashim, 2016). That is, urban areas that already have populations above one million can benefit from public transportation improvements due to the possibility of people shifting from heavy carbon-emitting transport (e.g., personal uses of traditional cars) to mass transit or active transport options (e.g., cycling to work). Yet, improving fuel efficiencies and electric vehicle use could have better co-benefit and carbon reduction potential than employment of public transportation (Kwan and Hashim, 2016). This example shows that planning or employing public transportation can be challenging, considering the dynamic infrastructure, technological, and social factors at play.[1]

3.1.4 Renewable energy and efficiency

Urban growth impacts energy use. In fact, it is estimated that a 1% increase in urban population translates into a 2.2% increase in energy use in urban environments, in that, the rate of change in energy use is twice that of urbanization (Santamouris, 2013). Therefore, with increasing urbanization there is a need for effective energy efficiency measures. Improving urban energy efficiencies assist in reducing carbon emissions.

Energy efficiency strategies are expansive and warrant detailed explanation. These strategies range from solar energy use to building automation system (BAS) management. This section focuses on popular energy efficiency strategies that tend to be used in urban areas, and it excludes power plant activities that generally occur outside of the urban boundary.

3.1.4.1 Solar collectors and energy storage

An abundant amount of solar energy from the sun hits the earth's atmosphere. According to Tsao et al. (2006), more energy hits the earth's surface in an hour and a half than is needed to energize the world for one year even after accounting for flux scattering, absorption and other environmental elements. There is potential for this radiation to be collected by solar collectors. Solar collectors are devices that collect, store, and move solar thermal energy (Capehart et al., 2008). These devices are useful in urban environments that receive enough solar energy from the sun, but they can be inhibited by capital costs,

[1] This chapter section is a very brief overview of public transportation strategies. Further information can be found in these book chapters (What Key Low Carbon Technologies are Needed to Meet Serious Climate Mitigation Targets and What is Their Status; Charging Strategies for Electric Transport; and Advancements, Challenges and Opportunities of Li-ion Batteries for Electric Vehicles.

energy storage challenges, and energy demand timing. For instance, energy needs are relatively high during the week when many people return from work around evening time. However, the sun does not provide much solar energy in the evening and at night. Useful energy storage and transfer techniques must then be integrated with solar collectors in order to improve efficiency and reduce energy losses. Yet, the development of highly-efficient energy storage systems is a work in progress (Branz et al., 2015; Hill et al., 2012; Khanafer and Vafai, 2018).

Some viable energy storage systems have been developed but can be dependent on location and environmental conditions. For instance, in some parts of the world, solar-heated hot water has been used as a way to store solar energy while also providing societal water needs in hotels, laundry facilities, and in the industrial sector (Capehart et al., 2008). Another kind of cost that is important when considering the use of solar collectors is operation and maintenance costs, which could include periodic cleaning of the collector surface from dust and debris, heat exchanger maintenance and building integration calibration.[2]

3.1.4.2 Waste to energy

Waste to energy approaches are also ideal for energy conservation in urban environments, considering these environments can produce copious amounts of useable waste. One of these approaches has been to convert food waste into energy. If there is enough food waste that can be delineated into functional waste streams, anaerobic digestion can be used to produce locally generated electricity and biogas. The approach not only reduces urban energy needs but also improves soil fertility by diverting volatile organics that would otherwise infiltrate SOM (Curry and Pillay, 2012).

Biomass is another kind of waste that can be used to produce energy in urban environments. Biomass, such as wood, can be used as fuel in combined heat and power (CHP) systems for residents and businesses (Henkel et al., 2016; Muley et al., 2016). CHP systems are co-generative, in that they can produce power as well as recover excess heat for integrative purposes, like producing steam for space heating (Capehart et al., 2008). Furthermore, CHPs can use different kinds of co-generative energy sources, such as fuel cells, gas turbines with heat recovery, and steam backpressure turbines, with varying levels of efficiency (Gambini and Vellini, 2015; Gvozdenac et al., 2017).[3]

3.1.4.3 Building Automation System (BAS)

Energy efficiency measures and other building management components require varying levels of operations management in order to maintain desired outcomes. A BAS serves this function by compiling data from one or many of these system components into a useable framework. That is, a BAS uses a network of microcomputers connected by high-capacity communication lines, networks, and the internet in order to control air conditioning, lighting, and other system components for optimal comfort and operation (Capehart et al., 2008; Ghaffarianhoseini et al., 2016). Some BASs aggregate data in a way that allows its users to be notified of system efficiency lapses. For instance, if an urban system contains a biomass CHP plant producing power and steam, the system can notify a user that a number of steam traps are operating below standard and reducing overall power plant efficiency.

Other BASs can notify users of system inefficiencies while also adjusting system components in real time. A popular example of these kinds of BASs are ones that control space air conditioning and heating functionality based on environmental conditions. For instance, a BAS may be programmed to increase the amount of conditioned air entering a space by compiling occupancy data from that space's carbon sensors with ambient temperature and humidity data. BASs that can adjust building components in real time using sensor data and the like are referred to as smart buildings or intelligent buildings (Ghaffarianhoseini et al., 2016).

---

[2] Further information on issues involving solar and other kinds of renewable energy sources can be found in Section 4 of this book (Wind/Solar/hydro/nuclear).

[3] Further explanation and details for how biomass can be used to generate electricity can be found in Section 3 of this book (Biomass sector).

It is important to note that current BAS technologies tend to use proprietary information structures which can present system integration and modularity challenges (Bu et al., 2015; Capehart et al., 2008). Coalescing data and controlling building components of proprietary BASs in a single building or facility can occur through synergistic approaches, including the use of Multi-agent architecture design and the employment of BACnet communication protocol (González-Briones et al., 2018; Zhou et al., 2012).

Urban developers, workers, and dwellers are not the only groups that can employ urban carbon management strategies. A utility entity also has effective ways of managing urban carbon emissions. One of these ways is for utilities to incorporate cost premiums into their rate structures in order to encourage reductions in energy use at peak-load periods (Newsham and Bowker, 2010). Governance structures can also have major influence on urban carbon emissions through tax-incentive and cap-and-trade policy implementation (Shammin and Bullard, 2009).

3.1.5 Urban agriculture

Urban agriculture is the production of crops and livestock in urban and peri-urban (i.e., an area or location adjacent to an urban area that is not considered entirely rural) environments (Hampwaye, 2013). The innovative potential of urban agricultural system designs has led some to propose an important niche, in which a critical proportion of a city's food (e.g., ~ 50% of fresh fruits and vegetables) is sourced from within the urban environment or within peri-urban regions adjacent to the city (Grewal and Grewal, 2012). Urban agriculture's range and intensity of benefits continue to be debated, but the general consensus is that it should play a part in managing carbon emissions and improving economic access (Ferreira et al., 2018; Hampwaye, 2013). Furthermore, these urban agricultural activities can displace a portion of food imported from distant growing regions which are often water-stressed (see Figure 3).

There is a policy focus on 'relocalization' of food systems in urban areas as a means of building in greater sustainability, such as the reduction of carbon emissions and resilience (Bloom and Hinrichs, 2011). Such a focus generally aims to alter the relationship of food production to urban systems through greater emphasis on innovations in local food production, harnessing of synergies with waste and energy byproducts, developing new models for the distribution and acquisition of healthy food, understanding the social and cultural contexts that influence dietary choices, and reducing policy disincentives to the consumption of healthy food (Johnson et al., 2012). Furthermore, studies have documented the many co-benefits (e.g., economic, environmental, security, social, and cultural benefits) of urban growing systems that reconnect people with their food sources (Bellows et al., 2003; Chance et al., 2018; Kulak et al., 2013; Martinez et al., 2010).

Urban agriculture can take many forms (Ferreira et al., 2018). These range from smaller community and home gardens, rooftop gardens, to relatively large farming operations located in peri-urban areas. Classifying these forms in some organized fashion is key in explaining the strategic importance and benefits of each. Rather than describe implications for employing each urban agriculture type individually, we follow the taxonomy established in Goldstein et al. (2016) which classifies urban agriculture into four types: Ground-based-non-conditioned (GB-NC), ground-based-conditioned (GB-C), building-integrated-non-conditioned (BI-NC), and building-integrated-conditioned (BI-C). This allows for a more streamlined explanation for urban developers and scientists to use in their planning and research activities. A qualitative synthesis of the ensuing information in this section can be found in Table 1.

GB-NC types of urban agriculture are generally characterized as requiring low external energy and capital resources; however, they are also characterized by relatively low crop yields (Goldstein et al., 2016). The term non-conditioned, in this context, means that no external forms of temperature control and conditioning are used as a part of these agricultural systems. They are susceptible to deleterious environmental conditions, like frost, floods, wind, drought, and extreme heat. This type of urban agriculture is considered to have a small carbon footprint due to its low use of energy and capital resources. Moreover, it is ideal for controlling carbon emissions while providing nutritional or social needs for urban communities. Examples of GB-NCs include home gardens, community gardens, low-tech greenhouses and food social enterprises. Although home gardens, community gardens and greenhouses may already be well understood, food social enterprises are not. Social enterprises based on urban agriculture tend

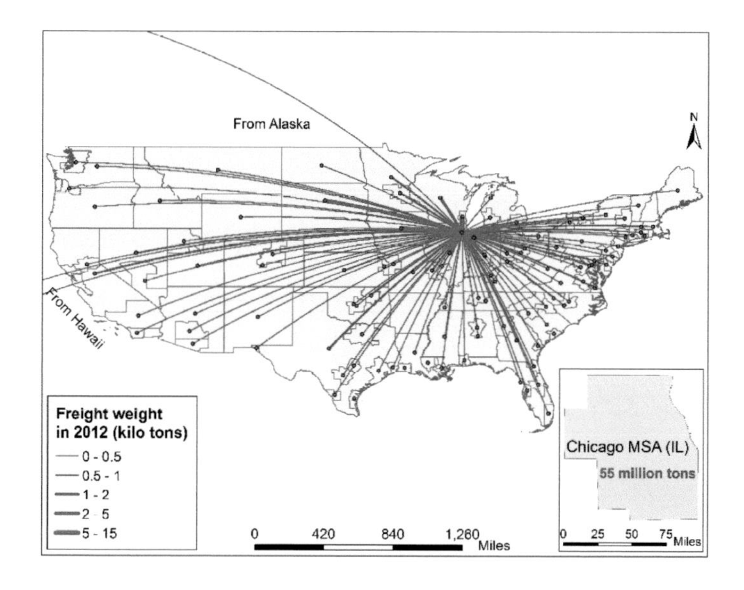

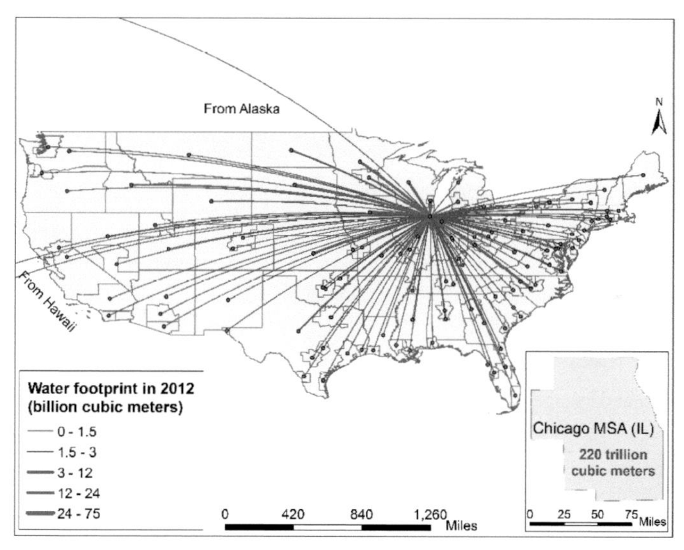

Figure 3. Inflows of agriculture products (upper) and the associated water footprint (lower) for the Chicago Metropolitan Statistical Area in 2012 (FAO, 2016; FAO, 2017a; FAO, 2017b; FHWA, 2016). We acknowledge Professor Bo Zou, Civil and Materials Engineering, University of Illinois at Chicago, and Professor Weslynne Ashton, Stuart School of Business, Illinois Institute of Technology, for figure development.

to serve as hubs for small-scale food production and food education (see Figure 4). In some cases, these social enterprises provide fresh produce to local food pantries, provide nutrition programming and facilitate educational opportunities to urban and peri-urban dwellers (City Grange, 2019; PGP, 2019).

GB-C types differ from GB-NCs in many ways. GB-Cs are characterized as requiring relatively high energy and capital resources while tending to produce higher crop yields with efficient water uses. The high energy and capital resources are generated by the use of temperature control and irrigation equipment, sensors and computer equipment. This type of urban agriculture system has its own subcategories to classify how these spaces are conditioned: Active and passive. Active GB-Cs tend to utilize more traditional conditioning systems, which require high operational energy to maintain ideal system conditions. Higher energy use of this kind tends to be linked to higher carbon emissions. Passive

Table 1. Basic characteristics of urban agriculture typologies: ground-based-non-conditioned (GB-NC), ground-based-conditioned (GB-C), building-integrated-non-conditioned (BI-NC), and building-integrated-conditioned (BI-C).

Type/Subcategory		Crop yields	Energy use	Capital requirements	Environmental susceptibility	Carbon impact	Symbiosis potential
GB-NC		L	L	L	H	L	L
GB-C	Passive	M	M to H	M to H	L to M	L to M	L to M
	Active	M to H	H	H	L to M	M	M
BI-NC		L	L	L to M	M	L	M
BI-C	Passive	M to H	M to H	M to H	L to M	L to M	M to H
	Active	H	H	H	L	H	H

Note: This table was adapted from Goldstein et al. (2016), where 'L' denotes low, 'M' denotes medium, and 'H' denotes high.

Figure 4. Picture of the grand opening of an urban food social enterprise located in Chicago, Illinois (Photo Credit: Joe F. Bozeman III; Permission: President of City Grange).

GB-Cs tend to use more capital resources to ensure passive forms of space conditioning as a function of the built environment. For example, these would include spaces that employ evaporative space cooling or sun-facing building design. The use of evaporative space cooling or sun-facing building design in passive GB-Cs tends to use less energy when compared to the traditional conditioning systems used in active GB-Cs. Nonetheless, there are not enough studies that compare active and passive conditioning resource uses to be able to fully compare them with quantitative detail (Goldstein et al., 2016).

BI-NC types exhibit many of the same characteristics as GB-NCs. That is, BI-NCs are also non-conditioned and tend to exhibit lower energy and capital resource use with lower crop yields, suggesting lower carbon emissions. BI-NCs differ from GB-NCs in that there is potential to increase efficiencies through symbiotic exchanges as a function of building integration. Whereas GB-NCs have little symbiotic potential, BI-NCs can impact carbon emissions at the building envelope through the provision of insulation and roof albedo, mitigating energy use through the improvement of stormwater runoff and the capture of building energy to lengthen growing season (Goldstein et al., 2016). Rooftop gardens, green roofs and hydroponic grows are examples of approaches that could integrate into a building as a non-conditioned urban agriculture system. As aforementioned, capital resources tend to be relatively low for BI-NCs, but this may not necessarily be the case for all forms of BI-NCs. For instance, retrofitting the building structure may be necessary in order to reinforce the roof of a green roof project, irrigation networks may need to be constructed to facilitate the watering of crops, and sensors and computer systems may need to be purchased in order to manage building components.

BI-C types tend to have higher operational energy and capital costs than BI-NCs, and tend to have higher crop yields. Similar to GB-Cs, BI-Cs have passive and active conditioning as subcategories. Improved symbiosis with other building system components can increase resource use efficiencies in water and energy. For example, Plant Chicago (2019) uses a BI-C style of design and the concept of the circular economy to recycle and reuse heat, energy and nutrient resources within the confines of the building. An approach such as this has the potential to reduce carbon emissions in the long-term but can require sizeable upfront costs if added building infrastructure is needed for irrigation or conditioning purposes. Furthermore, operational energy costs, and the associated carbon emissions, will be contingent upon how much of the BI-C space is passively or actively conditioned. This type of urban agriculture tends to be linked to for-profit, farming operations (Goldstein et al., 2016). That is, a well-resourced and managed BI-C space can produce high enough crop yields to offset or supplant certain peri-urban crop production.

3.2 Tools

In this section, we highlight 12 tools, used to estimate carbon stocks and flux, that are still being used by urban developers and the scientific community. This means that tools that are no longer being updated or are considered a previous version of a more current tool are excluded from this section. Furthermore, these tool descriptions and names are not an exhaustive list. That is, a certain tool's functionality and example based from one country or region may share qualities of similar tools from another country or region under a different name or title. The goal here is not to provide an exhaustive list of urban carbon management tools. Instead, the goal is to provide examples of what kinds of tools are readily available. Qualitative synthesis of the ensuing information in this section can be found in Table 2.

Table 2. Basic characteristics of urban carbon management tools.

Tool	Spatial scale descriptor (generalized)	Ease of use
BLCC	Building component to building	L
CCT	State to national	L
Carbon footprint calculators	Individual to national	L to M
CMS	National to global	L to M
CASA	Forest to national	L
FVS	Forest stand	L
GIS	Local to global	M to H
i-Tree	Tree to community	L
LANDIS	National forest to landscape	H
LIDAR	Local to national	M to H
NED	Forest stand	M to H
Scenario analysis	Localized to global	M to H

Note: This table was modelled after USDA (2019); where 'L' denotes low difficulty (e.g., a two hour or less learning curve), 'M' denotes medium difficulty (e.g., a one week or less learning curve), and 'H' denotes high difficulty (i.e., a major time investment that may require collaboration with a qualified research scientist).

3.2.1 Building life cycle cost programs

The National Institute of Standards and Technology developed a computational analysis tool for equipment and buildings called the building life cycle cost (BLCC). The U.S. Department of Energy has lauded it as a useful life cycle assessment (LCA) or costing (LCC) tool (US DOE, 2019). By incorporating energy escalation rates and other LCA factors, the BLCC enables the comparison of building components or building designs. This tool is relatively easy to use, but may require some initial training.

It is important to note that the BLCC is only one LCA or LCC tool of many. These kinds of tools are vast and range in capabilities. Listing and describing them all exceeds the scope of this chapter and section. Nonetheless, it is important to note that the BLCC, and other tools like it, can be used to assist urban planners and scientists to assess the viability and economical benefit of various urban agriculture and energy efficiency strategies. As with any data computational tool, it is vital to input accurate and comprehensive information. For instance, ignoring energy costs over time and focusing mostly on capital costs in an LCC analysis could result in purchasing more expensive system components.

3.2.2 Carbon calculation tool

The Carbon Calculation Tool (CCT) reads data from the U.S. Forest Service's Forest Inventory and Analysis (FIA) Program in order to generate state-level estimates of carbon stocks (Smith et al., 2007). CCT also has the ability to produce national carbon stocks and flux data. This tool's functionality is dependent on FIA file types, and it requires that data be imported. It is estimated that CCT requires minimal time investment to learn and use (USDA, 2019).

3.2.3 Carbon footprint calculators

Carbon footprint calculators have been popular for some time. They generally take self-reported information about a person's or group's behavior and calculate the amount of GHGs emitted. These tools can calculate individual to country level carbon footprints (Mulrow et al., 2019). There are many available online which have been developed by small and large organizations, including nonprofit organizations and the U.S. federal government.

These tools can be used to help urban planners and scientists encourage behavior that reduces urban carbon emissions. For instance, an individual could use the calculator developed by the Environmental Protection Agency (EPA) to assess their household carbon footprint (US EPA, 2016). Such information is designed to encourage activities that reduce carbon emissions. However, being aware of one's household or individual carbon footprint does not necessarily lead to behavior change. This is an important point for urban practitioners and scientists who are trying to manage or reduce carbon emissions.[4]

3.2.4 Carbon monitoring system

The National Aeronautics and Space Administration's (NASA's) Carbon Monitoring System (CMS) takes data from various sources to estimate carbon net flux for the 48 continental states of the U.S. This tool and dataset go further in that they help in predicting the evolution of global carbon sinks (NASA CMS, 2019). In other words, CMS has a global focus but also provides granularity to regional and county level carbon estimates by using NASA's satellite observations and modelling capabilities (Hagen et al., 2016). Carbon sequestration dynamics are assessed while also providing map displays at 100-meter resolution. This tool is stated to require minimal time investment to use, however, the user must incorporate their own data (USDA, 2019).

3.2.5 Carnegie ames stanford approach

The Carnegie Ames Stanford Approach (CASA) project uses NASA satellites to collect carbon stock and exchange data. It takes multi-year global data and runs it through simulation models at a local to global scale. CASA is capable of more than just carbon estimation, it can also analyze dynamics of the nitrogen cycle as driven by climate change and land use (US NASA, 2019). It is considered to require minimal time investment to use successfully (USDA, 2019).

[4] Challenges involving individual behavior change and the use of carbon footprint calculators are further explained later in this chapter, in the 3.3 *Unique Challenges in Managing Individual Carbon Emissions* section.

3.2.6 Forest vegetation simulator

The Forest Vegetation Simulator (FVS) is a useful carbon calculation tool for ecosystem and wood products. It serves as a growth simulation tool that can model forest vegetation at an individual-tree and forest stand level in the U.S. (US FS, 2019b). Some of the specific benefits are that it recognizes all major tree species, can facilitate environmental disturbances within simulation, and yields useful outputs for tree canopy cover (US FS, 2019a). It is important to note that FVS tends to require a moderate level of time investment to learn and use successfully (USDA, 2019).

3.2.7 Geographic information system

The geographic information system (GIS) is a framework that incorporates many types of geographical data. It has the ability to layer spatial information at small and large scales, and can even project two-dimensional and three-dimensional imagery. GIS also has robust synergistic capabilities, i.e., it can be incorporated into many software platforms and can link to other platforms to assess historical data and to project outcomes. However, the system and its capabilities are vast. It, therefore, can take a considerable amount of time to learn how to use if prior experience is lacking. Nonetheless, it is a powerful tool that could be used by scientists and urban planners, for example, to locate zoning space for urban infrastructure or to identify urban locations for afforestation.

3.2.8 i-Tree

i-Tree is an urban forest management tool that is peer reviewed from the USDA Forest Service. It uses tree inventory data to quantify environmental factors, such as energy conservation, air quality, carbon emission reductions, and stormwater control (USDA FS, 2019b). USDA FS (2019b) lauds i-Tree's reach, stating that it has been used around the world to report on the benefits trees provide at the individual tree level up to the state level (Endreny et al., 2017; Nowak et al., 2018). It is free to use and has supplemental tools for analyses and utility programming. The supplemental analysis tools include Landscape, which facilitates the exploration of geospatial data, Hydro, which can simulate the effects stream flow and water quality have on trees and land cover, Canopy, which uses Google Maps to produce statistically valid estimates of tree and land cover, and Species, which helps urban planners assess tree species, given environmental and geographic factors, and more. The supplemental utility programming includes the Pest Detection Module, which is a protocol for assessing trees for insect or disease problems, and Storm, which gives communities the ability to assess storm damage following an actual severe storm event. This tool is considered easy to use and tends to take only a few hours of time investment to learn (USDA, 2019).

3.2.9 LANDIS

LANDIS is a landscape tool designed to simulate forest growth and disturbances, such as climate change effects at the national forest and landscape level. It displays landscapes as a grid of cells and does not allow for analysis at the individual tree level (USDA FS, 2019a). LANDIS has variants available, each of which specializes in different simulation tasks. For example, LANDIS-II is a decadal to multi-century forecasting tool that is highly customizable.

Many have used LANDIS in analyses, assessments, and published works spanning from at least the late 1990s (USDA FS, 2019a). However, this tool tends to require a major time investment to learn and use successfully. It is suggested to partner with a qualified research scientist (USDA, 2019).

3.2.10 Light detection and ranging

Light Detection and Ranging (LIDAR) is a remote sensing method that uses pulsed laser light to create profiles of natural and manmade environments (NOAA, 2018). LIDAR allows for the examination of these environments at a topographical and bathymetric level. In other words, the LIDAR system can yield images and data for surface level environments (i.e., aboveground) as well as for seabed and riverbed environments (i.e., underwater or below sea level). This system could be used in carbon management strategies when needed. For example, urban planners could use LIDAR information to identify natural or manmade bodies of water when trying to locate water sources that could supply an urban farm or cooling fluids for a CHP. Many of the datasets and display interfaces are readily available for download and use online.

3.2.11 NED

NED is a decision-support software that leverages the strengths of several USDA Forest Service software products. It aids in developing strategic and management plans. NED is ideal for use at the landscape level, meaning that it is not ideal at the forest stand or individual tree level. Rather, it helps to assess the complex ecosystem dynamics (e.g., aesthetics, wildlife, water dynamics, and wood production) that emerge at the landscape level for meeting objectives and goals. It is important to note that this system utilizes very traditional growth and yield modeling, and requires a moderate amount of time investment to use successfully (USDA, 2019).

3.2.12 Scenario analysis

Scenario analysis allows for the structured assessment of several contingencies in qualitative or quantitative forms. The scale of assessment spans the spectrum of urban planning, i.e., it can be used to analyze a single building or component to national or global scales. There is an abundance of scenario analysis techniques and tools, some of which have already been described. A few other popular scenario analysis techniques and tools are computational in nature and include machine learning and agent-based modelling. Not all scenario analysis approaches are driven by computer software packages. For instance, in-person surveys could be used to generate responses of what actions organizations or people could take given a scenario.

Generally, these are used to forecast future outcomes. In some cases, historical data is used to help forecast future events. For the purposes of urban carbon management, scenario analysis could help assess which individual or hybrid strategies mitigate carbon emissions the greatest. Scenario analysis is largely dependent on the data that is used in its modelling. For this reason, among others, this technique often requires a relatively large amount of time to master. It may be prudent to work with an expert (e.g., an experienced computer coder and/or scientist) when employing scenario analysis to ensure critical factors are incorporated into the model(s).

3.3 Unique challenges in managing individual carbon emissions

One way to approach the problem of carbon management is from the bottom up (see Figure 5). Taking this perspective, human society's contribution to carbon flux is simply an aggregation of individual decisions. The carbon effects of deforestation, for example, can be linked to the demand for agricultural products, which require cleared land and sun exposure. This demand for land, in turn, arises from individual tastes for ingredients, such as beef, palm oil and soy beans (Hosonuma et al., 2012), that will be grown there. No matter the scale or the industry, carbon emissions can be traced back to an individual need or desire.

The primary tool for taking a bottom-up approach to carbon management is the individual carbon footprint calculator. Such calculators have proliferated in the form of mobile phone applications and online tools (Mulrow et al., 2019). The typical calculator instructs the user to measure a range of personal activities that ultimately generate, or in rare cases, sequester carbon emissions. The simplest calculators derive carbon emissions based on energy-related activities alone, while more detailed calculators consider lifestyle and consumption choices, such as food and travel (Padgett et al., 2008). Some even include the individual's share of collective emissions from government activities (depending on citizenship) in the final carbon footprint tally (West et al., 2015).

There is a famous business and policy maxim that says "What gets measured gets managed", but whether the measurement of personal carbon footprints will prompt individuals to better manage their

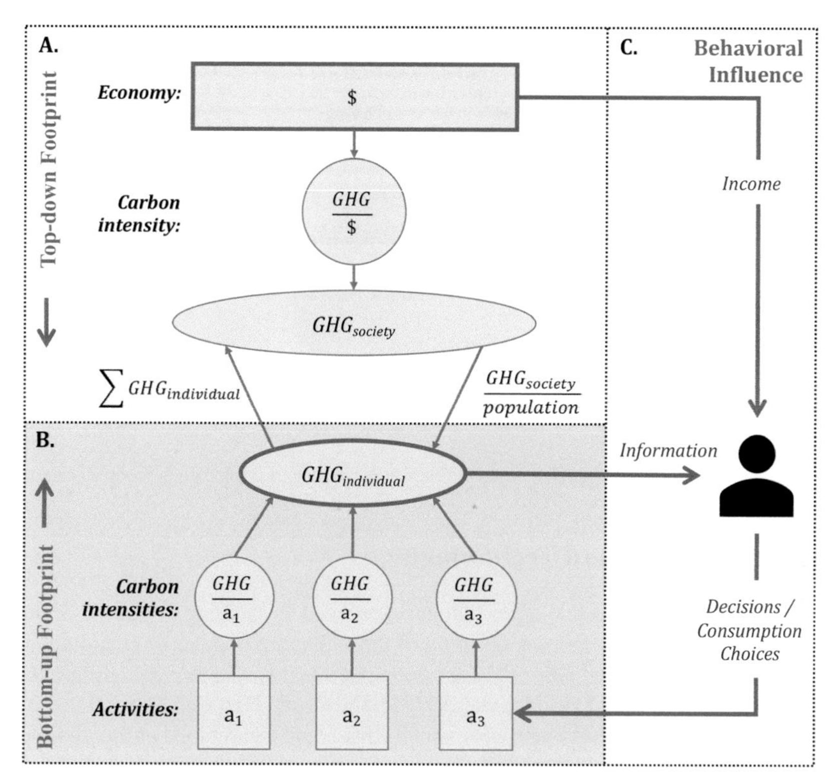

Figure 5. Different ways of approaching and calculating carbon footprints: (A) Top-down footprint: Divides economic activity and the resultant carbon emissions of a society among the population, yielding an individual-scale carbon footprint; (B) Bottom-ip footprint: The carbon intensity and activity levels of an individual are surveyed and summed, yielding a unique individual carbon footprint (used in individual carbon footprint calculators); and (C) Behavioral influence: Individuals can learn from their carbon footprint and be empowered to choose activities that will lower their footprint.

emissions (i.e., reduce them) is in doubt. In a recent survey on carbon footprint calculator usage, Mulrow et al. (2019) found that of the individuals who had previously used a calculator, less than 16% could recall the actual results they were given. Furthermore, 67% of respondents who had used a calculator noted that the exercise did not motivate them to make any changes in their energy consumption activities. Finally, when asked "Do you know what 1 kg of CO_2 emissions represents in terms of daily activity?", only 9% of respondents felt confident that they could provide an activity level (e.g., miles driven or hours of computer use) equivalent to this level of carbon emissions (see Figure 6).

This survey, together with a comprehensive review of 31 existing online calculators, revealed that there is much room for improvement in carbon calculator design and implementation (Mulrow et al., 2019). Further footprinting efforts should aim to create memorable and action-inducing outcomes from the calculator experience. It is worth asking whether widespread carbon footprint knowledge is even necessary for reducing emissions at a societal scale.

It is unclear whether better and more precise carbon measurement at the individual scale could deliver robust emissions reduction. Furthermore, there are powerful proxies for carbon emissions that are already widely measured and managed. These are energy demand and economic activity. A study of environmental beliefs, behaviors and carbon footprints revealed that income levels were the best predictor of carbon footprint, ranking above stated environmental beliefs and dietary habits (Boucher, 2016). Rather than attempt to measure and manage carbon emissions directly, a more effective strategy may be to reduce energy demand or to facilitate cost increases for carbon-intensive activities through policy. Nevertheless, individual carbon footprint calculators should continue to improve and be researched in order to identify tool effectiveness and strategic importance.

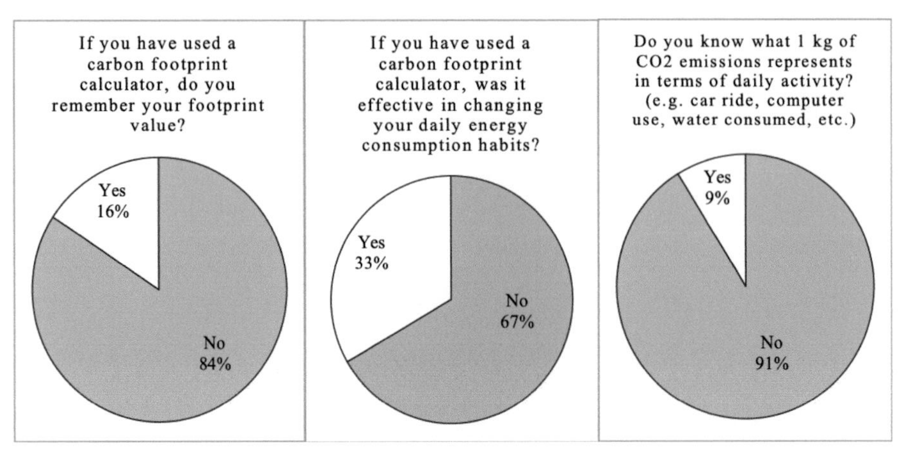

Figure 6. Survey responses, regarding behavior/knowledge impact of using carbon footprint calculators.

4. Emerging Strategies and Technologies

There are several strategies and technologies that have emerged as showing promise in improving urban carbon management. This section focuses on some of those that are popular or emerging, and that have particular applicability to urban and peri-urban environments. In this section, we highlight three: (1) BAS Integration Strategies, (2) Social Densification, and (3) Solid Oxide Fuel Cell Technology. It is important to disclaim that there are individual system components that could be adjoined or incorporated with these strategies and technologies. However, not all system components are described or explained as part of this section. This is a generalization and not an exhaustive list of all emerging strategies and technologies.

4.1 BAS integration strategies

Harnessing the data gathered in a BAS can be challenging. This can occur when either several kinds of BASs have been employed in a single building or system over time, or when components of a BAS use different computer languages as a primary means for data transfer. For example, a building may have employed several legacy BAS components that do not use standard open protocols for data transfer. Building operators within this same building may now want to upgrade their operations into a singular BAS using direct digital controls as their main data transfer medium. Doing so, however, may require that the data being transferred from these legacy components be translated into a useable format for integration with the new BAS. This kind of challenge requires the integration of an innovative, modular BAS, one that can effectively translate various kinds of legacy and open protocol data points into a useable format which is compatible with the overarching BAS framework. The development of these integrative, modular systems is still developing and have taken on different names and forms over the years (e.g., proportional-integral controls, energy optimization software, and smart building systems). Practitioners and scientists continue to analyze and assess the effectiveness of these integrative, modular strategies and technologies (Plageras et al., 2018; Rathore et al., 2018; Stergiou et al., 2018).

4.2 Social densification

With the worldwide migration towards urban centers comes a concentration of diverse populations within a relatively small area. This means that understanding intergroup relations of various demographics is ever more important for effective urban carbon management. Recent research has begun to analyze demographic differences as it relates to consumption of products and foods that have implications for urban and peri-urban carbon emissions.

Here are three example studies that have implications for social densification. In a study that explored air pollution exposure of whites, blacks and Hispanics in the U.S., it was found that non-Hispanic whites experience 17% less air pollution exposure than is caused by their consumption (Tessum et al., 2019). This contrasts with the experience of blacks and Hispanics, who tend to endure 56% and 63% excess exposure, respectively. A second study, which explored climate change implications and carbon emissions that come from white, black, and Latinx food consumption in the U.S., concluded that whites had the greatest per-capita impact on GHG emissions and water use, whereas blacks had the greatest per-capita impact on land use (Bozeman et al., 2019). In a third study, which investigated the relationship between racial attitudes and public opinion about climate change, results showed that reduced agreement with scientific consensus on climate change strongly correlated with high levels of racial resentment (Benegal, 2018).

These study findings are important when trying to design and develop urban areas that may develop or employ carbon emission policies, since issues involving carbon or GHG emissions are linked to climate change (Benegal, 2018; Bozeman et al., 2019). Understanding the behavior tendencies of demographics can assist urban planners in designing more effective policies and systems that result in carbon emission reductions. These example studies are at the country level, so future studies of these kind may need to be framed with greater spatial granularity in order to gain insight into the behavior tendencies of urban dwellers. Nonetheless, it is expected that urban planners and scientists will continue to investigate matters of social densification for the foreseeable future with demographic and socioeconomic specificity.

4.3 Solid oxide fuel cell technology

Solid Oxide Fuel Cells (SOFCs) have emerged as a viable option for baseline energy demand reductions in urban and peri-urban environments. A SOFC operates like a traditional battery, in that it has an anode and cathode which facilitate the transfer of electrons for electricity production. It differs from traditional batteries in some unique ways. For instance, SOFCs can theoretically operate for very long periods of time with minimal or no carbon emissions, all by oxidizing fuel. Efficient SOFC use can reduce energy demand requirements and improve efficiencies by obviating the need for complex power distribution systems. These kinds of improvements offset energy production from what would otherwise be from heavy carbon emitting, fossil energy sources. Nevertheless, there are operational and design challenges that persist. Continued research and design are needed for cost reduction and widespread use. There have, however, been successful examples of SOFC use in urban businesses and as hybrid components of CHPs (FCB, 2017). This technology is promising, and it is expected that further research and design will attempt to bring down costs and to increase hybridization capabilities.

Acknowledgements

This research and review is partly supported by the National Science Foundation (NSF) CAREER award #155173, NSF Graduate Research Fellowship Program award #1444315, and the Bayer-Monsanto Science, Technology, Engineering, and Mathematics Fellowship.

Nomenclature

Afforestation: The process of planting trees in an area of land without forest cover.

Building Automation System (BAS): A network of microcomputers connected by high-capacity communication lines, networks, and the internet that provides controls for air conditioning, lighting, and other systems in each building or facility.

Deforestation: The removal of a forest or stand of trees from an area of land for purposes other than re-forestation.

Ecosystem Services: The benefits humans receive from ecosystems.

Least Developed Country (LDC): A lower-income country with low levels of human assets; are vulnerable to economic and environmental shocks, and have structural impediments to sustainable development.

More Developed Country (MDC): A higher-income country with relatively high levels of human assets that are capable of withstanding economic and environmental shocks.

Urbanization: The gradual shift in residence of the human population from rural to urban areas.

References

Alam, M.M., Murad, M.W., Noman, A.H.M. and Ozturk, I. 2016. Relationships among carbon emissions, economic growth, energy consumption and population growth: Testing Environmental Kuznets Curve hypothesis for Brazil, China, India and Indonesia. Ecological Indicators 70: 466–479.

Ananda, J. 2018. Productivity implications of the water-energy-emissions nexus: An empirical analysis of the drinking water and wastewater sector. Journal of Cleaner Production 196: 1097–1105.

Auch, R., Taylor, J. and Acevedo, W. 2004. Urban Growth in American Cities: Glimpses of U.S. Urbanization. United States Geological Survey, Circular: 1252.

Bellows, A., Brown, K. and Smit, J. 2003. Health benefits of urban agriculture. Community Food Security Coalition. Retrieved from: http://www.foodsecurity.org/pubs.html#urban_ag. Last Accessed June 18, 2019.

Benegal, S.D. 2018. The spillover of race and racial attitudes into public opinion about climate change. Environmental Politics 27(4): 733–756.

Birch, E.L. and Wachter, S.M. 2011. World urbanization: The critical issue of the twenty-first century. Global Urbanization 1: 3–23.

Bot, A. and Benites, J. 2005. The importance of soil organic matter: Key to drought-resistant soil and sustained food production. Food and Agriculture Organization of the United Nations, Soils Bulletin 80.

Boucher, J.L. 2016. Culture, carbon, and climate change: A class analysis of climate change belief, lifestyle lock-in, and personal carbon footprint. Socijalna Ekologija 25(1-2): 53–80.

Bozeman, J.F., Bozeman, R. and Theis, T.L. 2019. Overcoming climate change adaptation barriers: A study on food-energy-water impacts of the average American diet by demographic group. Journal of Industrial Ecology. Available at: https://doi.org/10.1111/jiec.12859.

Bloom, J. and Hinrichs, C. 2011. Moving local food through conventional food system infrastructure: Value chain framework comparisons and insights. Renewable Agriculture and Food Systems 26(1): 13–23.

Branzm H.M., Regan, W., Gerst, K.J., Borak, J.B. and Santori, E.A. 2015. Hybrid solar converters for maximum exergy and inexpensive dispatchable electricity. Energy & Environmental Science 8: 3083–3091.

Brundtland, G.H. 1987. World commission on environment and development: Our common future. The United Nations 5–20.

Bu, S., Shen, G., Anumba, C.J., Wong, A.K.D. and Liang, X. 2015. Literature review of green retrofit design for commercial buildings with BMI implication. Smart and Sustainable Built Environment 4(2): 188–214.

Capehart, B.L., Turner, W.C. and Kennedy, W.J. 2008. Guide to Energy Management: Sixth Edition. CRC Press, Taylor & Francis Group.

Chance, E., Ashton, W., Pereira, J., Mulrow, J., Norberto, J., Derrible, S. and Guilbert, S. 2018. The Plant—An experiment in urban food sustainability. Environmental Progress and Sustainable Energy 37: 82–90.

Cheng, L., Zhang, N., Yuan, M., Xiao, J., Qin, Y., Deng, Y., Tu, Q., Xue, K., Nostrand, J.D.V., Wu, L., He, Z., Zhou, X., Leigh, M.B., Konstantinidis, K.T., Schuur, E.A.G., Luo, Y., Tiedje, J.M. and Zhou, J. 2017. Warming enhances old organic carbon decomposition through altering functional microbial communities. The ISME Journal: Multidisciplinary Journal of Microbial Ecology 11: 1825–1835.

City Grange. 2019. What We Believe. https://citygrange.com/pages/what-we-believe. Last Accessed April 19, 2019.

Conticelli, E., Proli, S. and Tondelli, S. 2017. Integrating energy efficiency and urban densification policies: Two Italian case studies. Energy and Buildings 155: 308–323.

Curry, N. and Pillay, P. 2012. Biogas prediction and design of a food waste to energy system for the urban environment. Renewable Energy 41: 200–209.

Derrible, S., Saneinejad, S., Sugar, L. and Kennedy, C. 2010. Macroscopic model of greenhouse gas emissions for municipalities. Journal of the Transportation Research Board 2191(1): 174–181.

Derrible, S. 2018. An approach to designing sustainable urban infrastructure. MRS Energy & Sustainability 5(e15): 1–15.

Duranton, G. and Turner, M.A. 2012. Urban growth and transportation. Review of Economic Studies 79: 1407–1440.

Endreny, T., Santagata, R., Perna, A., De Stefano, C., Rallo, R.F. and Ulgiati, S. 2017. Implementing and managing urban forests: A much needed conservation strategy to increase ecosystem services and urban wellbeing. Ecological Modelling 360: 328–335.

Federal Highway Administration (FHWA). 2016. Freight analysis framework. Available at: http://ops.fhwa.dot.gov/FREIGHT/freight_analysis/faf/index.htm. Last Accessed June 18, 2019.

Ferreira, A.J.D., Guilherme, R.I.M.M., Ferreira, C.S.S. and Oliveira, M. 2018. Urban agriculture, a tool towards more resilient urban communities? Current Opinion in Environmental Science and Health 5: 93–97.
Food and Agriculture Organization of the United Nations (FAO). 2016. FAO Water databases and software. Available at: http://www.fao.org/nr/water/infores_databases.html. Last Accessed June 18, 2019.
Food and Agriculture Organization of the United Nations (FAO). 2017a. Crops. Available at: http://www.fao.org/faostat/en/#data/QC. Last Accessed June 18, 2019.
Food and Agriculture Organization of the United Nations (FAO). 2017b. Water use and reuse for urban agriculture. Available at: http://www.fao.org/fcit/upa/water-urban-agriculture/en/. Last Accessed June 18, 2019.
Fuel Cells Bulletin (FCB). 2017. SOLIDpower SOFC units commissioned at Microsoft data centre. Fuel Cells Bulletin 2017(11): 6.
Gambini, M. and Vellini, M. 2015. High efficiency cogeneration: Electricity from cogeneration in CHP plants. Energy Procedia 81: 430–439.
Ghaffarianhoseini, A., Berardi, U., AlWaer, H., Chang, S., Halawa, E. and Clements-Croome, D. 2016. What is an intelligent building? Analysis of recent interpretations from an international perspective. Architectural Science Review 59(5): 338–357.
Goldstein, B., Hauschild, M., Fernández, J. and Birkved, M. 2016. Urban versus conventional agriculture, taxonomy of resource profiles: A review. Agronomy for Sustainable Development 36(9): 1–19.
González-Biones, A., Prieto, J., Prieta, F., Herrera-Viedma, E. and Corchado, J.M. 2018. Energy optimization using a case-based reasoning strategy. Sensors 18(865): 1–27.
Grewal, S. and Grewal, P. 2012. Can cities become self-reliant in food? Cities 29(1): 1–11.
Grote, R., Samson, R., Alonso, R., Amorim, J.H., Cariñanos, P., Churkina, G., Fares, S., Thiec, D., Niinemets, Ü., Mikkelsen, T.N., Paoletti, E., Tiwary, A. and Calfapietra, C. 2016. Functional traits of urban trees: Air pollution mitigation potential. Frontiers in Ecology and the Environment 14(10): 543–550.
Guinard, P. and Margier, A. 2018. Art as a new urban norm: Between normalization of the City through art and normalization of art through the City in Montreal and Johannesburg. Cities 77: 13–20.
Gvozdenac, D., Urošević, B.G., Menke, C., Urošević, D. and Bangviwat, A. 2017. High efficiency cogeneration: CHP and non-CHP energy. Energy 135: 269–278.
Hagen, S.N., Harris, N., Saatchi, S.S., Pearson, T., Woodall, C.W., Ganguly, S., Domke, G.M., Braswell, B.H., Walters, B.F., Jenkins, J.C., Brown, S., Salas, W.A., Fore, A., Yu, Y., Nemani, R.R., Ipsan, C. and Brown, K.R. 2016. CMS: Forest Carbon Stocks, Emissions, and Net Flux for the Coterminous US: 2005–2010. ORNL DAAC, Oak Ridge, Tennessee, USA.
Hampwaye, G. 2013. Benefits of urban agriculture: Reality or illusion? Geoforum 49: R7–R8.
Henkel, C., Muley, P.D., Abdollahi, K.K., Marculescu, C. and Boldor, D. 2016. Pyrolysis of energy cane bagasse and invasive Chinese tallow tree (*Triadica sebifera* L.) biomass in an inductively heated reactor. Energy Conversion and Management 109: 175–183.
Hewko, J. 2017. How much does it cost to power the world's fastest growing economies? International Trade Forum 2: 14015.
Hill, C.A., Such, M.C., Chen, D., Gonzalez, J. and Grady, W.M. 2012. Battery energy storage for enabling integration of distributed solar power generation. IEEE Transactions on Smart Grid 3(2): 850–857.
Hoesly, R.M., Smith, S.J., Feng, L., Klimont, Z., Janssens-Maenhout, G., Pitkanen, T., Seibert, J.J., Vu, L., Andres, R.J., Bolt, R.M., Bond, T.C., Dawidowski, L., Kholod, N., Kurokawa, J., Li, M., Liu, L., Lu, Z., Moura, M.C.P., O'Rourke, P.R. and Zhang, Q. 2018. Historical (1750–2014) anthropogenic emissions of reactive gases and aerosols from the Community Emissions Data System (CEDS). Geoscientific Model Development 11: 368–408.
Hong, Y., Xinyue, H., Qun, R., Tao, L., Xinhu, L., Guoqin, Z. and Shi, L. 2017. Effect of urban micro-climatic regulation ability on public building energy usage carbon emission. Energy and Buildings 154: 553–559.
Hosonuma, N., Herold, M., Sy, V., Fries, R.S., Brockhaus, M., Verchot, L., Angelsen, A. and Romijn, E. 2012. An assessment of deforestation and forest degradation drivers in developing countries. Environmental Research Letters 7(044009): 1–12.
Ignatieva, M., Stewart, G.H. and Meurk, C. 2011. Planning and design of ecological networks in urban areas. Landscape and Ecological Engineering 7: 17–25.
Ito, K. 2017. CO_2 emissions, renewable and non-renewable energy consumption, and economic growth: Evidence from panel data for developing countries. International Economics 151: 1–6.
Jandl, R., Lindner, M., Vesterdal, L., Bauwens, B., Baritz, R., Hagedorn, F., Johnson, D.W., Minkkinen, K. and Byrne, K.A. 2007. How strongly can forest management influence soil carbon sequestration? Geoderma 137: 253–268.
Johnson, R., Cowan, T. and Aussenburg, R. 2012. The role of local food systems in U.S. farm policy. Congressional Research Service: R42155.
Kaygusuz, K. 2012. Energy for sustainable development: A case of developing countries. Renewable and Sustainable Energy Reviews 16: 1116–1126.
Khanafer, K. and Vafai, K. 2018. A review on the applications of nanofluids in solar energy field. Renewable Energy 123: 398–406.

Knorr, W., Prentice, I.C., House, J.I. and Holland, E.A. 2005. Long-term sensitivity of soil carbon turnover to warming. Nature 433: 298–301.

Kulak, M., Graves, A. and Chatterton, J. 2013. Reducing greenhouse gas emissions with urban agriculture: A life cycle assessment perspective. Landscape and Urban Planning 111: 68–78.

Kwan, S.C. and Hashim, J.H. 2016. A review on co-benefits of mass public transportation in climate change mitigation. Sustainable Cities and Society 22: 11–18.

Lal, R. and Augustin, B. (eds.). 2012. Carbon Sequestration in Urban Ecosystems. Springer.

Li, C., Liu, M., Hu, Y., Shi, T., Qu, X. and Walter, M.T. 2018. Effects of urbanization on direct runoff characteristics in urban functional zones. Science of the Total Environment 643: 301–311.

Martinez, S., Hand, M., Da Pra, M., Pollock, S., Ralston, K., Smith, T., Vogel, S. and Newman, C. 2010. Local Food Systems: Concepts, Impacts, and Issues. United States Department of Agriculture, Washington, D.C.

McPherson, G., Simpson, J.R., Peper, P.J., Maco, S.E. and Xiao, Q. 2005. Municipal forest benefits and costs in five U.S. cities. Journal of Forestry 103(8): 411–416.

Mohareb, E., Derrible, S. and Peiravian, F. 2016. Intersections of Jane Jacob's conditions for diversity and low-carbon urban systems: A look at four global cities. Journal of Urban Planning and Development 142(2): 1–14.

Muley, P.D., Henkel, C., Abdollahi, K.K., Marculescu, C. and Boldor, D. 2016. A critical comparison of pyrolysis of cellulose, lignin, and pine sawdust using an induction heating reactor. Energy Conversion and Management 117: 273–280.

Mulrow, J., Machaj, K., Deanes, J. and Derrible, S. 2019. The state of carbon footprint calculators: An evaluation of calculator design and user interaction features. Sustainable Production and Consumption 18: 33–40.

National Aeronautics and Space Administration Carbon Monitoring System (NASA CMS). 2019. About CMS Data & Products for Applications. Available at: https://carbon.nasa.gov/data.html. Last Accessed June 18, 2019.

National Oceanic and Atmospheric Administration (NOAA). 2018. What is LIDAR. Available at: https://oceanservice.noaa.gov/facts/lidar.html. Last Accessed June 18, 2019.

Navamuel, E.L., Morollón, F.R. and Cuartas, B.M. 2018. Energy consumption and urban sprawl: Evidence for the Spanish case. Journal of Cleaner Production 172: 3479–3486.

Newsham, G.R. and Bowker, B.G. 2010. The effect of utility time-varying pricing and load control strategies on residential summer peak electricity use: A review. Energy Policy 38: 3289–3296.

Nowak, D.J., Crane, D.E. and Stevens, J.C. 2006. Air pollution removal by urban trees and shrubs in the United States. Urban Forestry & Urban Greening 4(3-4): 115–123.

Nowak, D.J., Greenfield, E.J., Hoehn, R.E. and Lapoint, E. 2013. Carbon storage and sequestration by trees in urban and community areas of the United States. Environmental Pollution 178: 229–236.

Nowak, D.J., Maco, S. and Binkley, M. 2018. I-Tree: Global tools to assess tree benefits and risks to improve forest management. Arboricultural Consultant 51(4): 10–13.

Ochoa, J.J., Tan, Y., Qian, Q.K., Shen, L. and Moreno, E.L. 2018. Learning from best practices in sustainable urbanization. Habitat International 78: 83–95.

Padgett, J.P., Steinemann, A.C. , Clarke, J.H. and Vandenbergh, M.P. 2008. A comparison of carbon calculators. Environmental Impact Assessment 28(2-3): 106–115.

Peterson Garden Project (PGP). 2019. About: Grow2Give and food security. Available at: https://petersongarden.org/about/. Last Accessed February 25, 2019.

Plant Chicago. 2019. WHO WE ARE. https://plantchicago.org/who-we-are/. Last accessed February 25, 2019.

Plageras, A.P., Psannis, K.E., Stergiou, C., Wang, H. and Gupta, B.B. 2017. Efficient IoT-based sensor big data collection-processing and analysis in smart buildings. Future Generation Computer Systems 82: 349–357.

Quéré, C.L., Andrew, R.M., Canadell, J.G., Sitch, S., Korsbakken, J.I., Peters, G.P., Manning, A.C., Boden, T.A., Tans, P.P., Houghton, R.A., Keeling, R.F., Alin, S., Andrews, O.D., Anthoni, P., Barbero, L., Bopp, L., Chevallier, F., Chini, L.P., Ciais, P., Currie, K., Delire, C., Doney, S.C., Friedlingstein, P., Gkritzalis, T., Harris, I., Hauck, J., Haverd, V., Hoppema, M., Goldewijk, K.K., Jain, A.K., Kato, E., Körtzinger, A., Landschützer, P., Lefévre, N., Lenton, A., Lienert, S., Lombardozzi, D., Melton, J.R., Metzl, N., Millero, F., Monteiro, P.M.S., Munro, D.R., Nabel, J.E.M.S., Nakaoka, S., O'Brien, K., Olsen, A., Omar, A.M., Ono, T., Pierrot, D., Poulter, B., Rödenbeck, C., Salisbury, J., Schuster, U., Schwinger, J., Séférian, R., Skjelvan, I., Stocker, B.D., Sutton, A.J., Takahashi, T., Tian, H., Tilbrook, B., Laan-Luijkx, I.T., Werf, G.R., Viovy, N., Walker, A.P., Wiltshire, A.J. and Zaehle, S. 2016. Global Carbon Budget 2016. Earth System Science Data 8: 1–45.

Rathore, M.M., Paul, A., Hong, W.-H., Seo, H., Awan, I. and Saeed, S. 2018. Exploiting IoT and big data analytics: Defining smart digital city using real-time urban data. Sustainable Cities and Society 40: 600–610.

Raupach, M.R., Marland, G., Ciais, P., Quéré, C.L., Canadell, J.G., Klepper, G. and Field, C.B. 2007. Global and regional drivers of accelerating emissions. Proceedings of the National Academy of Sciences of the United States of America 104(24): 10288–10293.

Robinson, S.L. and Lundholm, J.T. 2012. Ecosystem services provided by urban spontaneous vegetation. Urban Ecosystems 15: 545–557.

Santamouris, M. 2013. Energy and Climate in the Urban Built Environment: Chapter 7. The energy impact of the urban environment. Routledge, London – eBook edition. Available at: https://learning.oreilly.com/library/view/energy-and-climate/9781873936900/014_9781315073774_chapter7.html#ch7. Last Accessed June 18, 2019.

Shammin, M.R. and Bullard, C.W. 2009. Impact of cap-and-trade policies for reducing greenhouse gas emissions on U.S. households. Ecological Economics 68: 2432–2438.

Smith, J.E., Heath, L.S. and Nichols, M.C. 2007. U.S. forest carbon calculation tool: Forest-land carbon stocks and net annual stock exchange. USDA FS, Northern Research Station NRS 13: 1–34. Available at: https://www.nrs.fs.fed.us/pubs/gtr/gtr_nrs13R.pdf. Last Accessed June 18, 2019.

Steckel, J.C., Edenhofer, O. and Jakob, M. 2015. Drivers for the renaissance of coal. Proceedings of the National Academy of Sciences of the United States of America 112(29): E3775–E3781.

Stergiou, C., Psannis, K.E., Kim, B.-G. and Gupta, B. 2018. Secure integration of IoT and cloud computing. Future Generation Computer Systems 78: 964–975.

Tamiya, N., Noguchi, H., Nishi, A., Reich, M.R., Ikegami, N., Hashimoto, H., Shibuya, K., Kawachi, I. and Campbell, J.C. 2011. Population ageing and wellbeing: Lessons from Japan's long-term care insurance policy. The Lancet 378: 1183–1192.

Tessum, C.W., Apte, J.S., Goodkind, A.L., Muller, N.Z., Mullins, K.A., Paolella, D.A., Polasky, S., Springer, N.P., Thakrar, S.K., Marshall, J.D. and Hill, J.D. 2019. Inequity in consumption of goods and services adds to racial-ethnic disparities in air pollution exposure. Proceedings of the National Academy of Sciences of the United States of America 116(13): 6001–6006. https://doi.org/10.1073/pnas.1818859116.

Tillie, N., Beurden, J.B., Doepel, D. and Aarts, M. 2018. Exploring a stakeholder based urban densification and greening agenda for rotterdam inner city—Accelerating the transition to a liveable low carbon city. Sustainability 10(1927): 1–27.

Tsao, J., Lewis, N. and Crabtree, G. 2006. Solar FAQs. U.S. Department of Energy. Available at: http://citeseerx.ist.psu.edu/viewdoc/download?doi=10.1.1.721.6459&rep=rep1&type=pdf. Last Accessed: April 19, 2019.

United Nations–Conference on Trade and Development (UN CTAD). 2017. Map of the Least Developed Countries (LDCs). Available at: https://unctad.org/en/Pages/ALDC/Least%20Developed%20Countries/LDC-Map.aspx. Last Accessed June 18, 2019.

United Nations–Department of Economic and Social Affairs (UN DESA). 2017. World Population Prospects: The 2017 Revision, Key Findings and Advance Tables. Population Division. ESA/P/WP/248.

United Nations–Department of Economic and Social Affairs (UN DESA). 2018a. 2018 Revision of World Urbanization Prospects. Publications. Available at: https://www.un.org/development/desa/publications/2018-revision-of-world-urbanization-prospects.html. Last Accessed June 18, 2019.

United Nations–Department of Economic and Social Affairs (UN DESA). 2018b. World Urbanization Prospects 2018. Population Division. Available at: https://population.un.org/wup/Country-Profiles/. Last Accessed June 18, 2019.

United Nations–Economic Analysis and Policy Division (UN EAPD). 2019. Least Developed Countries (LDCs). Available at: https://www.un.org/development/desa/dpad/least-developed-country-category.html. Last Accessed June 18, 2019.

United Nations–Statistical Division (UN SD). 2017. Population density and urbanization. Available at: https://unstats.un.org/unsd/demographic/sconcerns/densurb/densurbmethods.htm. Last Accessed June 18, 2019.

United Nations–World Economic Situation and Prospects (UN WESP). 2014. Country Classification 143–150.

United States Census Bureau (US CB). 2019. U.S. and World Population Clock. Available at: https://www.census.gov/popclock/.

United States Department of Agriculture (USDA). 2019. Carbon Estimation Tools: A Primer. Available at: https://www.fs.usda.gov/ccrc/tools/carbon-primer. Last Accessed June 18, 2019.

United States Department of Agriculture–Forest Service (USDA FS). 2019a. About LANDIS Landscape Disturbance and Succession model. Available at: https://www.nrs.fs.fed.us/tools/landis/. Last Accessed June 18, 2019.

United States Department of Agriculture–Forest Service (USDA FS). 2019b. i-Tree Streets: Overview. Available at: https://www.itreetools.org/tools/i-tree-streets/i-tree-streets-application-overview. Last Accessed June 18, 2019.

United States Department of Energy (US DOE). 2019. Building Life Cycle Cost Programs. Available at: https://www.energy.gov/eere/femp/building-life-cycle-cost-programs. Last Accessed June 18, 2019.

United States Environmental Protection Agency (US EPA). 2017. Global Greenhouse Emissions Data. Available at: https://www.epa.gov/ghgemissions/global-greenhouse-gas-emissions-data. Last Accessed June 18, 2019.

United States Forest Service (US FS). 2019a. Forest Vegetation Simulator (FVS). Available at: https://www.fs.fed.us/fvs/. Last Accessed June 18, 2019.

United States Forest Service (US FS). 2019b. Forest Vegetation Simulator (FVS). Available at: https://www.fs.fed.us/fvs/whatis/index.shtml. Last Accessed June 18, 2019.

United States National Aeronautics and Space Administration (US NASA). 2019. NASA-CASA Project: Overview of Global Ecosystem Research in Earth Science Division. Available at: http://cquest.arc.nasa.gov:8399/casa/. Last Accessed June 18, 2019.

Vergnes, A., Pellissier, V., Lemperiere, G., Rollard, C. and Clergeau, P. 2014. Urban densification causes the decline of ground-dwelling anthropods. Biodiversity and Conservation 23: 1859–1877.

Wang, J., Bai, J., Zhao, Q., Lu, Q. and Xia, Z. 2016. Five-year changes in soil organic carbon and total nitrogen in coastal wetlands affected by flow-sediment regulation in a Chinese delta. Scientific Reports 6(21137): 1–8.

Wang, L., Omrani, H., Zhao, Z., Francomano, D., Li, K. and Pijanowski, B. 2019. Analysis on urban densification dynamics and future modes in southeastern Wisconsin, USA. PLoS One 14(3): 1–22.

Wang, Y. and Zhao, T. 2015. Impacts of energy-related emissions: Evidence from under developed, developing and highly developed regions in China. Ecological Indicators 50: 186–195.

West, S.E., Owen, A., Aexelsson, K. and West, C.D. 2015. Evaluating the use of a carbon footprint calculator: Communicating impacts of consumption at household level and exploring mitigation options. Journal of Industrial Ecology 20(3): https://onlinelibrary.wiley.com/doi/full/10.1111/jiec.12372.

World Atlas. 2019. North America Facts. Fast Facts. Available at: https://www.worldatlas.com/webimage/countrys/nafacts.htm#page. Last Accessed June 18, 2019.

Zhou, P., Lei, X. and Lv, Z. 2012. Study on integrating BACnet with IPv6-based wireless sensor networks. Procedia Engineering 29: 275–279.

CHAPTER 14

Adaptive Lean and Green (L&G) Manufacturing Approach in Productivity and Carbon Management Enhancement

Wei Dong Leong,[1] *Hon Loong Lam*,[1,*] *Chee Pin Tan*[2] and *Sivalinga Govinda Ponnambalam*[3]

1. Introduction

Global trade has increased cross border opportunities for many manufacturing companies around the world. The increase in global competition has driven the industry players to seek out alternative options to strive for higher competitiveness. Cross-border trading has also created challenges in the manufacturing industry. The factors such as technological advancement, political influence, economic stability, regulatory restructuring, monetary fluctuation and environmental pressure are consistently changing the competitive landscape (Issa et al., 2010). One of the easier options for an industry player to remain sustainable is through reduction of operation waste.

Anand and Kodali (2008) highlighted that many large industry players (e.g., Toyota, Ford, Boeing, etc.), have applied the lean approach in order to maintain their competitiveness in the global market. Lean manufacturing (LM) is known as an approach which eliminates all non-value-added activity in a production process (Womack and Jones, 1994). Lean is also an integrated socio-technical system where the main objective is to eliminate waste by concurrently reducing or minimizing supplier, customer and internal variability (Rodríguez et al., 2017). As lean focuses on the operational aspect, green highlights the environmental aspect. Bhattacharya et al. (2011) stated that the green represents ecological sustainability and encompasses many different concerns, including waste generation and recycling, air, water and land pollution, energy usage and efficiency. On top of that, the European Commission has stated that the implementation of the green application into other related areas, such as green economy, will create more value with fewer resources (European Commission, 2015). With the development in terms of green growth, the World Bank (2012) defined the green growth as growth that is efficient in the consumption of natural resources, minimises pollution and environmental impacts as well as resilient towards natural hazards. Green element focuses on minimizing environmental impact and pollution by creating more value with fewer resources.

[1] University of Nottingham Malaysia Campus, Dep. of Chemical Engineering, Jalan Broga, 43500 Semenyih, Selangor, Malaysia.

[2] Monash University Malaysia, School of Engineering, Jalan Lagoon Selatan, 47500 Bandar Sunway, Selangor, Malaysia.

[3] Faculty of Manufacturing Engineering, University Malaysia Pahang, 26600 Pekan, Pahang, Malaysia.

* Corresponding author: HonLoong.Lam@nottingham.edu.my

The operation performance can be further enhanced through synergy between lean and green (L&G) approaches, which results in value creation and waste reduction. The implementation of L&G will not only improve the performance efficiency but also the environmental performance. Generally, the initiation of L&G approach will reduce the overall carbon emission from the manufacturing sector.

To effectively implement the L&G approach, an adaptive model is introduced into the L&G framework to ease implementation of L&G. The adaptive model is a dynamic model that can cope with constant updating data and provide performance analysis. Based on the priority and input from the organisation, the adaptive model can analyse the process data and provide potential improvement recommendations. The adaptive model utilises the back-propagation method, which provides improvement recommendation based on the error percentage between actual and expected performance.

2. Lean Manufacturing

Since the first industrial revolution, the lean approach has been portrayed as a significant element in manufacturing. Lean manufacturing (LM) is generally used to eliminate activities and procedures that do not add value to the final product (Nawanir et al., 2018). The evolution of traditional manufacturing to lean approach allows the organisation to operate at higher efficiency and in a competitive manner. The five key principles (e.g., define value, identify value stream, create smooth value flow, implement pull-based production and strive for excellence) of LM have laid down a systematic framework to ease lean implementation (Leong et al., 2019a). The principles of LM mainly focus on identifying the value-added and non-value-added elements in an organisation. The identification and reduction of non-value-added elements in LM will be aided by a list of well-established lean tools, such as Just in time, Kaizen, key performance indicator, 5S, etc.

The productivity improvement of an organisation can be done through value creation by reducing 7 wastes (e.g., over-production, waiting, inventory, motion, transportation, over-processing and defects) from an operation. The industry players' confidence in implementing lean practice can be developed through appropriate guidance and selecting the right lean tools to tackle the existing operation challenges. Lean tools can be used in every organisation, provided appropriate understanding and modification are applied. On top of that, the misapplication of lean tools may result in wastage. The dependency on an experienced lean mentor to implement lean tools is necessary.

The implementation of the LM approach and tools has shown positive results in terms of operation performance. Zimmer (2000) highlighted that the successfully implementation of LM has delivered positive operation outcome with 50% or greater increase in capacity, 50% improvement in product quality and 60% reduction in production time cycle in different sector. The LM outcome is expected to result in a smooth, high quality and well-organised operation that can produce customer demanded product on time with minimal waste. On top of that, LM has also reflected the impact on cost-performance resulting from waste reduction that contributes to improving the competitiveness of the organisation. Besides that, the increase in operational efficiency will contribute indirectly to lower carbon footprint per unit of production. The organisation's leadership is critical in influencing followers' attitude and behaviour in practicing LM. The implementation of LM with a top-down approach can have better outcome.

3. Green Manufacturing

Green manufacturing (GM) is a business strategy that focuses on profitability through proactive and reactive environmental friendly operation. GM is introduced in order to enhance and compensate the organisation performance in the environment aspect. An organisation that emphasizes green innovation can improve environmental performance and corporate image among customers and regulators.

GM is defined as an environmental and economically-driven approach to minimizing the waste through process design, consumption of products and materials (Maruthi and Rashmi, 2015). Mittal and Sangwan (2014) highlighted that the GM approach helps the industry player to achieve better economy without compromising the environment. The International Standard Organization (ISO) published the Environmental Management System (EMS) standard internationally, known as the ISO 14001, as a systematic continuous improvement tool to improve an organization's environmental-related matters (Agan et al., 2013). The organisations need to explore alternatives in order to achieve a balance between operation and environmental performance and, hence, remain competitive.

Helu and Dornfeld (2012) highlighted that GM consists of 5 principles, including (i) evaluation and improvement of the manufacturing process from a green perspective, (ii) system assessment across both the vertical and horizontal directions, (iii) reduction or removal of harmful input and output of the system (iv) reduction in net resource consumption and (v) consideration of the temporal effect on the process. These principles underline the assessment of the manufacturing process based on product lifecycle, resources, energy and environmental emissions.

In many situations, 80% of the economic, social and environmental impacts have already been decided in the process during product design stage (Kim and Kara, 2012). Thus, Design for Environment (DfE) and life cycle assessment (LCA) need to include the green element during process design. Generally, DfE can identify and design the product based on the environmental impact of the product through the LCA (Eibel, 2017). On top of that, the LCA evaluates the potential environmental impact of the entire life cycle of a product.

Energy consumption is also a major factor in manufacturing. Green technology, such as photovoltaic, co-generation, biogas, etc., can improve the green aspect of a factory (Leong et al., 2019a). To achieve GM, many changes and modifications are required at the technology, system perspective, paradigm shift and system knowledge level. Moreover, the lack of expertise and financial scheme in the organisation have delayed the implementation of the green approach. In many occasions, the management is not fully alert of the potential improvement in the operation. Due to the lack of energy-efficient experts, the industrialist has switched the focus to smaller equipment which only contributes to a potential improvement of 2–5% (UNIDO, 2010).

Generally, the driving forces to implement GM are from an external driving force, such as legislation, corporate image and tax incentive. The government role in supporting green project financing and incentives will further encourage the industry players to practise GM. Maloni and Brown (2006) stated that ISO 14001 certified companies are 50% likely to require their suppliers to undertake specific environmental practices and 40% likely to assess their suppliers' environmental performances. With that, an ISO certified company will have to maintain the accreditation by practising the green approach continuously. This will create a chain effect in demanding the supply chain to adapt to green practices.

4. Lean and Green (L&G) Approach

Factories that practice with lean principles tend to have greater enthusiasm towards higher operation efficiency and better environmental performance (Leong et al., 2019b). Boeing's lean manufacturing (LM) program has shown improvement in quality efficiency and a reduction in environmental waste (EPA, 2000). Dües et al. (2013) emphasized that lean is a catalyst in green implementation while green helps to maintain best-practice in lean. Generally, the synergy of lean and green (L&G) approach develops the efficiency and effectiveness for both operation and environmental outcome as compared to an individual approach. A study has discovered that the L&G approach shares the same objectives in improving organisation performance (Hallam and Contreras, 2016).

According to Pampanelli et al. (2014), there are limited approaches that integrate L&G with the fundamental principles of the lean approach and green approach. The industry players are facing the main challenge in the lack of L&G expertise in the industry. The need for a systematic approach should be developed in order to support the organisation and reap the benefits of L&G. With that, the top-down approach from the management is the key to ensure that the L&G practices are being executed successfully. The L&G approach can be used as an execution strategy for continuous improvement.

With proper aiding tools, the benefits of L&G can impact the operation and environmental performance through an effective management strategy. The adoption of new technologies not only enhances the industry competition but also improves carbon management performance. The successful implementation of L&G approach will reflect directly in the operation cost.

Cherrafi et al. (2017) pointed out a list of the main hurdles faced during the implementation of L&G, such as poor quality of the human resource, lack of training and guidance, lack of regulatory and policy enforcement, fund constraints and resistance to change. However, the self-initiative and commitment from the organisation can be beneficial to improve themselves in the competitive market.

Dües et al. (2013) highlighted the similarity of both the lean approach and green approach that show a strong commitment to efficiency-driven practice and zero waste. The waste reduction in L&G approach has a direct impact on the total carbon emission in an organisation. As waste is the common denominator in the L&G approach, the lean practises can be implemented into green practises (Porter and Linde, 1995). The strong focus on the continuous development of the lean approach requires consistent development of human resource in order to avoid future mistakes. This is also advocated in the green approach (Johansson and Winroth, 2009). Verrier et al. (2014) stated that the L&G approach will lead to better competitive advantages and profitability. Despite the fact that the L&G approach shares a similar goal in reducing waste, the implementation method can be different. For example, investment in technology is inevitable to reduce environmental emission from the green aspect. However, from the lean perspective, investment does not contribute directly to customer value.

5. Lean and Green Analytic Model

A systematic lean and green (L&G) analytic model is critical in accelerating industrialist progress towards sustainable manufacturing. Moradi and Huang (2018) introduced a two-layer hierarchy to implement strategic and tactical control for dynamic sustainability problems. Azapagic et al. (2016) added that the framework should include 3 stages: Problem structuring, analysis and resolution. Figure 1 demonstrates the L&G framework that addresses the challenges in achieving L&G.

This framework will be able to guide the practitioner in identifying areas that require improvement. The L&G approach can assist the industrialist to make better carbon management decisions within the production boundary. The L&G approach assesses the facility based on manpower, money, material, machine and environment (4M1E). The 4M1E model identifies areas that can maximise both operational and environmental performance concurrently. On top of that, the L&G index (LGI) is used as a benchmarking tool for the organisation to track their progress. The back-propagation (BP) optimisation method is used to complement the L&G continuous improvement.

Generally, manpower (MP) is the foundation of an organisation. The continuous development of talent is critical in improving their competitiveness and striving for innovation (Leong et al., 2018). In this era, the manufacturing workforce is decreasing as the younger generation lost interest in the manufacturing sector. Kar (2018) discovered that the younger generation has the misconception for the manufacturing sector with lack of competitive wages, innovation, unsafe, etc. Williamson et al. (2018) highlighted that human behaviour is one of the main components of carbon management. Effective and efficient carbon management practices will lead to a better, sustainable organisation.

Machine (MC) is one of the major components in converting raw material to product. Availability efficiency, performance efficiency and quality efficiency are the critical criteria that the industrialists monitor on MC (Leong et al., 2019a). These three main factors will contribute to overall equipment efficiency (OEE) and will reflect the performance of the MC (Singh et al. 2013). The OEE can be translated into the performance of carbon footprint per product produced from the processing facility.

Material (MT) reflects the amount of resources that are being converted from the raw material into product. It takes into consideration the recycling of defective products as well. In MT, the inventory system ensures that the storage area is being utilized efficiently. This is to ensure that the production flow will not be interrupted due to lack of resources. Inventory handling of MT can contribute to higher carbon footprint due to the different handling method. Thus, carbon management must be considered when dealing with inventory management.

Money (MY) is often classified as the financial evaluating tool for an organisation. Many times, the fluctuation in monetary policy creates a huge financial challenge to the industrialists. However, there are some internal factors that can be controlled in order to maximise the efficiency of human resources

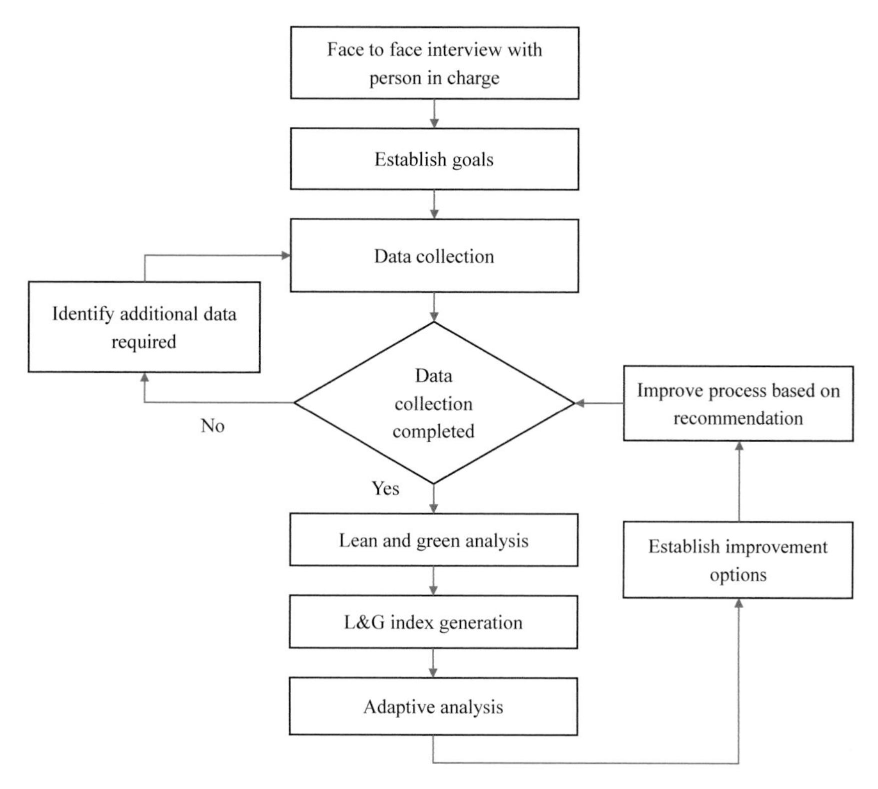

Figure 1. Systematic approach of the lean and green framework.

and equipment that will contribute directly to higher profit. On top of that, MY can act as a benchmark in reflecting the organisation financial performance in terms of carbon footprint with respect to their production.

The fifth component is known as the environment (EV). The carbon footprint overall carbon management pattern of an organisation. Generally, EV combines all the carbon footprint from the other 4 components (i.e., MP, MC, MT, MY) in order to reflect the carbon performance of the organisation. On top of that, it is the responsibility of the organisation to ensure that environmental emissions, such as air, solid and water waste, are treated according to environmental regulation before being emitted into the environment.

5.1 Data collection and analysis procedures based on L&G framework

The methodology demonstrated in Figure 1 is explained as below:

1. Conduct interview with the plant manager/operation manager/maintenance manager/human resource manager to understand the operation behaviour of the factory.
2. Establish an independent data collection team. L&G questionnaires are used to gain the management team's feedback/opinion on their standard operating procedure (SOP) and decision-making hierarchy. The objectives and goals of the organisation will be identified prior to data collection. Figure 2 illustrates a guideline for data collection.
3. The data collection schedule shall be developed in order to avoid any unnecessary delay. Data can be collected through supervisory control and data acquisition (SCADA), distributed control system (DCS) or manual logbook. Data collection shall be over a period of six months or one operation year.

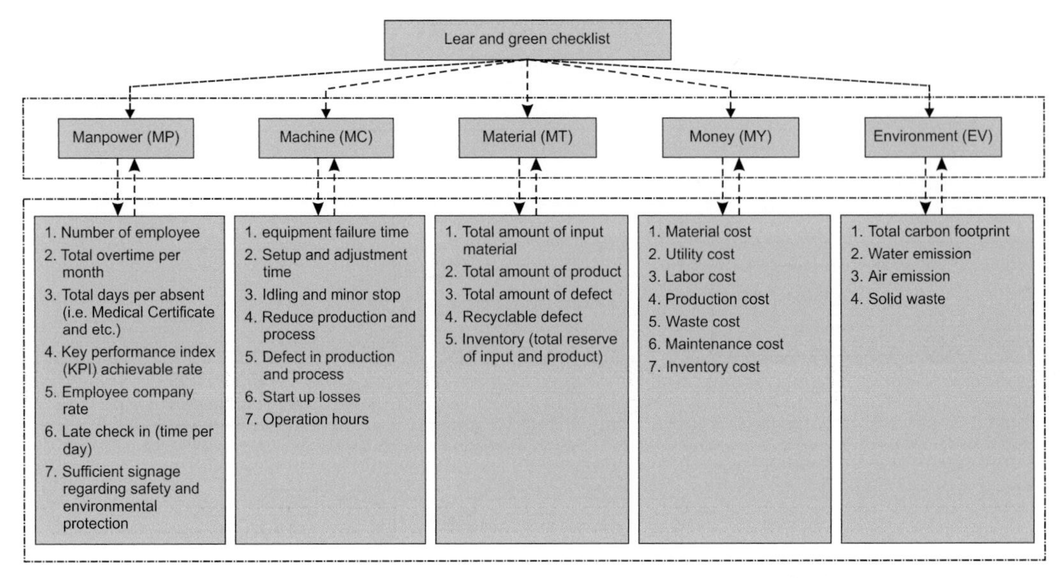

Figure 2. L&G checklist and guidelines for data collection (Leong et al., 2019a).

4. Once the data collection is completed, review and discuss the collected data before analysis. Observe data behaviour on special events.
5. Perform L&G analysis to generate the lean and green index (LGI). The initial LGI for the organisation performance will be used as the benchmarking index.
6. Based on the benchmarked LGI, the organisation can set a new target to achieve for the following month. The back-propagation optimiser will analyse the LGI outcome for the following month to provide an updated weight. The updated weight will act as guidance to the industrialist on the areas that require improvement.
7. Based on the updated weight for the components, the industry players can plan their improvement areas based on the ranking of the components.
8. Repeat step 3 for continuous data collection.

5.2 Multi-criteria decision method—Analytic hierarchy process (AHP)

After the establishment of the lean and green (L&G) framework, multi-criteria decision making (MCDM) tools shall be selected to complement the model. Despite there being many different MCDM methods available, the analytic hierarchy process (AHP) method is selected due to its simple but powerful method in decision making. The AHP method is used to develop a decision-making model based on goals and criteria (i.e., main components and indicators). Figure 3 illustrates an AHP structure model. The downward arrow, which connects the upper level to the lower level, shows the dependency of the lower level group on the upper level.

Based on the interview with the management and plant personnel, a set of questionnaires will be developed and distributed to the industry experts in order to obtain their input on the related sector. In this AHP analysis, Saaty's pairwise comparison scale is adopted in order to represent the intensity of the relationship between the pairing (Saaty, 2012). The 9-point fundamental scale is described in Table 1.

The pairwise comparison matrix will use the industry experts' input as the primary input. Equation (1) demonstrates the input from the experts that are filled from the second upper left part of the matrix (i.e., w_{12}), while the lower left is the inverse of upper-right value (i.e., $1/w_{12}$). Ngan et al. (2018) emphasized that the geometric mean method is used for multiple inputs. Saaty (1980) highlighted that the consistency

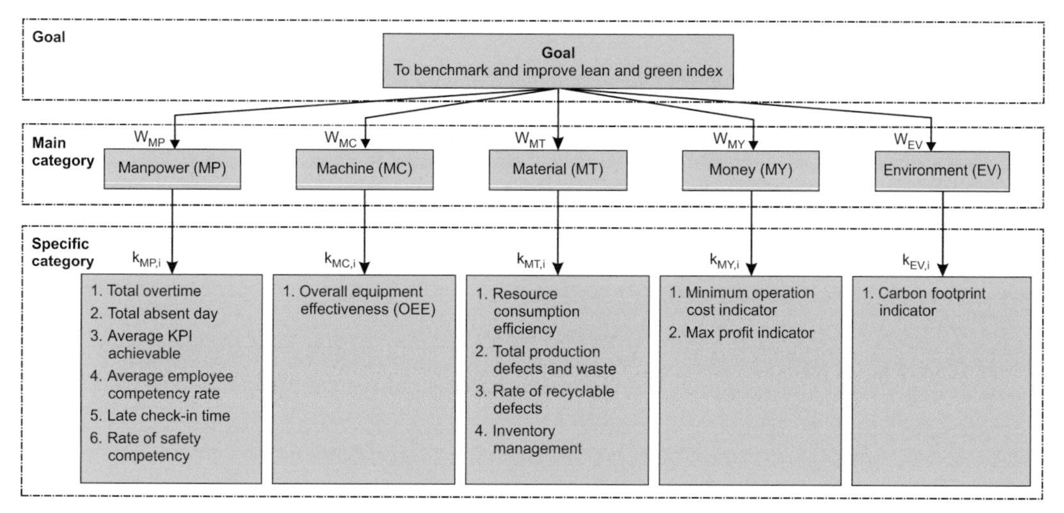

Figure 3. Analytic hierarchy process model structure (Leong et al., 2019a).

Table 1. Saaty pairwise comparison scale (Saaty, 2012).

Verbal judgement	Numeric value
Extremely important	9
	8
Very strongly more important	7
	6
Strongly more important	5
	4
Moderate more important	3
	2
Equally important	1

ratio (CR) of 0.10 or less is acceptable. The judgment of the analysis needs to be re-evaluated if CR is more than 0.10, which reflects inconsistency.

$$m = \begin{matrix} & C_1 & C_2 & C_{...} & C_n \\ C_1 \\ C_2 \\ C_{...} \\ C_n \end{matrix} \begin{bmatrix} 1 & w_{12} & w_{1...} & w_{1n} \\ 1/w_{12} & 1 & w_{2...} & w_{2n} \\ 1/w_{1...} & 1/w_{2...} & 1 & w_{...n} \\ 1/w_{1n} & 1/w_{2n} & 1/w_{...n} & 1 \end{bmatrix} \quad (1)$$

5.3 *Lean and green index (LGI)*

Lean and green index (LGI) is used for benchmarking and evaluation of lean and green (L&G) progress. Figure 2 reflects the data required for analysis. LGI is a composite index contributed by 4M1E in Equation (2). The w_{MP}, w_{MT}, w_{MC}, w_{MY}, and w_{EV} represent the weights of the components obtained from the industrialists' feedback.

$$LGI = w_{MP} \times MP + w_{MT} \times MT + w_{MC} \times MC + w_{MY} \times MY + w_{EV} \times EV \quad (2)$$

5.3.1 Manpower index, MP

The main indicators in the manpower (MP) component are attendance rate, key performance indicator (KPI) and competency. In MP, an excellent employee is competent personnel who fulfils his responsibility with maximum attendance rate and KPI. More importantly, the employee must be constantly updated and trained with sufficient safety knowledge regarding the performed task. In the development of the MP index, the indicators required are as follows:

a. Total overtime per employee, MP_{OT} (hr/year.pax)

$$MP_{OTi} = \frac{\textit{Total employee over time per year}}{\textit{Average number of employee working in a year}} \tag{3}$$

Normalise the indicator by comparing with total operation hours per annum.

$$MP_{OT} = \frac{hr_{opt.year} - MP_{OTi}}{hr_{opt.year}} \tag{4}$$

b. Total absent day per employee, MP_{AB}(day/year.pax)

$$MP_{ABi} = \frac{\textit{Total employee over time per year}}{\textit{Average number of employee working in a year}} \tag{5}$$

Normalise the indicator by comparing with total operation hours per annum.

$$MP_{AB} = \frac{hr_{opt.year} - MP_{ABi}}{hr_{opt.year}} \tag{6}$$

c. Average KPI achievable per employee, MP_{KPI}(%/year.pax)

$$MP_{KPIi} = \frac{\textit{Total employee KPI achievable per year}}{\textit{Average number of employee working in a year}} \tag{7}$$

Normalise the indicator by 100% achievable

$$MP_{KPI} = \frac{MP_{KPIi}}{100} \tag{8}$$

d. Average employee competency rate per employee, MP_{CR}(%/year.pax)

$$MP_{CRi} = \frac{\textit{Total employee competency rate per year}}{\textit{Average number of employee working in a year}} \tag{9}$$

Normalise the indicator relative to the full rate.

$$MP_{CR} = \frac{MP_{CRi}}{100} \tag{10}$$

e. Late check-in time per employee, MP_{LT}(hr/year.pax)

$$MP_{LTi} = \frac{\textit{Total employee late check in time rate per year}}{\textit{Average number of employee working in a year}} \tag{11}$$

Normalise the indicator by comparing with total operation hours per annum.

$$MP_{LT} = \frac{hr_{opt.year} - MP_{LTi}}{hr_{opt.year}} \tag{12}$$

f. The rate of safety competency per employee, MP_{SC}(%/year.pax)

$$MP_{SCi} = \frac{\textit{Total rate of safety competency per year}}{\textit{Average number of employee working in a year}} \tag{13}$$

Normalise the indicator relative to full rate.

$$MP_{SC} = \frac{MP_{SCi}}{100} \tag{14}$$

MP index is represented as below:

$$MP = k_{MP,OT} \times MP_{OT} + k_{MP,AB} \times MP_{AB} + k_{MP,KPI} \times MP_{KPI} + k_{MP,CR} \times MP_{CR} + k_{MP,LT} \times MP_{LT} + k_{MP,SC} \times MP_{SC} \quad (15)$$

5.3.2 Machine index, MC

The machine (MC) index is represented by the overall equipment effectiveness (OEE). The intention of OEE is to numerically illustrate the production efficiency through a simple and clear metric. Through data analysis, the OEE is able to identify the bottleneck of the process. The six main losses (e.g., equipment failures, set up and adjustment, idling and minor stoppage, reduced speed, defect in the process and start-up losses) are used to calculate the metric for OEE. Figure 4 demonstrates the relationship of the six main losses.

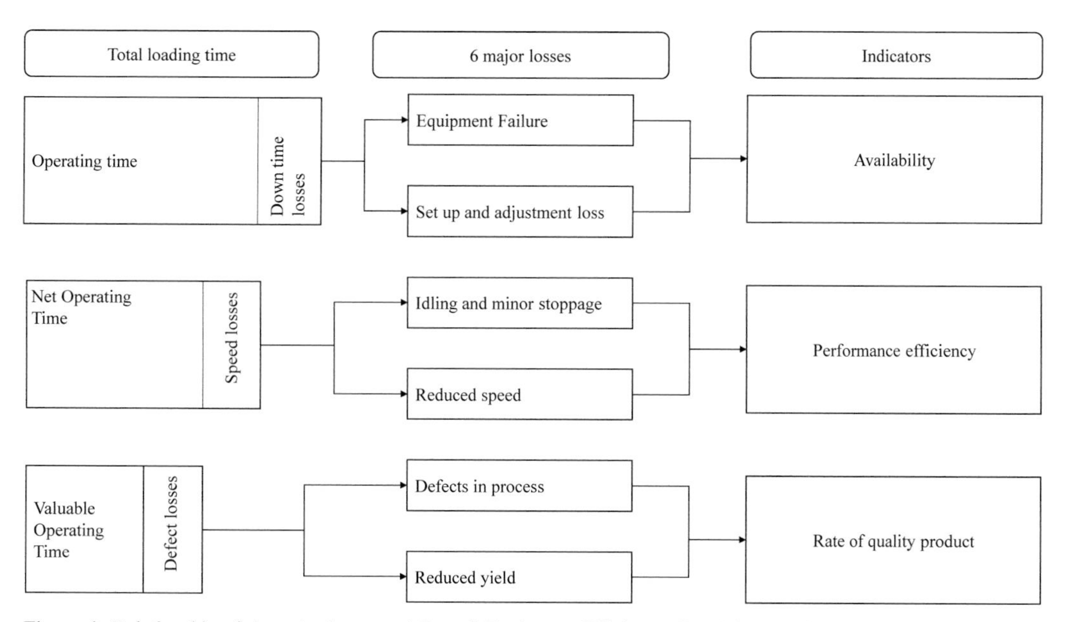

Figure 4. Relationship of six major losses and Overall Equipment Efficiency (OEE) (Batumalay and Santhapparaj, 2009).

Thus, availability (A) is calculated as below:

$$A = \frac{\textit{Planned operation time} - (\textit{equipment failure time-set up and adjustment loss})}{\textit{Planned operation time}} \quad (16)$$

Performance efficiency (P) indicator is shown below:

$$P = \frac{\textit{Total processed amount} \times \textit{Ideal production rate}}{\textit{Actual operation time}} \quad (17)$$

The rate of quality product (Q) is expressed as below:

$$Q = \frac{\textit{Processed amount} - \textit{Defect amount}}{\textit{Processed amount}} \quad (18)$$

Globally, the OEE is widely accepted as a quantitative tool to measure the performance of the equipment. Any organisation that can achieve an overall OEE scoring of 85% is considered world-class.

$$OEE = A \times P \times Q \quad (19)$$

5.3.3 Material index, MT

The material (MT) component focuses on raw material, produced product and inventory management. As sustainable development goal (SDG) 12 emphasizes responsible consumption and production, MT can contribute by improving the rate of resources conversion to quality products (SDG, 2019). The indicators for MT are illustrated as below:

Resource consumption efficiency, MT_{RE} reflects the relationship between total product produced to total consumed resources.

$$MT_{RE} = \frac{\textit{Total amount of product}}{\textit{Total amount of material input}} \tag{20}$$

Product to defect indicator, MI_{DI} is to assess the ratio of defects to total production. A value closer to 1 is highly favourable as it indicates less defects are produced.

$$MT_{DI} = \frac{\textit{Total amount of product} - \textit{Total product defects}}{\textit{Total amount of product}} \tag{21}$$

The rate of recyclable defects, MT_{RD} to evaluate the total amount of recycling goods. The reduction in recyclable goods will improve the overall operation cost.

$$MT_{RD} = \frac{\textit{Total amount of defects} - \textit{Recycable defects}}{\textit{Total amount of defects}} \tag{22}$$

The inventory, MT_{IN} ensures resource and product storage area are sufficient to support smooth production. To identify the optimum inventory space required, statistical process control (SPC) method can be used to evaluate the storage behaviour of inventory.

$$MT_{IN} = \frac{\textit{(Minimum monthly material input + average monthly production)}}{\textit{(Average monthly input material + maximum monthly production)}} \tag{23}$$

Finally, the performance of resource consumption is represented by MT index.

$$MT = k_{MT,RE} \times MT_{RE} + k_{MT,DI} \times MT_{DI} + k_{MT,RD} \times MT_{RD} + k_{MT,IN} \times MT_{IN} \tag{24}$$

5.3.4 Money index, MY

In money (MY), total operation cost (MY_{OC}) and total profit (MY_{TP}) are used to reflect the financial performance of a process. The performance for both MY_{OC} and MY_{TP} are evaluated on a per-unit-of-production basis.

MY_{OC} illustrates the operation expenditure (OpEx), which includes the costs required to maintain the process, such as labour cost, maintenance cost, energy cost, raw material cost, etc.

$$MY_{OC} = \frac{\textit{Minimum operation cost/unit of product}}{\textit{Average operation cost/unit of product}} \tag{25}$$

MY_{TP} reflects the performance total profit per unit of product from the production.

$$MY_{TP} = \frac{\textit{Maximum profit cost/unit of product}}{\textit{Average profit cost/unit of product}} \tag{26}$$

Thus, MY index is generated by both MY_{OC} and MY_{TP}, as follows:

$$MY = k_{MY,OC} \times MY_{OC} + k_{MY,TP} \times MY_{TP} \tag{27}$$

5.3.5 Environment index, EV

From the environmental (EV) perspective, global warming and climate change have been an inevitable issue for every industry. Čuček et al. (2012) emphasized the importance of including global warming potential (GWP) as one of the EV indicators. According to EPA (2017), GWP reflects the heat-absorbing capacity of the emitted gas compared to a unit of CO_2 at that time. Moreover, Ng et al. (2014) added that the proper management and reduction in production waste (solid and water) can further reduce the environmental impact. As illustrated in Figure 2, EV targets several indicators, such as water emission, air emission and solid waste.

For air emission, the evaluation of the carbon footprint (tCO_2/unit of product) of the exhaust gas component (i) is highly dependent on the flowrate and composition of the exhaust air. The GWP can be obtained from EPA (2017).

$$\textit{Carbon emission of } i = \textit{Total operation hours} \times \textit{exhaust gas mass flowrateof } i \textit{ gas} \times \textit{GWP of component } i \quad (28)$$

$$CO_2 \textit{ footprint indicator, } EV_{CO_2} = \frac{\textit{minimum monthly } CO_2 \textit{ emission/total } i \textit{ monthly production}}{\textit{average monthly } CO_2 \textit{ emission/total } i \textit{ monthly production}} \quad (29)$$

As for water emission, the targeted wastewater components j should be identified. Specific wastewater component j will be assessed and identified accordingly as per the organisation's requirement. In this context, the flow rate and composition of wastewater components are needed.

$$\textit{Water emission of } j = \textit{Total operation hours} \times j \textit{ mass flowrate} \quad (30)$$

$$\textit{Waste water indicator, } EV_{WW} = \frac{\textit{minimum monthly wastewater emission/total monthly production}}{\textit{average monthly wastewater emission/total monthly production}} \quad (31)$$

Solid wastes varies between organisations. The industry players can evaluate specific solid waste s according to their requirements.

$$\textit{Solid waste, } s = \textit{Total operation hours} \times \textit{rate of waste, } s \textit{ generation per unit of product} \times \textit{rate of production} \quad (32)$$

$$\textit{Solid waste indicator, } EV_{SW} = \frac{\textit{minimum monthly generated solidwaste/total monthly production}}{\textit{average monthly generated solidwaste/total monthly production}} \quad (33)$$

In many organisations, the carbon footprint is used as an indicator in environmental performance. The EV index can be used as an indicator to reflect the environmental performance of the organisation.

$$EV = k_{EV,CO_2} \times EV_{CO_2} + k_{EV,WW} \times EV_{WW} + k_{EV,SW} \times EV_{SW} \quad (34)$$

The 4M1E model can be modified to suit the operational behaviour of the individual organisation. It is important that the plant manager and engineer have a good understanding of their process indicator and parameter prior to data collection.

5.4 Adaptive lean and green approach for continuous improvement

The static analytical method has been widely illustrated in many case study associated with output prediction. However, the static analytic method is unable to cope with the dynamic changes in the real world. The need for an adaptive analytical model with an updating algorithm with time step can be beneficial in addressing the dynamic challenges. The constant feed-in of data from the organisation can be beneficial to the model in predicting a better outcome. The continuous updating algorithm is also known as the adaptive feature that responds over time. The back-propagation (BP) algorithm, which operates based on the reverse mode of differentiation (Griewank, 2012), is incorporated in the adaptive model. The BP method is then popularized with an application on neural networks by Rumelhart et al. (1986).

In the analytical lean and green (L&G) approach, the expectations for the lean and green index (LGI) can change with time. The accessed LGI is constantly being corrected with respect to the expected index. The update rules are being utilised by the BP algorithm to compare the difference between the accessed

LGI and expected LGI for every time step. The error used is the mean squared error, which is defined as the following:

$$E = \frac{1}{2}(LGI_{accessed} - LGI_{expect})^2 \tag{35}$$

where E is the error and LGI can be an accessed or expected index. As an improvement strategy is planned, the expectation index can be targeted by the industry player prior to implementation. On the other hand, the processing data are evaluated in order to obtain the accessed LGI after the improvement strategy has been executed. The gradient descent method with chain rule is applied with the use of BP. Equation (36) indicates that the updating of AHP weight for the intermediate layers (4M1E) is computed:

$$w_{i+1} = w_i - \eta \frac{E_i - E_{i-1}}{w_i - w_{i-1}} \tag{36}$$

w_i represents the weight of the five main components (4M1E) on the current month. w_{i+1} is the new weight (target) for the next month, determined by the organisation. w_{i-1} represents the weight for the weight of the previous month, while E reflects the error having subscripts of the same notation. A proportional constant, which known as the learning rate η, is used to update the weights of each components. Wilson and Martinez (2001) discovered that the learning rate of 0.05 reflects the smallest error, indicating that this is the fastest learning rate in terms of reduction in total sum square (tss) error.

$$k_{i+1} = k_i - \eta \frac{E_i - E_{i-1}}{w_i - w_{i-1}} \times \frac{w_i - w_{i-1}}{k_i - k_{i-1}} \tag{37}$$

k_i are the weights for the each of the indicators under the 5 main components, where i is the time step for current month. The calculated error is passed down to the analytical model with the chain rule and updated for each time step.

The weight update for the components and sub-components are critical in performing performance iteration. The continuous iteration on the component's weight will reflect on the areas that can be further improved by the organisation. Thus, the updating process is important for the organisation and the analytical model to achieve the target LGI.

5.5 Lean and green case study

An application of the lean and green (L&G) model is demonstrated with a cogeneration (COGEN) case study.

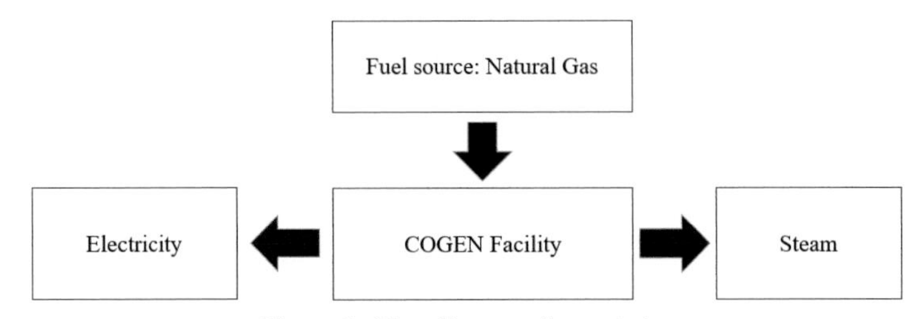

Figure 5. Flow diagram of case study.

In year 2015, the COGEN plant started its operation independently without synchronising with the national grid. The requirements of the COGEN plant are shown below:

a. Main fuel source : Natural Gas
b. Maximum electricity generation : 6.5MW @ 32degC
c. Maximum steam recovery : 16 tonnes/hour @ 10barG saturated steam
d. Total operation days : 351 days
e. Total operation hours : 8424 hours
f. Annual shutdown days : 14 days

5.5.1 Data collection and discussion

Figure 6 demonstrates the battery limit of the COGEN case study. The 5 main components (i.e., manpower, machine, material, money and environment) are reflected in Figure 6. The data collection is executed as per Figure 2. In this study, operational data of the plant from the year 2015 to the year 2018 are collected. According to the operation manager, some of the indicators (i.e., rate of recyclable defects and inventory management) in level three (in Figure 2) are not applied in this study, as reflected in Figure 2.

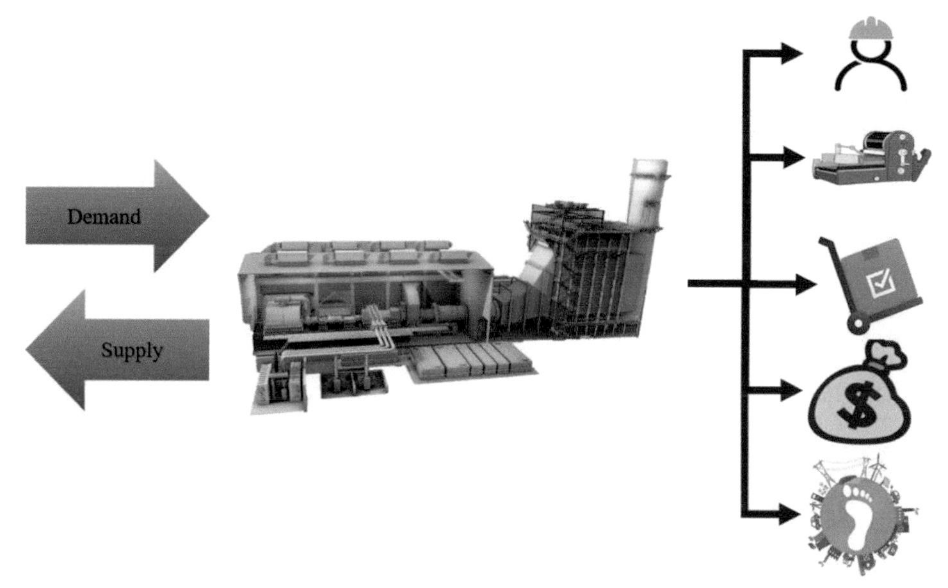

Figure 6. Boundary limit of COGEN case study (Leong et al., 2019a).

5.5.2 Analytic hierarchy process (AHP) analysis

Table 2 demonstrates the distribution of the 5 main components' (4M1E) weights. Based on the industry expert's input, money (MY) was ranked first in improving LGI of COGEN plant (26.10%), followed by environment (EV) (21.54%), manpower (MP) (18.7%), machine (MC) (18.28%) and material (MT) (15.38%). The industry expert emphasizes the importance of experienced personnel in managing the COGEN facility due to the lack of experienced talent in the COGEN industry. This has increased the challenges for an organisation in retaining the talent.

Table 3 illustrates the priorities of manpower (MP) indicators. According to the industry expert, the compliance of occupational health and safety of the employee are the utmost priority of an organisation. The operation manager added that continuous improvement training can improve the level of competency of the employee in order to allow the employee to perform their duty safely.

Table 2. Consolidated pairwise comparison matrix of 4M1E with respect to the goal.

	MP	MC	MT	MY	EV	Eigenvector	Ranking
MP	1.00	1.00	1.17	0.70	1.04	0.1870	3
MC	1.00	1.00	1.43	0.67	0.80	0.1828	4
MT	0.85	0.70	1.00	0.41	1.11	0.1539	5
MY	1.42	1.50	2.46	1.00	0.76	0.2610	1
EV	0.96	1.25	0.90	1.32	1.00	0.2154	2

CR: 0.0283.

Table 3. Pairwise comparison matrix of manpower (MP) indicators.

	MP-OT	MP-AB	MP-KPI	MP-CR	MP-LC	MP-SC	Eigenvector	Ranking
MP-OT	1.00	0.36	0.91	0.54	0.78	0.51	0.0963	6
MP-AB	2.79	1.00	3.47	1.72	1.12	0.44	0.2340	2
MP-KPI	1.10	0.29	1.00	0.67	2.05	0.80	0.1359	4
MP-CR	1.85	0.58	1.50	1.00	2.35	0.83	0.1837	3
MP-LC	1.28	0.89	0.49	0.43	1.00	0.47	0.1084	5
MP-SC	1.97	2.27	1.25	1.20	2.11	1.00	0.2417	1

CR = 0.063.

Table 4. Pairwise comparison matrix for money (MY) indicators.

	MY-OC	MY-PT	Eigenvector	Ranking
MY-OC	1.00	0.58	0.3678	2
MY-PT	1.72	1.00	0.6322	1

On top of that, Table 4 shows the dominance factor of money (MY) where profit indicators are deemed to be more significant in comparison to operation cost. Based on the operating team, the maintenance of the system has been contracted to a qualified contractor at a monthly fixed cost.

5.5.3 Lean and green index (LGI)

Figure 7 reflects the historical trend of manpower (MP) performance. The lower competency and key performance index (KPI) score are mainly due to the initiation of the new process where the MP has minimal experience in handling the COGEN process. To ensure the COGEN team can manage the operation effectively, extensive training and guidance are provided, ensuring the competency of the employee. In Figure 7, it can be observed that the improvement in competency rate has also improved the rate of achievable KPI of the employee. This reflects the overall improvement in MP performance.

Figure 8 reflects the indicator for the machine (MC). It is observed that the MC index has been improved substantially in the year 2017, as the COGEN output demand has been maximised. As informed by the operation team, the COGEN facility operated at lower demand rate in the years 2015 and 2016 due to lack of energy demand from the end-user. In the year 2017, the MC index also achieved an overall equipment effectiveness (OEE) of higher than 85%, which is considered as a world-class performance.

The input resource and output resources performance of the COGEN plant are reflected in Figure 9. The MT index tends to be lower in the years 2015 and 2016 because the demand of the COGEN facility was low. On top of that, the COGEN facility requires a minimum fuel input to maintain the idle condition. In this case, the storage and inventory requirement for the COGEN facility is not required because the fuel source (i.e., natural gas) is supplied directly from the gas pipeline.

Based on the expert's input, money (MY) index is the top priority in the COGEN facility. Referring to Figure 10, the profit rating of the year 2015 is relatively low compared to other years. Based on the operation manager's explanation, the performance in the year 2015 is mainly due to a lack of experience in operating the plant.

Lastly, Figure 11 demonstrates the performance indicator of the environment (EV). The scoring for carbon footprint indicator has improved. The overall EV index has increased as the performance of the COGEN system has improved. The EV performance is highly dependent on the MC performance, as the dry low emission (DLE) mode depends highly on the load demand of the prime mover in order to maintain the low NOx emission.

Figure 12 and Table 5 show the L&G outcome of the case study. Generally, a consistent improvement in LGI can be observed from 2015 onwards. The carbon performance reflected from EV takes time to reflect after improvements to the other components have been made. Therefore, the improvement on 4M1E

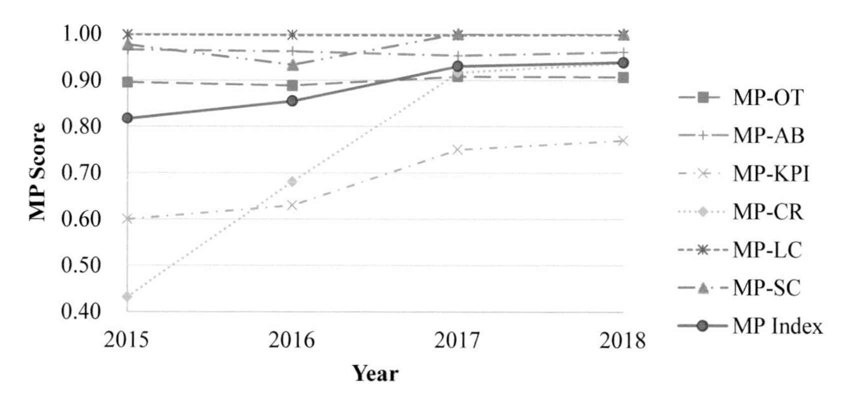

Figure 7. Manpower (MP) indicators.

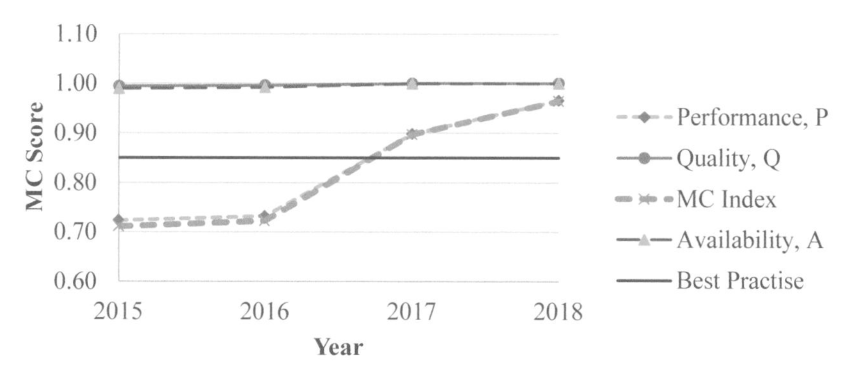

Figure 8. Machine (MC) indicators.

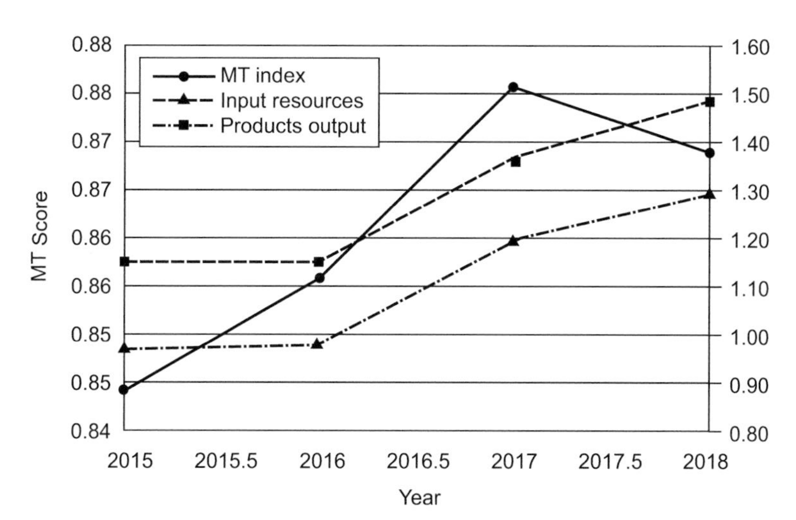

Figure 9. Material (MT) performance indicator.

shall be monitored progressively in order to ensure that the L&G approach is effectively implemented. In this case study, carbon management is challenged by the competency of human resource. Figures 7, 8, 9, 10 and 11 have shown that the operational and environmental performance improved with better employee competency rate. Thus, the carbon management strategy is highly dependent on MP planning.

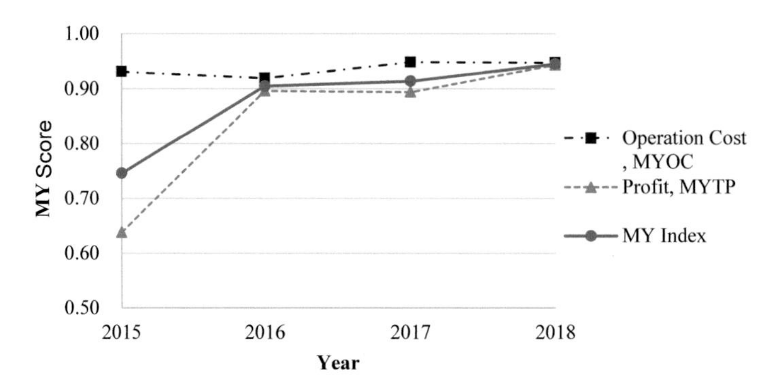

Figure 10. Money (MY) performance indicator.

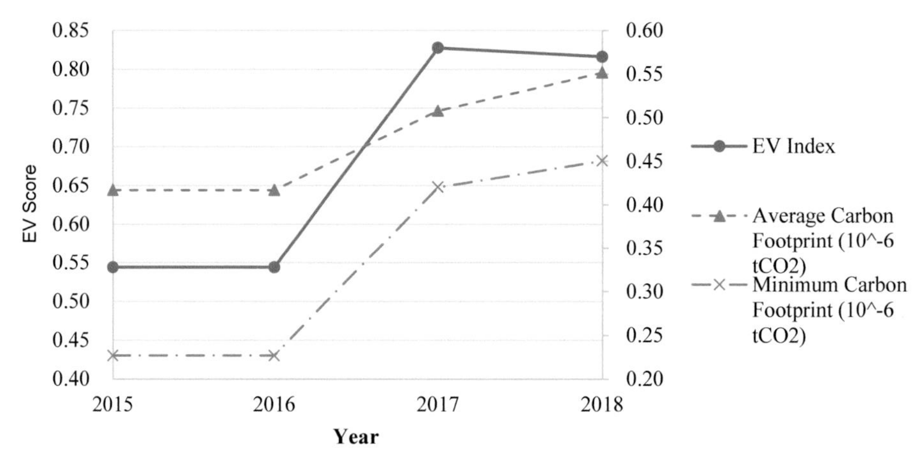

Figure 11. Environment (EV) performance indicator.

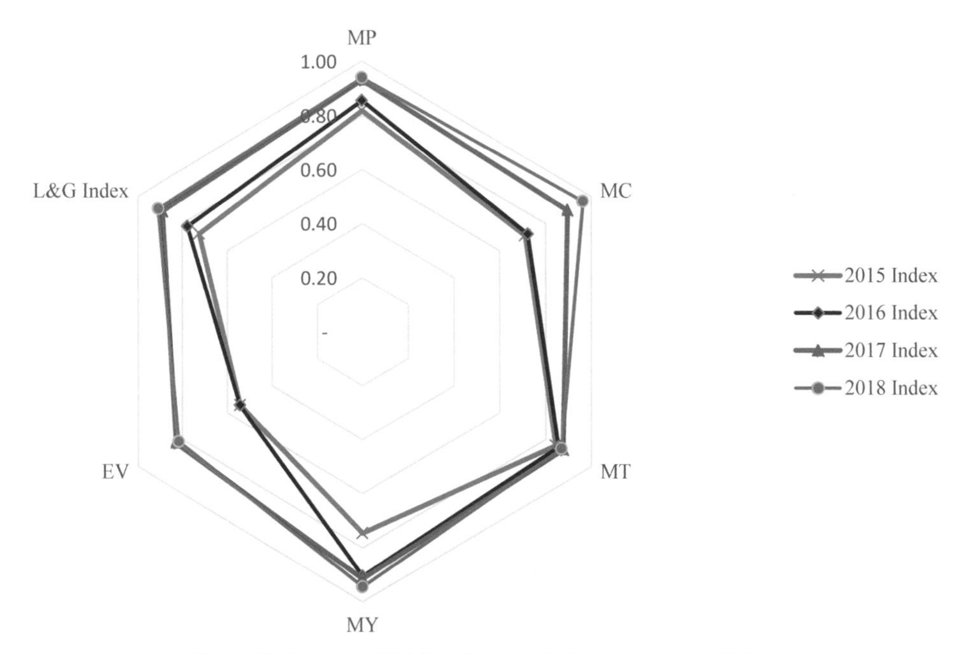

Figure 12. Progress of L&G performance index (Leong et al., 2019a).

Table 5. Summary of L&G index.

	MP	MC	MT	MY	EV	L&G index
2015 Index	0.8173	0.7108	0.8442	0.7456	0.5439	0.7244
2016 Index	0.8551	0.7219	0.8559	0.9045	0.5439	0.7767
2017 Index	0.9306	0.8967	0.8758	0.9138	0.8276	0.8894
2018 Index	0.9390	0.9640	0.8689	0.9448	0.8162	0.9078

5.5.4 Analytical continuous improvement in the lean and green index

According to the operation team, the COGEN facility has been operating under the same operation strategy since 2015. In July 2018, the backpropagation (BP) optimisation, where the COGEN facility is required to established a monthly target, was introduced. Figure 13 reflects the LGI performance after the implementation of BP optimiser. The BP optimiser evaluates the feedback on improvement done based on the five main components and its sub-indicators. The performance of new changes or action done in the operation will be reflected in the LGI in Figure 13. It can be observed that, starting from August 2018, the monthly LGI indicates an oscillation pattern as the BP optimiser provides feedback to the industrialist for improvement.

Figures 14, 15 and 16 show the weight update of 4M1E components, MP indicators and MY indicators based on BP optimisation, respectively. The figures are indicating a similar graph pattern as the BP optimise the L&G model. Referring to Figures 14, 15 and 16, the BP optimiser explores the potential improvement area by using the previous data points in the months of September and October 2018. With the continuous feedback from BP optimiser and operation team, the BP optimiser has shown signs of data convergence in the months of November and December. The convergence of the indicators is evaluated by the gradient descent method, where the optimum point of the system is identified. The outcome of the converged values is similar to the initial weight, this indicates that the experts have provided useful insights on the operation. The generated results from BP optimiser are entirely reliant on the input of the process parameter data. Thus, BP optimiser can enhance an organisation's performance through continuous process improvement.

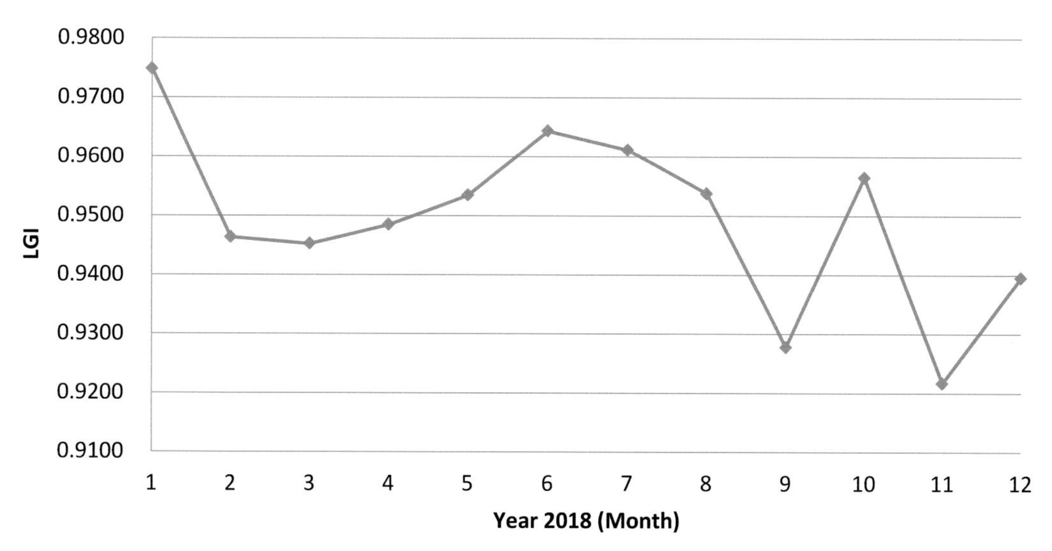

Figure 13. Monthly LGI update (BP optimisation initiated in July).

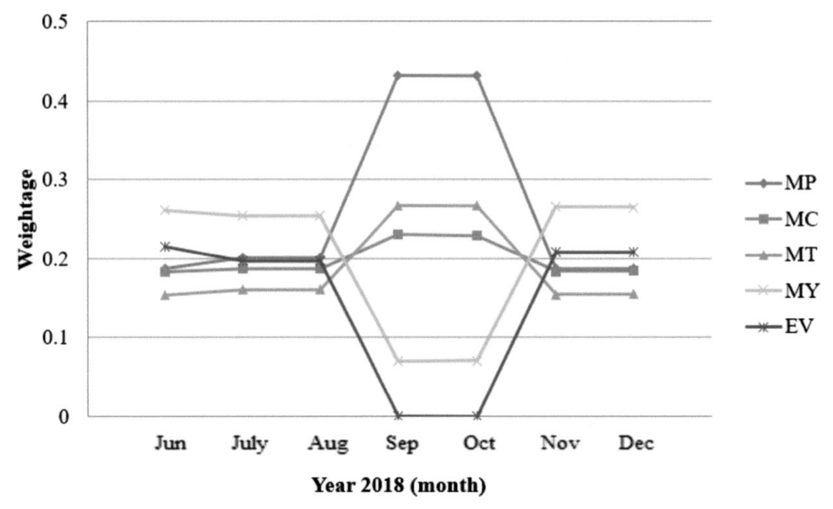

Figure 14. Backpropagation optimisation on 4M1E main components.

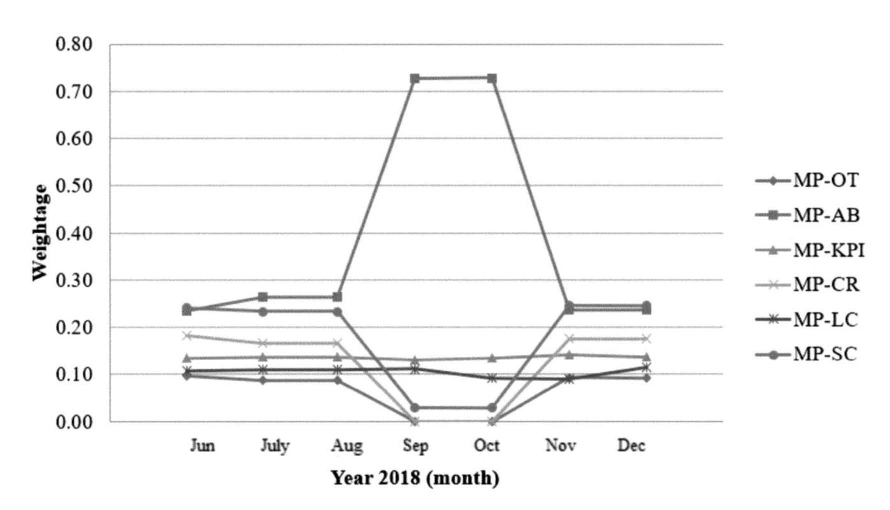

Figure 15. Backpropagation optimisation on MP indicator.

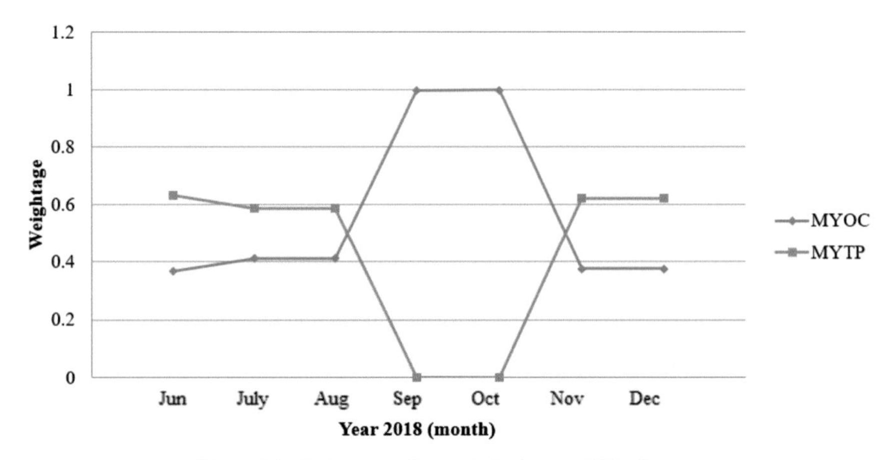

Figure 16. Backpropagation optimisation on MY indicator.

6. Conclusions

The efficiency and effectiveness of traditional manufacturing practices are no longer competitive in this global trading world. Adding on to global warming and climate change, the enforcement of stringent regulations and policies have challenged the industrialists to seek out alternative technology in order to improve their environmental performance. With the implementation of the adaptive lean and green approach, the industrialists will be assisted in the reduction of non-value-added product from the operation and environmental based on 4M1E. This will not only enhance global competitiveness, but also improve the carbon management system of the organisation. The adaptive method allows the organisation to practice continuous improvement and track the improvement rate by constantly exploring potential improvement areas. The integration of L&G approach not only reflects a positive outcome in carbon management, but also performance improvement. The adaptive approach will be able to predict and provide a better performance improvement indication with the increase in the operation database.

Acknowledgement

The authors would like to acknowledge financial support from the Ministry of Higher Education (FRGS/1/2016/TK03/MUSM/01/1). Research funding and support from Newton Fund and the EPSRC/RCUK (Grant Number: EP/PO18165/1) is also gratefully acknowledged.

Nomenclature

4M1E	Manpower, Machine, Money, Material, Environment
A	Availability
AHP	Analytical hierarchy process
BP	Backpropagation
COGEN	Co-generation
CR	Consistency ratio
DCS	Distributed control system
DfE	Design for environment
EV	Environment
EV_{CO_2}	Annual total overtime per employee
EV_{SW}	Annual total overtime per employee
EV_{WW}	Annual total overtime per employee
GM	Green manufacturing
GWP	Global warming potential
ISO	International standard organisation
KPI	Key performance index
L&G	Lean and green
LCA	Life cycle assessment
LGI	Lean and green index
LM	Lean manufacturing
MC	Machine
MP	Manpower
MP_{AB}	Annual total absent day per employee
MP_{CR}	Annual average competency rate of employee
MP_{KPI}	Annual average KPI achievable per employee
MP_{LC}	Annual late check-in time per employee
MP_{OT}	Annual total overtime per employee
MP_{SC}	Annual average safety competency rate of employee
MT	Material

MT_{DI}	Annual total overtime per employee
MT_{IN}	Annual total overtime per employee
MT_{RD}	Annual total overtime per employee
MY	Money
MY_{OC}	Total operation cost
MY_{TP}	Total operation profit
OEE	Overall equipment effectiveness
P	Performance efficiency
Q	Rate of quality product
SCADA	Supervisory control and data acquisition
SOP	Standard operating procedure

Recommended Reading List

1. Saaty, T.L. 1980. The Analytic Hierarchy Process. McGraw-Hill, New York.
2. Rumelhart, D.E., Hinton, G.E. and Williams, R.J. 1986. Learning representations by backpropagating errors. Nature 323(6088): 533–536. Doi: 10.1038/323533a0.
3. Roos, D., Womack, J. and Jones, D. 2014. The Machine that Changed the World. USA: Free Press.

References

Agan, Y., Acar, M. and Borodin, A. 2013. Drivers of environmental processes and their impact on performance: A study of Turkish SMEs. Journal of Cleaner Production 51: 23–33.

Anand, G. and Kodali, R. 2008. Selection of lean manufacturing systems using the PROMETHEE. Journal of Modelling in Management 3(1): 40–70. Doi: 10.1108/17465660810860372.

Azapagic, A., Stamford, L., Youds, L. and Barteczko-Hibbert, C. 2016. Towards sustainable production and consumption: A novel DEcision-Support Framework IntegRating Economic, Environmental and Social Sustainability (DESIRES). Computers & Chemical Engineering 91: 93–103. Doi: 10.1016/j.compchemeng.2016.03.017.

Batumalay, K. and Santhapparaj, A.S. 2009. Overall Equipment Effectiveness (OEE) through Total Productive Maintenance (TPM) practices—A study across the Malaysian industries. 2009 International Conference for Technical Postgraduates (TECOGENOS). Doi: 10.1109/tecogenos.2009.5412049.

Bhattacharya, A., Jain, R. and Choudhary, A. 2011. Green manufacturing energy, products and processes. The Boston Consulting Group, pp. 1–24.

Čuček, L., Klemeš, J.J. and Kravanja, Z. 2012. A review of footprint analysis tools for monitoring impacts on sustainability. Journal of Cleaner Production 34: 9–20. Doi: 10.1016/j.jclepro.2012.02.036.

Dües, C.M., Tan, K.H. and Lim, M. 2013. Green as the new Lean: How to use lean practices as a catalyst to greening your supply chain. Journal of Cleaner Production 40: 93–100. Doi: 10.1016/j.jclepro.2011.12.023.

Eibel, D. 2017. Green Manufacturing—An Essential Success Factor in a Globalization World.

European Commission. 2015. Managing resources. [online] European Commission–Basics–Managing resources. Available at: http://ec.europa.eu/environment/basics/green-economy/resources/index_en.htm [Accessed 31 May 2018].

[EPA] Environmental Protection Agency. 2017. Understanding Global Warming Potentials | *US EPA*. [online] US EPA. Available at: https://www.epa.gov/ghgemissions/understanding-global-warming-potentials [Accessed 14 Jan. 2019].

Gilbreth, F. 1921. Process Charts. The American Society of Mechanical Engineers. New York. USA.

Griewank, A. 2012. Who invented the reverse mode of differentiation? Documenta Mathematica, Extra Volume ISMP, 389–400.

Hallam, C. and Contreras, C. 2016. The interrelation of Lean and green manufacturing Practices: A case of push or pull in implementation. 2016 Portland International Conference on Management of Engineering and Technology (PICMET).

Helu, M. and Dornfeld, D. 2012. Principles of green manufacturing. Green Manufacturing, pp. 107–115.

Issa, T., Chang, V. and Issa, T. 2010. Sustainable business strategies and PESTEL framework. GSTF International Journal on Computing 1(1): 73–80.

Johansson, G. and Winroth, M. 2009. Lean vs. Green manufacturing: Similarities and differences.

Kar, S. 2018. 6 Challenges for HR in the Manufacturing Industry. [online] Blog.gethyphen.com. Available at: https://blog.gethyphen.com/blog/6-challenges-for-hr-in-the-manufacturing-industry [Accessed 3 Jan. 2019].

Kim, S. and Kara, S. 2012. Impact of technology on product life cycle design: Functional and environmental perspective. Leveraging technology for a sustainable world, pp. 191–196. https://doi.org/10.1007/978- 3-642-29069-5_33.

Leong W.D., Lam, H.L., Tan, C.P. and Ponnambalam, S.G. 2018. Development of multivariate framework for lean and green process. Chemical Engineering Transactions 70: 2191–2196. Doi: 10.3303/CET1870366.

Leong, W.D., Teng, S.Y., How, B.S., Ngan, S.L., Lam, H.L., Tan, C.P. and Ponnambalam, S.G. 2019a. Adaptive analytical approach to lean and green operations. Journal of Cleaner Production, https://doi.org/10.1016/j.jclepro.2019.06.143.

Leong, W.D., Lam, H.L., Ng, W.P.Q., Lim, C.H., Tan, C.P. and Ponnambalam, S.G. 2019b. Lean and green manufacturing—a review on application and impact. Process Integration and Optimisation for Sustainability, 5–23.

Maloni, M. and Brown, M. 2006. Corporate social responsibility in the supply chain: An application in the food industry. Journal of Business Ethics 68(1): 35–52.

Maruthi, G. and Rashmi, R. 2015. Green manufacturing: It's tools and techniques that can be implemented in manufacturing sectors. Materials Today: Proceedings 2(4-5): 3350–3355. https://doi.org/10.1016/j. matpr.2015.07.308.

Mittal, V. and Sangwan, K. 2014. Prioritizing drivers for green manufacturing: Environmental, social and economic perspectives. Procedia CIRP 15: 135–140.

Moradi-Aliabadi, M. and Huang, Y. 2018. Decision support for enhancement of manufacturing sustainability: A hierarchical control approach. ACS Sustainable Chemistry & Engineering 6(4): 4809–4820. Doi: 10.1021/acssuschemeng.7b04090.

Nawanir, G., Fernando, Y. and Teong, L. 2018. A second-order model of lean manufacturing implementation to leverage production line productivity with the importance-performance map analysis. Global Business Review 19(3_suppl): S114–S129.

Ng, W.P.Q., Lam, H.L., Varbanov, P.S. and Klemeš, J.J. 2014. Waste-to-energy (WTE) network synthesis for municipal solid waste (MSW). Energy Conversion and Management 85: 866–874. Doi: 10.1016/j.enconman.2014.01.004.

Ngan, S.L., Promentilla, M.A.B., Yatim, P. and Lam, H.L. 2018. A novel risk assessment model for green finance: The case of malaysian oil palm biomass industry. Process Integration and Optimization for Sustainability. Doi: 10.1007/s41660-018-0043-4.

Pampanelli, A., Found, P. and Bernardes, A. 2014. A lean & green model for a production cell. Journal of Cleaner Production 85: 19–30.

Porter, M.E. and van der Linde, C. 1995. The daily local news: Serving chester county. Harv. Bus. Rev. 119–134. https://doi.org/10.1016/0024-6301(95)99997-E.

Rodríguez, D., van Landeghem, H., Lasio, V. and Buyens, D. 2017. Determinants of job satisfaction in a lean environment. International Journal of Lean Six Sigma 8(2). Doi: 10.1108/ijlss-01-2016-0002.

Rumelhart, D.E., Hinton, G.E. and Williams, R.J. 1986. Learning representations by back-propagating errors. Nature 323(6088): 533–536. Doi: 10.1038/323533a0.

Saaty, T.L. 1980. The Analytic Hierarchy Process. McGraw-Hill, New York.

Saaty, T.L. 2012. Decision Making for Leaders: The Analytic Hierarchy Process for Decisions. RWS Publications, PA, USA.

Singh, R., Shah, D., Gohil, A. and Shah, M. 2013. Overall Equipment Effectiveness (OEE) calculation—automation through hardware & software development. Procedia Engineering 51: 579–584.

[SDG] Sustainable Development Goals. 2019. Sustainable Development Goals. Sustainable Development Knowledge Platform. [online] Sustainabledevelopment.un.org. Available at: https://sustainabledevelopment.un.org/?menu=1300 [Accessed 2 Jan. 2019].

[UNIDO] United Nation Industrial Development Organisation. 2010. Motor Systems Efficiency Supply Curves. United Nations Industrial Development Organisation (UNIDO), Vienna, Austria.

Verrier, B., Rose, B., Caillaud, E. and Remita, H. 2014. Combining organizational performance with sustainable development issues: The Lean and Green project benchmarking repository. J Clean Prod. https://doi.org/10.1016/j.jclepro.2013.12.023.

Williamson, K., Satre-Meloy, A., Velasco, K. and Green, K. 2018. Climate change needs behavior change. Yale Progr Clim Chang Commun.

Wilson, D.R. and Martinez, T.R. 2001. The need for small learning rates on large problems. IJCNN'01. International Joint Conference on Neural Networks. Proceedings (Cat. No.01CH37222). Doi: 10.1109/ijcnn.2001.939002.

Womack, J. and Jones, D.T. 1994. From lean production to the lean enterprise. Harvard Business Review 72: 93–104.

World Bank. 2012. Inclusive Green Growth: The Pathway to Sustainable Development. Washington, DC. © World Bank. https://openknowledge.worldbank.org/handle/10986/6058 License: CC BY 3.0 IGO.

Wu, L., Subramanian, N., Abdulrahman, M., Liu, C., Lai, K., Pawar, K., Wu, L., Subramanian, N., Abdulrahman, M.D., Liu, C., Lai, K. and Pawar, K.S. 2015. The impact of integrated practices of lean, green, and social management systems on firm sustainability performance—evidence from chinese fashion auto-parts suppliers. Sustainability 7: 3838–3858. https://doi.org/10.3390/su7043838.

Zimmer, L. 2000. Get Lean to boost profits. Forming and Fabricating 7: 12–23.

CHAPTER 15

Advancements, Challenges and Opportunities of Li-ion Batteries for Electric Vehicles

Qianran He[1] and *Leon Shaw*[2,*]

1. Introduction

Since Karl Benz designed and built the first motorized tricycle powered by an internal-combustion engine in 1885, the time of automobiles has begun. From then on, human life was modernized by gasoline-powered vehicles, and cars were labelled as "the machine that changed the world". It is believed that electric vehicles will again reshape the world, just like what internal combustion engine and gas vehicles did in the second industrial revolution.

Integrated with big data and new technology, like autonomous driving, the electric vehicles are acting not only as high performance transportation tools but also basic units of distributed energy, playing an important role in the smart grid system, and at the same time providing an interactive access port for human beings to connect with the smart city, IoT (Internet of Things), and other future scenarios.

Most importantly, battery powered electric vehicle is also one of the most promising clean transportation solutions for the energy and environmental crises faced by our next generation. Based on the fact of limited fossil fuel reservation and the heavy pollution with the consumption of traditional energy, it becomes very urgent to have an alternative energy source to power automobiles that is clean, safe and efficient. There are several types of new energy vehicles other than electric vehicles, like hydrogen vehicles and biofuel vehicles. However, electric vehicles powered by lithium ion batteries as well as hybrid and plug-in hybrid vehicles have been demonstrated to be the practical candidates and have already dominated the new energy vehicles market.

The use of electric vehicles can reduce CO_2 emission by improving energy efficiency and increasing the use of renewable energies. Electric vehicles convert 59–62% of the electrical energy from the grid to power at the wheels, whereas gasoline vehicles only convert 17–24% of the gasoline energy to power at the wheels (Hekkert et al., 2005). Because of the higher energy efficiency of electric vehicles and the possibility to increase the use of electricity produced from renewable resources (i.e., from hydro-, solar-, or wind-powered plants), the life cycle greenhouse gas emission can be reduced. Therefore, broad market penetration of electric vehicles can move society in a more sustainable direction.

[1] 2801 S. King Dr. APT 1314.
Email: qhe6@hawk.iit.edu
[2] 10 W. 32nd St.
* Corresponding author: lshaw2@iit.edu

Because of the aforementioned advantages, the development of electric vehicles is highly emphasized worldwide by many countries and leading companies. There are more than 15 countries that have announced their timelines of banning fossil vehicles before 2050 or even earlier. Several major car manufacturers, like Toyota, Volkswagen and GM, have also announced their deadlines of phasing out the traditional gas engine vehicles production line.

While the governments and companies are making schedules of welcoming the all-electric vehicles era, we are still facing some technological issues in electric vehicles that need further improvement. For instance, the relatively high cost of production, slow charging rate, limited battery life, safety in extreme conditions, inadequate distribution of charge stations and the corresponding services are examples of the issues that need to be solved in order to enable broad market penetration of electric vehicles. In spite of these challenges, we believe that the future route of electric vehicles is becoming clear and that research/ development of battery technologies will sooner or later bring the era of electric vehicles.

2. Working Principles of Li-ion Batteries

Li-ion batteries (LIBs), the key device to power electric vehicles, have attracted attention, enthusiasm and intensive development activity throughout recent years because of their high output voltage, high specific energy, long cycle life and absence of memory effect (Tarascon and Armand, 2001). It's important for one to understand the mechanism of how batteries provide the energy to power the wheels and keep up with the latest advancements made in LIBs and their applications in the area of electric vehicles. The following briefly highlights the operation principles of LIBs.

A typical Li-ion battery is composed of two main building blocks: The anode and cathode, with other supporting components which are the separator, electrolyte, and current collectors. The anode (negative electrode) is the electrode that stores lithium ions during charging and releases lithium ions during discharging. Usually, carbon (graphite) is used as the anode material, providing intercalation sites for lithium ions and allowing lithium de-intercalation without much change in the structure (Deng, 2015). The cathode (positive electrode) is the electrode that gives up lithium ions to the anode during charging and collects lithium ions back during discharging. Lithium metal oxides (e.g., $LiCoO_2$) and multi-metal oxides, such as $Li(Ni_xMn_yCo_z)O_2$, where $x + y + z = 1$ (NMCs), and $Li(Ni_xCo_yAl_z)O_2$, where $x + y + z = 1$ (NCAs), have been studied extensively and are the most commonly used cathode materials, acting as the lithium source of LIBs (Ashuri et al., 2019; Sahni et al., 2019). The separator is a porous membrane, basically made of polypropylene (PP) and polyethylene (PE), which works as an insulator that prevents direct contact between the anode and cathode while allowing the lithium ions to pass through. The electrolyte is composed of lithium salt(s) dissolved in an organic solvent and plays the role of transporting lithium ions back and forth between the anode and cathode (Goodenough and Kim, 2010). Current collectors are metal foils that allow electrons to flow from the cathode to anode via the external circuit during charge and vice versa during discharge. Due to the difference in the working potential, a copper foil is used as the anode current collector and an aluminum foil is utilized for the cathode current collector.

Figure 1 provides an illustration on how a lithium ion battery works. In the charging process, the external electrical power applied on the battery will force lithium ions to leave the cathode material crystal lattice, pass through the separator, migrate to the anode side, and embed into the underlying structure of the carbon (graphite) layer. While lithium ions are moving from the cathode to anode through the separator, the electrons released from the oxidation of the cathode material flow in the same direction but through the external circuit (because the electrically insulated separator blocks electron transfer within the battery). In the discharging process, the electric potential difference between the anode and cathode will drive the anode side lithium ions move back to the cathode side. As the positively charged lithium ions leave the carbon host structure, the electrons released from the oxidation of the anode material also flow back to the cathode through the external circuit, which provide electricity for the usage of electric vehicles.

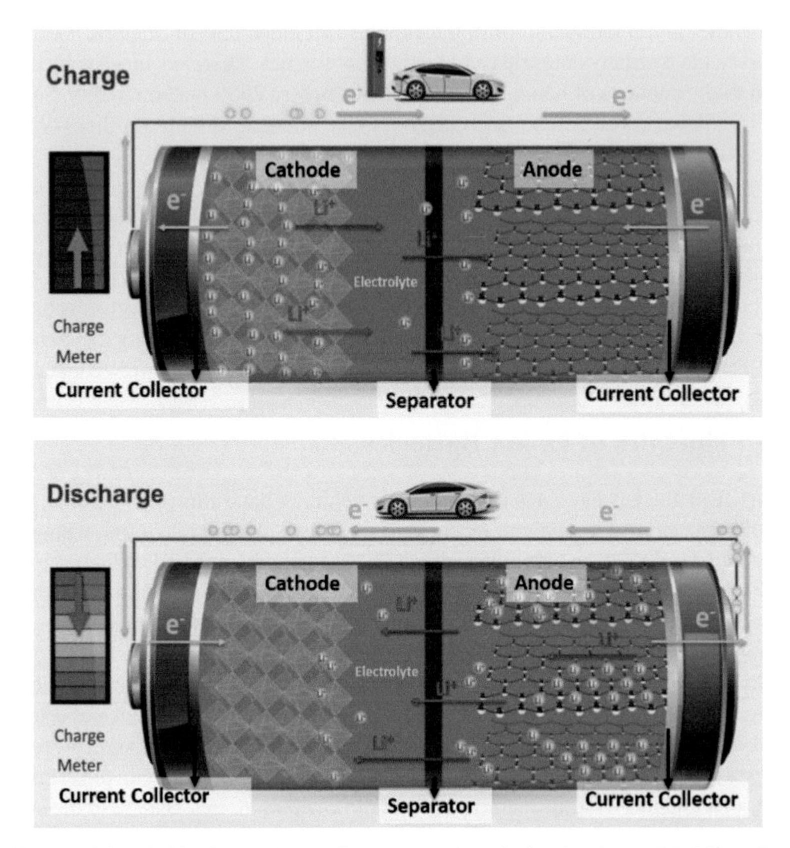

Figure 1. Working principles of Li-ion battery (annotation on screenshot of animation from DOE-Office of Energy Efficiency & Renewable Energy: https://www.energy.gov/eere/articles/how-does-lithium-ion-battery-work animation created by Sarah Harman and Charles Joyner).

3. Challenges Faced by Li-ion Batteries for Electric Vehicles

The application of Li-ion batteries (LIBs) in electric vehicles is at an early stage of development compared to the internal combustion engine in conventional vehicles. Although significant breakthroughs have been made in technologies of LIBs in the past decades, LIBs that can provide higher energy density, lower cost, faster charging ability as well as safer and better performance at low temperature are in constant demand from consumers. In particular, the following advancements are needed to lower the barriers for widespread acceptance of electric vehicles by consumers.

First, the energy density of LIBs must be doubled in order to compete with the current market of fossil fuel vehicles. The-state-of-the-art automotive LIB packs only have 130–140 Wh/kg of gravimetric energy density and 210 Wh/l of volumetric density, while the U.S. Department of Energy projects a need of 235 Wh/kg and 500 Wh/l at the battery pack level to achieve 500 km driving distance for electric vehicles (Schmuch et al., 2018). Thus, the relatively low energy density of LIBs is still a major limitation in its massive application. Possible solutions include introducing Ni-rich and Li-rich oxides for cathode materials (Myung et al., 2017) and incorporate high specific capacity materials (such as Si) in the anode (Ashuri and Shaw, 2016).

Second, the cost along the LIBs value chain for electric vehicles is still at a high level. The cost could be divided into three major parts: Material cost, fabrication cost and storage/supply cost. To compete against conventional fossil fuel vehicles with combustion engines, the cost of the LIB pack, currently priced at 200–250 U.S. dollar per kWh, needs to fall below 120–150 U.S. dollar per kWh (Nykvist and Nilsson, 2015). It is encouraging to see a continuous decline of the battery pack cost for electric vehicles.

Efforts have been made to decrease the material cost, for example, by producing low cobalt (Co) or Co-free LIBs, since Co is the most costly raw material for the cathode (Shaw and Ashuri, 2019). Besides, the scale-up automation production line has also greatly lowered the fabrication cost of LIBs.

Third, the fast charging capability of electric vehicles is one of the most critical demands from the user side. The desired rate that car manufactures are pursuing is 5–20 minutes of charge to achieve 80% of the charge state, however, current super charge stations need about 40 minutes to fulfill the requirement. At the same time, as the charge rate is increased, the risk of battery capacity deterioration has also increased along with the concern regarding several safety issues caused by the fast charge (Somerville et al., 2016). One of the biggest problems is the formation of Li dendrites, which is highly probable to happen due to the high current polarization at the anode (Uhlmann et al., 2015). The continuous growth of Li dendrites might penetrate through the separator, causing battery internal circuit short or even triggering explosion of the cell and vehicle. In addition, low temperatures, especially below –20 ºC, will also increase the possibility of Li plating because of the sluggish kinetics of Li diffusion at low temperature (Liu et al., 2016).

In summary, the application of LIBs in electric vehicles is facing several significant challenges. LIBs with higher energy density, lower pack cost, faster charging capability as well as safer performance during fast charging and at low temperatures are needed in order to satisfy the current market growth of electric vehicles.

4. Status and Advancements in Anode Materials

There are three categories of anode materials for LIBs, grouped according to their lithiation mechanisms: (1) insertion materials, (2) alloying materials, and (3) conversion materials. The most widely-used material for current electric vehicles is the insertion type, such as graphite. Graphite holds a low lithium intercalation potential, at about 0.2 V vs. Li^+/Li, and yields a theoretical specific capacity of 372 mAh/g in its fully lithiated state (LiC_6) (Deng, 2015). The reaction taking place during the lithiation and delithiation of graphite is (Yata et al., 1993):

$$yC + xLi^+ + xe^- \leftrightarrow Li_xC_y \tag{1}$$

Actually, the LIBs using graphitic carbon as anode material didn't provide 372 mAh/g of specific capacity because of the shielding effects of the electrons intercalated in the graphene layers, which prohibit further entry of following lithium ions to the structure (Inagaki et al., 2003). The limited lithium intercalation in graphite, thus, controlled the volume expansion during lithiation to only 12%, which helps to keep the host structure stable through the cycles and results in a long cycle life and high capacity retention (Yata et al., 1993). In addition, graphite also has good electron conductivity and thermal stability, making it an excellent anode material for application in LIBs (Amine et al., 2016).

However, with the growing demand for higher energy density LIBs, people are looking for alternative anode materials because the limited intercalation mechanism of graphite also results in a limited specific capacity. In contrast, silicon (Si) has attracted tremendous attention as one of the most promising anode materials for the next-generation LIBs because of its extremely high specific capacity and abundant amount on earth. The lithiation mechanism of Si is different from the intercalation mechanism of graphite. It is an alloying process during lithiation and de-alloying process in de-lithiation. Equations (2a) and (2b) illustrate the lithiation and de-lithiation chemistry of Si at room temperature (Obrovac and Chevrier, 2014).

$$(c)Si + xLi^+ + xe^- \leftrightarrow (\alpha)Li_xSi \tag{2a}$$

$$(\alpha)Li_xSi + yLi^+ + ye^- \leftrightarrow (c)Li_{15}Si_4 \tag{2b}$$

where the prefix "(α)" and "(c)" refer to amorphous phase and crystalline phase, respectively, suggesting the crystallinity and phase changes in the lithiation/de-lithiation process. At room temperature, the fully lithiated Si could deliver 3,572 mAh/g of specific capacity, which is almost 10 times that of graphite

(Ashuri and Shaw, 2016). However, a large volume expansion of Si (> 300%) comes along with the large specific capacity (Ashuri and Shaw, 2016). The huge volume expansion and shrinkage in each cycle will result in continuous formation and collapse of the solid electrolyte interface (SEI), causing problems of particle disconnection, continuous consumption of electrolyte and, thus, the decay of the capacity (Wu et al., 2012). To solve the problems of the huge volume expansion and low intrinsic electric conductivity of Si, a wide range of strategies have been investigated. These strategies include nanostructure design of Si material (Kim et al., 2010; Liu et al., 2012; Yang et al., 2015; Graetz et al., 2003; Liu et al., 2014), Si porous structures (Yi et al., 2013; Liu et al., 2013; Song et al., 2014; Hwang et al., 2015), Si-based nanocomposites (Li et al., 2015; Lee, Shim and Kim, 2014; Han et al., 2015), conductive coatings (Wang et al., 2014; Wan et al., 2014; Pan et al., 2014), addition of electrolyte additives (Choi et al., 2006), and use of novel binders (Magasinski et al., 2010; Kovalenko et al., 2011). Among these strategies, the best performance of Si anodes with high specific capacity and long cycle life is attained through well-designed internal structures by combining multiple strategies, including nanoscale building blocks, conductive coatings and engineered void space. The examples of such Si material designs are pomegranate-inspired Si design (Liu et al., 2014) and carbon-coated, graphene-supported, micro-sized porous Si (Yi et al., 2014). Both of these designs not only have high specific capacities (1,160 mAh/g after 1,000 cycles for pomegranate-inspired Si design and 1,150 mAh/g after 100 cycles for carbon-coated, graphene-supported, micro-sized porous Si) with long life, but also possess high areal capacities (> 3.0 mAh/cm^2) at high mass loading. For example, Liu et al. (Liu et al., 2014) reports that the Si pomegranate anodes have high specific capacity (> 1,000 mAh/g) for the entire 1,000 cycles when the mass loading is ~ 0.2 mg/cm^2 and the current density is at 2.1 A/g. In addition, at high mass loading (such as 3.12 mg/cm^2) the areal capacity is higher than 3.0 mAh/cm^2 while the specific capacity is still maintained at ~ 1,000 mAh/g when the current density is 0.22 A/g or 0.7 mA/cm^2 in terms of the current per unit area (Liu et al., 2014). Note that the areal capacity of 3.0 mAh/cm^2 reported by Liu et al. (2014) is comparable to those of commercial LIBs with graphite anodes. Thus, the use of Si-based anodes will result in lighter batteries.

Figure 2 illustrates why the Si anode design with nanoscale building blocks, conductive coatings and engineered void space could provide the best electrochemical performance among various strategies. Here, the conductive carbon shell outside the Si nanoscale building blocks serves two functions: (i) offering a super highway for both electron and Li-ion transport during cycling and, thus, providing high power densities and (ii) confining the Si volume expansion and shrinkage inside the shell during cycling, thereby providing an interface for the formation of stable SEI layers, leaving no stresses to the binder, avoiding electrode pulverization and maintaining good contact with the current collector for long-lasting cycle life. The presence of engineered void space permits Si nanoscale building blocks to expand inwards without causing too much stress to the conductive shell which, in most cases, is made of brittle carbon. The Si nanoscale building blocks inside the conductive shell will reduce volume expansion per particle, minimizing the tensile stress at the conductive shell during lithiation. Therefore, a combination of the nanoscale Si building blocks, outer conductive shell and inner engineered voids will offer excellent electrochemical performance, as demonstrated by Liu et al. (2014) and Yi et al. (2014).

Our recent study (Ashuri et al., 2016) has offered direct evidence for the superior performance derived from a combination of Si nanoscale building blocks, conductive coatings and engineered void space. As shown in Figure 3 (Ashuri et al., 2016), the anode made of micro-Si particles exhibits a very high first discharge capacity (3,600 mAh/g). The anode made of nano-Si particles have a similar first discharge capacity. However, the specific capacity of the micro-Si cell drops to below 100 mAh/g only after 20 cycles, while the nano-Si cell has a better performance than the micro-Si cell, but its capacity drops below 500 mAh/g after 40 cycles. In contrast, hollow Si nanoparticles coated with a carbon shell, denoted as HSi@C nanoparticles (without MgO), have a far better cycle stability than both micro-Si and nano-Si cells. For example, at the 60th cycle, the specific capacity of the micro-Si cell is near zero and that of the nano-Si cell is about 200 mAh/g, whereas the specific capacity of HSi@C (without MgO) is at 750 mAh/g, much higher than both micro-Si and nano-Si cells (Ashuri et al., 2016). The rapid capacity fading of the micro-Si cell is likely due to Si particle cracking, whereas the nano-Si cell has a better performance because Si nanoparticles (70–130 nm) have a lower propensity for cracking. However, the

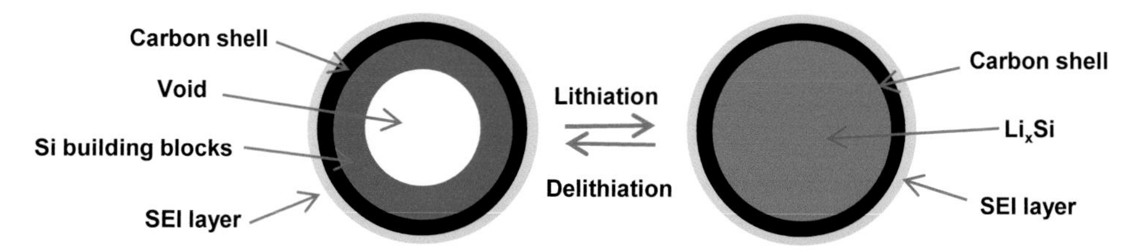

Figure 2. Schematic of a Si particle design with nanoscale building blocks, a carbon shell and engineered void space inside. The volume expansion and shrinkage of the Si anode will take place inside the carbon shell, leading to a stable SEI layer and elimination of continuous Li consumption.

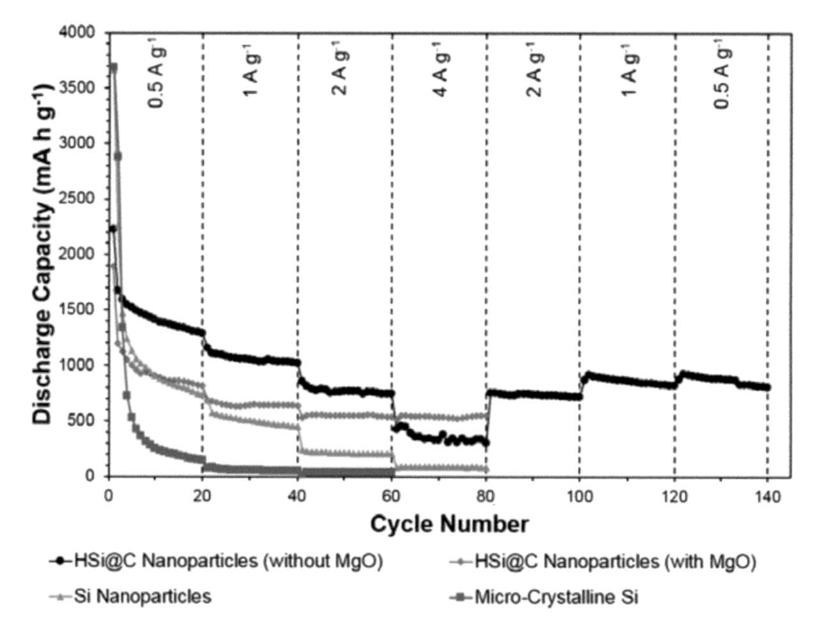

Figure 3. Discharge capacity vs. cycle number for various Si anodes as indicated. HSi@C half cells have the best cyclic stability, whereas micro-Si half cells have the worst stability. See the text for detailed discussion. Reproduce with permission (Ashuri et al., 2016) ©2019, Elsevier.

nano-Si cell still exhibits inferior performance to the HSi@C cell (without MgO) because nanoparticles, although they may not crack, still undergo volume expansion and shrinkage during lithiation and delithiation, which can result in repeated formation and fracture of the SEI layer and, thus, increased impedance, continued consumption of the electrolyte and capacity fading. In sharp contrast to micro-Si and nano-Si cells, the void space inside the HSi@C particle can allow Si to expand during lithiation with little or no change in the volume of the entire particle, thereby providing a stable SEI layer without continuous Li consumption of the electrolyte while keeping the particle in contact with conductive pathways for continuous charge/discharge operation (Ashuri et al., 2016). Hollow Si nanoparticles coated with a carbon shell and containing MgO impurity, denoted as HSi@C nanoparticles (with MgO) in Figure 3, have poorer performance than HSi@C nanoparticles (without MgO) because of the presence of MgO impurity from the synthesis process (Ashuri et al., 2016).

It should be noted that, in spite of successful lab scale modifications of Si anodes, as reported in many publications (Kim et al., 2010; Liu et al., 2012; Yang et al., 2015; Graetz et al., 2003; Liu et al., 2014; Yi et al., 2013; Liu et al., 2013; Song et al., 2014; Hwang et al., 2015; Li et al., 2015; Lee, Shim and Kim, 2014; Han et al., 2015; Wang et al., 2014; Wan et al., 2014; Pan et al., 2014; Choi et al., 2006; Magasinski et al., 2010; Kovalenko et al., 2011; Yi et al., 2014; Ashuri et al., 2016), industry so far is only using Si as small fraction additive (< 10 wt%) in the state-of-the-art graphite anodes (Blomgren,

2017). The high synthesizing cost and low mass loading of nanostructured Si hindered its way of mass application. Efforts were made to reduce the cost of synthesizing Si anode. The most recent Si-anode work using inexpensive micron size particles has shown reversible capacity of 1050 mAh/g after 200 cycles (Jana and Singh, 2019). In this work, polycrystalline micron-sized silicon powders were used as the starting material and ball milled, followed by subsequent mixing with superconducting carbon, dispersing them in a liquid carbon precursor and then freezing the constituents for the final carbonization. Another work started with micron silicon powders has used etching to create silicon nanofibers followed by carbonization, showing ~ 778 mAh/g specific capacity even after 200 cycles (Jana, Ning and Singh, 2018). However, further breakthroughs are expected in the application of high Si loading in the anode to fully harvest the enormous potential of Si anodes for LIBs.

5. Status and Advancements in Cathode Materials

Since the commercialization of LIBs in 1990s, $LiCoO_2$ has acted as the dominant cathode material for LIBs in most of the application of portable devices. With a reversible Li extraction voltage at 4.3 V vs. Li^+/Li, $LiCoO_2$ is giving about 150 mAh/g of specific capacity (Ashuri et al., 2019; Chen et al., 2019). The lithium intercalation and de-intercalation of $LiCoO_2$ as well as other layered oxides $LiMO_2$ with M = Cr, Co, Ni or other transition metals are suggested as Equation (3) below (Goodenough and Park, 2013).

$$LiMO_2 - xLi^+ - xe^- \leftrightarrow Li_{1-x}MO_2 \tag{3}$$

However, the unstable structure of $LiCoO_2$ at high potentials and the high cost of cobalt (Co) as raw material make it no longer suitable for large scale Li-ion battery applications, especially for electric vehicles (Shaw and Ashuri, 2019). Thus, $LiMO_2$-type layered metal oxides that incorporate with nickel, cobalt and manganese (NMC) or nickel, cobalt and aluminum (NCA) have been explored and successfully used for electric vehicles (Shaw and Ashuri, 2019). These blended transition metal oxides with different metal contents not only lower the total cost of raw material, but also achieve higher capacity by increasing Ni content due to the larger charge transfer contribution from Ni^{2+}/Ni^{4+} redox couple (Noh et al., 2013). For example, NMC811 ($LiNi_{0.8}Mn_{0.1}Co_{0.1}O_2$) gives about 200 mAh/g of specific capacity, while NMC333 ($LiNi_{1/3}Co_{1/3}Mn_{1/3}O_2$) typically offers 160 mAh/g (Noh et al., 2013). However, one drawback of the Ni rich layered cathodes is the poor thermal stability and structural stability in the charged state. Composition arrangements and core-shell structures have been designed to reduce the safety hazard of these high Ni content materials (Myung et al., 2017). Coatings have also been widely pursued in order to improve the cycle stability of NMC and NCA cathodes (Shaw and Ashuri, 2019).

Many investigations have established that proper coatings can provide multiple functions in improving the cycle stability of NMC and NCA cathodes. It is known that $LiPF_6$-based electrolyte always contains a small amount of water which can cause breakdown of the electrolyte accompanying with HF generation (Myung et al., 2005). The generated HF attacks the NMC, leading to dissolution of the NMC surface into the electrolyte, whereas LiF is deposited on the surface of the NMC, resulting in an increased impedance of the cell (Myung et al., 2005). Thus, a well-designed coating is critical to achieve multi-functionalities, including: (i) scavenging HF in the electrolyte, (ii) scavenging water molecules in the electrolyte and, thus, suppressing HF propagation during charge/discharge cycles, (iii) serving as a buffer layer to minimize HF attack on NMCs and suppress side reactions between NMCs and the electrolyte, (iv) hindering phase transitions and impeding loss of lattice oxygen, (v) preventing microcracks in NMC particles to keep participation of most NMC material in lithiation/de-lithiation, and (vi) enhancing the rate capability of NMC cathodes (Shaw and Ashuri, 2019). Our recent study on nano-$LiCoO_2$ with tunable $LiAlO_2/Al_2O_3$ coating reveals that integration of nano-$LiCoO_2$ and $LiAlO_2/Al_2O_3$ coating can lead to improvements in three electrochemical properties simultaneously, i.e., high specific capacity, good cycle stability, and fast charge rate (Chen et al., 2019). As shown in Figure 4, nano-$LiCoO_2$ with $LiAlO_2/Al_2O_3$ coating can be charged/discharged at the 3C rate (1C = 145 mAh/g) for more than 400 cycles with a final specific capacity of 128 mAh/g. This result represents a significant improvement

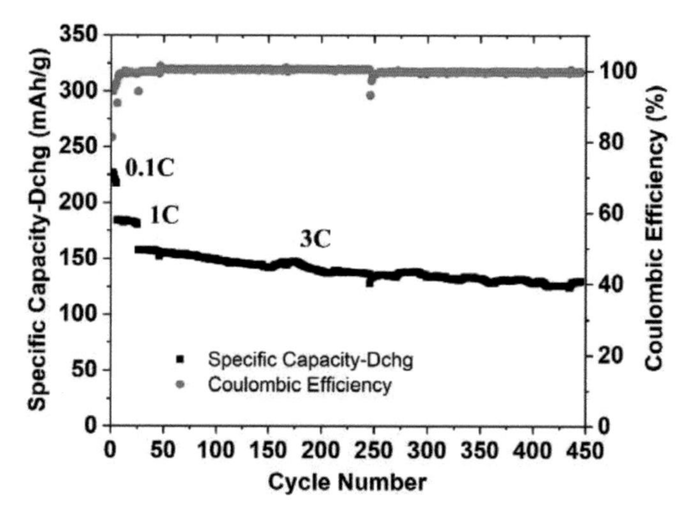

Figure 4. Discharge capacity and coulombic efficiency of $LiAlO_2/Al_2O_3$-coated nano-$LiCoO_2$ versus cycle number. The cell was charged/discharged for 5 cycles at 0.1C, then 20 cycles at 1C, and finally 425 cycles at 3C. Reproduce with permission (Chen et al., 2019) ©2019, American Chemical Society.

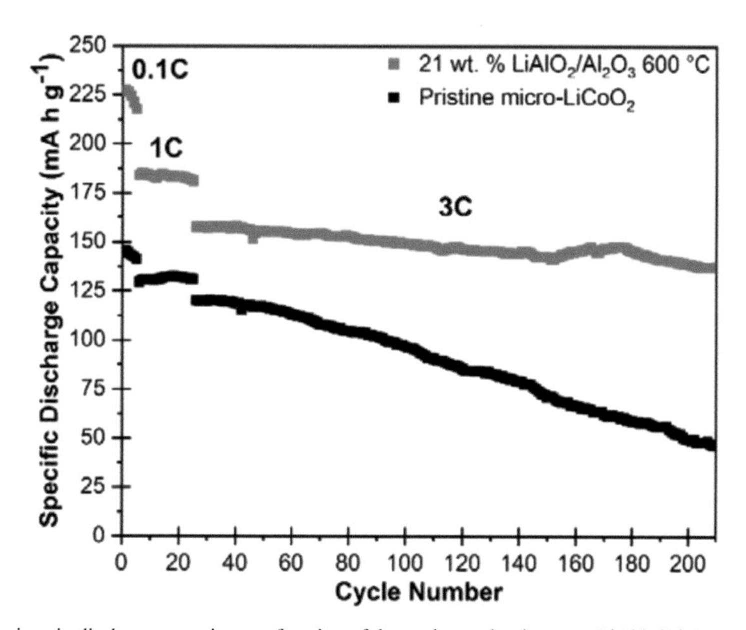

Figure 5. Comparison in discharge capacity as a function of the cycle number between $LiAlO_2/Al_2O_3$-coated nano-$LiCoO_2$ and micro-$LiCoO_2$. The cells were charged/discharged for 5 cycles at 0.1C, then 20 cycles at 1C, and finally 185 cycles at 3C. Reproduce with permission (Chen et al., 2019) ©2019, American Chemical Society.

over uncoated nano-$LiCoO_2$ which exhibits extremely fast capacity decay with a capacity about 90 mAh/g for just 20 cycles at 3C (Blomgren, 2017). The result of nano-$LiCoO_2$ with $LiAlO_2/Al_2O_3$ coating also represents significant advancement over micro-$LiCoO_2$ when the charge/discharge rate is high. As shown in Figure 5, $LiAlO_2/Al_2O_3$-coated nano-$LiCoO_2$ not only possesses higher specific capacities, but also has much better capacity retention than micro-$LiCoO_2$. The higher specific capacity of the $LiAlO_2/Al_2O_3$-coated nano-$LiCoO_2$ is due to its substantially larger electrode/electrolyte interface area and, thus, faster lithiation and de-lithiation when compared to micro-$LiCoO_2$ (Chen et al., 2019). The better capacity retention is due to the protective function of the $LiAlO_2/Al_2O_3$ coating.

Another type of the most successful alternative cathode material that has been applied in electric vehicles is $LiFePO_4$, which is much cheaper than $LiCoO_2$ and has significantly improved the safety and lifetime of LIBs (Yamada and Hinokuma, 2002). Nevertheless, there are also some challenges that remain for $LiFePO_4$, for example, its relative low energy density and low conductivity. Thus, the $LiFePO_4$ cathode is more utilized in short range electric vehicles, often with the use of a conductive carbon coating or conductive polymers that improve the electron transport efficiency.

6. Status and Advancements in Electrolytes

Besides the anode and cathode, the electrolyte is the third key component of the lithium-ion battery, serving as the medium that transports Li^+ ions back and forth between the anode and cathode. Main requirements of the electrolyte are: (1) large voltage window of thermodynamic stability, (2) high Li^+ conductivity (with ionic conductivity $\sigma_{Li} > 10^{-3}$ S/cm), (3) low electronic conductivity ($\sigma_e < 10^{-10}$ S/cm), (4) good chemical stability with the electrodes over the operation temperature range and power range, and (5) low toxicity, low cost, nonflammable, and nonexplosive (Goodenough and Kim, 2010).

Nowadays, most of the commercialized LIBs use organic liquid electrolytes with lithium hexafluorophosphate ($LiPF_6$) as the conducting salt dissolved in various mixtures of carbonate solvents. The most commonly-used carbonate solvents are ethylene carbonate (EC), diethyl carbonate (DEC), dimethyl carbonate (DMC) and ethylmethyl carbonate (EMC). These carbonates have large electrolyte voltage windows, possess low viscosities that enable better Li^+ diffusion, and are oxidatively stable enough to be used for high voltage cathode materials (Aurbach, 2000).

However, one big challenge of organic liquid electrolytes is the safety issue, since the carbonates are highly flammable and have very low flash points (usually < 30 ºC) (Vogdanis et al., 1990). Furthermore, $LiPF_6$ could autocatalytically decompose into toxic and corrosive products that deteriorate the battery and result in safety hazards. Besides, a separator part is needed to pair with the organic liquid electrolyte in order to avoid shortage between the electrodes. In contrast, solid electrolyte, that has recently been extensively investigated, provides a very promising solution to the safety concern. There are two major types of solid electrolyte: Solid polymer electrolytes and inorganic solid electrolytes. Both types of solid electrolytes act as the electrolyte and separator simultaneously. A solid polymer electrolyte made up with polyethylene oxides (PEOs) with lithium salt (for example $LiPF_6$) could remain in contact with the electrode surface while the electrode material has moderate volume change during different states of charge. However, the room temperature conductivity of PEOs ($\sigma_{Li} < 10^{-5}$ S/cm) is too low to drive high power battery application (Meyer, 1998). In contrast, the inorganic solid electrolyte has much higher Li^+ conductivity ($\sigma_{Li} > 10^{-4}$ S/cm), but is unable to keep a good surface contact due to the volume change of the electrode during cycling; it has, therefore, only been used for thin-film battery applications (Besenhard, Winter and Biberacher, 1995).

7. Concluding Remarks

It is widely accepted that the market of electric vehicles will grow continuously in the future, powered by advanced lithium-ion batteries (LIBs) with higher energy densities, lower costs and better thermal stability. To achieve the desired level of capacity and safety for large scale applications, several strategies have been explored and executed in order to deliver better performance LIBs for electric vehicles.

At the anode side, graphite is widely used for stable lithium intercalation and de-intercalation while offering good conductivity and long cycle life at the same time. To further boost the energy level, Si is being incorporated into current carbonaceous anodes. Although the quantity of Si added in the graphite anode is currently still small (about 5%), the content of Si is expected to increase continuously and eventually replace graphite.

At the cathode side, $LiMO_2$-type layered metal oxides containing nickel, cobalt and manganese (NMC) or nickel, cobalt and aluminum (NCA) with different compositions have been explored and successfully used for electric vehicles. New generation electric vehicles are pursuing high Ni content in the blended layered oxides in order to improve the capacity and reduce the cost along with designs to improve the thermal stability of Ni-rich cathodes at their charged state.

The state-of-the-art electrolyte has also been optimized in order to maximize the ionic conductivity of $LiPF_6$ salt in the carbonates, form stable SEI layers and avoid further degradation of the electrolyte. However, safety issues remain a big concern for liquid electrolytes due to their intrinsic flammable property and unstable chemical stability. Solid electrolytes have been investigated lately in order to address the conductivity and contact issues and provide promising solutions for future solid-state batteries.

In summary, LIBs as the most dominant energy solution for electric vehicles in the coming decades are continuously making progress in improving the energy density and cycle performance with low costs, while concerns of safety and fast charging capability remain big challenges in meeting the growing requirements. New technologies, like lithium-metal-based solid state batteries, have promised to offer solutions to the safety concerns regarding electric vehicles, but fast charging requirement could present a barrier for widespread application of solid-state batteries in the foreseeable future. In spite of some near-term challenges, we are optimistic and believe that with investment from the public and private sectors, lithium ion batteries with high performance and minimum safety issues will soon bring about the era of electric vehicles and play a critical role in reducing CO_2 emissions and helping to create a more sustainable society.

References

Amine, Khalil, Zifeng Ma, Feng Pan, Larry A. Curtiss, Zonghai Chen and Jun Lu. 2016. The role of nanotechnology in the development of battery materials for electric vehicles. Nature Nanotechnology 11(12): 1031–38. https://doi.org/10.1038/nnano.2016.207.

Ashuri, Maziar, Qianran He, Yuzi Liu, Kan Zhang, Satyanarayana Emani, Monica S. Sawicki, Jack S. Shamie and Leon L. Shaw. 2016. Hollow silicon nanospheres encapsulated with a thin carbon shell: An electrochemical study. Electrochimica Acta 215: 126–41. https://doi.org/10.1016/j.electacta.2016.08.059.

Ashuri, Maziar, Qianran He and Leon Shaw. 2016. Silicon as a potential anode material for li-ion batteries: Where size, geometry and structure matter. Nanoscale 8(1): 74–103. https://doi.org/10.1039/c5nr05116a.

Ashuri, Maziar, Qianran He, Zhepu Shi, Changlong Chen, Weiliang Yao, James Kaduk, Carlo Segre and Leon Shaw. 2019. Long-term cycle behavior of nano-$LiCoO_2$ and its postmortem analysis. J. Phys. Chem. C 123: 3299–3308. https://doi.org/10.1021/acs.jpcc.8b10099.

Aurbach, Doron. 2000. Review of selected electrode-solution interactions which determine the performance of Li and Li ion batteries. Journal of Power Sources Vol. 89.

Besenhard, Jo, Winter, M. and Biberacher, W. 1995. Filming mechanism of lithium-carbon anodes in organic and inorganic electrolytes. Journal of Power Sources Vol. 54.

Blomgren, George E. 2017. The development and future of lithium ion batteries. Journal of the Electrochemical Society 164(1): A5019–25. https://doi.org/10.1149/2.0251701jes.

Chen, Changlong, Weiliang Yao, Qianran He, Maziar Ashuri, James Kaduk, Yuzi Liu and Leon Shaw. 2019. Tunable $LiAlO_2/Al_2O_3$ coating through a wet-chemical method to improve cycle stability of nano-$LiCoO_2$. ACS Applied Energy Materials, in Press. https://doi.org/10.1021/acsaem.8b02079.

Choi, Nam-Soon, Kyoung Han Yew, Kyu Youl Lee, Minseok Sung, Ho Kim and Sung-Soo Kim. 2006. Effect of fluoroethylene carbonate additive on interfacial properties of silicon thin-film electrode. Journal of Power Sources 161: 1254–59. https://doi.org/10.1016/j.jpowsour.2006.05.049.

Deng, Da, 2015. Li-ion batteries: Basics, progress, and challenges. Energy Science and Engineering 3(5): 385–418. https://doi.org/10.1002/ese3.95.

Goodenough, John B. and Youngsik Kim. 2010. Challenges for rechargeable li batteries. Chem. Mater. 22: 587–603. https://doi.org/10.1021/cm901452z.

Goodenough, John B. and Kyu-Sung Park. 2013. The li-ion rechargeable battery: A perspective. J. Am. Chem. Soc. 135: 1167–76. https://doi.org/10.1021/ja3091438.

Graetz, J., Ahn, C.C., Yazami, R. and Fultz, B. 2003. Highly reversible lithium storage in nanostructured silicon. Electrochemical and Solid-State Letters 6(9): A194. https://doi.org/10.1149/1.1596917.

Han, Hyoung Kyu, Chadrasekhar Loka, Yun Mo Yang, Jae Hyuk Kim, Sung Whan Moon, Jong Soo Cho and Kee-Sun Lee. 2015. High capacity retention Si/silicide nanocomposite anode materials fabricated by high-energy mechanical milling for lithium-ion rechargeable batteries. Journal of Power Sources 281: 293–300. https://doi.org/10.1016/j.jpowsour.2015.01.122.

Hekkert, Marko P., Franka H.J.F. Hendriks, Andre P.C. Faaij and Maarten L. Neelis. 2005. Natural gas as an alternative to crude oil in automotive fuel chains well-to-wheel analysis and transition strategy development. Energy Policy 33: 579–94. https://doi.org/10.1016/j.enpol.2003.08.018.

Hwang, Gaeun, Hyungmin Park, Taesoo Bok, Sinho Choi, Sungjun Lee, Inchan Hwang, Nam Soon Choi, Kwanyong Seo and Soojin Park. 2015. A high-performance nanoporous Si/Al_2O_3 foam lithium-ion battery anode fabricated by selective chemical etching of the Al-Si alloy and subsequent thermal oxidation. Chemical Communications 51(21): 4429–32. https://doi.org/10.1039/c4cc09956g.

Inagaki, Michio, Kaneko, K., Endo, M., Oya, A. and Tanabe, Y. 2003. Carbon Alloys: Novel Concepts to Develop Carbon Science and Technology. Elsevier.

Jana, M., Tianxiang Ning and Raj N. Singh. 2018. Hierarchical nanostructured silicon-based anodes for lithium-ion battery: Processing and performance. Materials Science and Engineering B: Solid-State Materials for Advanced Technology 232–235 (November): 61–67. https://doi.org/10.1016/j.mseb.2018.11.002.

Jana, M. and Raj N. Singh. 2019. A facile route for processing of silicon-based anode with high capacity and performance. Materialia Vol. 6. https://doi.org/10.1016/j.mtla.2019.100314.

Kim, Hyejung, Minho Seo, Mi Hee Park and Jaephil Cho. 2010. A critical size of silicon nano-anodes for lithium rechargeable batteries. Angewandte Chemie-International Edition 49(12): 2146–49. https://doi.org/10.1002/anie.200906287.

Kovalenko, Igor, Bogdan Zdyrko, Alexandre Magasinski, Benjamin Hertzberg, Zoran Milicev, Ruslan Burtovyy, Igor Luzinov and Gleb Yushin, 2011. A major constituent of brown algae for use in high-capacity Li-ion batteries. Science 334(6052): 75–79. https://doi.org/10.1126/science.1209150.

Lee, Duk-Hee, Hyun-Woo Shim and Dong-Wan Kim. 2014. Facile synthesis of heterogeneous Ni-Si@C nanocomposites as high-performance anodes for Li-ion batteries. Electrochimica Acta 146 (November): 60–67. https://doi.org/10.1016/j.electacta.2014.08.103.

Li, Mingqi, Jingwei Gu, Xiaofang Feng, Hongyan He and Chunmei Zeng. 2015. Amorphous-silicon@silicon oxide/chromium/carbon as an anode for lithium-ion batteries with excellent cyclic stability. Electrochimica Acta 164(May): 163–70. https://doi.org/10.1016/j.electacta.2015.02.224.

Liu, Nian, Hui Wu, Matthew T. McDowell, Yan Yao, Chongmin Wang and Yi Cui. 2012. A yolk-shell design for stabilized and scalable li-ion battery alloy anodes. Nano Letters 12(6): 3315–21. https://doi.org/10.1021/nl3014814.

Liu, Nian, Kaifu Huo, Matthew T. Mcdowell, Jie Zhao and Yi Cui. 2013. Rice husks as a sustainable source of nanostructured silicon for high performance Li-ion battery anodes. Scientific Reports 3: 1919. https://doi.org/10.1038/srep01919.

Liu, Nian, Zhenda Lu, Jie Zhao, Matthew T. Mcdowell, Hyun Wook Lee, Wenting Zhao and Yi Cui. 2014. A pomegranate-inspired nanoscale design for large-volume-change lithium battery anodes. Nature Nanotechnology 9(3): 187–92. https://doi.org/10.1038/nnano.2014.6.

Liu, Qianqian, Chunyu Du, Bin Shen, Pengjian Zuo, Xinqun Cheng, Yulin Ma, Geping Yin and Yunzhi Gao. 2016. Understanding undesirable anode lithium plating issues in lithium-ion batteries. RSC Advances 6(91): 88683–700. https://doi.org/10.1039/c6ra19482f.

Magasinski, Alexandre, Bogdan Zdyrko, Igor Kovalenko, Benjamin Hertzberg, Ruslan Burtovyy, Christopher F. Huebner, Thomas F. Fuller, Igor Luzino and Gleb Yushin. 2010. Toward efficient binders for Li-ion battery Si-based anodes: Polyacrylic acid. ACS Applied Materials and Interfaces 2(11): 3004–10. https://doi.org/10.1021/am100871y.

Meyer, Wolfgang H. 1998. Polymer electrolytes for lithium-ion batteries. Advanced Materials 10(6): 439–48. https://doi.org/10.1002/(SICI)1521-4095(199804)10:6<439::AID-ADMA439>3.0.CO;2-I.

Myung, Seung-taek, Kentarou Izumi, Shinichi Komaba and Yang-kook Sun. 2005. Role of Alumina Coating on Li-Ni-Co-Mn-O Particles 10: 3695–3704.

Myung, Seung-Taek, Filippo Maglia, Kang-Joon Park, Chong Seung Yoon, Peter Lamp, Sung-Jin Kim and Yang-Kook Sun. 2017. Nickel-rich layered cathode materials for automotive lithium-ion batteries: Achievements and perspectives 2: 196–223. https://doi.org/10.1021/acsenergylett.6b00594.

Noh, Hyung-Joo, Sungjune Youn, Chong Seung Yoon and Yang-Kook Sun. 2013. Comparison of the structural and electrochemical properties of layered Li[NixCoyMnz]O_2 (x = 1/3, 0.5, 0.6, 0.7, 0.8 and 0.85) cathode material for lithium-ion batteries. Journal of Power Sources 233: 121–30. https://doi.org/10.1016/j.jpowsour.2013.01.063.

Nykvist, Björn and Måns Nilsson. 2015. Rapidly falling costs of battery packs for electric vehicles. Nature Climate Change 5: 329–332.

Obrovac, M.N. and Chevrier, V.L. 2014. Alloy negative electrodes for Li-ion batteries. Chemical Reviews 114(23): 11444–502. https://doi.org/10.1021/cr500207g.

Pan, Lei, Haibin Wang, Lei Tan, Shengyang Chen, Dacheng Gao and Lei Li. 2014. Facile synthesis of yolk-shell structured Si-C nanocomposites as anode for lithium-ion battery. Chemical Communications 50: 5878–80.

Sahni, Karan, Maziar Ashuri, Qianran He, Ritu Sahore, Ira D. Bloom, Yuzi Liu, James A. Kaduk and Leon L. Shaw. 2019. H_3PO_4 treatment to enhance the electrochemical properties of Li(Ni1/3Mn1/3Co1/3)O_2 and Li(Ni0.5Mn0.3Co0.2)O2 Cathodes. Electrochimica Acta 301(April): 8–22. https://doi.org/10.1016/j.electacta.2019.01.153.

Schmuch, Richard, Ralf Wagner, Gerhard Hörpel, Tobias Placke and Martin Winter. 2018. Lithium-based rechargeable automotive batteries. Nature Energy 3: 267–78. https://doi.org/10.1038/s41560-018-0107-2.

Shaw, Leon and Maziar Ashuri. 2019. Coating—a potent method to enhance electrochemical performance of Li(NixMnyCoz)O_2 cathodes for Li-ion batteries. Advanced Materials Letters 10(6): 369–80. https://doi.org/10.1021/acs.chemmater.8b03827.

Somerville, L., Bareño, J., Trask, S., Jennings, P., McGordon, A., Lyness, C. and Bloom, I. 2016. The effect of charging rate on the graphite electrode of commercial lithium-ion cells: A post-mortem study. Journal of Power Sources 335(December): 189–96. https://doi.org/10.1016/j.jpowsour.2016.10.002.

Song, Jiangxuan, Shuru Chen, Mingjiong Zhou, Terrence Xu, Dongping Lv, Mikhail L. Gordin, Tianjun Long, Michael Melnyk and Donghai Wang. 2014. Micro-sized silicon-carbon composites composed of carbon-coated Sub-10 Nm Si primary particles as high-performance anode materials for lithium-ion batteries. Journal of Materials Chemistry A 2(5): 1257–62. https://doi.org/10.1039/c3ta14100d.

Tarascon, J.M. and Armand, M. 2001. Issues and challenges facing rechargeable lithium batteries. Nature 414: 359–67. https://doi.org/10.1038/35104644.

Uhlmann, C., Illig, J., Ender, M., Schuster, R. and Ivers-Tiffée, E. 2015. *In situ* detection of lithium metal plating on graphite in experimental cells. Journal of Power Sources 279(April): 428–38. https://doi.org/10.1016/j.jpowsour.2015.01.046.

Vogdanis, Lazaros, Barbara Martens, Hermann Uchtmann, Friedrich Hensel and Walter Heitz. 1990. Synthetic and thermodynamic investigations in the polymerization of ethylene carbonate. Die Makromolekulare Chemie 191(3): 465–72. https://doi.org/10.1002/macp.1990.021910301.

Wan, Jiayu, Alex F. Kaplan, Jia Zheng, Xiaogang Han, Yuchen Chen, Nicholas J. Weadock, Nicholas Faenza et al. 2014. Two dimensional silicon nanowalls for lithium ion batteries. Journal of Materials Chemistry A 2(17): 6051–57. https://doi.org/10.1039/c3ta13546b.

Wang, Dingsheng, Mingxia Gao, Hongge Pan, Junhua Wang and Yongfeng Liu. 2014. High performance amorphous-Si@SiOx/C composite anode materials for Li-ion batteries derived from ball-milling and *in situ* carbonization. Journal of Power Sources 256: 190–99. https://doi.org/10.1016/j.jpowsour.2013.12.128.

Wu, Hui, Gerentt Chan, Jang Wook Choi, Ill Ryu, Yan Yao, Matthew T. Mcdowell, Seok Woo Lee et al. 2012. Stable cycling of double-walled silicon nanotube battery anodes through solid-electrolyte interphase control. Nature Nanotechnology 7(5): 310–15. https://doi.org/10.1038/nnano.2012.35.

Yamada, A., Chung, S.C. and Hinokuma, K. 2002. Optimized LiFePO[Sub 4] for lithium battery cathodes. Journal of the Electrochemical Society 148(3): A224–29. https://doi.org/10.1149/1.1348257.

Yang, L.Y., Li, H.Z., Liu, J., Sun, Z.Q., Tang, S.S. and Lei, M. 2015. Dual yolk-shell structure of carbon and silica-coated silicon for high-performance lithium-ion batteries. Scientific Reports 5: 10908. https://doi.org/10.1038/srep10908.

Yata, S., Kinoshita, H., Komori, M., Ando, N., Anekawa, A. and Hashimoto, T. 1993. Extended Abstract of 60th Annual Meeting of the Electrochemical Society of Japan, Tokyo. In 2G09.

Yi, Ran, Fang Dai, Mikhail L. Gordin, Shuru Chen and Donghai Wang. 2013. Micro-sized Si-c composite with interconnected nanoscale building blocks as high-performance anodes for practical application in lithium-ion batteries. Advanced Energy Materials 3(3): 295–300. https://doi.org/10.1002/aenm.201200857.

Yi, Ran, Jiantao Zai, Fang Dai, Mikhail L. Gordin and Donghai Wang. 2014. Dual conductive network-enabled graphene/Si-C composite anode with high areal capacity for lithium-ion batteries. Nano Energy 6: 211–18. https://doi.org/10.1016/j.nanoen.2014.04.006.

CHAPTER 16

Charging Strategies for Electrified Transport

*Sheldon Williamson,** *Deepa Vincent,*
AVJS Praneeth and *Phouc Hyunh Sang*

1. Introduction

Conventional vehicles (CV) or internal combustion engine (ICE) vehicles represent the majority of vehicles on the road and the high demand for petroleum is a major cause for concern. The shortage of petroleum is one of the most critical global issues and increasingly costly fuel is one of the most critical issues for CV users. CVs have detrimental effects on the environment as they emit greenhouse gases (GHGs), Figure 1 (Environment and Climate Change Canada, 2019); as a result, it is becoming difficult for CVs to satisfy increasingly stringent environmental regulations. One of the most promising and attractive alternatives to CVs are electric vehicles (EVs) or zero-emission vehicles that only consume electrical energy. Replacing conventional ICE vehicles with EVs on a global scale could result in immense benefits (of environmental sustainability), saving our world from dangerous and ever-increasing rates of pollutant emissions. The key benefits of moving towards (or adopting) EVs include a decrease in oil consumption and reduction in pollution.

The aforementioned benefits are based on using batteries as a green source of energy. The chemical nature of batteries being highly non-linear makes them dependent on key factors, such as chemistry, temperature, aging, load profile, charging algorithm and so on. In addition, initially, due to the limited energy densities of commercial battery packs, the performance of EVs was restricted with various limitations, including slow charging rate, short driving distance, heavy battery packs and high costs.

A typical EV drivetrain topology is shown in Figure 2. The charger constitutes power electronic converters that transform AC from grid to DC in order to charge the batteries. Generally, batteries have much lower specific energy than fossil fuels. Thus, to provide a reasonable mileage range, heavy battery packs are required. The battery management system (BMS) ensures a balanced and efficient energy flow between the cells of a battery. As shown in Figure 2, a higher power density ultra-capacitor bank is required in order to form a hybridized energy storage system. This provides an added power capability in addition to energy capacity for the EV. At the end of discharge, batteries are replenished by chargers. The frequent charging opportunities will reduce the battery pack size and increase the life of batteries. The electric drive feeds electricity into the motor in required amounts and at variable frequencies, thereby indirectly controlling the motor's speed and torque.

University of Ontario Institute of Technology, Oshawa, Ontario, Canada.
* Corresponding author: sheldon.williamson@uoit.ca

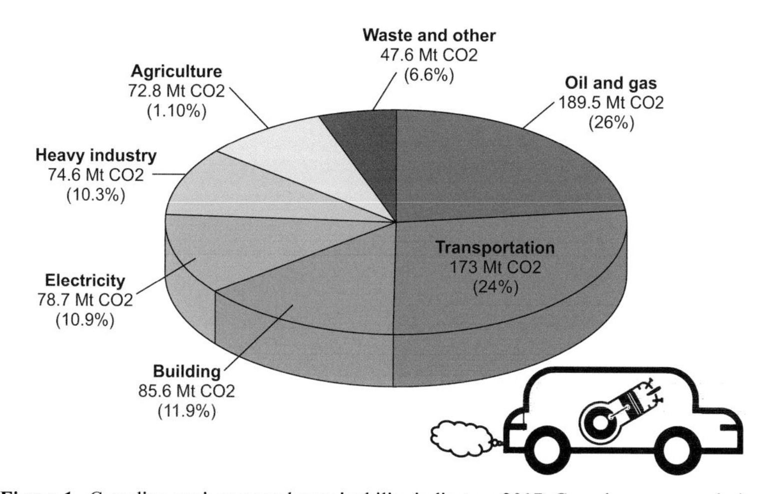

Figure 1. Canadian environmental sustainability indicators 2017: Greenhouse gas emissions.

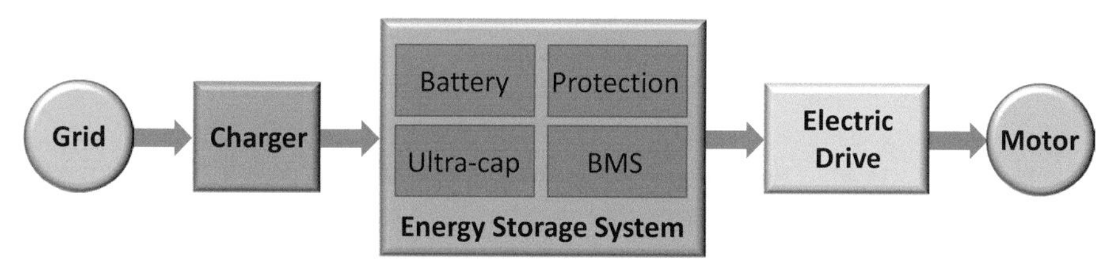

Figure 2. EV drive train topology.

The chapter is organized as follows: Section 2 of this chapter presents innovative energy storage techniques for transportation electrification and autonomous e-mobility applications. Section 3 deals with advanced plugged charging infrastructures, both onboard and offboard chargers, for electrified transport. Next, DC fast charging power electronic converters for Level 3 SAE J1772 standards and future universal voltage high-power charging levels are discussed in section 4. Section 5 introduces wireless charging using inductive power transfer (IPT) technology and PV/grid-connected charging infrastructures for EV charging are discussed in section 6. A brief comparison of various charging strategies is presented in the concluding remarks.

2. Energy Storage Systems

Electric energy storage systems (ESS) have been the most researched area in transportation electrification (Williamson, 2007; Sidhu et al., 2016). Within the domain of ESS, research revolved mostly around batteries. The battery chemistries explored during this time period were better than the lead-acid batteries used in the first electric vehicle. These new chemistries were NiMH (Nickel metal hydride), NiCd (Nickel Cadmium), and Li-ion (Lithium Ion). Within Li-ion, various different chemistries of the cathode and anode have been explored with varying results regarding energy density, safety, cycle life, nominal cell voltage, etc. A Li-ion cell basically consists of an anode and cathode immersed in an electrolyte. The battery storage is obtained when a number of these cells are stacked together. However, batteries have limitations on their power handling capabilities and cycle life. Cycle life of Li-ion batteries varies from 1000–1500 cycles for different chemistries, and, if stressed to high pulsed power, the electrodes of batteries can be severely impacted, reducing the life of a battery.

On the other end of the energy storage spectrum is the ultra-capacitor, or EDLC (Electric Double Layer Capacitors), also commonly known as super-capacitor. These storage elements are made up on principles similar to capacitor storage, i.e., two conducting electrodes separated by a dielectric. The energy storage mechanism in super-capacitors is electrostatic energy. Super-capacitors have complimentary attributes when compared to batteries. They have high power density, over 500,000 charge-discharge cycles and very low ESR (equivalent series resistance), which makes them suitable for high power pulsed operations.

Batteries and super-capacitors with their complimentary attributes, provide an opportunity to get the best from both. SCs and battery, combined through an electrical interface and controlled via a power and energy management technique, can deliver a battery-super-capacitor hybrid system. This HESS (Hybrid Energy Storage System) has an energy element in the form of battery and a power element in the form of a super-capacitor. The characteristics of super capacitor and battery are given in Table 1.

The hybrid topologies can either have both the devices arrangeed in series or both battery and super-capacitor arranged in parallel, Figure 4. The series architecture does not decouple the battery from load transients and, hence, is not explored any further. The parallel topology is the one which is implemented in hybrid electrical energy storage systems. Within the parallel architecture, there are different configurations of connecting battery and super-capacitor to common DC bus. These configurations can be broadly divided into active and passive. The passive topology does not include any additional electronic interfacing to connect super-capacitor to DC bus, this makes the voltage across the super-capacitor fixed to the battery voltage. As energy stored in the super-capacitor is equivalent to the voltage across its terminals, by fixing it to an almost constant DC bus voltage, the available energy from the super-capacitor is severely restrained. In the active hybrid topologies, power electronics-based DC/DC converter would be used to interface super-capacitors to the DC bus.

Table 1. Typical characteristics of super-capacitors and batteries.

Parameters	Super-capacitors	Li-ion batteries
Cell voltage (V)	2.2–3.3	2.5–4.2
Charge-discharge cycles	10^5–10^6	500–2000
Temperature range (°C)	–40–70	–20 to 60
Specific power (kW/kg)	3–10	0.3–1.5
Specific energy (Wh/kg)	4–9	100–265

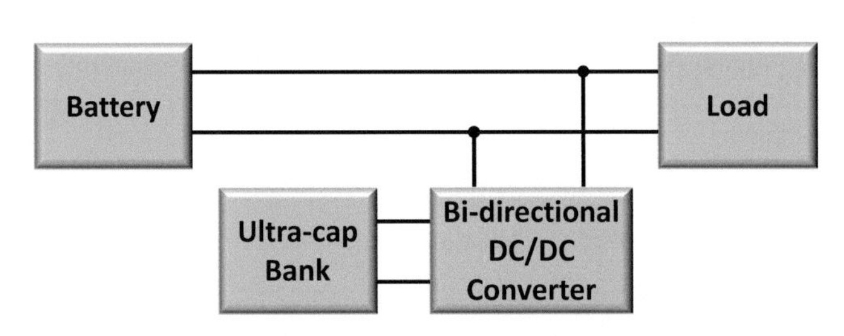

Figure 3. Hybrid energy storage topology.

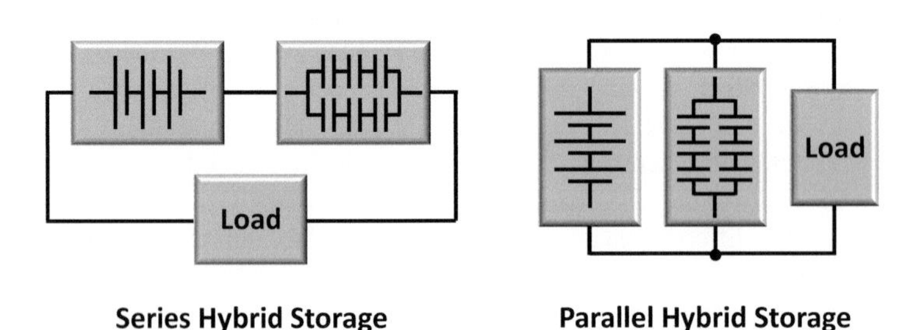

Figure 4. Battery and super-capacitor hybridization topologies.

3. Charging Methodologies I—Wired/Plug-in Charging

Battery cells have a limited energy density and can afford to discharge loads of their capacity. The time taken to dissipate energy and to reach the minimum threshold level depends on the load conditions, which can be either light loads or heavy loads. At the end of discharge, a cell will regain its capacity through a power electronic device known as a 'charger'. These chargers can accept broad input levels and different types of power supplies (AC or DC) to charge the battery cells. The devices which connect input grid supply of different voltage and frequencies levels to charge the cells are termed as AC-DC converters (Praneeth et al., 2018). The devices which can also accept the inputs from renewables, fossil fuels or DC sources to charge the battery cells or packs are termed as DC-DC converters. A typical EV charger layout is given in Figure 5. Irrespective of the input supply, the conversion process from one form of supply to another remains the same in these converters. Typical structure of these converters could be either isolated or non-isolated, depending on the type of application used. The most interesting application is grid-connected battery chargers, which need to ensure safe charging of battery packs while maintaining power quality at the grid. Maintaining power quality is one of the key requirements for any electric vehicle (EV) battery charger. Poor power quality results in huge penalty on the customers. The front-end rectifiers available in the grid connected battery chargers of the vehicle address all the power quality problems. The major classifications of the battery chargers for EV applications are on-board or off-board, based on whether the equipment is located inside or outside the vehicle, as shown in Figure 6.

In general, off-board chargers, also known as fast chargers (FCs), are equipped outside the vehicle because of their bulky size and high power intake required to charge the battery pack at a much faster rate (typically of charge rate 1C). The time taken by these chargers are in minutes, depending on the battery capacity. The typical ratings of the FC are in several kW (> 50 kW). For example, the Tesla Super charger takes about 45–60 minutes to charge a 90 kWh battery pack and about one hour and 40 minutes with ABB DC fast charger and even less than 60 minutes with ABB and Siemens ultra-fast chargers. The high input power supply to these chargers restricts users in terms of safety aspects and it also requires high infrastructural maintenance. Moreover, charging at such a fast rate may involve thermal constraints and can degrade the life of the battery. Many experts and researchers are still focusing on the exact prediction and estimation of the life cycles for different chemistries of Li-ion batteries when subjected to fast charging. However, at present, due to the lack of fast charging infrastructure requisites, the automotive manufactures prefer on-board chargers. These on-board chargers have an upper limit for their power level, based on cost, weight and space constraints (Praneeth et al., 2018). There are two broad categories of on-board chargers for electric vehicles: Conductive and inductive. Conductive chargers have a wired

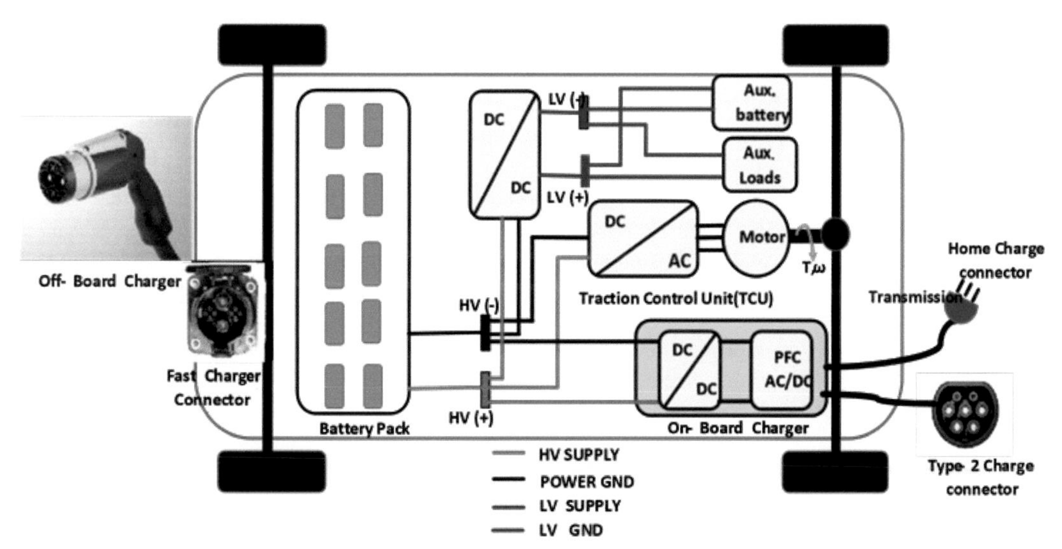

Figure 5. Typical EV charger layout.

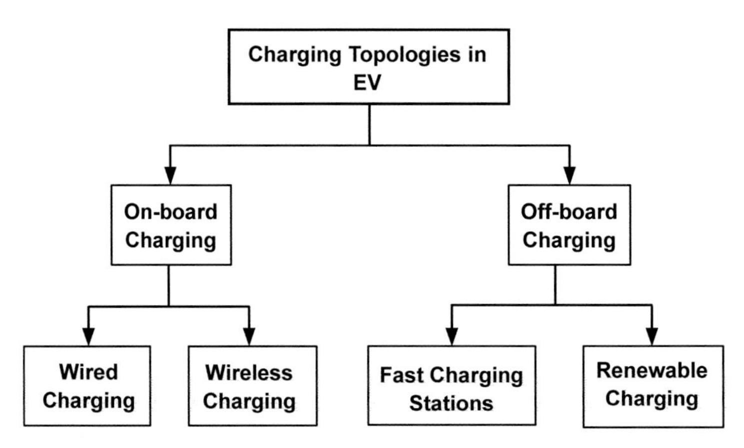

Figure 6. Classifications of electric vehicle charging.

connection from the power supply to the power converter in a vehicle. An inductive charger does not have direct contact from the supply through wire; instead, it transfers power wirelessly through a high frequency link consisting of transmitter and receiver coils. Inductive charging, also known as contactless or wireless charging, has limitations in power levels, operating frequency and emission levels.

As the cost of utility power is cheaper at night, most EV charging happens overnight at homes with Level-I charging. Typical time taken by Level-I charging is about 12–15 hours. To reduce the time, Level-II charging (known as semi-fast charging) at 208–240 V is used. The AC charging levels as per SAE standards are summarized in Table 2. The customers usually have the range anxiety problem for long drives due to the required long charging times at the infrequent EV charging stations. To resolve the curiosity and have a safe journey, automotive manufactures, with the support from local governments and OEMs, have initiated projects for quick charging of the vehicles. A high power charging infrastructure with safety considerations has been installed in major cities and in major commercial places. The DC charging levels as per SAE standards are summarized in Table 3.

Battery chargers need to ensure safe charging of battery packs while maintaining power quality. Power converter topologies with charge control algorithms have significant effect on charge currents and efficiency of the battery chargers. The existing on-board battery chargers from manufactures are shown in Figure 7. They can accept a broad range of input power supply, ranging from single-phase 120 V to three-phase 400 V. Standard single-phase household power sockets are suitable to charge an electric vehicle. Layouts of on-board battery chargers are mainly classified as two-stage converter topologies and single-stage converter topologies, based on output ripple reduction and isolation requirements.

Typically, a lead acid battery needs only single stage conversion (AC-DC) to charge, even though there exists a large low frequency ripple component (2f) from the input to the output current of the battery during charging. However, a two stage (AC-DC/DC-DC) conversion provides an inherent low frequency rejection, which is preferable for lithium ion battery charging, as it is sensitive to low frequency current ripples. Figure 8 shows the single-phase on-board two-stage conversion for lithium battery charging.

A bulky electrolytic capacitor at the output of power factor correction (PFC) stage is used to maintain stiff voltage and to provide energy during the hold-up time. The effect of the ripple current along with the power loss generated in the equivalent series resistance (ESR) causes a temperature rise on the electrolytic capacitor and eventually degrades its lifetime, sometimes resulting in failure of the components.

The limited lifetime of the electrolytic capacitors restricts the automotive manufactures from operating at higher DC link capacitance. The input current of a single-phase PFC converter when operated in continuous conduction mode (CCM) contains both the high frequency and low frequency ripple components. A solution to eliminate the electrolytic capacitor is to absorb the high frequency switching ripple using film capacitors and allow the low frequency components to charge the battery packs. This method of charging is termed as sinusoidal charging. A sinusoidal charging with the low frequency ripple uses film capacitors instead of bulky capacitors as the DC link in battery charging.

Table 2. SAE EV AC charging power levels.

Charge method	Supply voltage (V)	Maximum current (A)	Output power level (kW)	Estimated charge time (h)
AC Level-I	120 V AC, single phase	12	1.4	7–17
	120 V AC, 1 single phase	16	1.9	7–17
AC Level-II	208 to 240 V AC, single phase	≤ 80	3.3–7	3–7
AC Level-III	230 to 400 V AC, three phase	32 to 63	14–43	< 0.8

Table 3. SAE EV DC charging power levels.

Charge method	Supply voltage (V)	Maximum current (A)	Output power level (kW)	Estimated charge time (h)
DC Level-I	200–450	80	< 36	0.5–1.2
DC Level-II	200–450	200	90	0.3–0.4
DC Level-III	200–450	≤ 400	250	< 0.1

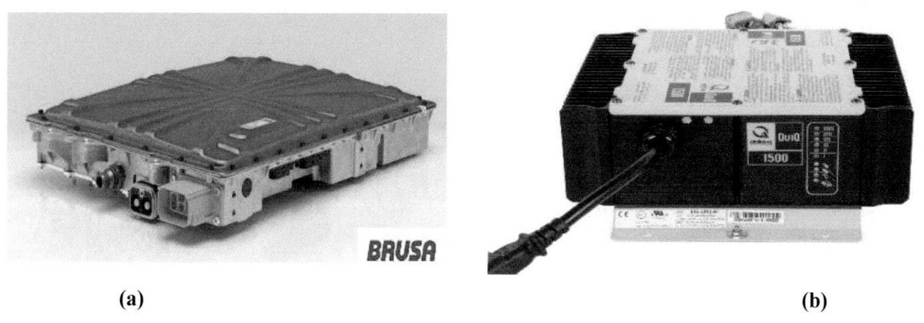

Figure 7. EV wired on-board battery charger (a) BRUSA® (b) Delta-Q.®

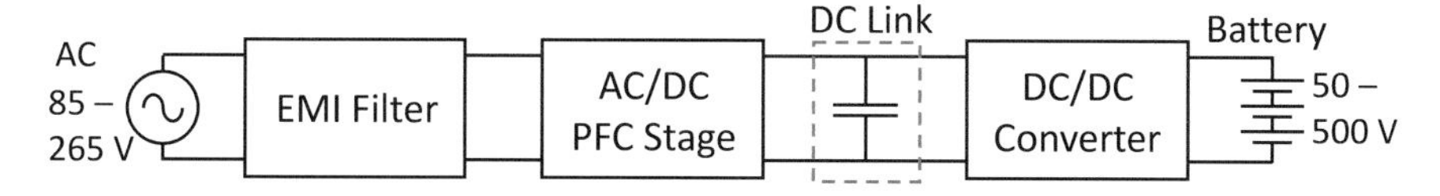

Figure 8. Single-phase on-board two-stage conversion.

The various capacitance and voltage ratings of the different types of capacitors are shown in Figure 9. However, a deep study is required in order to understand the impacts of the sinusoidal current charging on the lithium-ion battery over 5000 cycles and also to elaborate the impacts on the capacity fade.

The schematic of two-stage single-phase 3.3 kW plugged on-board battery charger is shown in Figure 10. The first stage is the AC/DC converter with a power factor correction (PFC) circuit and the second stage is the DC/DC converter with high frequency transformer. The DC/DC stage provides the isolation for the system as well as the control required at the output battery. There are numerous PFC topologies to improve the power quality, efficiency and power density of the converters, which suit the battery charging application.

The second stage in the on-board battery charger is the DC/DC converter stage. The most important safety requirement for the automotive manufacturers is to provide the isolation between the high voltage (HV) battery pack and the grid supply. The configurations of the DC/DC converters for all applications use an isolated DC/DC converter. The non-isolated DC/DC converters have the advantage of simple structure, high efficiency, high reliability, low cost and small size, but don't provide a galvanic isolation. The non-isolated converters, like buck, boost, cuk, buck-boost converters, can refill the auxiliary battery

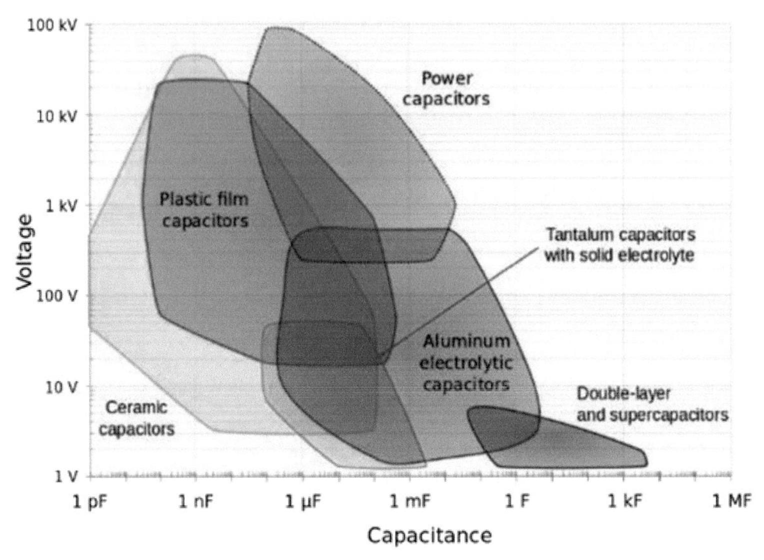

Figure 9. Different types of capacitors with voltage and capacitance levels (Wang, 2016).

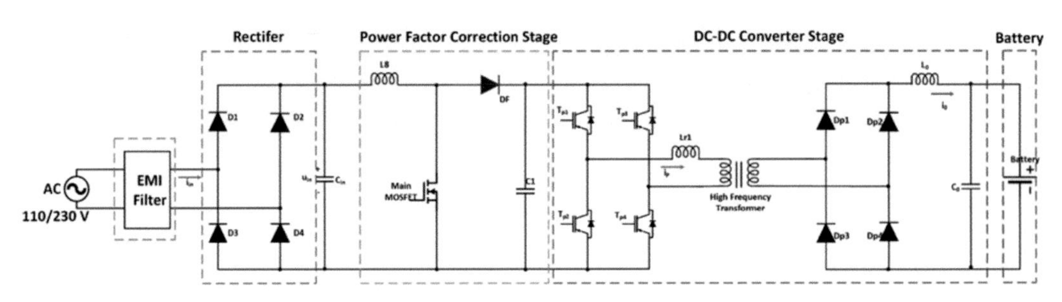

Figure 10. Schematic of two-stage on-board battery charger.

Table 4. Power ratings of on-board battery chargers.

Manufacturer	Input voltage (V) (Single Phase-AC)	Input voltage (V) (Three Phase-AC)	Power rating (kW)	Maximum current (A)	Output voltage range (V)	Efficiency (%)	Cooling
BRUSA-NLG513-U1	85–265	–	3.3	12.5	200–520	93	Liquid
BRUSA-NLG514-U1	100–265	–	3.3 k	9	300–720	93	Air
BRUSA-NLG664	200–250	360–440	7/22	30	200–450	> 90	Liquid
Delta-Q technologies	85–265	–	3.3	11	200–450	94	Liquid
Delta-Q technologies	100–265	–	1.5	20	72–100	94	Air
Delphi	85–265	–	3.3	12	200–440	> 94	Liquid/Air
Delta-power electronics	85–265	–	3.3	12	180–430	98.1	Air/Liquid

(LV) pack through the available solar energy. In isolated topologies, fly back and forward converters can be used in electric vehicle applications with limited power levels. The most common topologies for battery chargers with high power levels (greater than 3 kW) use half bridge or full bridge topologies. Half bridge converter configurations with LLC topology are used in the design of the battery chargers for power ratings up to 1.5 kW. Table 4 provides the information on the existing on-board battery chargers for electric vehicle applications.

4. Charging Methodologies II—DC Fast Charging Topologies

Presently, the SAE J1772 standard specifies three levels of charging (Yilmaz et al., 2013). The charging levels are classified by AC and DC charging levels (Hybrid-EV Committee, 2010; Rubino et al., 2017). The parameters of the charging levels are listed in Tables 1 and 2. The major features of each charging level are listed as follows:

- AC Level 1 Charging: The most common household slow-charging method. For Level 1 charging, no extra facilities are needed. Total initial cost is about $800 USD.
- AC Level 2 Charging: This can be an on-board equipment. Level 2 charging is most preferred today, since it is quicker and has a dedicated connector. The initial cost is about $3,000 USD.
- DC Level 3 or DC fast Charging: This option provides the customer the option to charge in < 1 hour and as little as 15 minutes of charge time. The power system for Level 3 charging is completely off-board. This is purely a commercial fast charging station and cannot be set up in residential areas. Total initial cost of such a power system would range between $40,000 USD to $70,000 USD, including initial investment costs, infrastructure costs, operating and maintenance costs.

From these standards, it is clear that building multiple level 3 or DC fast charging stations is required in order to enable EVs to compete with commercial gasoline vehicles. The DC fast charging stations will also help alleviate the EV users' range anxiety. In order to find practical charging times, the Nissan Leaf® 24 kWh Li-ion battery pack is considered (Channegowda et al., 2015). The time to fully charge the battery pack with different charging levels is shown in Table 5. Since the available charging levels have been established, it is necessary to review the power converters being employed in the charging infrastructure. The following sections provide a brief overview of the topologies meant for fast charging the batteries of EVs. There are a number of power converter topologies available for the purpose of rapid charging of batteries or super capacitors. A few feasible options include:

- Unidirectional Boost Converters

  The unidirectional boost converter is shown in Figure 11, these are employed in situations where the output voltage has to be boosted for loads which require higher voltage (Singh et al., 2004).

  The primary goal of using such a boost converter instead of a traditional diode bridge rectifier is to provide better power factor, remove harmonics at the input end and to have an unvarying DC voltage at the output if unwanted perturbations occur at the AC end. The boost converter has been a popular choice for EV charging due to its Power Factor Correction (PFC) stage, there have been many control strategies developed to reduce input current harmonics as well (Mallik et al., 2017).

- Vienna Rectifier

  Another popular power converter topology is the Vienna rectifier, shown in Figure 12. The Vienna rectifier is a popular choice due to its ability to achieve a high power factor and attain lower harmonic distortion (Liu et al., 2017). As shown in Figure 12, there is only one active switch per phase, making the Vienna rectifier easier to control and more dependable. This is basically a pulse width modulated converter, the boost inductors present at the input help in power factor correction (Kedjar et al., 2014).

  Essentially, the stored energy acquired by the inductor when the switch is OFF is transmitted to the load via the diodes whenever the switch is turned ON. The advantages of employing this topology include the absence of a neutral point connection and the lack of auxiliary commutation circuits, which eliminates dead time problems.

- AC/DC Reduced-Switch Buck-Boost Converter

  The main highlight of this topology is that it is inexpensive, has fewer switches and, most importantly, since this is a buck-boost converter, the output voltage can be varied over a wide range. The topology is shown in Figure 13 (Wijeratne et al., 2012). There have been a few three-phase front end rectifiers proposed, however they are mostly boost converters which do not allow variation of voltage over

Table 5. A 24 kWh batterypack charged with standard EV charging levels (SAE J1772).

Level	Voltage (V)	Phase	Power (kW)	Time(h)
AC Level 1	120	1 phase	1.4	17
AC Level 2	240	1 and split phase	4	6
DC Level 3	208/415	3 phase	50	0.5

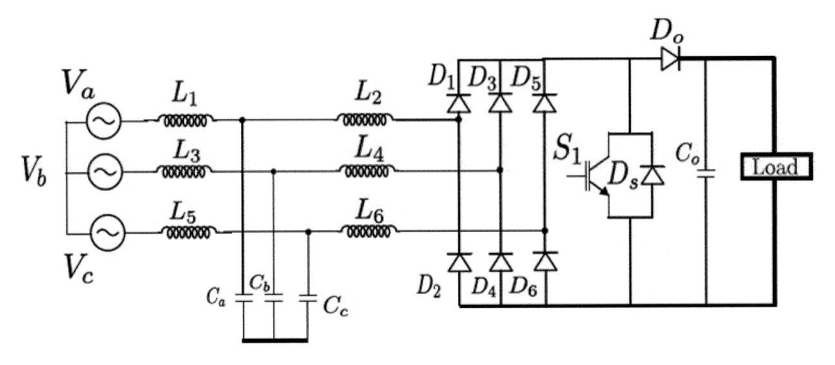

Figure 11. Unidirectional boost converter (Singh et al., 2004).

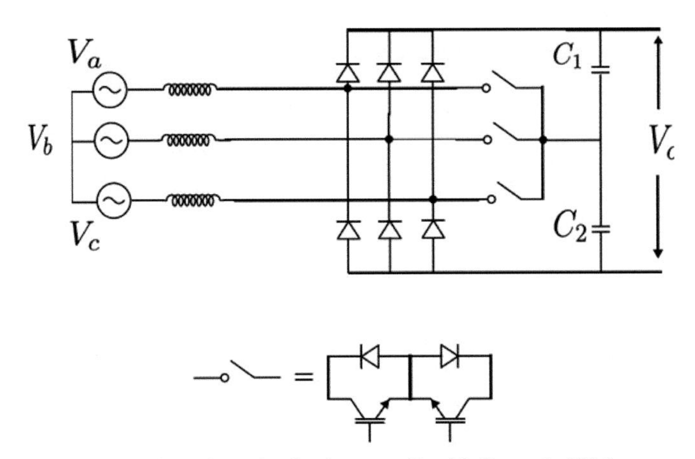

Figure 12. Schematic of a vienna rectifier (Kedjar et al., 2014).

wide ranges. This converter topology can operate in buck mode when the duty ratio is below 0.5 and in boost mode when the duty ratio is above 0.5.

- Swiss Rectifier

The Swiss rectifier is a three-phase buck derived power converter, Figure 14. This topology has been considered for EV battery charging (Soeiro et al., 2012). This topology enables wide output voltage range along with the inherent Power Factor Correction (PFC) capability. The Swiss rectifier converter topology combines a buck derived DC/DC converter along with a third harmonic current introducing circuit (Schrittwieser et al., 2017). For a power converter of 7.5 kW rating, an efficiency of 96.5% was achieved.

The Three Phase Three Switch Rectifier (TPTS) is yet another buck derived three-phase rectifier converter topology, Figure 15 (Baumann et al., 2000). The TPTS only has three active switches and four diodes in each leg. A TPTS rectifier is a very good candidate for fast charging applications, due to its high efficiency and wide output voltage range (Nussbaumer et al., 2007).

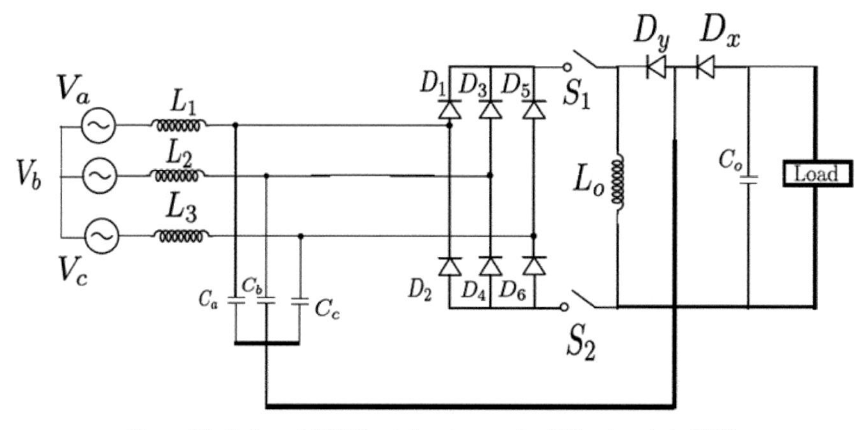

Figure 13. 3-phase AC/DC buck-boost converter (Wijeratne et al., 2012).

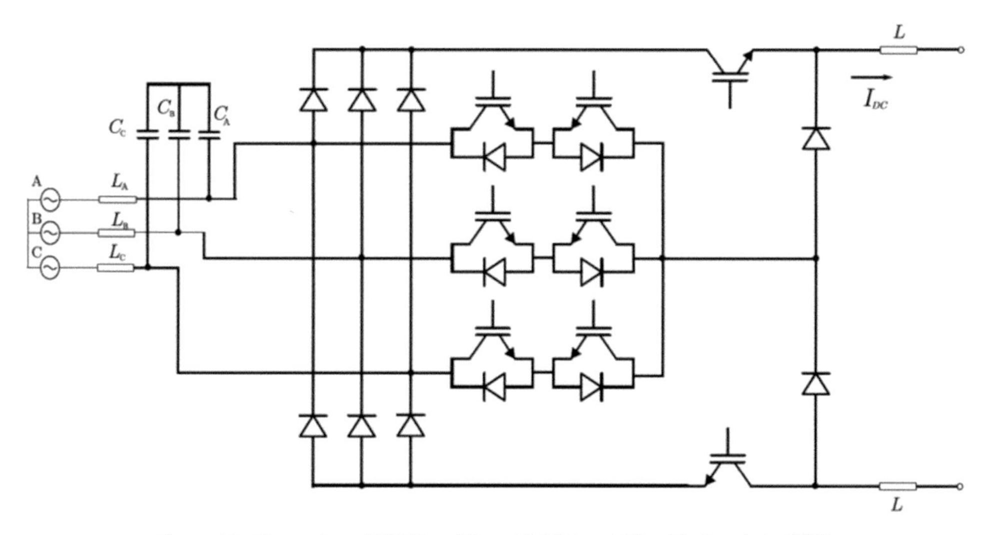

Figure 14. Three-phase SWISS rectifier with LC input filter (Soeiro et al., 2012).

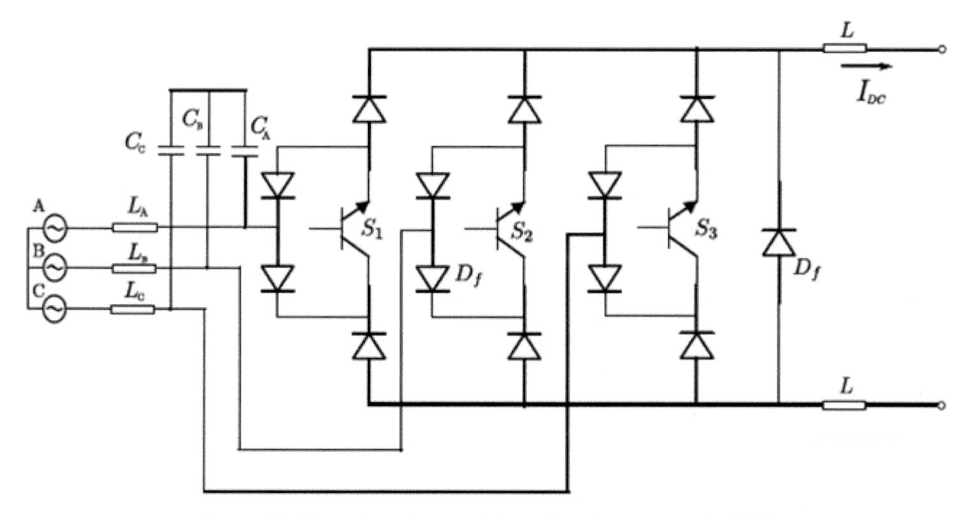

Figure 15. Three-phase three switch rectifier (Baumann et al., 2000).

5. Charging Methodologies III—Wireless Charging

Traditionally, EVs usually use a simple plug-in charging method, also known as conductive charging or plug-in charging, in which a copper connected cable forms the power link. Plug-in charging is a widely accepted method on the market, and it is available for most EVs: Chevrolet Volt, Tesla Roadster, Nissan Leaf and Mitsubishi i-MiEV. However, there are disadvantages to plug-in charging, such as safety issues caused by exposed plugs and damaged cables. To avoid the drawbacks of plug-in charging, wireless charging methods have been widely studied in recent years (Covic et al., 2007; Vincent et al., 2019; Aditya et al., 2019).

It has been established in the literature that only an inductive WPT system has the potential to be applied in medium and high-power applications, and particularly for the charging of EV batteries. In the literature, this method of power transfer has also been referred, contactless power transfer (CPT), contactless energy transfer (CET), inductively coupled power transfer (ICPT), resonant inductive power transfer (RIPT) and inductive power transfer (IPT). The EVs adopting wireless charging technology can provide the following advantages over conventional wired charging:

- Range extension: Wireless charging has the scope for 'opportunity charging', i.e., charging the vehicle little and often during the day when the EV is not in use (Birrell et al., 2015). For example, in addition to residential garage, public and private parking areas, the wireless charger can also be installed at: Traffic lights, bus stops, high traffic congestion area where vehicles are slow moving, and taxi ranks that move forward as taxis are hired (Fisher et al., 2014). This opportunity charging is possible since wireless charging does not require human intervention and charging can, therefore, be carried out automatically. This, in turn, leads to significant improvements in range compared to that available from a single overnight plugin charge.
- Safety and convenience: Wireless charging provides galvanic isolation between load and source. Therefore, it eliminates some disadvantages of plug-in charging technology, such as risk of electrocution from aging wiring and bad connections, especially in wet and hostile environments, failure to plug in, trip hazard from a long connecting wire, poor visual appeal due to hanging cords, contactor wear caused by excessive use and thermal cycling and, most importantly, discomfort in handling a plug-in charger in a harsh climate that commonly has snow and where the charge point may become frozen onto the vehicle (Wu et al., 2011).
- Battery volume reduction: Due to the scope of opportunity charging, as shown in Figure 16, charging can take place more frequently. Therefore, EVs can travel the same distance with a reduced battery pack (Huh et al., 2011) This, in turn, can lower the price of EVs and make them more efficient due to the reduced weight. Frequent charging also extends the battery life by reducing the depth of discharge in the battery.
- Weather proof: In a wireless charger, power transfer takes place via an electromagnetic link, therefore, charging is not affected by the presence of snow, rain or dust storms. Besides, a transmitter is embedded underground and is, therefore, safe from extreme weather conditions and requires less frequent maintenance or replacement than a plug-in charger.

EV wireless power transfer principle is based on well-established Ampere's circuital law and Faraday's law of induction. Figure 17 shows two coils linked by inductive coupling. Here, the subscript P and S refers to primary and secondary coil, respectively. The terms ϕ_M, ϕ_{IP}, and ϕ_{IS} are mutual flux, primary leakage flux, and secondary leakage flux, respectively. Let M, L_P, L_S be mutual inductance, self-inductance of primary and self-inductance of secondary coil, respectively. When a time varying current is applied to the primary coil, a time varying flux of the same frequency is produced in the region surrounding the primary. The strength of the magnetic field around a closed path is directly proportional to the current carried by the coil and is given by Ampere's law (for the case when the displacement current is neglected) by equation (1):

$$\oint H.\, dl = \int J.\, ds \tag{1}$$

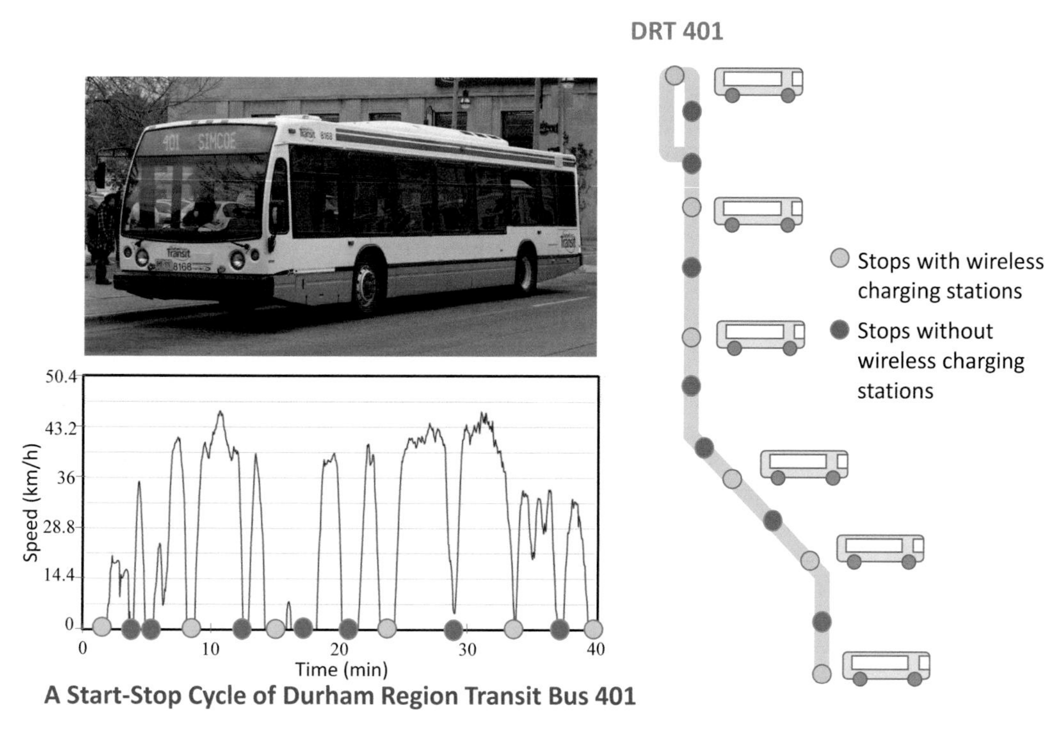

Figure 16. Opportunity charging of an urban bus route (Vincent et al., 2017).

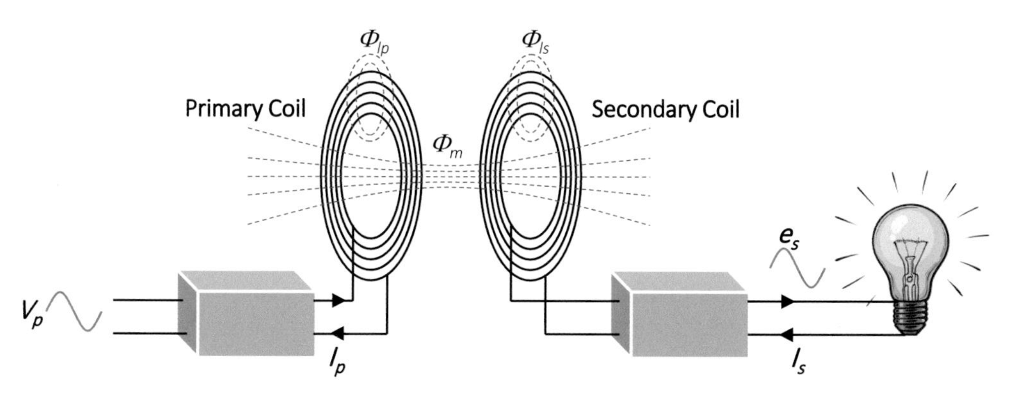

Figure 17. Power transfer via two mutually coupled air cored coils (Aditya, 2016).

If the currents are carried by wires in a coil with N turns, then equation (2) can be simplified as:

$$\oint H.dl = N_p I_P \tag{2}$$

Here, H is the magnetic field strength, N_P is number of turns while I_P is the current flowing in the primary coil, and l is the length of the circumference of the closed path. This time varying magnetic flux links the secondary coil and emf is induced in the secondary coil by the principle of Faraday's law of electromagnetic induction given by equation (3):

$$e_s = N_s \, d\phi_m/dt \tag{3}$$

Here, e_s is the emf induced in the secondary coil and ϕ_m is the flux linking the secondary coil. This emf is capable of driving a current to the load if the circuit is closed. From equation (2), one can understand that the greater the magnitude of current in the primary, the greater the magnetic field strength. From equation (3), one can understand that the higher the rate of change of mutual flux, the greater is the magnitude of emf induced in the secondary. This rate of change of mutual flux which is equal to supply frequency signifies that a high frequency current in the primary is needed to establish high emf in the secondary. However, all the flux does not link the secondary; coupling coefficient k relates common flux to total flux.

$$k = \phi_m/\phi_m + \phi_{lP} \tag{4}$$

In equation (4), $\phi_p = \phi_m + \phi_{lP}$ is the total flux produced by the primary coil. Induced emf in terms of the coupling coefficient is given by equation (5):

$$e_s = kN_s d\phi_p/dt \tag{5}$$

If all the flux link the secondary, then k is 1, and if none of the flux links the secondary, k is 0, i.e., $0 \leq k \leq 1$. Based on the values of k, the magnetically coupled system can be classified into two categories: Tightly coupled systems and loosely coupled systems. In tightly coupled systems, such as transformer and induction motor, the primary is placed in proximity of secondary and flux is shaped by placing windings on the core of high magnetic permeability. Therefore, they have mutual inductance greater than the leakage inductance. Because of tight coupling, k usually lies between 95% to 98% for the transformer and approximately 92% for the induction motor (Zaheer et al., 2011). In the case of using magnetically coupled system for powering EVs, a large air gap is required to allow for inconsistency in the road surface and better clearance between the road and vehicle. Because of this large air gap, the leakage flux is very high and the coefficient of coupling is from 1% to 3% only. Such applications are classified under loosely coupled systems. Poor coupling in loosely coupled systems leads to poor transfer of power. To improve coupling and compensate leakage inductance, capacitive compensation in primary and secondary windings is required (Wu et al., 2012).

Figure 18 shows the global energy chain for a typical RIPT-based EV battery charger. On the way-side, power is usually provided by a utility supply that may be DC from a battery or low frequency (LF) 60 Hz AC from a grid. If the utility is supplied from a grid, then a power factor correction stage may also be included in this block in order to reduce the harmonic pollution of the grid. A high-frequency (HF) voltage at a few tens of kHz is then generated by a high-frequency converter, which is simply an inverter if the utility is DC or a power factor correction rectifier followed by an inverter if the utility is 60 Hz AC. The power level and operating frequency standards are given by SAE J2954 in Table 6.

This high-frequency voltage generates energy in the form of a high-frequency current through a compensation network and primary coil. A primary compensation circuit is added so as to have the primary input voltage and the current in phase to minimize the VA-rating and, thus, the size of the high-frequency power converter. Moreover, primary compensation also acts as a band pass filter, blocking

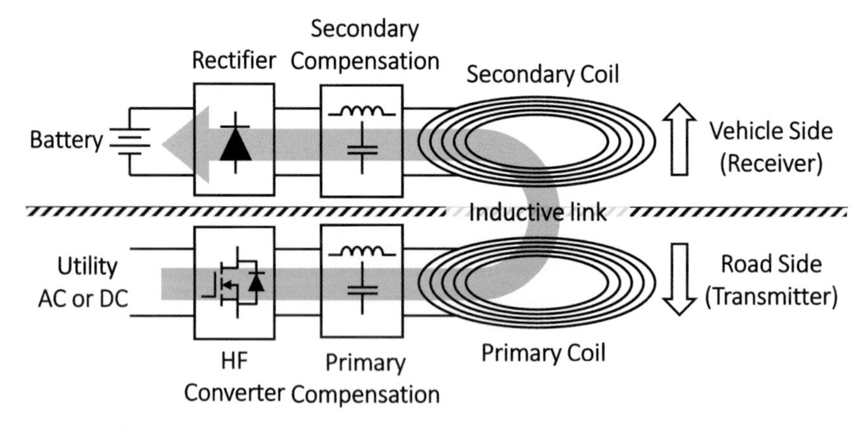

Figure 18. Generalized block diagram of IPT system for EV battery charging (Aditya, 2016; Vincent et al., 2019).

Table 6. Power level and operating frequency standards for static wireless power transfer charging (SAE J2954).

Classification	**WPT1**	**WPT2**	**WPT3**	**WPT4**
Frequency band	81.39 kHz – 90 kHz (typical 85 kHz)			
Maximum AC input power (Volt Amps)	3.7 kVA	7.7 kVA	11.1 kVA	22 kVA
Minimum target efficiency at nominal x, y alignment	> 85%	> 85%	> 85%	> 85%
Minimum target efficiency at offset position	> 80%	> 80%	> 80%	> 80%

undesirable frequency components generated from the power electronic converter feeding the primary. Therefore, an almost sinusoidal current flows in the primary coil and this enables soft switching operation of the converter feeding the primary.

Energy is then transferred to the vehicle side through the secondary coil which is mutually coupled to the primary coil through the flux generated in the air-gap by the primary coil current. The energy received by the secondary coil is then processed by the secondary compensation circuit which is added to improve the power transfer capability of the system. Finally, the voltage received is rectified so as to make it exploitable by the load (batteries). Depending upon the control, an additional DC-DC converter is sometimes included between the rectifier and the load. Apparently, the DC-DC converter brings in more components and corresponding losses.

Compensating networks, which are capacitor arrays, are made to resonate with coil inductance, thus forming a resonant inductive link. Depending on the connection of the compensating capacitor in the primary and secondary coils, four types of resonant inductive links can be defined: Series-series (SS), series-parallel (SP), parallel-series (PS) and parallel-parallel (PP). Primary parallel compensation, such as PP and PS resonant inductive links, allows for a higher primary current as only a small part of the current flows through the semiconductor. However, PP and PS have several fundamental drawbacks. First, they require an additional series inductor to regulate the inverter current flowing into the primary resonant tank. This series inductor, in turn, increases the converter size and, therefore, the total cost of the RIPT system. Secondly, due to the circulating current in the primary resonance tank, the partial load efficiency of the parallel compensated primary system is lower. Thirdly, in PP and PS resonant inductive links, the value of the primary compensation capacitor is not constant but varies with varying mutual coupling and load. Therefore, PP and PS resonant inductive links will require sophisticated control strategies to maintain unity power factor operation in the primary power supply, irrespective of load and mutual coupling variation.

Primary series compensation enables canceling of the significant voltage drop of a primary coil, therefore, the required voltage rating of the power supply is reduced. In SS and SP resonant inductive links, no extra inductor is needed. Moreover, primary compensation is independent of load. However, in an SP resonant inductive link, primary compensation depends upon mutual coupling and, therefore, needs consideration in dynamic charging applications. The SP resonant inductive link requires a higher value of capacitance for stronger magnetic coupling, and its peak efficiency is inferior to an SS resonant inductive link. Therefore, an SS resonant inductive link is theoretically the best in terms of efficiency, component count, complexity of control, and cost.

6. Photovoltaic/Grid Interconnected DC Charging Applications

Currently, charging electric vehicles heavily involves the use of the AC grid. Integrating renewable energy sources into a charging infrastructure reduces the dependency on the AC grid. Photovoltaic (PV)/ Grid interconnected systems have been used for commercial charging infrastructure. The use of a solar and grid interconnected system is an attractive solution for residential charging systems for EVs. For systems up to 10 kW, single phase inverters can be used for residential applications. For interconnection of the residential solar PV to the grid, various isolated and non-isolated topologies are available with multiple stages. Residential photovoltaic systems for EV charging require features such as bidirectional power flow capability, isolation and voltage boost capability to match the solar PV array voltage to the grid voltage requirements.

6.1 Photovoltaic/grid integrated DC charging architecture

The most common architectures that are used for photovoltaic grid connected charging infrastructures have a common AC Link or a DC Link, shown in Figures 19 and 20, respectively.

A detailed comparison of the various architectures has been presented in (Mouli et al., 2015). The AC link structure provides isolation to the EV from the grid as well as to the PV from the grid through a high frequency transformer. It consists of multiple power conversion stages, which can be costly, and the reliability can be an issue. The DC link architecture is a more common structure in which the PV is connected to the grid through a DC/DC stage followed by an inverter stage.

In (Gurkaynak, 2009), a solar powered residential home with AC charging of PHEVs was proposed. The topology has single phase photovoltaic/grid integrated H-bridge inverter integrated with a bidirectional DC-DC converter for energy storage. The PHEV to be charged is connected as an AC load on the grid side. The isolation requirement was not taken into consideration in this topology.

In (Traube, 2013), the architecture used is similar to the one shown in Figure 20. A 10 kW PV/grid connected charger topology was proposed for mitigating the intermittency of the photovoltaic power system using the charger functionality. The PV/Grid connection was achieved through a MPPT-based DC/DC converter. The 25 kW PV systems consisted of two strings of PV arrays connected to individual DC/DC. The outputs of the DC/DC converters are connected together, forming a common DC bus voltage which is then connected to the grid through a three-phase inverter. A 10 kW charger has been connected to the DC link. The charger consists of four 2.5 kW soft switched bidirectional DC-DC converters.

EV charging through a residential battery energy storage system (BESS) through a PV/Grid connected 2.2 kW system has been proposed in (Saxena, 2017). The topology consists of a common DC link, to which the PV is connected through a DC/DC converter.

The DC link serves as the input to the single-phase inverter tying the system to the AC grid. The BESS is connected to the DC link of the inverter through a bidirectional DC/DC buck-boost converter. The single-phase inverter has multiple functions, such as to operate as an active power filter and as harmonics mitigation along with reactive power compensation. However, the bidirectional capability of the voltage source inverter has not been addressed. In this topology, for each mode of power flow, there are at least two stages. Any EV connected is charged from the BESS.

As can be seen, solar inverters have used multi-stage systems for grid integration of solar photovoltaics. There has been a great interest in increasing the efficiency and the reliability of single stage inverter systems. When such inverters are used in applications such as a charging infrastructure, the multiple-stages can result in lower efficiencies and power densities in different modes of operations. The

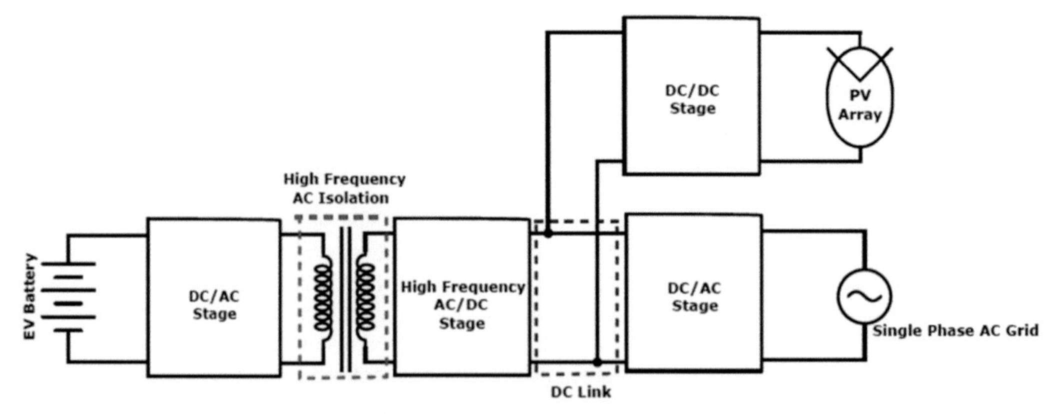

Figure 19. Block diagram of a typical DC charging architecture with DC link (Singh, 2018).

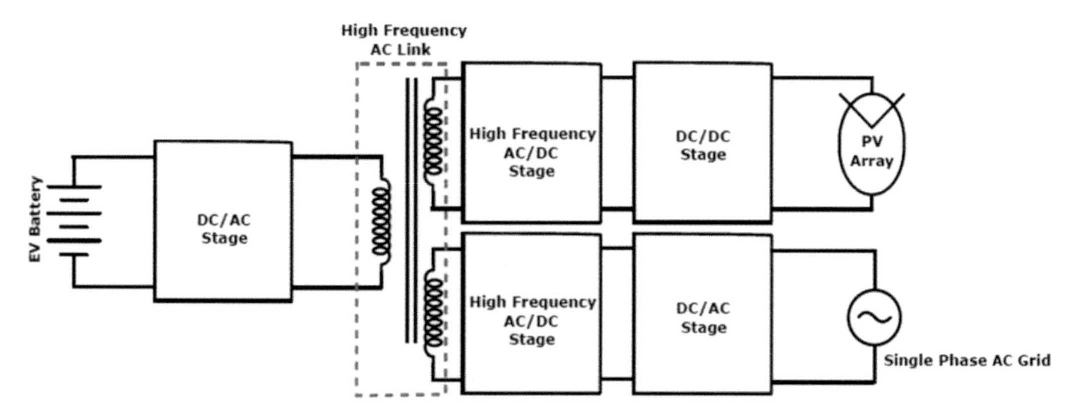

Figure 20. Block diagram of a typical DC charging architecture with an AC link (Singh, 2018).

need for high-efficiency solar converter chargers has resulted in generating particular interest in single stage solar inverters and, more specifically, in the z-source inverter topology proposed in (Singh et al., 2018). It has gained tremendous interest in photovoltaic-grid connected and residential applications.

The operation of a ZSI is heavily dependent on the passive components. It presents an opportunity to integrate energy storage units into such a system. In Shi et al. (2012), the Z-source DC/DC converter topology was used to charge an EV from the PV directly, but the reverse power flow of the EV to the grid was achieved through a separate PWM rectifier topology. The ZSI topology was modified to develop a DC charging infrastructure in (Carli et al., 2013) as a centralized photovoltaic grid connect inverter/charger. Some of the advantages of a Z-source inverter for PV/grid/EV connected charging applications include:

- It has a unique ability to behave as a buck and boost converter in a single stage.
- With the presence of the "shoot through" state, the reliability improves under accidental short-circuit conditions due to Electromagnetic Interference (EMI) from the surrounding equipment.
- Compact high-frequency ZSI systems can be designed by reducing the component sizing.
- The PV, EV and the AC grid operate at three different voltage levels. Two of them are different DC voltages and one of them is an AC voltage. This is an advantage that ZSI topology offers without an additional stage.

7. Concluding Remarks

To conclude, a comparison between wired/plugged-in and wireless/inductive chargers/research prototypes is provided in Table 7. The selected chargers are classified based on input and output voltage ranges, single-phase (1Φ) or three-phase (3Φ), power levels and efficiency. Table 8 gives an overview of the type of charging (slow/fast), charge time and battery capacities of existing selected EV models. The full recharge time in hours is calculated as the ratio of battery capacity in kWh to the power in kW. For DC fast charging, time taken to reach 80% state of charge is given instead of full recharge time.

The electrified transport is safe, reliable, has lower environmental impact and convenient technology. The developments that will happen in the coming years will determine how significant the role of charging strategies in advancing the transportation electrification and the sustainability of autonomous e-mobility. The future implications of electrified transport that need to be investigated thoroughly are listed below:

- Reliable grid connectivity and management during wired or wireless EV charging to balance the demand and supply.
- Optimized deployment of charging infrastructure and energy storage systems.

Table 7. Comparison of wired/wireless chargers/prototypes.

Charging technologies	Selected types	Input voltage (V)	Output voltage	Power (kW)	Efficiency (%)
Conductive/ Wired charger	Brusa-NLG513-U1 (2010)	100–265 (1Φ)	200–520	3.3	93
	Brusa-NLG514-U1 (2010)	100–265 (1Φ)	300–720	3.3	93
	Brusa-NLG664-U1 (2018)	200–250 (1Φ) 360–440 (3Φ)	200–450 310–430	7 22	> 90 > 94
	Continental (2017)	120–240 (1Φ) 208–416 (3Φ)	220–470 450–835	7 22	-----
Inductive/ Wireless charger	(Li et al., 2015)	<= 450 (DC)	300–500	6	95.3
	(Bosshard et al., 2016)	<= 800 (DC)	500–700	50	95.8
	Brusa ICS115 (2014)	85–265 V (1Φ)	170–440	3.3	> 90

Table 8. EV models comparison.

Selected EV models	Battery size (kWh)	Type of charging/ Max power level (kW)	Full or 80% recharge time (Hrs)	Range per full charge (km)	Sample calculation	
					Power level (kW)	Range (km) per hour of charging
Nissan LEAF® (2015)	24	AC level 1/1.92	12.5	~ 135	3.3	~ 19
		AC level 2/19.2	1.25		6.6	~ 37
Volkswagen e-Golf® (2019)	35.8	AC level 1/1.92	18.64	~ 201	3.6	~ 20
		AC level 2/19.2	1.864		7.2	~ 40
		DC fast/50	0.72			
Tesla Model S® (2019)	85	AC level 2/19.2	4.42	~ 400	11.5	~ 54
		DC fast/120 (Tesla Supercharger)	0.7		19.2	~ 90

- Government policies and standards to promote the growth and development of EV technology.
- Vehicle to grid (V2G) and grid to vehicle (G2V) bidirectional power transfer that transforms EVs as mobile energy storage which can regulate the grid by storing excess generation from uncontrolled renewable energy sources, like PV/solar, wind, etc.

Acknowledgements

Authors would like to acknowledge Mr. Navbir Sidhu, Dr. Janamejaya Channegowda, Dr. Kunwar Aditya, and Dr. Siddhartha A. Singh for their contributions.

References

Aditya, K. 2016. Design and Implementation of an Inductive Power Transfer System for Wireless Charging of Future Electric Transportation, Ph.D. diss., Ontario Tech University.

Aditya, K. and Williamson, S.S. 2019. Design guidelines to avoid bifurcation in a series-series compensated inductive power transfer system. IEEE Transactions on Industrial Electronics 66(5): 3973–3982.

Baumann, M., Drofenik, U. and Kolar, J.W. 2000. New wide input voltage range three-phase unity power factor rectifier formed by integration of a three switch buck-derived front-end and a DC/DC boost converter output stage. Presented at the Twenty-Second International Telecommunications Energy Conference.

Birrell, S.A., Wilson, D., Yang, C.P., Dhadyalla, G. and Jennings, P. 2015. How driver behaviour and parking alignment affects inductive charging systems for electric vehicles. Transportation Research Part C Emerging Technology 58: 721–731.

Bosshard, R. and Kolar, J.W. 2016. Multi-objective optimization of 50 kW/85 kHz IPT system for public transport. IEEE Trans. Emerg. Sel. Topics Power Electron 4(4): 1370–1382.
BRUSA. 2010. NLG513-on-board-charger [Online]. Available: https://www.brusa.biz/_files/drive/05_Sales/Datasheets/BRUSA_DB_EN_NLG513.pdf.
BRUSA. 2010. NLG514-on-board-charger [Online]. Available: https://www.brusa.biz/_files/drive/05_Sales/Datasheets/BRUSA_DB_EN_NLG514.pdf.
BRUSA. 2014. ICS–inductive charging system ICS115 [Online]. Available: https://www.brusa.biz/fileadmin/Diverses/Download/Datenblaetter/BRUSA_DB_EN_ICS115.
BRUSA. 2018. NLG664-on board fast charger [Online]. Available: https://www.brusa.biz/fileadmin/template/Support-Center/Datenbl%C3%A4tter/BRUSA_DB_EN_NLG664.pdf.
Carli, G. and Williamson, S.S. 2013. Technical considerations on power conversion for electric and plug-in hybrid electric vehicle battery charging in photovoltaic installations. IEEE Transactions on Industrial Electronics 28(12): 5784–5792.
Channegowda, J., Pathipati, V.K. and Williamson, S.S. 2015. Comprehensive review and comparison of DC fast charging converter topologies: Improving electric vehicle plug-to wheels efficiency. Presented at the IEEE 24th International Symposium on Industrial Electronics (ISIE), pp. 263–268.
Channegowda, J. 2018. Analysis and Design of Computationally Efficient Modulation Schemes for Three Phase Three Switch Rectifier for Transportation Electrification, Ph.D. diss., Ontario Tech University.
Continental. 2017. On-board charger [Online]. Available: https://www.continental-automotive.com/en-gl/Passenger-Cars/Powertrain/Electrification/Charging-Technologies/On-Board-Charger.
Covic, G.A., Boys, J.T., Kissin, M.L.G and Lu, H.G. 2007. A three-phase inductive power transfer system for roadway-powered vehicles. IEEE Transactions on Industrial Electronics 54(6): 3370–3378.
Environment and Climate Change Canada. 2019. Canadian environmental sustainability indicators: Greenhouse gas emissions, Consulted on 8/21/2019. Available at: www.canada.ca/en/environment-climate-change/services/environmentalindicators/greenhouse-gas-emissions.html.
Fisher, T.M., Farley, K.B., Gao, Y., Bai, H. and Tse, Z.T.H. 2014. Electric vehicle wireless charging technology: A state-of-the-art review of magnetic coupling systems. Wireless Power Transfer 1(2014): 87–96.
Gurkaynak, Y., Li, Z. and Khaligh, A. 2009. A novel grid-tied, solar powered residential home with plug-in hybrid electric vehicle (PHEV) loads. Presented at the IEEE Vehicle Power and Propulsion Conference.
Huh, J., Lee, W., Cho, G.H., Lee, B. and Rim, C.T. 2011. Characterization of novel inductive power transfer systems for on-line electric vehicles. Presented at the Proc. IEEE Applied Power Electronics Conference and Exposition (APEC).
Hybrid-EV Committee. 2010. J1772-SAE Electric Vehicle and Plug in Hybrid Electric Vehicle Conductive Charge Coupler, SAE International.
Kedjar, B., Kanaan, H.Y. and Al-Haddad, K. 2014. Vienna rectifier with power quality added function. IEEE Transactions on Industrial Electronics 61(8): 3847–3856.
Li, W., Zhao, H., Li, S., Deng, J., Kan, T. and Mi, C.C. 2015. Integrated LCC compensation topology for wireless charger in electric and plugin electric vehicles. IEEE Trans. Ind. Electron 62(7): 4215–4225.
Liu, J., Ding, W., Qiu, H., Zhang, C. and Duan, B. 2017. Neutral-point voltage balance control and oscillation suppression for Vienna rectifier. Presented at the 3rd International Future Energy Electronics Conference and ECCE Asia (IFEEC 2017-ECCE Asia).
Mallik, A., Ding, W., Shi, C. and Khaligh, A. 2017. Input voltage sensorless duty compensation control for a three-phase boost PFC converter. IEEE Transactions on Industry Applications 53(2): 1527–1537.
Mouli, G.R.C., Bauer, P. and Zeman, M. 2015. Comparison of system architecture and converter topology for a solar powered electric vehicle charging station. Presented at the 9th International Conference on Power Electronics and ECCE Asia (ICPE-ECCE Asia).
Nussbaumer, T., Baumann, M. and Kolar, J.W. 2007. Comprehensive design of a three-phase three-switch buck-type PWM rectifier. IEEE Transactions on Power Electronics 22(2): 551–562.
Praneeth, A.V.J.S. and Williamson, S.S. 2018. A review of front end AC-DC topologies in universal battery charger for electric transportation. pp. 293–298. *In*: 2018 IEEE Transportation Electrification Conference and Expo (ITEC).
Praneeth, A.V.J.S., Patnaik, L. and Williamson, S.S. 2018. Boost-cascaded-by-buck power factor correction converter for universal on-board battery charger in electric transportation. pp. 5032–5037. *In*: IECON 2018—44th Annual Conference of the IEEE Industrial Electronics Society.
Rubino, L., Capasso, C. and Veneri, O. 2017. Review on plug-in electric vehicle charging architectures integrated with distributed energy sources for sustainable mobility. Applied Energy 207: 438–464.
Saxena, N., Hussain, I., Singh, B. and Vyas, A.L. 2017. Implementation of grid integrated PV battery system for residential and electrical vehicle applications. IEEE Transactions on Industrial Electronics 99: 1–1.
Schrittwieser, L., Kolar, J.W. and Soeiro, T.B. 2017. Novel SWISS rectifier modulation scheme preventing input current distortions at sector boundaries. IEEE Transactions on Power Electronics 32(7): 5771–5785.
Shi, L., Xu, H., Li, D., Zhang, Z. and Han, Y. 2012. The photovoltaic charging station for electric vehicle to grid application in smart grids. Presented at the IEEE 6th International Conference on Information and Automation for Sustainability.

Sidhu, N., Patnaik, L. and Williamson, S.S. 2016. Power electronic converters for ultra-capacitor cell balancing and power management: A comprehensive review. pp. 4441–4446. *In*: IECON 2016—42nd Annual Conference of the IEEE Industrial Electronics Society.
Sidhu, N. 2017. Hybrid Electric Energy Storage with Lithium-Ion Battery and Bank Switched Super-capacitors, Master's thesis, Ontario Tech University.
Singh, B., Singh, B.N., Chandra, A. and Al-Haddad, K. 2004. A review of three-phase improved power quality AC-DC converters. IEEE Transactions on Industrial Electronics 51(3): 641–660.
Singh, S.A. 2018. Design and Implementation of a Single Phase Modified Z-Source Inverter Topology for Photovoltaic/Grid Interconnected DC Charging Applications, PhD diss., Ontario Tech University.
Singh, S.A., Carli, G., Azeez, N.A. and Williamson, S.S. 2018. Modeling, design, control, and implementation of a modified Z-source integrated PV/Grid/EV DC Charger/Inverter. IEEE Transactions on Industrial Electronics 65(6): 52135220.
Soeiro, T.B., Friedli, T. and Kolar, J.W. 2012. Swiss rectifier—A novel three-phase buck type PFC topology for Electric Vehicle battery charging, presented at the 2012 Twenty-Seventh Annual IEEE Applied Power Electronics Conference and Exposition (APEC).
Traube, J. et al. 2013. Mitigation of solar irradiance intermittency in photovoltaic power systems with integrated electric-vehicle charging functionality. IEEE Transactions on Power Electronics 28(6): 3058–3067.
Vincent, D., Huynh, P.S., Patnaik, L., Capano, D. and Williamson, S.S. 2017. Accelerating the autonomous electric transportation revolution: wireless charging and advanced battery management systems—IEEE transportation electrification community [Online]. Available: https://tec.ieee.org/newsletter/december-2017/accelerating-the-autonomous-electric-transportation-revolution-wireless-charging-and-advanced-battery-management-systems. [Accessed: 22-Aug-2019].
Vincent, D., Chakraborty, S., Huynh, P.S. and Williamson, S.S. 2018. Efficiency analysis of a 7.7 kW inductive wireless power transfer system with parallel displacement. pp. 409–414. *In*: 2018 IEEE International Conference on Industrial Electronics for Sustainable Energy Systems (IESES).
Vincent, D., Huynh, P.S., Azeez, N.A., Patnaik, L. and Williamson, S.S. 2019. Evolution of hybrid inductive and capacitive AC links for wireless EV charging—a comparative overview. IEEE Transactions on Transportation Electrification 5(4): 1060-1077.
Wang, H. 2016. Capacitors in Power Electronics Applications—Reliability and Circuit Design, IECON (tutorial), Florence, Italy.
Wijeratne, D.S. and Moschopoulos, G. 2012. A novel three-phase buck boost AC DC converter. IEEE Transactions on Power Electronics 29(3): 1331–1343.
Williamson, S.S. 2007. Electric drive train efficiency analysis based on varied energy storage system usage for plug-in hybrid electric vehicle applications. pp. 1515–1520. *In*: 2007 IEEE Power Electronics Specialists Conference.
Wu, H.H., Gilchrist, A., Sealy, K., Israelsen, P. and Muhs, J. 2011. A review on inductive charging for electric vehicles. Presented at the Proc. IEEE International Electric Machines and Drives Conference.
Wu, H.H., Gilchrist, A., Sealy, K.D. and Bronson, D. 2012. A high efficiency 5 kW inductive charger for EVs using dual side control. IEEE Transactions on Industry Informatics 8(3): 585–595.
Yilmaz, M. and Krein, P.T. 2013. Review of battery charger topologies, charging power levels, and infrastructure for plug-in electric and hybrid vehicles. IEEE Transactions on Power Electronics 28(5): 2151–2169.
Zaheer, A., Budhia, M., Kacprzak, D. and Covic, G.A. 2011. Magnetic design of a 300 W underfloor contactless Power Transfer system. Presented at the Proc. Annual Conference on IEEE Industrial Electronics Society.

Section 3

Electricity and the Grid

CHAPTER 17

The Role of Microgrids in Grid Decarbonization

Md Rejwanur Rashid Mojumdar, *Homam Nikpey Somehsaraei* and *Mohsen Assadi**

1. Introduction

Modern civilization is heavily dependent on a secure supply of electricity as it is the most efficient medium for energy transmission and distribution. For the last 200 years, the enormous electricity grid, which is now the biggest infrastructure that humanity has ever built, has been developed. However, that grid infrastructure has grown old and been challenged to meet expected security, reliability and, quality of supply. The decentralized demand for electricity has been reaching beyond the centralized transmission capacity and the primary sources of centralized energy production, such as coal and oil, have been responsible for dramatic increases in atmospheric CO_2 concentrations. In every regard, the electricity infrastructure needs to be renewed and revised. This will eventually require a very significant amount of investment. International Energy Agency (IEA) estimated that an investment of $16 trillion would be required for the energy sector over the period 2003–2030 (Birol, 2004). For this much further investment, it was found wise to revise the operational philosophies of conventional power system so that the modernized infrastructure meets the expected system security, environmental protection, energy efficiency, cost of supply, operational safety and power quality.

Certainly, one of the major objectives of electricity grid transformation is to achieve a decarbonized power grid, which will be highly utilized to limit the global temperature rise from carbon emissions. At the historic Paris Agreement on climate change, countries committed to keeping global warming no higher than 2 ºC (3.6 ºF) above pre-industrial levels while trying to limit the temperature increase to 1.5 ºC (2.7 ºF). In the recent report, the Intergovernmental Panel on Climate Change (IPCC) claims that even a 1.5 ºC increase will have large effects on the climate. IPCC also claims that 1.5 ºC to 2 ºC, this half a degree more of warming, will induce drastic consequences of climate change (Intergovernmental Panel on Climate Change, 2018). In the last decade, though many countries have taken large initiatives and the share of renewable energies in the energy mix has been increasingly visible, these efforts are not even close to being sufficient. Under current policies and development, energy-related emissions only restrain their upward slope and remain flat around 35 Gt/year, whereas these emissions should be gradually decreased to 9.7 Gt/year by 2050 in order to keep the temperature rise just below 2 °C (Remap

University of Stavanger, Dept. of Energy and Petroleum, Postboks 8600 Forus, 4036 Stavanger, Norway.
Emails: md.r.mojumdar@uis.no; homam.nikpey@uis.no
* Corresponding author: mohsen.assadi@uis.no

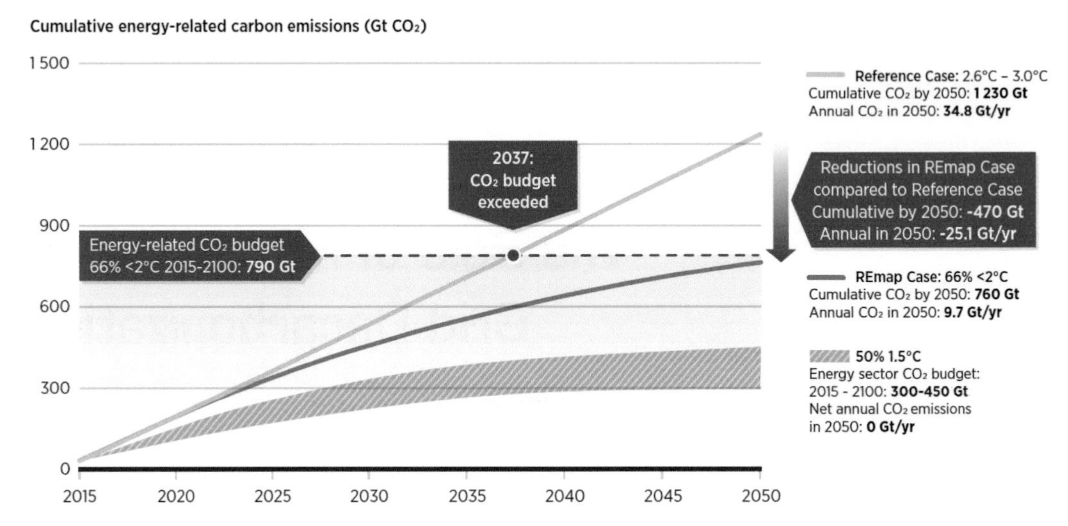

Figure 1. Cumulative energy-related CO_2 emissions and emissions gap, 2015–2050 (Gt CO_2) (International Renewable Energy Agency, 2018).

Case) (International Renewable Energy Agency, 2018), as shown in Figure 1. In addition, to keep the temperature rise at 1.5 °C, energy-related emissions should be gradually decreased to zero Gt/year by 2050, which seems quite difficult to achieve. In fact, according to the International Renewable Energy Agency (IRENA)'s global energy transformation roadmap to 2050 (International Renewable Energy Agency, 2018), renewable energy (RE) and energy efficiency schemes (EES) in combination can provide over 90% of the necessary energy-related CO_2 emission reductions. There, it was also suggested that the deployment of renewable energy sources (RESs) and EES needs to be scaled up at least six times faster for the world to meet the decarbonization and climate mitigation goals set out in the Paris Agreement.

Therefore, to limit the global temperature rise from carbon emissions, along with other major concerns, worldwide electricity grids require modernization, decarbonization and transformation to smart grid (SG). The notion of SG refers to a thoroughly modernized electricity delivery system that appends communication and control systems to the existing power delivery infrastructure. According to the European Technology Platform of Smart Grids (European Smart Grids Technology Platform, 2006), an SG is an electricity network that can intelligently integrate the actions of all users connected to it—generators, consumers and those that assume both roles—in order to deliver sustainable, economic and secure electricity supplies efficiently. There, the key idea is to integrate a great amount of renewable and cleaner energy in the electrical power system, make that system energy efficient and then serve the all potential energy demands for all the sectors (buildings, transport, district heating, industry and power sector itself) utilizing the SG.

It is worth noting that the power systems have always been "smart" at the transmission level. In fact, it is the power distribution level which requires more "smartness". Firstly, there are many micro-RESs available at the distribution level, which are distributed in nature. In addition, small-scale distributed generations (DGs) from less carbon-emitting gas/biofuel based combined heat and power (CHP) or combined cooling, heat and power (CCHP) plants, fuel cells, different storages and other micro-turbines can be deployed in the distribution side. As well as being mostly RES and relatively lower emitters of carbon, these distributed energy resources (DERs) can contribute to EES in two significant ways. DERs can save energy, avoiding significant transmission losses. Then, being locally generated, CHP and CCHP plants can supply local heat and cooling demands, which, in turn, increases energy efficiency. As more of the energy demands can be supplied locally, the DG philosophy also relieves the requirement of extended transmission infrastructure and saves major capital investments. On top of all, most of the energy customers are at the distribution level. Active participation from the customers in terms of controllable loads can be significantly utilized in demand response (DR) schemes. Considering all of these, it is obvious that the distribution level needs more control and management capabilities to be active and smart for the integration of DERs and DR.

Over the years, with their higher share in the generation mix, DERs will have much more impact on power balance and grid frequency. Therefore, control, aggregation and management architecture are required to integrate DERs and DR resources at the distribution level and realize an SG. There comes the concept of microgrid (MG), the building block of the active distribution network and, hence, the SG. The formation of MGs arises from the aggregation and control capability over a smaller distribution network, which possesses DERs, such as DGs, different storages and controllable loads.

Of course, there are many other concepts in SG that work outside or in parallel with MGs and play a vital role in the grid decarbonization. For example, advanced metering infrastructure can enable major energy efficiency programs by monitoring, feedback and diagnosis, which are estimated to save a lot of energy (Pratt et al., 2010). However, beyond monitoring, an SG needs distributed control and management platforms to aggregate and control DERs. In that regard, MG is a vital concept of the SG paradigm because it can aggregate as well as control DERs located in a particular geographical area. Other concepts in this regard, such as the virtual power plants (VPPs) (Hatziargyriou, 2014), can only aggregate output of power plants located at different geographical locations to take part in the electricity market but they do not have any energy management system (EMS) and control over its resources. Therefore, the role of MGs is crucial and unique in the SG paradigm.

In this chapter, section 2 discusses the major technical challenges of transformation towards the decarbonized grid and the corresponding role of MGs. Section 3 discusses the MG concept and its position in the hierarchical structure of SG and electricity markets. Section 4 articulates MGs' role in the facilitation of different grid decarbonization schemes and finally, section 5 concludes the chapter with a summary.

2. Major Technical Challenges in the Transformation towards Decarbonized Grid and MGs' Role

The transformation from the conventional power grid to a decarbonized, smart and modernized power grid brings many challenges. In SGs, the expected higher share of RE will cause the system net demand (demand minus generation from RE) to be largely variable, uncertain, and highly ramping (Thatte and Xie, 2016). As we know, the demand for energy is always varying; however, conventional power plants are very flexible in regulating their output, so, in a conventional grid, balancing the supply to the demand is straightforward. In contrast, in modernized SGs, a large share of RE production, specifically regarding solar and wind, are intermittent. Furthermore, the outputs from these resources are either direct current (DC) or alternating current (AC) with harmful harmonics, so their output needs well-designed conversion and conditioning to conform to the power quality standard (Qi et al., 2007; Tande, 2002). These intermittent resources currently contribute roughly 40% of the global RE production and their share in the energy mix is expected to grow exponentially (International Renewable Energy Agency, 2018). Moreover, for economic reasons, an SG is not built from scratch, but rather uses much of the conventional grid infrastructure which was not designed to serve the greatly increased generation from renewables and multi-sector loads. Particularly, in the paradigm of the SG, the power distribution system, which was almost unsupervised conventionally, becomes highly active. Considering these issues, the main challenges of energy system transformation, from the technical and engineering perspective, are:

i) the integration of renewables at all the levels of the power grid, especially in the distribution side
ii) the implementation of novel energy efficiency measures and programs
iii) the formation of novel electricity market structures to include DERs
iv) the active management of unbalanced, radial and conventionally invisible distribution system
v) the equilibration of ever-varying demands with intermittent RE
vi) the assurance of power quality, reliability, protection and, security in the grid with renewables
vii) the solution of line congestion to serve increasing demands from all the sectors, especially plug-in hybrid electric vehicles (PHEVs)

Exclusively, the well-designed operation and control functions of MGs, the building blocks of SGs, are contrived to play a crucial role in tackling these challenges and many more. Importantly, MGs can continually facilitate the optimized operation of the enclosed power network that they maintain. In fact, MGs, with their aggregation, control, power conditioning, and marketing capabilities, make it possible to safely integrate DERs, specifically renewables, storages, and CHPs, at the distribution side. MGs also largely contribute to the aggregation, control, and management of PHEVs and DR applications. Altogether, MGs play a crucial role in the transformation towards a decarbonized grid.

However, with the emergence of the MG concept, the operational philosophy, control hierarchy and market operation of SGs will be fundamentally different from the conventional grid practices. In general, intelligence will be distributed. Consequently, the operation and control will be much more complex. Therefore, to emphasize MGs' crucial role in grid decarbonization, it is most important to realize the MG concept with its placement in the hierarchical structure of SG for control, optimization and marketing. Furthermore, a brief discussion regarding the mechanism of a single MG unit is necessary in order to comprehend MGs' role in grid decarbonization. Hence, the following two sections present these essential descriptions regarding MGs with aiding illustrations.

3. The Concept of MG and Multi-MG Systems

According to the definition from the EU research projects (Hatziargyriou et al., 2006; Schwaegerl, 2009) MGs comprise low voltage (LV) distribution systems with DERs (micro-turbines, fuel cells, PV, etc.), together with storage devices (flywheels, capacitors, batteries and heat storages) and flexible loads (electrical, heat). Such systems can be operated in a non-autonomous mode if interconnected to the grid, or in an autonomous mode if disconnected from the main grid. If aggregated and controlled efficiently, MGs can provide distinct benefits.

It is important to note that there are three mandatory MG features: Local loads, local DERs and intelligent control. Environmental protection schemes in terms of carbon credit will also be an essential feature of MGs if the specific country promotes that provision (Hatziargyriou, 2014). Another important point about MGs is that they are built at a specific geographic location. Therefore, generation and load aggregation ideas like VPP are not MGs as their sources are located in diverse locations and they do not have control over them. Then, it is worth mentioning that, though MGs can be operated in autonomous or islanded mode when there are disturbances in the main grid, the majority of them will be operated in the grid-connected mode for most of the time (Hatziargyriou, 2014). A longer period of islanded operation will require the MGs to have larger storage systems. Moreover, as grid-connection offers the opportunity of participation in the larger electricity markets, the main benefits of the MG concept are seen from the grid-connected operation. In some conceptions, especially for cold countries, a heat grid is also an important part of MGs, where heat from CHP plants can be utilized to supply local heat needs and, therefore, the overall generation efficiency can be improved (Hatziargyriou, 2014). Distributed storages (DSs) and distributed heat storages (DHSs) play the vital role in MGs of balancing out generation and demand, smoothing out the ripples and shifting the peak load to a period of lesser load density.

MGs can be of various grid sizes, and, generally, they range in 100 kW to 1–2 MW of power. At large sizes, an entire LV grid can be defined as a single MG. On the other hand, an MG can be an LV feeder or even an LV house, meeting the mandatory features. In Figure 2, an MG formed of an LV feeder is shown. It can be seen that there exists an MG central controller (MGCC), assigned for central optimization, aggregation and control. In addition, there must be local controllers (LCs) for the local control of different DERs in the MG. MGCC and LCs have bidirectional communication (as denoted by the green dashed arrows) between them and these schemes vary with the assigned control strategy.

MGs in the hierarchical structure of an SG is illustrated in Figure 3, which primarily bounds the four levels of SG hierarchical structure for control and optimization. A concise and introductory description of each level and their interaction is given here. This overview and Figure 3 will be referenced when key concepts are explained in the ensuing sections.

The top level is the level of the transmission network operator (TNO), with big generation plants and associated markets such as the capacity market, day-ahead market and, intra-day market. TNO manages

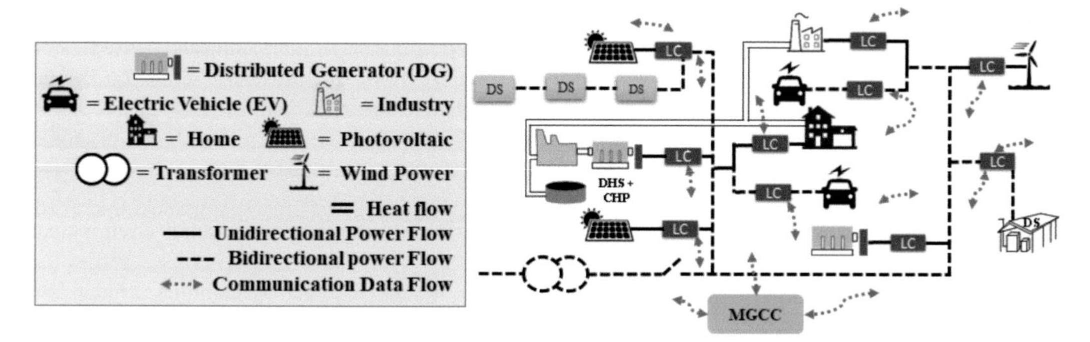

Figure 2. A typical MG on LV feeder.

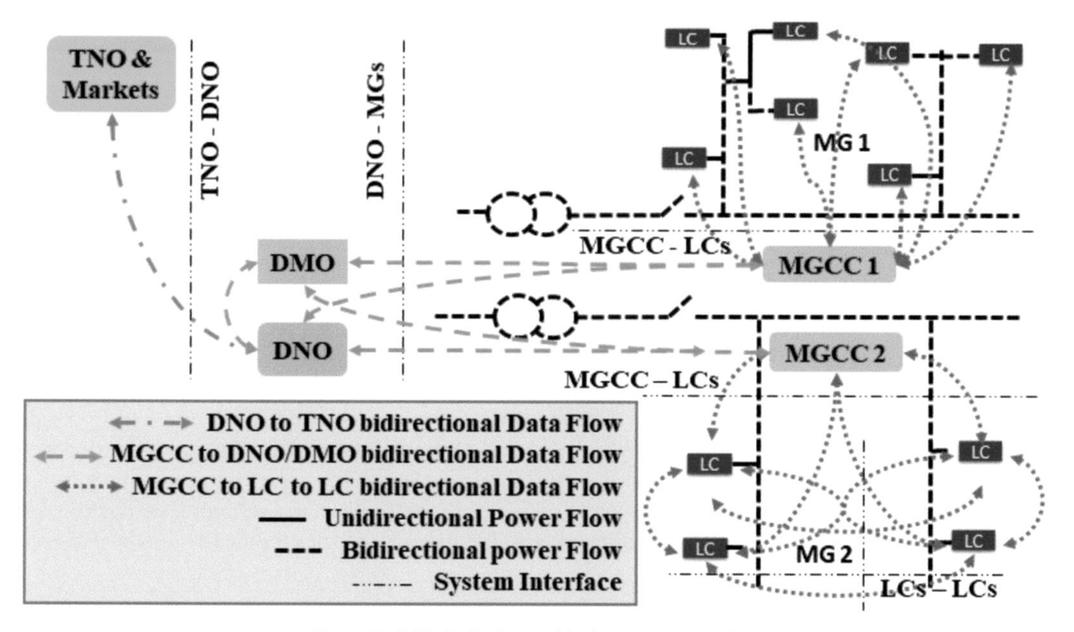

Figure 3. MGs in the hierarchical structure of an SG.

and supervises the wide-area power transmission line through the transmission supervisory control and data acquisition (SCADA) system. The second level is formed with the distribution network operator (DNO) and distribution market operator (DMO). The DNO will possess the distribution SCADA system to acquire real-time states of the distribution system. The DMO will be a platform to host an electricity market in the TNO-DNO boundary. Following this is the third level, constituted with MG laterals with MGCCs. Finally, there is the level of local controllers (LCs) for DERs and bundled loads of building and industry. Importantly, Figure 3 depicts general schemes of bidirectional data flow (denoted by dashed arrows of different colors) between different components like LC, MGCC, DNO, DMO and TNO. Dashed arrows of the same color mean that they serve data between the same pair of control or optimization level. For example, the green dashed arrows represent the data exchange between LCs and a corresponding MGCC in charge. The MGCC receives bids, forecasts and states from the LCs' sensors. Then, the MGCC utilizes these data in an optimization algorithm to send control references back to the LCs and bids up to the DNO/DMO for market participation. Similarly, the data exchanges between MGCCs and DNO/DMO as well as between the DNO/DMO and TNO is shown. At each of those interfaces, some optimization is done as the data is exchanged.

It is worth mentioning that, in some propositions, a multi-MGs level is suggested, generally applied at higher voltage levels (i.e., medium voltage), implying the coordination of interconnected but separated MGs collaborating through a multi-MG controller (Hatziargyriou, 2014). Then, if the multi-MG optimization, where several MGs can also be optimized together, is included, then there will be five levels in the hierarchical structure of control and optimization for an SG. There, the multi-MG level would also have data exchange and optimization. In multi-MG optimization, different DGs and controllable loads will be aggregated and controlled by an intermediate control level between MGs and DNO/DMO. This level will optimize the operation of the multi-MG network, under a real market environment (Hatziargyriou, 2014). This level can contribute in the tertiary control level of MGs (discussed in the next section). However, multi-MGs level for SG is not a generally accepted or common scheme, so, for the sake of simplicity, this level is not illustrated in Figure 3. For the same reason, this level is not explicitly mentioned further in the discussion below. However, if this level is operational in a zone of SG, its placement in the hierarchy, data exchange, optimization, and market participation is straightforward, considering another level of optimization between MGs and DNO/DMO.

3.1 Control, optimization and market mechanisms of a single MG unit

An MG works both as an aggregator of DERs and DR resources and as a controller of the resources that it encloses. It has a complex mechanism for control, resource optimization and market participation. The mechanism is summarised below.

3.1.1 Control hierarchy of MGs

As the mechanism of microgrid control is discussed with great detail in (Bidram et al., 2017; Bidram, Member and Davoudi, 2012; Hatziargyriou, 2014), a single MG has four levels in its own hierarchical control structure, as shown in Figure 4. Zero level control with constant active power and reactive power (PQ) mode control or constant voltage and frequency (VF) mode control is the fasted level of inner loop control. Primary level control with droop characteristics is the next level of fast controls that provide voltage and frequency stability, load sharing and plug-and-play capability. In fact, droop characteristics of any generator are the linear relationships that exists between active power versus frequency (P-Q droop) and reactive power versus voltage (Q-V droop) (Bidram et al., 2012). Primary and zero level controls are implemented in the LCs of the MG where different grid following topologies for the LCs can be designed

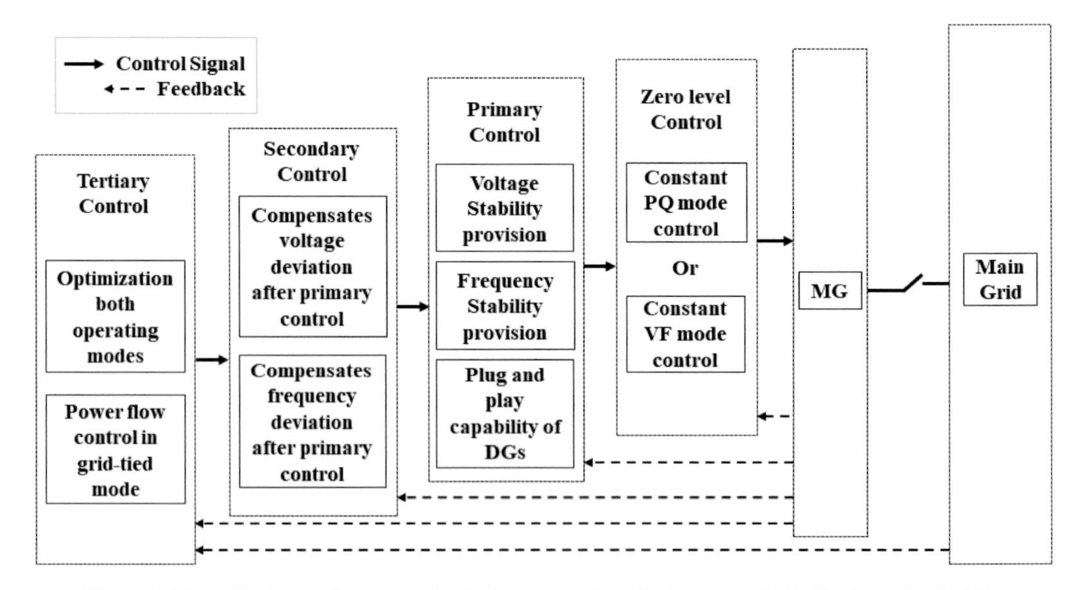

Figure 4. Hierarchical control structure of an MG (concepts from (Bidram et al., 2017; Hatziargyriou, 2014)).

by a combination of droop characteristics with constant PQ or constant VF control. Secondary control is managed by the EMS of MG, which is at the MGCC in the case of the centralized scheme and formed by a combination of MGCC and LCs in the case of the distributed scheme, explained in next subsection. At the secondary control level, references are set by a vital optimization algorithm. The optimization has other objectives, however, for the secondary control, it sets references to compensate voltage and frequency deviations by balancing demand and generation in a grid-connected or islanded market situation. This optimization keeps running in the time resolution of 5–15 minutes and the output of this is used as the references of the primary and zero level control. Above the secondary control, the tertiary control levels are managed by both the DNO/DMO operators and the EMS of MG. The mechanism of the tertiary control also varies depending on the MGs' mode of operation (grid-connected or islanded). When an MG is operating in a grid-connected way, the tertiary control references come from the market dispatch decisions. There exist several markets with several objectives and time horizon where data of bids, forecasts and states from many agents, including MGs, are used and dispatch schedules are made. Here, MG EMSs run optimization to bid optimally to different markets. The sequence of markets is discussed in a later subsection. However, in the case where MGs are run in isolated mode, the tertiary control references are made solely from the MG EMSs by solving the related constrained optimization problem, briefed at the next subsection.

3.1.2 The optimization

The vital role of MGs is to resolve the conflicting interests of all the stakeholders. The EMS of an MG plays an important role in that resolution, being in the middle of two hierarchical levels—the DNO/DMO and LCs. Firstly, there are information and requests from the LCs of each component within an MG. These requests can be in the form of bids for generation, loads and controllable loads from each stakeholder. In some alternative schemes, instead of bids, information for optimization can also be forecasts of demands and generations (Hatziargyriou and Tsikalakis, 2005). Secondly, MGs have communications with the DNO/DMO. When an MG runs in grid-connected mode, it communicates to get utility requirements, the forecast of market prices and the larger market platforms to bid from MG side. If an MG runs in islanded mode, though it does not consider the grid-side market, it still forms grid quality supply (voltage and frequency) and balance power by dispatching its generation and load. Therefore, in both grid-connected mode and islanded mode, the EMS of an MG needs to resolve many conflicting interests of stakeholders by optimization in three axes of concern—economical, technical and environmental (Hatziargyriou, 2014). A few concerns of technical and environmental axes also have an effect on the economical axes. For example, greenhouse gas (GHG) emissions in the environmental axes incur emission costs and power losses in the technical axes cause power loss costs, both on the economical axes. Finally, the outputs of this very optimization in MG EMS level are communicated to the LCs of DERs to set control references of secondary level control, which has been discussed in the previous subsection. Furthermore, optimal bidding from the MG side is also an output of the optimization which are communicated to different markets. This information flow and the optimization functions of EMS for an MG are shown in Figure 5.

For this vital optimization, centralized or distributed, either of the control schemes is applied in MG EMSs (Su and Wang, 2012). As an example, in Figure 3, MG 1 is built with a centralized scheme and MG 2 is built with a distributed scheme. These schemes are described below.

3.1.2.1 The centralized scheme

In the centralized control scheme, as shown for MG 1 of Figure 3, the EMS of the MG is placed solely in a central controller, as detailed further in (Chaouachi et al., 2013; Daniel E. Olivares and Claudio A., 2014; Su and Wang, 2012). That central controller has bi-directional communication with all the LCs of the MG and DNO/DMO level. Here, the LCs have no autonomy and intelligence. The LCs are obliged to follow the outcome references from the central controller's optimization. This scheme is simple to implement, easy to maintain and requires a lower cost. However, having a single controller, the EMS has a huge computational burden, requires high-bandwidth links and possesses a single point of failure. This

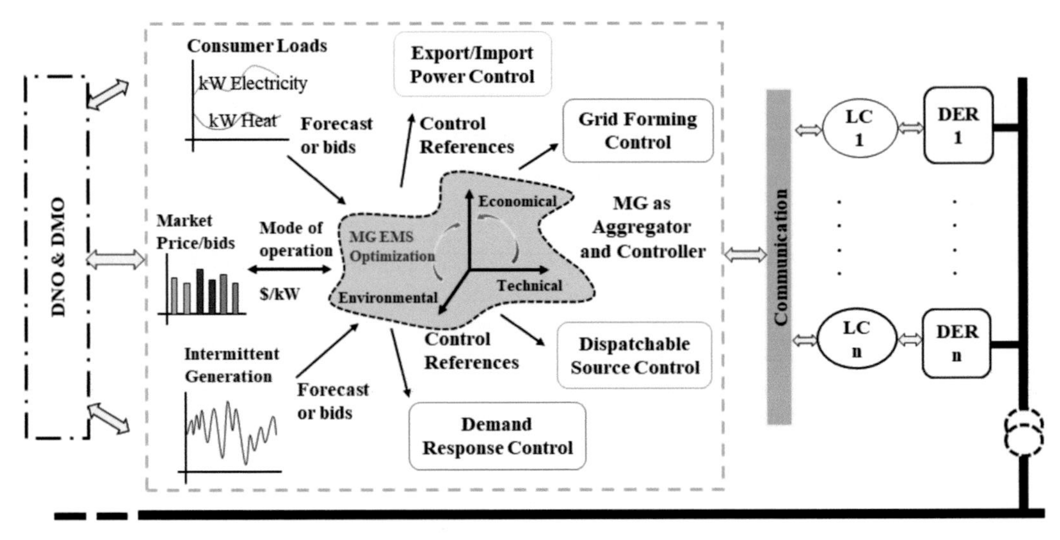

Figure 5. Optimization in three axes at the MG EMS between DNO/DMO and LCs.

scheme is not easy to expand and has very limited plug-and-play functionality, which is required in order to deploy PHEVs and for the ever-growing installation of DERs (Su and Wang, 2012).

3.1.2.2 The distributed scheme

In the distributed scheme, as shown for the MG 2 of Figure 3, each MG component is regulated by one or more LCs, rather than being governed by a central master controller. Every LC communicates with other local controllers and possesses the intelligence to make operational decisions on their own. This scheme accommodates more autonomy and intelligence for the individual stakeholders. However, it is worth noting that a central controller and EMS might still be necessary for this scheme, which exchanges energy price information, utility requirements and bids with the DNO/DMO. That central controller may take over the control of the local controller in the event of grid contingencies or equipment failure. In normal operations, the control is given back to the LCs and they operate autonomously using primary control loops, like frequency droop control, Volt-VAR control and local information (Bidram et al., 2017; Huang et al., 2011; Kantamneni et al., 2015; Logenthiran et al., 2012; Su and Wang, 2012). This scheme has better plug-and-play capabilities, system expansion ability and a lower computational burden in each controller. This scheme also avoids a single point of failure. However, this scheme requires a larger investment in the communication facility and more time for convergence between local controllers (Huang et al., 2011; Su and Wang, 2012).

3.2 Electrical market structure and MG's participation

In a concise description, an MG's two main functions are aggregation and control. As it aggregates DERs and DR resources, its position and participation mechanism in the electrical market is an area of great interest. The decarbonized grid of the future will run with the same three levels of regulations in a conventional power system. As the regulations and markets in power system operation are depicted in Figure 6, the primary regulation is the automatic energy variation of a unit by local proportional control to stabilize frequency. It is mandatory for every generator and not remunerated. Primary regulation is the fastest regulation and its response time is in seconds. As can be seen in Figure 4, the zero and primary level of control in the hierarchical control of MGs takes care of this primary regulation.

Errors still exist after proportional primary regulation. The secondary regulation, which is also called automatic generation control (AGC), is there to minimize remaining errors. It also returns tie-line powers

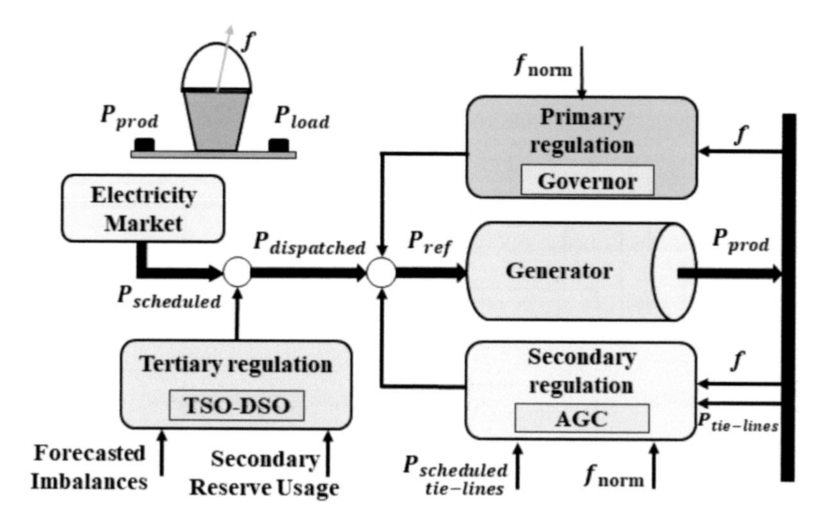

Figure 6. Regulations and markets in power system operating philosophy.

of an area to their scheduled values. The secondary level of MGs' hierarchical control does a similar job, as can be seen in Figure 4. However, from the main grid's operating philosophy, AGC is optional for generators and it is allotted to generators after they bid in the capacity/reserve market. Selected generators need to adjust their generation within 5 minutes. This market is highly paid by the TNO and the penalty of failure is also high, so only technically robust generators bid here. The acceptance of aggregated renewables to bid in this market is just beginning, so whether MGs can participate in this market depends on the market regulation of a specific country. However, in the islanded mode of operation, MGs must maintain this secondary regulation for the network it encircles, through its secondary level of control.

Between secondary and tertiary regulations, two wholesale electricity markets—the day-ahead and intraday market are operated. These are energy markets, so generators sell and consumers buy. TNO only sees who fails and who should pay penalties. Most generators initially bid in the day-ahead market. If a bid of a generator gets accepted and then there are some problems due to forecasting errors or plant failures, that generator buys the lacking generation from other generators in the intra-day market—a market that runs every 15 min in a day. Also, any generator may decide to bid only in the intra-day market, without taking part in the day-ahead market, if this strategy suits the generator. Now MGs, of course, can take part in these markets, optimizing their aggregated DERs and DR resources.

After all of these, there still remains imbalances and the capacity reserves also require refills. There comes the tertiary regulation. The flexibility market to serve the tertiary regulation is the solution of still existing imbalances and restoration of secondary regulation reserves. Flexibility market is also a market with bids where flexibility/shift from generators already assigned schedule are sold again. In this market, the response time varies from minutes to hour. With the increasing inflexibility due to the spread of RESs, it is expected that flexibility markets will be formed to run every 15 minutes. It is contrived that new flexibility market will be formed in the TNO-DNO interface where DERs, DR applications and, consequently, the MGs as aggregators can play a vital role.

Now, the day-ahead and intra-day energy market are called financial markets and the flexibility market for tertiary regulation is called a service market. As the cash flow interaction of these markets is shown in Figure 7, it can be seen that DERs, including micro-generators and storage units, will participate in wholesale financial markets mainly with the aid of MGs' aggregation. In some cases, VPPs can be formed for aggregation if the case requires, but it would not have controllability. Then, to take part in the ancillary service market, DERs and DRs have both options—through aggregation by MGs or individually with some distributed agent-based method having blockchain-based format for privacy, as in (Guan et al., 2018; Pop et al., 2018). In any of these markets, when the participation is through MG aggregation, there will be two modes of operation: Grid-connected or islanded and the

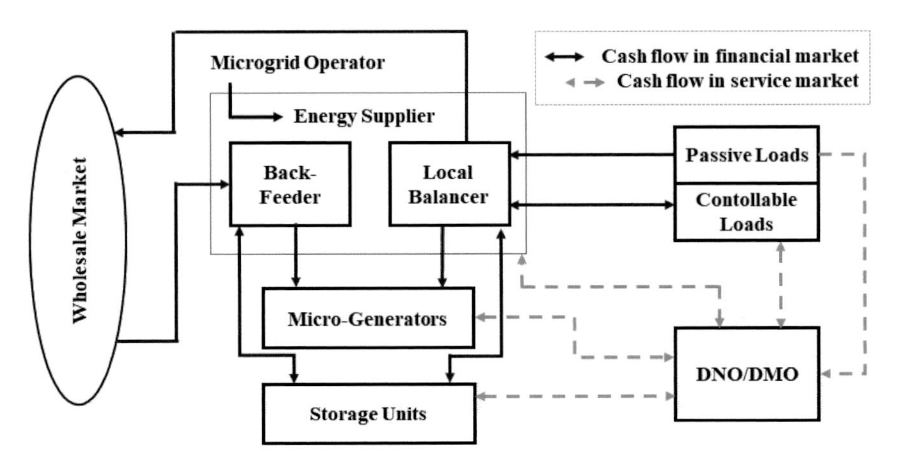

Figure 7. Cash flow interaction of financial and service markets from the distribution side (concept from (Hatziargyriou, 2014)).

optimization in three axes, as stated in subsection before. The optimization there can also be formed as an internal market of the MG itself. Operation of MGs in markets is described in detail in (Celli et al., 2005; Hatziargyriou, 2014; Hatziargyriou et al., 2005; Liu et al., 2016; Logenthiran et al., 2008; Sinha et al., 2008).

4. MGs' Roles to Facilitate Decarbonization Schemes

MGs can play a vital role in grid decarbonization. The reasoning behind the formation of MGs is that they have both aggregation and control capability over resources located in a common geographical location. Therefore, for the power system operation, they can be regarded as a controllable unit—generation or load. MGs' role in major decarbonization schemes are discussed below.

4.1 Integration of renewables

RESs are dispersed in nature and they need to be integrated into the distribution systems. However, conventional power distribution systems were not designed to integrate those intermittent RESs. Therefore, novel power distribution system clusters with superior control and aggregation capabilities were needed to integrate those micro-RESs. This is what MG effectively facilitates. MG's capability of integrating micro-RESs smoothly in the distribution system comes as the first reason to form them. MGs facilitate the integration in the following four ways:

i) Foremostly, the EMS of an MG resolves the conflicting interests of all the stakeholders at its range—RESs, storages, other generations, distribution grid and demand. Hence, the optimization in the MG EMS level is crucial in integrating renewables.
ii) An MG allows a part of the distribution grid to be operated as an islanded grid. This facilitation greatly enables the integration of local renewables in many scenarios.
iii) The LCs embedded in MG have fast and dynamic control, which adapt with the state changes in order to maintain the grid voltage and frequency. Through power conditioning, they also ensure the quality of supply from intermittent renewables, which is a great enabler.
iv) Being an entity in the electricity market, MGs provide the market power to their constituent DERs, which are mostly RESs.

MGs vital role to integrate more RESs is quite evident by now. Numerous research studies have been published around the integration of renewables utilizing MG concept. An off-line optimization approach towards real-time energy storage management for renewable integration in MG was presented in (Rahbar et al., 2015). In (Xuan and Bin, 2008), the advantages and mechanism of MGs to integrate renewables were presented. A detailed study of an MG, its control, architecture, and consequent advantages to integrate renewables were discussed in (Hatziargyriou, 2014).

4.2 Reduction of power line losses

Energy loss in transmission and distribution lines is a major issue when evaluating the performance of a power transmission network. International Electrotechnical Commission (IEC) reports that the overall electrical energy losses between the power plant and consumers are in the range of 8–15% (International Electrotechnical Commission (IEC), 2007). They estimate, of the total electricity generated, between 1–2% is lost in the step-up transformers between generators and transmission lines, around 2–4% is lost in the transmission lines, about 1–2% is lost in step-down transformers between transmission lines and distribution networks and around 4–6% is lost in the distribution network transformers and cables. In reality, 8 to 15% loss of electrical energy in the transmission and distribution is a great amount of loss. MGs can aid in reducing energy losses in various ways, the prominent ones being listed here:

i) As MGs are enabling the integration of DERs into the distribution side, this, in turn, deducts the transmission losses in long transmission lines and meets the demand locally (Hatziargyriou, 2014). Optimal location and sizing of DERs aid further in the loss reduction (Bhumkittipich and Phuangpornpitak, 2013).
ii) Considering time-varying demand and generation, the multi-period optimal power flow in the EMS of an MG can determine the optimal allocation of loads between DERs in a way that minimizes losses (Ochoa and Harrison, 2011).
iii) Voltage control and dispatchable power factor control of DGs, coordinated from EMSs of MGs can reduce distribution energy losses.
iv) Network reconfiguration, such as supplying load from a different feeder (Mesut E. Baran and Felix F. Wu, 1989) or meshing the radial distribution network (Murty and Kumar, 2014) can reduce distribution losses. Back-to-back (BTB) loop power flow controllers (LPFCs) can also be utilized to mesh the radial network inside an MG and provide reactive power support, which, in turn, reduces power losses (Cano et al., 2015). These reconfiguration schemes are coordinated from the EMSs of MGs, as depicted in Figure 8, as an example of distribution loss reduction by an MG.
v) Conservation voltage reduction (CVR) and advanced voltage control (AVC) from the EMSs of MGs can be a great way of reducing losses, as detailed in (Pratt et al., 2010). Concisely speaking, with the reduced electricity service voltage, end-use energy consumption drops in certain end-use loads like incandescent lights and certain electronics. This scheme is called CVR. Contrarily, electric losses in a distribution system increase with voltage drop as the motors and constant power loads draw more currents. So, there exists an opportunity to continually optimize tradeoffs in service voltage and energy use by precisely controlling voltage within acceptable limits. This is called AVC and MG

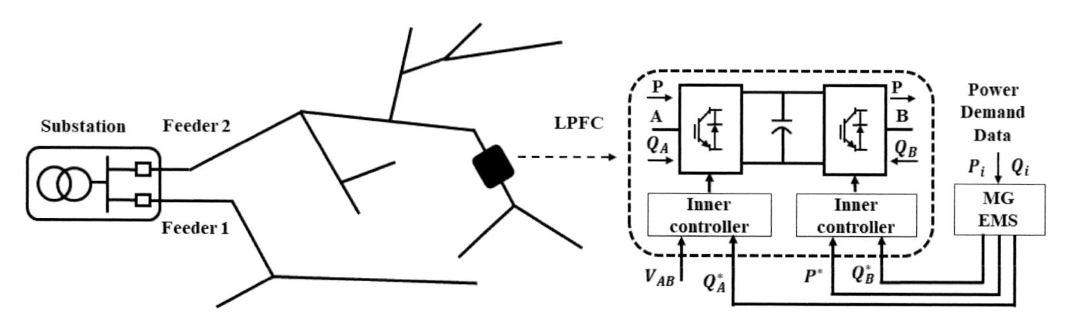

Figure 8. Distribution loss reduction by MG EMS controlled BTB LPFC (concept from (Cano et al., 2015)).

EMS can effectively activate this trade-off. AVC to minimize distribution power loss of MGs was discussed in (Ahn and Peng, 2013). Evaluation of CVR has been carried out in (Pacific Northwest National Laboratory (PNNL), 2010) and (Pasha et al., 2017) on a national level and at MG scale, respectively.

4.3 Facilitation of CHP

Cogenerations or CHP plants provide much higher conversion efficiency than plants for only electricity generation. CHP plants are a manifestation of the second law of thermodynamics as they produce electricity from high-exergy heat and thermal energy from the waste heat at a lower temperature. Consequently, the total energy efficiency can reach as high as 90% (Wang et al., 2015). However, as transmission of heat is not an efficient option, there should be enough local demand for heat in order to utilize CHP plants. MGs facilitate the spread of CHP plants as they accommodate installation of DG, like CHP, in the distribution grid. In contrast to the transmission network, distribution networks are normally placed in urban areas with sufficient thermal energy demand. Therefore, MGs can embed CHP plants, which can play a significant role in providing energy with much higher efficiency, consequently reducing GHG emissions. This scheme is expected to contribute greatly in the colder climate countries (Hatziargyriou, 2014). For the warmer countries, tri-generation plants, such as CCHP, are contrived to play a similar role (Wu and Wang, 2006). A detailed review of modeling, planning and optimal energy management of a combined cooling, heating and power MG is presented in (Gu et al., 2014). In (Aghaei and Alizadeh, 2013b), the multi-objective self-scheduling of CHP-based MGs considering DR programs and DS systems is discussed.

4.4 Aggregation, control and management of PHEVs

The spread of PHEV will require the integration of a major part of the transportation sector into the electric power system. This trend will gradually reduce transportation sectors' total dependency on fossil fuels and consequent emission of CO_2 and other particulates. Then, from the power systems' point of view, integration of a considerable number of (in millions) PHEVs into the electricity grid poses both great opportunities and challenges. Opportunities arise from the fact that a large number of PHEVs will have enormous storage capacity in their batteries, which can be utilized to store electricity at the time of abundance and then provide ancillary services for balancing the grid in the time of scarcity (Kempton and Tomić, 2005). On the other hand, PHEVs might cause many problems for power system operation, such as overloading, reduced efficiency, decreases in voltage, power quality and system load factor, particularly at the distribution level (Clement and Driesen, 2009; Mwasilu et al., 2014; Pieltain Fernández et al., 2011; Sortomme et al., 2011; Taylor et al., 2009). However, it is the management of the charging and discharging of PHEVs that determines whether they are opportunities or threats to the future power system. Hence, an energy management platform like an MG that can optimize the charging and discharging is a possible solution for overcoming the integration problem of PHEVs.

The voltage level and concept for charging PHEVs vary with countries and continents (Su et al., 2012). It is important to note that, among different types of charging (Kempton and Tomić, 2005), most of the PHEV charging is expected to take place in public charging facilities of level 2 (from a single-phase branch circuit with typical voltage ratings from 208 to 240 V, which allows for the maximum current of up to 80 amp AC with a 100 amp circuit breaker) and level 3 (with higher voltage and a maximum current of 400 amp, fast-rate DC charging for commercial and public applications) (Su et al., 2012).

Smart controllers with intelligence and optimization capability will be required for PHEV charging and discharging, especially in the public facilities where thousands of them will be connected in a short period. Those controllers will regulate clusters of PHEVs considering real-world constraints of line capacity, transformer capacity and infrastructure variations among vehicles. The controller should explore the relationship between feeder losses, load factor and load variance for coordinated PHEV charging in order to mitigate problems in power system operation. Now, if the situation suggests, sometimes

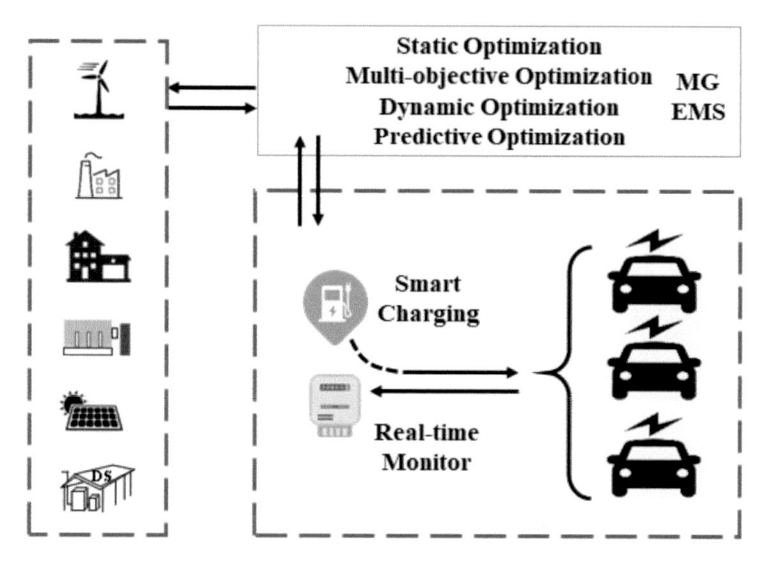

Figure 9. MG EMS controlled PHEV charging structure (concept from (Su et al., 2012)).

this controller can be a dedicated controller for managing a cluster of PHEVs and a VPP type separate aggregator might be assigned for market participation. However, in most of the cases, the charging section in the power distribution system will be formed into MGs as they can work as both aggregator and controller and facilitate more functions in a better way without further investments. Figure 9 shows an exemplary structure for PHEV charging station controlled from MG EMS where it provides the intelligence and optimization for the integration and charging/discharging of PHEVs considering overall resources and network. In the figure, a set of black arrows represent two-way electrical energy flow and communications. There, MG EMS facilitates the penetration of PHEV by providing (Su et al., 2012):

i) Static optimization or single-objective optimization for certain given constraints.
ii) Multi-objective optimization, like multi-objective energy scheduling, for minimizing energy usage, minimizing peak demand, minimizing charging cost, maximizing customer preference, etc.
iii) Dynamic optimization for plug-and-play operation.
iv) Predictive optimization for time steps ahead prediction.

With this structure, a number of research studies have been carried out in order to improve the efficiency and reliability of the power grid, as well as reduce overall cost and carbon emission with the utilization of PHEV assigned into MGs. In (Jian et al., 2013), the possibility of smoothing out the load variance in a household MG by regulating the charging patterns of family PHEVs is investigated. Different issues concerning the integration of single-phase charging devices for electric vehicles (EVs) in low-voltage MGs are addressed in (Peças Lopes et al., 2010) by developing and testing different control strategies. In (Tushar et al., 2014), an optimal centralized scheduling-method to jointly control the electricity consumption of home appliances and consumption/discharge of PHEVs is proposed for increasing the reliability and stability of MGs and giving lower electricity prices to customers.

4.5 Aggregation and control of DR resources

As discussed in section II, a large portion of RESs is inherently intermittent and non-dispatchable. Hence, with the spread of RESs and decrease of conventional generators, management of flexibility and ancillary services is a major concern for the operation and control of future power systems. Different storage technologies, like high energy density batteries, high power density super-capacitors, compressed air storages, heat storages and others, will be deployed throughout the power grid to tackle the flexibility issues of an SG. However, storage technologies are costly and inefficient in some cases. There, the DR scheme is a much cheaper and efficient alternative to storage systems. Therefore, their implementation is prioritized and they are contrived to be a major provider of ancillary services and flexibility for the future grid (Aghaei and Alizadeh, 2013a; Palensky et al., 2011).

Demand resources are specifically defined as a subset of non-spinning reserves which can be accessed by controlling the load and must be available within 10 minutes from the time they are called upon (Kathan and Daly, 2009). The DR strategy is called direct load control (DLC) if the demand resources can be controlled immediately upon the occurrence of disturbances. DR can be achieved through responsive electric devices and a popular candidate of such electric devices is residential electric water heaters (EWHs) equipped with hot water storage tanks (Heffner and Moezzi, 2006; Jia and Pierre, 2007; Kondoh and Hammerstrom, 2011). Residential EWHs have huge potentials for DR applications as they account for about 11% of the total electricity consumption in houses on average and their share on total household demand increases to over 30% during peak demand hours (Kondoh et al., 2011; Ning Lu and Katipamula, 2005). Other than EWHs, lesser priority loads, like heating, ventilation and air-conditioning (HVAC), washing machines and dryers, are also suitable candidates for DR applications. With these resources and direct or market-based controllability, DR applications are capable to provide ancillary services for primary, secondary and tertiary frequency regulations.

Other than ancillary services, DR applications also induce peak shaving/shifting effect due to the fact that, in a DR allowed market and its time-based pricing structure, customers automatically intend to switch their lesser priority loads towards the period of less costly electricity. Thus, the peak load of a distribution grid can be shaved and shifted to a less demanding period. This phenomenon is illustrated in Figure 10.

Peak shaving has two advantages to the utilities' and environmental concerns. First, utilities delay investments on the excessive generation and save money from paying the most expensive spinning reserves to meet the peak load. Second, environmental objectives are met because a large portion of the peak loads would otherwise be met with fossil fuel-based flexible reserves and, due to peak shaving, these loads can be served with more RESs. Other than peak shaving, DR applications will have energy conservation effects on consumers. When DR and energy efficiency programs are advertised, time-varying pricing is introduced; hence, consumers are automatically induced to be economic in their electricity consumption (Pratt et al., 2010). DR application is, therefore, a significant enabler of SGs and decarbonization.

Now, MGs have a special role to play for accommodating this significant enabler—DR application. DR application requires both aggregation into a large amount of resources and controllability to curtail a specific number of loads in short notice. In fact, the aggregated nature and control capability makes DR applications more reliable than conventional generation, where the failure of one generator to start can cause the loss of considerable spinning reserve capacity (U.S., 2006). MGs can manage both aggregation and control of DR applications.

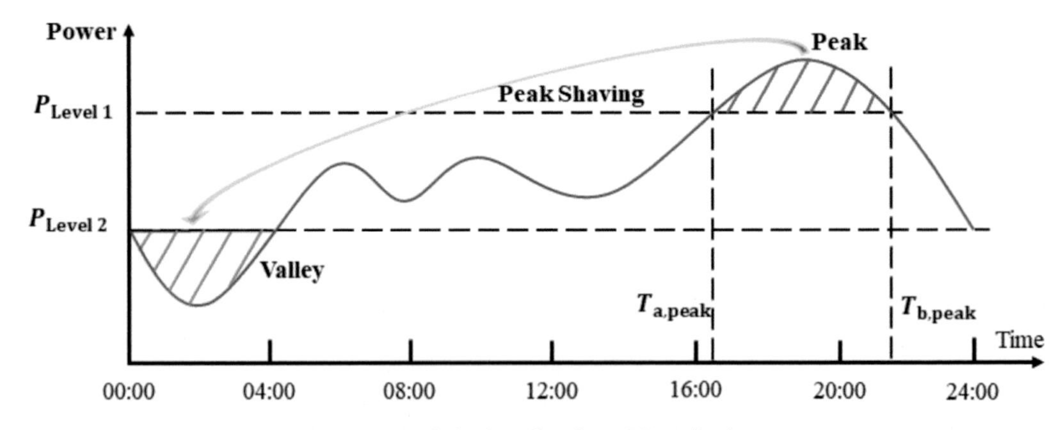

Figure 10. Peak shaving effect due to DR applications.

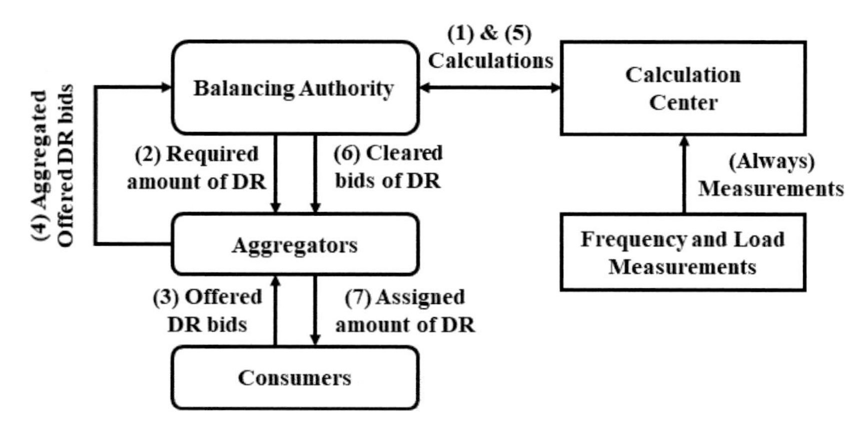

Figure 11. Steps of DR applications process in any level of regulation (concepts from (Hatziargyriou et al., 2005; Pourmousavi and Nehrir, 2012)).

Figure 11 illustrates the steps of DR application for any level of regulation. In general, there are four stakeholders/participants in DR applications (Hatziargyriou et al., 2005; Pourmousavi and Nehrir, 2012):

i) The balancing authority, which will be the combination of DNO and any market operator, like the tertiary market operator in a TNO-DNO interface or an intra-day wholesale market operator. MGs can also be the balancing authority if they run in islanded mode and optimize the generation and demand. Having calculation from the calculation center, the balancing authority initiates the DR process and later, having bids and the aid of the calculation capability, this center finds the bids to be cleared.
ii) A DR aggregator integrates the bids from the individual DR resources and offers aggregated bids to the balancing authority. MGs can play this role of aggregator when grid-connected.
iii) The customers who participate in the DR program. In many cases, these customers will be inside the network that an MG encircles.

Of course, a DR scheme can be implemented, avoiding MGs and using VPPs, but that scheme would lack the capability of direct control and might fail to meet the regulation, especially in fast response cases. Therefore, many research studies have been conducted to establish MGs as the aggregator and controller of DR applications. A comprehensive central DR algorithm for frequency regulation, while minimizing the amount of manipulated load in an MG, was presented in (Pourmousavi and Nehrir, 2012). An integrated scheduling method in MG, where the reserve requirement for compensating renewable forecast errors is provided by both responsive loads and DG units, was presented in (Mazidi et al., 2014). In (Doolla and Nunna, 2012), an agent-based intelligent EMS is proposed to facilitate power trading among multiple MGs and allow customers to participate in DR applications. A risk-constrained profit maximization bidding structure for MG aggregators with DR was developed in (Nguyen and Le, 2015).

5. Conclusion

Transformation of conventional power grid towards SG is an urgent requirement to reduce our carbon footprint. With this transformation, the power grid will be able to safely accommodate DERs and energy efficiency measures. However, with SG's vision and its revolutionary philosophies, like distributed and intermittent generation at the distribution side as well as the bi-directional power flows, it will require a suitable building block, which will possess vital intelligence to optimize an array of complex operations. In this regard, the concept of MGs, which possesses vital optimization, aggregation and control capability, emerged. MGs can optimally manage a distribution grid network having high penetration of DERs, either in grid-connected or islanded mode. There are three main stakeholders of MG—the grid operator, the utility and the customer. In the grid operator's point of view, an MG is an aggregation concept where both supply-side and demand-side resources in distribution grid participate, given attractive remuneration as a small source of power or ancillary services supporting the network, in a way that they can be controlled as a single entity, like load or generator. At the utility's advantage, small DERs near the load center can reduce distribution and transmission requirements and power flows with three important impacts: RES integration, loss-reduction and substitution of network assets. Importantly, from the customers' point of view, MGs can simultaneously satisfy electrical and thermal needs, improve power quality, reduce emissions, enhance reliability and potentially lower costs of energy supply.

In this chapter, the significant contribution of MGs towards grid decarbonization is discussed from the environmental point of view. It was shown that MGs contribute to grid decarbonization by facilitating the integration of renewables and CHP plants, minimizing distribution losses, managing PHEVs and accommodating DR applications. There are other mechanisms and concepts where MGs can be an aid for decarbonization, however, this chapter only covers the main mechanisms. Overall, the role of MGs in grid decarbonization is well-established. This justifies why many MG projects are underway in developed countries. This also gives hope that the spread of intelligence in the distribution grid by means of MGs can possibly be scaled up to meet the goals delineated in the Paris Agreement.

Acronyms

AC	Alternating Current
AGC	Automatic Generation Control
AVC	Advanced Voltage Control
BTB	Back-to-Back
CCHP	Combined Cooling, Heat and Power
CHP	Combined Heat and Power
CVR	Conservation Voltage Reduction
DC	Direct Current
DER	Distributed Energy Resources
DG	Distributed Generation
DHS	Distributed Heat Storage
DLC	Direct Load Control
DMO	Distribution Market Operator
DNO	Distribution Network Operator
DR	Demand Response
DS	Distributed Storage
EES	Energy Efficiency Schemes
EMS	Energy Management System
EV	Electric Vehicle
EWH	Electric Water Heater
GHG	Greenhouse Gas
HVAC	Heating, Ventilation and Air-conditioning
IEA	International Energy Agency
IEC	International Electrotechnical Commission
IPCC	Intergovernmental Panel on Climate Change
IRENA	International Renewable Energy Agency
LC	Local Controller
LPFC	Loop Power Flow Controller
LV	Low Voltage
MG	Microgrid
MGCC	Microgrid Central Controller
PHEV	Plug-in Hybrid Electric Vehicle
PNNL	Pacific Northwest National Laboratory

PQ	Active Power and Reactive Power
RE	Renewable Energy
RES	Renewable Energy Sources
SCADA	Supervisory Control and Data Acquisition
SG	Smart Grid
TNO	Transmission Network Operator
VF	Voltage and Frequency
VPP	Virtual Power Plant

References

Aghaei, Jamshid and Mohammad Iman Alizadeh. 2013a. Demand response in smart electricity grids equipped with renewable energy sources: A review. Renewable and Sustainable Energy Reviews 18: 64–72.

Aghaei, Jamshid and Mohammad Iman Alizadeh. 2013b. Multi-objective self-scheduling of CHP (Combined Heat and Power)-based microgrids considering demand response programs and ESSs (Energy Storage Systems). Energy 55: 1044–54.

Ahn, Changsun and Huei Peng. 2013. Decentralized voltage control to minimize distribution power loss of microgrids. IEEE Transactions on Smart Grid 4(3): 1297–1304.

Baran, Mesut E. and Wu, Felix F. 1989. Network reconfiguration in distribution systems for loss reduction and load balancing. IEEE Transaction on Power Delivery.

Bhumkittipich, Krischonme and Weerachai Phuangpornpitak. 2013. Optimal placement and sizing of distributed generation for power loss reduction using particle swarm optimization. Energy Procedia 34: 307–17.

Bidram, Ali, Student Member and Ali Davoudi. 2012. Hierarchical structure of microgrids control system. IEEE Transactions on Smart Grid 3(4): 1963–76.

Bidram, Ali, Vahidreza Nasirian, Ali Davoudi and Frank L. Lewis. 2017. Cooperative Synchronization in Distributed Microgrid Control. Springer.

Birol, Fatih, 2004. World energy investment outlook to 2030. Geopolitics of Energy 26(3-4): 13–18.

Cano, José M., Norniella, Joaquín G., Rojas, Carlos H. and Orcajo, Gonzalo A. 2015. Application of loop power flow controllers for power demand optimization at industrial customer sites. *In*: IEEE Power & Energy Society General Meeting, 2015.

Celli, G., Pilo, F., Pisano, G., Student Member and Soma, G.G. 2005. Optimal participation of a microgrid to the energy market with an intelligent EMS. pp. 1–6. *In*: 2005 International Power Engineering Conference, IEEE.

Chaouachi, Aymen, Kamel, Rashad M., Ridha Andoulsi and Ken Nagasaka. 2013. Multiobjective intelligent energy management for a microgrid. IEEE Transactions on Industrial Electronics 60(4): 1688–99.

Clement, K., Haesen, E. and Driesen, J. 2009. Stochastic analysis of the impact of plug-in hybrid electric vehicles on the distribution grid. pp. 160–160. *In*: CIRED 2009–20th International Conference and Exhibition on Electricity Distribution-Part 1. IET. 2009. Vol. 25.

Doolla, Suryanarayana and Kumar Nunna, H.S.V.S. 2012. Demand response in smart distribution system with multiple microgrids. IEEE Transactions on Smart Grid 3(4): 1–9.

European Smart Grids Technology Platform. 2006. Vision and Strategy for Europe's Electricity Networks of the Future. Vol. 19.

Gu, Wei, Zhi Wu, Rui Bo, Wei Liu, Gan Zhou, Wu Chen and Zaijun Wu. 2014. Modeling, planning and optimal energy management of combined cooling, heating and power microgrid: A review. International Journal of Electrical Power and Energy Systems 54(2014): 26–37.

Guan, Zhitao, Guanlin Si, Xiaosong Zhang, Longfei Wu, Nadra Guizani, Xiaojiang Du and Yinglong Ma. 2018. Privacy-preserving and efficient aggregation based on blockchain for power grid communications in smart communities. IEEE Communications Magazine 56(7): 82–88.

Hatziargyriou, N.D., Dimeas, A. and Tsikalakis, A.G. 2005. Management of microgrids environment in market. *In*: 2005 International Conference on Future Power Systems. IEEE.

Hatziargyriou, N.D., Jenkins, N., Strbac, G., Peças Lopes, J.A., Alfred Engler, Oyarzabal, J., Kariniotakis, G. and Amorim, A. 2006. Microgrids—large scale integration of microgeneration to low voltage grids. Cigre C6-309 (Lv):Cigre 2006.

Hatziargyriou, N.D. 2014. Microgrids Architectures and Control. Wiley & IEEE Press.

Heffner, G.C., Goldman, C.A. and Moezzi. M.M. 2006. Innovative approaches to verifying demand response of water heater load control. IEEE Transactions on Power Delivery 21(1): 388–97.

Huang, Alex Q., Crow Mariesa, L., Gerald Thomas Heydt, Zheng Jim, P. and Dale Steiner, J. 2011. The future renewable electric energy delivery and management (FREEDM) system: The energy internet. Proceedings of the IEEE 99(1): 133–48.

Intergovernmental Panel on Climate Change (IPCC). 2018. Summary for Policymakers. An IPCC Special Report on the Impacts of Global Warming of 1.5 °C above Pre-Industrial Levels and Related Global Greenhouse Gas Emission Pathways.
International Electrotechnical Commission (IEC). 2007. Efficient Electrical Energy Transmission and Distribution.
International Renewable Energy Agency (IRENA). 2018. Global Energy Transformation: A Roadmap to 2050. 2018th ed. International Renewable Energy Agency.
Jia, Runmin, Nehrir, M. Hashem and Pierre, Donald A. 2007. Voltage control of aggregate electric water heater load for distribution system peak load shaving using field data. 2007 39th North American Power Symposium, NAPS 492–97.
Jian, Linni, Honghong Xue, Guoqing Xu, Xinyu Zhu, Dongfang Zhao and Shao, Z.Y. 2013. Regulated charging of plug-in hybrid electric vehicles for minimizing load variance in household smart microgrid. IEEE Transactions on Industrial Electronics 60(8): 3218–26.
Kantamneni, Abhilash, Brown, Laura E., Gordon Parker and Weaver, Wayne W. 2015. Survey of multi-agent systems for microgrid control. Engineering Applications of Artificial Intelligence 45: 192–203.
Kathan, David and Caroline Daly. 2009. Assessment of demand response and advanced metering. Federal Energy Regulatory Commission 4(8): 2–5.
Kempton, Willett and Jasna Tomić. 2005. Vehicle-to-grid power implementation: From stabilizing the grid to supporting large-scale renewable energy. Journal of Power Sources 144(1): 280–94.
Kondoh, Junji, Ning Lu and Hammerstrom, Donald J. 2011. An evaluation of the water heater load potential for providing regulation service. IEEE Transactions on Power Systems 26(3): 1309–16.
Liu, Guodong, Yan Xu and Kevin Tomsovic. 2016. Bidding strategy for microgrid in day-ahead market based on hybrid stochastic/robust optimization. IEEE Transactions on Smart Grid 7(1): 227–37.
Logenthiran, T., Dipti Srinivasan and David Wong. 2008. Multi-agent coordination for DER in microGrid. 2008 IEEE International Conference on Sustainable Energy Technologies 77–82.
Logenthiran, Thillainathan, Dipti Srinivasan, Khambadkone, Ashwin M. and Htay Nwe Aung. 2012. Multiagent system for real-time operation of a microgrid in real-time digital simulator. IEEE Transactions on Smart Grid 3(2): 925–33.
Mazidi, Mohammadreza, Alireza Zakariazadeh, Shahram Jadid and Pierluigi Siano. 2014. Integrated scheduling of renewable generation and demand response programs in a microgrid. Energy Conversion and Management 86: 1118–27.
Murty, V.V.S.N. and Ashwani Kumar. 2014. Electrical power and energy systems mesh distribution system analysis in presence of distributed generation with time varying load model. International Journal of Electrical Power and Energy Systems 62: 836–54.
Mwasilu, Francis, Jackson John Justo, Eun Kyung Kim, Ton Duc Do and Jin Woo Jung. 2014. Electric vehicles and smart grid interaction: A review on vehicle to grid and renewable energy sources integration. Renewable and Sustainable Energy Reviews 34: 501–16.
Nguyen, Duong Tung and Long Bao Le. 2015. Risk-constrained profit maximization for microgrid aggregators with demand response. IEEE Transactions on Smart Grid 6(1): 135–46.
Ning Lu and Katipamula, S. 2005. Control strategies of thermostatically controlled appliances in a competitive electricity market. IEEE Power Engineering Society General Meeting 2005: 164–69.
Ochoa, Luis F. and Harrison, Gareth P. 2011. Minimizing energy losses: Optimal accommodation and smart operation of renewable distributed generation. IEEE Transactions on Power Systems 26(1): 198–205.
Olivares, Daniel E., Cañizares, Claudio A. and Mehrdad Kazerani, 2014. A centralized energy management system for isolated microgrids. IEEE Transactions on Smart Grid 5(4): 1864–75.
Pacific Northwest National Laboratory (PNNL). 2010. Evaluation of CVR on a National Level.
Palensky, Peter, Senior Member, Dietmar Dietrich and Senior Member. 2011. Demand side management: Demand response, intelligent energy systems, and smart loads. IEEE Transactions on Industrial Informatics 7(3): 381–88.
Pasha, Aisha Munawer, Zeineldin, Hatem H., Ameena Saad Al-Sumaiti, Mohamed Shawky El Moursi and Ehab Fahamy El Sadaany. 2017. Conservation voltage reduction for autonomous microgrids based on V-I droop characteristics. IEEE Transactions on Sustainable Energy 8(3): 1076–85.
Peças Lopes, J.A., Polenz, Silvan A., Moreira C.L. and Rachid Cherkaoui. 2010. Identification of control and management strategies for LV unbalanced microgrids with plugged-in electric vehicles. Electric Power Systems Research 80(8): 898–906.
Pieltain Fernández, Luis, Tomás Gómez San Román, Rafael Cossent, Carlos Mateo Domingo and Pablo Frías. 2011. Assessment of the impact of plug-in electric vehicles on distribution networks. IEEE Transactions on Power Systems 26(1): 206–13.
Pop, Claudia, Tudor Cioara, Marcel Antal, Ionut Anghel, Ioan Salomie and Massimo Bertoncini. 2018. Blockchain based decentralized management of demand response programs in smart energy grids. Sensors (Switzerland) 18(1).
Pourmousavi, S. Ali and Nehrir, M. Hashem. 2012. Real-time central demand response for primary frequency regulation in microgrids. IEEE Transactions on Smart Grid 3(4): 1988–96.
Pratt, Robert G., Balducci, P., Clint Gerkensmeyer, Srinivas Katipamula, Kintner-Meyer, Michael C.W., Sanquist, Thomas F., Schneider, Kevin P. and Secrets, T.J. 2010. The smart grid: An estimation of the energy and CO_2 benefits. U.S. Department of Energy (January): 172.

Qi, Li, Lewei Qian, Stephen Woodruff and David Cartes. 2007. Prony analysis for power system transient harmonics. Eurasip Journal on Advances in Signal Processing 2007.

Rahbar, Katayoun, Jie Xu and Rui Zhang. 2015. Real-time energy storage management for renewable integration in microgrid: An off-line optimization approach. IEEE Transactions on Smart Grid 6(1): 124–34.

Schwaegerl, Christine. 2009. Advanced architectures and control concepts for more microgrids. More Microgrids Siemens AG, STREP Project (December): 1–145.

Sinha, Arup, Basu, A.K., Lahiri, R.N., Chowdhury, S., Chowdhury, S.P. and Crossley, Peter A. 2008. Setting of Market clearing price (MCP) in microgrid power scenario. pp. 1–8. *In*: 2008 IEEE Power and Energy Society General Meeting-Conversion and Delivery of Electrical Energy in the 21st Century. IEEE.

Sortomme, Eric, Student Member, Hindi, Mohammad M., Student Member, Macpherson, S.D. James, Student Member, Venkata, S.S. and Life Fellow. 2011. Coordinated charging of plug-in hybrid electric vehicles to minimize distribution system losses. IEEE Transactions on Smart Grid 2(1): 198–205.

Su, Wencong and Jianhui Wang. 2012. Energy management systems in microgrid operations. Electricity Journal 25(8): 45–60.

Su, Wencong, Habiballah Eichi, Wente Zeng and Mo-yuen Chow. 2012. A survey on the electrification of transportation in a smart grid environment. IEEE Transactions on Industrial Informatics 8(1): 1–10.

Tande, John Olav Giæver. 2002. Applying power quality characteristics of wind turbines for assessing impact on voltage quality. Wind Energy 5(1): 37–52.

Taylor, Jason, Arindam Maitra, Mark Alexander, Daniel Brooks and Mark Duvall. 2009. Evaluation of the impact of plug-in electric vehicle loading on distribution system operations. 2009 IEEE Power and Energy Society General Meeting, PES'09 1–6.

Thatte, Anupam A. and Le Xie. 2016. A metric and market construct of inter-temporal flexibility in time-coupled economic dispatch. IEEE Transactions on Power Systems 31(5): 3437–46.

Tushar, Mosaddek Hossain Kamal, Chadi Assi, Martin Maier and Mohammad Faisal Uddin. 2014. Smart microgrids: Optimal joint scheduling for electric vehicles and home appliances. IEEE Transactions on Smart Grid 5(1): 239–50.

U.S. Energy Dept. 2006. Benefits of Demand Response in Electricity Markets and Recommendations for Achieving Them. A Report to U.S. Congress (February).

Wang, Zhaoyu, Bokan Chen, Jianhui Wang, Begovic, Miroslav M. and Chen Chen. 2015. Coordinated energy management of networked microgrids in distribution systems. IEEE Transactions on Smart Grid 6(1): 45–53.

Wu, D.W. and Wang, R.Z. 2006. Combined cooling, heating and power: A review. Progress in Energy and Combustion Science 32(5-6): 459–95.

Xuan, Liu and Su Bin. 2008. Microgrids—an integration of renewable energy technologies. *In*: 2008 China International Conference on Electricity Distribution, CICED 2008.

CHAPTER 18

Storage of Fluctuating Renewable Energy

Daniel Fozer[1] and *Peter Mizsey*[1,2,*]

1. Introduction

The fluctuating energy supply at different places of the Earth is a natural phenomenon since the most important energy source on planet Earth is the Sun, that is, solar energy. Due to the rotation of the Earth, the Sun shines on different points of our planet at varying times and intensities. From our point of view, the Sun is shining always in a different way, that is, the solar energy supply is fluctuating at different places on the Earth. Nature has already provides storage of the fluctuating solar energy via photosynthesis, which has a modest efficiency of only 1% (Olah et al., 2005), and the result is vegetation. Therefore, the storage of energy is also a typical natural phenomenon.

Humans have been the beneficiary of this natural phenomenon. Making a fire with wood uses stored solar energy. Utilization of biomass, however, is limited by the finite area of arable lands (~ 10% share of the world (World Energy Council, 2016)). Nowadays, other renewable energy sources, like solar panels and wind turbines, are developing very rapidly. Production of renewables and their consumption is increasing in the world, as illustrated in Figure 1. These devices, however, produce electricity in a fluctuating manner due to weather conditions. Electricity consumption is also a fluctuating phenomenon due to times of day or night and seasons. The efficient and large-scale usage of renewables can be managed if there are efficient storage alternatives at every different level, that is, large, medium and/or small scale.

In this chapter, an overview about different energy storage alternatives of different scales and different principles is given, including:

- thermal way of storage,
- mechanical ways of storage, like water reservoirs, compressed air, and fly wheels,
- chemical ways of storage: (i) batteries (ii) reversible reactions (iii) hydrogen economy and (iv) CO_2-based circular economy producing energy carrier molecule, e.g., methanol, methane,
- biological way of storage, typically algae-based technologies.

1.1 Storage capacity and volume required for different energy systems

Figure 2 shows storage capacities and discharge times of the different kinds of energy storage alternatives. As it is presented, batteries and fly wheels have lower storage capacity. The highest value belongs to the

[1] Budapest University of Technology and Economics, Department of Chemical and Environmental Process Engineering, Environmental Process Engineering Research Group.
[2] University of Miskolc, Institute of Chemistry, Department of Fine Chemicals and Environmental Technologies.
* Corresponding author: kemizsey@uni-miskolc.hu

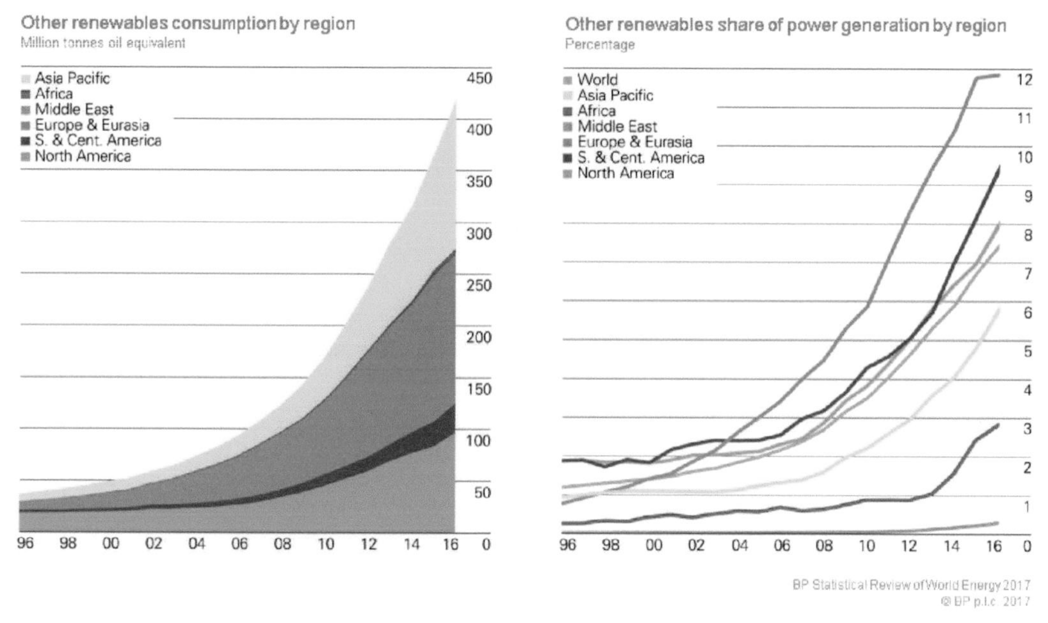

Figure 1. Increasing trend of renewable energy production in the World (BP, 2017; Deák et al., 2018).

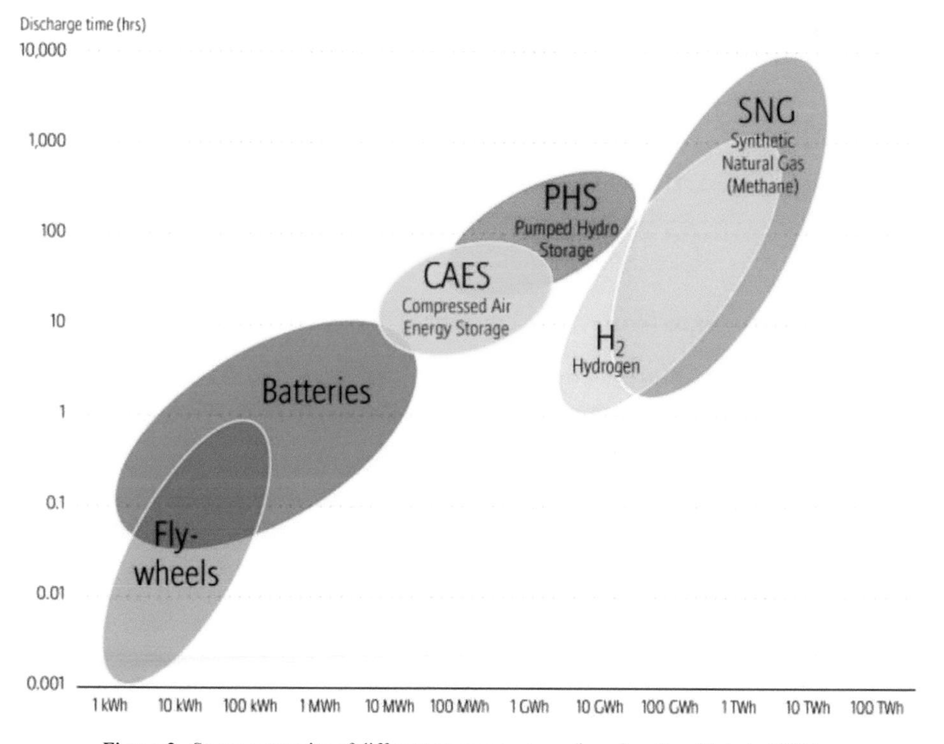

Figure 2. Storage capacity of different storage systems (based on Specht et al., 2009).

Table 1. The volume of storing 10 MWh electrical energy (Teichman et al., 2012).

Pumped hydro, 300 m	14.000 m^3
Compressed air, 20 bar	3.400 m^3
Li-ion battery	30 m^3
Chemical storage of hydrogen	5 m^3
Diesel	1 m^3

Synthetic Natural Gas (methane). Table 1 shows the volumetric energy density of energy storage for 10 MWh energy (Teichman et al., 2012). As can be seen, the pumped hydro needs the highest volume and the hydrocarbon needs the smallest volume due to its high carbon content.

2. Thermal Storage

The oldest form of energy storage. The heat can be stored as sensible heat or latent heat.

2.1 Storage using sensible heat

The housing structures do that job and protect us against external hot or cold temperatures. In the industry, e.g., steel industry, so-called recuperators are applied. These devices contain heat resistant solid material, e.g., shamot brick or fire brick, which is heated up with flue gases and then the accumulated heat is used for the preheating of feed gases.

2.2 Storage using latent heat of melting

The latent heat method is more powerful than the use of sensible heat. Molten material can store heat more efficiently: A large amount of heat can be stored in a small volume. Smaller volume but more heat.

A rapidly developing area of this heat storage option is a microcapsule-based solution where the melting material is closed in a microcapsule. There are many alternatives for the melting material to match its melting point to the stored heat features. To find or produce the appropriate material in the microcapsule is the subject of many research projects. Such materials can be paraffin, modified hydrocarbons, polymers, etc. The microcapsule can find many application areas, including tempering of buildings, especially in the desert, where significant money can be saved on the air conditioning or other utility requirements.

3. Mechanical Storage: Water Reservoirs

Pumped hydro storage is coupled with hydropower turbines for generating electricity. As can be seen in Table 1, the hydro needs the highest volume to store the same amount of energy. In pumped hydro storage, in the case of electric energy surplus, water is pumped to the upper reservoir and, in the case of energy requirement, the water is let down into the lower reservoir. Energy is released due to the change in position; water potential energy becomes kinetic energy. Kinetic energy turns a turbine which turns a generator which generates electricity. The efficient operation of such a pumped hydro power station requires pump-turbine/motor-generator assemblies, as shown in Figure 3.

Pumped hydro energy storage can be built in places where the geographical situation is suitable. This is the reason why the first pumped hydro power stations were built in Italy and Switzerland. It is important to note that the hydro power stations require a relatively high capital investment.

The current hydro power storage capacities represent about 2.5–10% of the total electricity production in countries where hydropower is used (e.g., USA, Europe, Japan). Considering, however, the advantages of the hydro power storage, more power stations are expected to be built.

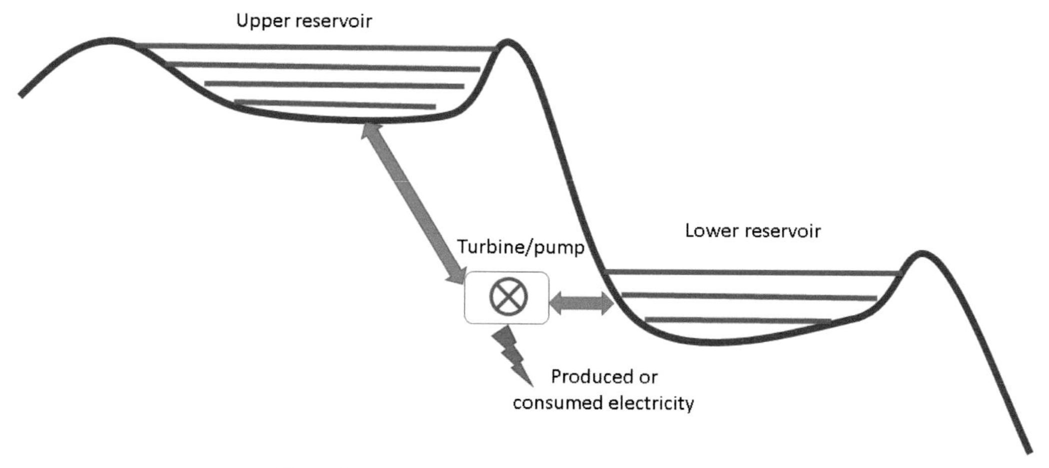

Figure 3. Schematic view of pumped hydro power station.

4. Chemical Storage

4.1 Batteries

Batteries are the best-known and most widespread tools for electrical energy storage. As shown in Figure 2, the batteries belong rather to lower or middle capacity storage devices compared to the other options. Many different batteries aiming to get a better and more suitable device have been developed. Typical features are the voltage, energy density, efficiency, cycles and price. The development is continuous, motivated by the promise of this method.

There are practically uncountable types of batteries. Some important ones are: Lead-acid battery, Nickel-Cadmium (Ni-Cd) battery, Nickel-Metal hydride (Ni-MeH) battery, Lithium-ion (Li-Ion) battery.

Advantages and disadvantages are listed in Table 2 (ADB, 2018).

Table 2. Comparison of batteries, advantages–disadvantages.

Battery	Advantages	Disadvantages
Lead-acid battery	The oldest battery, well known. Relatively cheap. Due to extremely low internal resistance, the discharge current can be very high (the reason why the car industry likes it). High capacity. Negligible maintenance.	Modest weight to energy ratio. It should be charged if storage. High charging time. Deep discharge and charge cycles reduce life time. Serious environmental effect at disposal. Acid spillage can happen.
Nickel-Cadmium (Ni-Cd) battery	Ultra-fast charge is possible. Needs maintenance. Long life. Good performance at low temperature.	Memory effect. Cadmium is toxic. High self-discharge. Low cell voltage.
Nickel-Metal hydride (Ni-MeH) battery	Good energy density performance. No environmental constraints, like Ni-Cd. Ni-Cd batteries can be simply replaced with Ni-MeH. Good low temperature operation.	Deep discharge limits the life cycle. Limited discharge current. Complex charge algorithm. High self-discharge.
Lithium-ion (Li-Ion) battery	Long life, no maintenance. Low internal resistance, high capacity. Simple charge and reasonable charge time.	Possible thermal runaway, protection is needed. Degradation at high temperature. Modest low temperature performance. Special regulations for large quantity transport.

Table 3. Data of different batteries (Rechargeable battery, 2019).

Battery	Cell voltage (V)	Energy by weight (Wh/kg)	Power (W/kg)	Efficiency (%)	Average energy cost ($/kWh)	Cycle number	Life time (year)
Lead-acid	2.1	30–40	180	70–92	150	< 350	3–20
Ni-Cd	1.2	40–60	150	70–90	400–800	1500	
Ni-Mh	1.2	30–80	250–1000	66	250	1000	
Li-ion	3.6	160	1800	99.9	200–350	1200	2–3

The different features of the batteries can be quantified if we define certain numbers that represent their performance. These numbers help to compare and rank the batteries. Such a comparison is shown in Table 3.

The comparison of the numbers explains why the Li-ion battery is selected for the electric cars. The partner of the Li in the cathode can be Cobalt Oxide (or Lithium Cobaltate), Manganese Oxide (or Lithium Manganate), Iron Phosphate, Nickel Manganese Cobalt (or NMC) and Nickel Cobalt Aluminum Oxide (or NCA).

It is an important issue for battery producers and researchers to create a battery that can store large amounts of electrical energy and can be charged in a very short time. Currently the charge time ranges from half an hour (quick charge to 80% of battery capacity) to 24 hours; in comparison, the "charge" time of gasoline cars is a few minutes.

A special issue is the problem posed by trucks. The issue can be summarized by the statement of an average truck driver: "*We do not want to transport batteries but goods*". That is, the energy requirement of trucks is much higher than that of the passenger cars and the energy content of the hydrocarbons is still much higher (30 times better, see Table 1) than that of the best battery. So, if we want to use electric cars/trucks on a large scale these are the challenges to be solved, that is, stored energy, energy density and charging time.

4.2 Reversible chemical reactions

There are reversible chemical reactions where the hydrogen is reversibly created and, if needed, released. Equation (1) shows an example of such a reaction (Mizsey et al., 1999). Methyl cyclohexane is reduced to toluene where 3 moles of hydrogen are released.

$$C_7H_{14} \leftrightarrow C_7H_8 + 3H_2 \quad \Delta H^0_{298} = \pm 216.35 \frac{kJ}{mol} \tag{1}$$

The reversible reaction has a positive balance, which means the reaction is capable of energy storage.

Teichman et al. (2012) proposed an efficient reaction system for hydrogen storage in a reversible reaction. N-ethylcarbazol is hydrogenated and dehydrogenated. The molecule is capable of storing 9 moles of hydrogen. Octene is also proposed (Teichman et al., 2012). The basic idea is the same: Solar cells are producing electricity which is used for water electrolysis. The produced hydrogen is chemically bound and released if needed. The hydrogen is then consumed in fuel cells to produce electricity and fulfill energy requirements. This method is patented but only for household scale.

The energy storage with reversible chemical reactions is capable of storing the fluctuating energy that is typical for certain renewable ones. It assumes that the renewable energy is available in the form of electricity. The electricity is transferred into hydrogen via water electrolysis. The system has advantages and disadvantages (Table 4).

As a conclusion, the energy storage of industrial scale is the SNG (Synthetic Natural Gas), see Figure 2, and this technology is the Power-to-gas method.

Table 4. Evaluation of energy storage with reversible chemical reactions.

Advantages	Disadvantages
The storage of fluctuating electrical energy is possible. There are different target-oriented reactions. Its economic features are competitive with the electricity production from gasoline in the house (Teichman et al., 2012).	The energy must be in the form of hydrogen. Huge installation cost. The outlook of the house changes (Figure 4). The operation with hydrogen needs special security actions. The storage capacity is in the range of a family house. It needs much secured space in the house.

Figure 4. Retrofit installation of solar cells on a private house (Mizsey's photo of his own home, 2019).

5. Power-to-Gas and Power-to-Liquid Energy Storage

5.1 Connections between chemical and biological ways of energy storage

As mentioned in the Introduction, there are four chemical storage methods to be discussed in our work:

- chemical ways of storage: (i) batteries (ii) reversible reactions (iii) hydrogen economy and (iv) CO_2-based circular economy producing energy carrier molecule, e.g., methanol, methane,

and the biological way:

- biological way of storage, typically algae-based technologies.

Two chemical ways are discussed in section 4, the hydrogen economy and the CO_2-based circular economy remain to be discussed. These two alternatives belong to the so-called Power-to-Gas (PtG) category.

The biological way, developed as a novel energy storage technology, also belongs to this category, since it is also a Power-to-Gas method. The CO_2-based storage way and the biological way can be applied and classified as the so-called Power-to-X (PtX) category, where the X can be energy carrier, any kind of platform molecule, etc.

Therefore, these solutions, having the same features and philosophy, all belong to the same category (PtG or PtX) and are discussed together in the following subsection.

5.2 Hydrogen economy, CO_2 based storage way, biological way

Storing energy by forming chemical bonds is a widespread and simple storage method.

Chemical ways of energy storage discussed here are based on the production of hydrogen via water electrolysis. Gaseous chemical energy vectors, such as methane (CH_4) and hydrogen (H_2), require suitable and economic storage facilities. Underground reservoirs (e.g., salt cavity, depleted oil and gas fields) can be applied for storage purposes, however, in the case of H_2, the application of certain storage alternatives depends on geological conditions and is not possible in all regions (Andersson and Grönkvist, 2019). The storage of hydrogen is a difficult task as it has low density (0.0893 g L^{-1} at standard temperature (273.15 K) and pressure (101.33 kPa)) and, thus, occupies high storage volume compared to other energy vectors (methane, methanol). There is strong criticism of hydrogen economy because of its low volumetric energy density (e.g., Olah et al., 2005). To achieve economic storage, the density of H_2 has to be increased. This can be done by applying chemical and mechanical methodologies, i.e., compression using metal containers (gas holders, spherical pressure vessels, pipe storage). Storing hydrogen is characterized by lower chemical energy density compared to methane. As shown in Figure 5, the highest hydrogen storage density can be obtained by the production of chemical hydrides with values ranging between 50–120 kg m^{-3}.

Another important issue, also against the hydrogen economy, is the production of hydrogen itself. If it is produced by electrolysis using renewable electricity to supply the electricity need of an airport such as Frankfurt, Germany, constant 50,000 MW electricity would have to be made available. This is technically possible. On the other hand, the volume of the wings of an airplane, which is the fuel tank, could not store so much hydrogen that would be needed for a long distance, e.g., transatlantic, flight.

The storage of CH_4 can be carried out in a more reliable way because it requires 4–5 times less storage volume than H_2. Methane can be suitably stored in underground reservoirs, metal vessels, in the form of methane hydrates and using adsorption process with various porous materials, such as metal organic frameworks (MOFs), porous organic polymers (POPs), covalent organic frameworks (COFs),

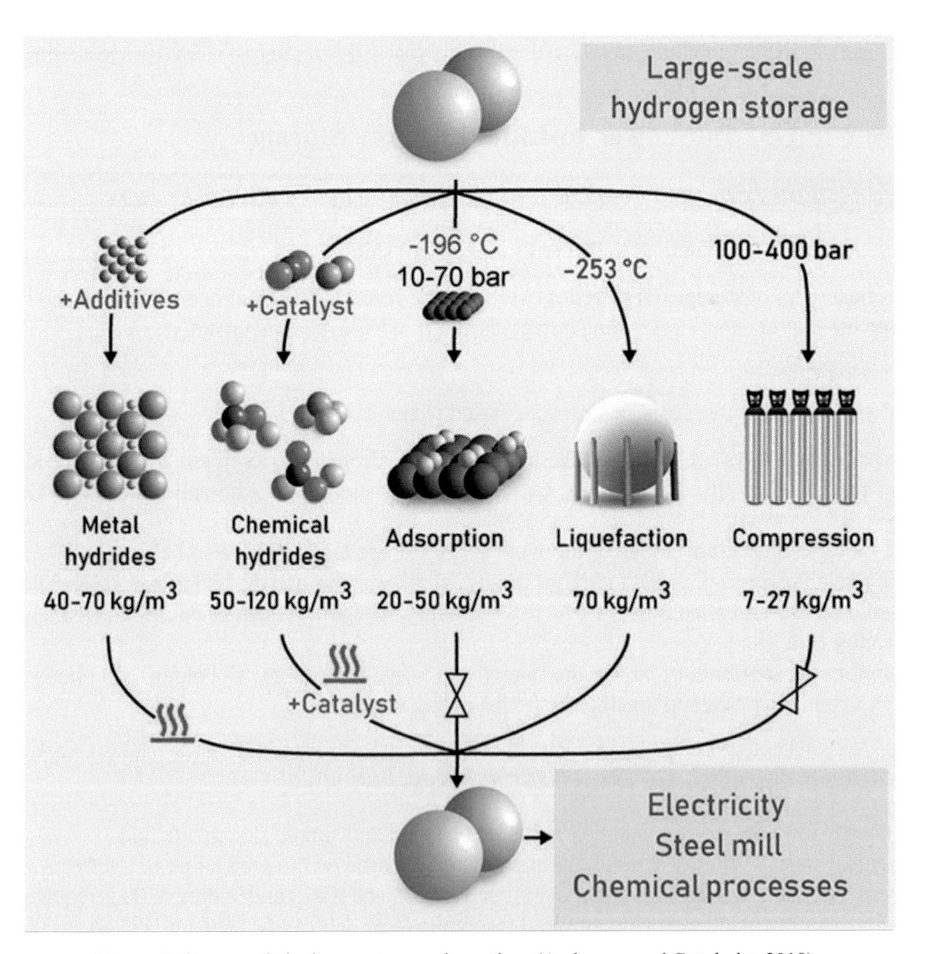

Figure 5. Large-scale hydrogen storage alternatives (Andersson and Grönkvist, 2019).

carbons and zeolites (Li et al., 2016). Natural gas, which contains mainly methane, can be stored in its liquid form below –162 °C. The advantage of the application of liquefied natural gas (LNG) is that it occupies 600 times less volume compared to its gaseous state, enabling the transportation over long distances via tanker ships.

The Power-to-X (PtX, X: synthetic fuels, chemicals, fertilizers, etc.), chain combines the storage of fluctuating renewable energy sources, carbon dioxide utilization and decarbonization of the energy sector. The common factor in PtX technologies is that intermittent renewable electricity is converted into chemical energy, producing hydrogen. The H_2 can be used to store energy, for fuel cells or to transform CO_2 into synthetic fuels such as methane or methanol. The great benefit of the technology is that it involves the circular economy concept since following the utilization of either methane or methanol, the emitted carbon dioxide is captured and used again as an intermediate in the storage chain. In this way, the circular economy aspect is fulfilled by closing the local carbon cycle.

The Power-to-Methane (PtM) technology combines the utilization of anthropogenic CO_2 and H_2 to produce gaseous methane. The electrolysis and methanation can be carried out in several ways: (1) alkaline electrolysis (AEL), (2) polymer electrolyte membranes (PEM) and (3) solid oxide electrolysis (SOEC). Concerning methanation, two different methodologies can be separated to reduce carbon dioxide: (1) chemical based catalytic methanation and (2) biological methanation using methanogen bacteria as a "biocatalyst". The biological methanation is characterized by low operational temperature (30–60 °C) and atmospheric pressure. On the other hand, the reaction kinetics are slow, the operation is complex and the mass transfer is inadequate. In both ways, the methane is formed by the reaction of CO_2 with H_2, which is known as the Sabatier reaction, Equation (2):

$$CO_2 + 4H_2 = \mathrm{CH_4} + 2H_2O \quad \Delta H^0_{298} = -165.1\ kJ\ mol^{-1} \tag{2}$$

The purity of carbon dioxide sources is an important factor regarding both of the methanation technologies. High oxygen, NO_x and SO_x concentrations prevent the metabolism of methanogenic bacteria and deactivate catalysts. Oxygen and siloxanes reduce the biological activity of methanogens, while certain impurities (such as ethylene) deactivate the catalyst that is used in chemical methanation. Depending on flue gas sources (e.g., power plants, biomass processes, industrial processes) the carbon dioxide concentration in exhaust gas ranges between 10–100 vol% (Ghaib and Ben-Fares, 2018). Absorption, adsorption, cryogenic distillation and membrane technologies can be applied to separate CO_2 from the other flue gas components.

In the PtM chain, the fluctuating variable renewable energy (VRE) is transformed into a gaseous energy carrier, methane, that can be injected and stored in the already existing natural gas infrastructure. The first step of the process is the production of hydrogen and oxygen via water electrolysis using renewable (solar, wind) electricity. In the second step, methanation is used to convert anthropogenic carbon dioxide into methane. Biomass and industrial processes can be used as carbon sources. The valuable co-products of the process are the evolving oxygen and heat.

The highly exothermic catalytic methanation is carried out at temperature ranging between 200 and 500 °C and pressure between 1 and 100 bar. Methanation requires sufficient temperature control; otherwise, catalyst sintering and thermodynamic limitations decrease the conversion of carbon dioxide. The methane yield can be increased by decreasing the reaction temperature and increasing pressure. During the methanation, co-products, such as CO, C_nH_{2n} and C_nH_{2n+2}, can be also produced in small quantities, which deactivates catalysts. The formation of hydrocarbons can be avoided by decreasing temperature and pressure. Decreasing temperature and increasing pressure is favorable for the evolution of CO (Klaus et al., 2010), thus, selecting and optimizing the applicable reaction conditions (temperature, pressure, catalysts) is crucial in achieving high methane selectivity and energy efficient operation.

The hydrogenation of carbon dioxide requires an effective catalyst due to the high kinetic barrier. Various metals can be used as catalysts in the methanation reaction (e.g., Ni, Ru, Co, Fe, Pd, Rh), while Al_2O_3, SiO_2, MgO, TiO_2 and ZrO_2 can be applied as support materials. Noble metals (Ru, Rh) show high activity and selectivity to methane but their high price limits their application. Ni is an appropriate catalytic agent with high activity, selectivity and low price but it can be oxidized in oxidizing atmosphere (Ghaib and Ben-Fares, 2018; Lunde and Kester, 1974; Götz et al., 2016).

The exothermic nature of methanation requires sufficient temperature control during the reaction. The evolving heat must be dissipated efficiently in order to avoid catalyst deactivation. Various types of reactor designs, such as fixed bed, fluidized bed, membrane and microchannel reactors, can be applied to methanation (Jia et al., 2019; Kreitz and Turek, 2018; Leonzio, 2019).

The biological methanation (BM) is based on the application of autotrophic hydrogenotrophic methanogen bacteria (e.g., *Methanothermobacter, Methanogenium, Methanopyrus* species) (Lecker et al., 2017; Guneratnam et al., 2017) that convert CO_2 and H_2 into methane. The decomposition of organic material to methane consists of several steps. These are (1) hydrolysis, (2) fermentation, (3) anaerobic oxidation, (4) syntrophic acetate oxidation, (5) hydrogen oxidation, (6) hydrogenotrophic methanogenesis (by methanogenic archea) (Benjaminsson et al., 2013). Bacteria can perform hydrogen oxidation, producing acetic acid from the substrates (CO_2 and H_2), as indicated in Equation (3). Acetotrophic methanogenesis can also be done using methanogens, converting the produced CH_3COOH to methane, Equation (4).

$$CH_3COO^- + 4H_2O = 2HCO_3^- + 4H_2 + H^+ \quad \Delta G = 104.6\ kJ\ mol^{-1} \tag{3}$$

$$4H_2 + HCO_3^- + H^+ = CH_4 + 3H_2O \quad \Delta G = 135.6\ kJ\ mol^{-1} \tag{4}$$

BM can be carried out in two ways: (1) the hydrogen is injected into an anaerobic digester to reduce the formed carbon dioxide (also known as *in situ* methanation) (2) a H_2 and CO_2 mixture is injected into a bioreactor according to the stochiometric ratio, this is known as *ex situ* methanation. Hydrogenotrophic methanogens can be applied in order to increase the calorific value of CO_2-rich anaerobic biogas.

CO_2 and H_2 are substrates for the methanogens during their metabolism. A small amount of the substrates is consumed to produce biomass. Wise et al. (Wise and Augenstein, 1978) demonstrated that the 0.28 to 0.43% of the inlet H_2 and approximately 3% of CO_2 is used in the growth phase of methanogens bacteria. Other studies found that the CO_2 uptake can be as high as 6.4 and 8.5% (Wilson, 2013; Battery University, 2019). Thus, metabolism is affecting the stochiometric ratio of the Sabatier reaction, resulting in an optimal H_2:CO_2 ratio of between 4 to 1.064 and 1.085. Substrate limitation causes decreased biomass growth rate and methane production.

The methane yield is also influenced by the temperature and pressure of the bioreactor. The activity of methanogens is temperature dependent and ranges between an optimal temperature of 15 and 98 °C. The temperature of the fermentation can influence the reaction rate and response time. Several studies show that raising the temperature can increase the reaction speed (Banach et al., 2018; Mickol, Laird and Kral, 2018; Blake et al., 2015; Thomas and Cavicchioli, 2000). The high temperature is not favorable in terms of the solubility of gases, however, it decreases the dissolved gas concentration and the gas-to-liquid mass transfer. On the other hand, the increased biological activity outperforms this effect.

The hydrogen pressure also has an important effect on methane yield. Low pressure of hydrogen is preferred by oxidizing organisms, but high pressure of H_2 is needed for methanogens.

Another important factor during the fermentation is the pH of the cultivation medium. Luo and Angelidaki (2013) showed that applying a pH range between 7.7 and 7.9 resulted in complete usage of H_2 throughout the thermophilic anaerobic fermentation. The methanation process consumes CO_2 during the operation, which may increase the broth pH above 8 (Zabranska and Pokorna, 2018). The increasing pH reduces biological activity but can be prevented by continuously adjusting the pH chemically or applying parallel fermentations with low pH substrates (Chae et al., 2010; Kim et al., 2004; Zhou et al., 2016).

In the case of *in situ* methanation, finding the optimal H_2 pressure is vital for efficient operation. The injection of hydrogen has to be fit to the CO_2 production rate in the digester, which requires continuous measurement of the gas composition and the setup of rigorous process control. On the other hand, these factors increase operating costs. *Ex situ* biogas upgrading bypasses the limitations that emerge in the case of existing biogas plants but it requires substrates streams that do not contain compounds hazardous to the culture.

One of the limitations regarding the technological side of biological methanation is the low hydrogen solubility in the media, i.e., water. This ultimately influences the applicable bioreactor designs for the methanation process. Several factors must be addressed in order to achieve enhanced gas-to-liquid mass transfer in bioreactors. These independent variables are the temperature, pressure, surface area for gas transfer and mixing. Other aspects of operation conditions, such as the recirculation of input gases, residence time, reactor volume and sparger pore size, should also be considered.

Different types of reactors can be applied for biological methanation, these include: (i) Continuously Stirred Tank Reactors (CSTR) (Zhang et al., 2015; Wandera et al., 2018; Khemkhao, Techkarnjanaruk and Phalakornkule, 2016), (ii) Fixed-Bed Reactors (FBR) (Alitalo and Aura, 2015) and (iii) Trickle-Bed Reactors (TBR) (Marko Burkhardt et al., 2019; Strübing et al., 2017). The advantages of CSTRs are high gas-to-liquid mass transfer rates due to the efficient mechanical agitation and a wide range of applicable impellers design. However, this design offers limited scale-up possibilities and the mixing in the vessel requires a high amount of energy, which increases the operational cost. In the case of fixed-bed reactors, the microbes are immobilized on the surface of particles and placed as a bed inside the reactor. The particles are characterized by high surface area-to-volume ratios, attaining high gas-to-liquid mass transfer rates without applying agitation. Similarly to fixed-bed reactors, TBRs also apply microbe immobilization on the surface of beds. The difference is that the reactor chamber is not filled with liquid medium but is sprinkled on the biofilm, forming a three-phase system (gas phase, liquid phase, biofilm). This type of reactor offers high volumetric efficiency and mass transfer rate.

The electricity production must meet some quality criteria, such as high efficiency and low GHG emission throughout the operation of a power plant. One of the main drawbacks that arises with the application of Power-to-Gas storage is the low overall electricity production efficiency. Hydrogen production can be done by 3 different water electrolysis technologies with different cell voltage efficiencies: (1) alkaline (AEL) (47–82%) (Ursúa et al., 2012), (2) polymer electrolyte membrane (PEM) (48–65%) (Ursúa, Gandía and Sanchis, 2012), (3) solid oxide electrolysis cells (SOEC) (83%) (Klotz, Leonide and Weber, 2014). The efficiency of catalytic methanation ranges between 70 to 85% (Blanco and Faaij, 2018), while the efficiency of biological methanation is above 74%, based on the higher heating value (HHV) (Electrochaea, 2018). The power production efficiency is estimated to be 30–40% in the case of an open cycle gas turbine, while this value ranges between 50 and 65 using combined cycle gas turbines (Blanco and Faaij, 2018). Regarding catalytic methanation, the overall efficiency of power-to-methane-to-power chain is assumed to be an average 36% (Zoss, Dace and Blumberga, 2016), while in the case of biological methanation this value is approximately the same. It means that, upon the storage process, only one third of the produced fluctuating renewable electricity can be gained back. The power production via burning fossil-based energy carriers (e.g., natural gas) has higher electric efficiency ($\eta_{electr.}$= 44.4%) and, thus, the PtG chain is currently not able to outperform the conventional technology in terms of efficiency. Therefore, to achieve a rentable and effective storage of the intermittent renewable energy produced by wind and solar farms, the efficiencies should be increased.

One of the main problems arising from the application of conventional fossil-based power plants (natural gas, coal) is that they are characterized by high global warming potential ranging GHG emissions between 0.490–1.183 kg $CO_{2,eq}$, respectively (WINGAS, 2018). The environmental aspect of the PtG chain is also an important factor that must be taken into consideration. The PtG technology becomes beneficial only if the cumulated greenhouse gas emission that is emitted throughout the operation is lower than fossil-based alternatives. The global warming potential of the Power-to-Gas technology is evaluated by several life cycle analyses. Collet et al. (Collet et al., 2017) found that 0.562 kg $CO_{2,eq}$ was emitted during the production of 1 kWh electricity involving carbon capture and PtG storage of fluctuating energy. It is also pointed out in this study, that GHG emission depends on the electricity mix, thus, the ratio of wind and photovoltaic influences the global warming potential. Parra et al. (Parra et al., 2017) investigated the life cycle emission profile of different Power-to-Gas alternatives. They determined that 0.309 kg $CO_{2,eq}$ kWh^{-1} is emitted through the Power-to-H_2 pathway, while 0.406 kg $CO_{2,eq}$ kWh^{-1} is associated with the Power-to-CH_4 storage option.

The Power-to-Methane storage chain offers the large-scale and long-term storage of fluctuating renewable energy in the form of methane. The synthetic methane can be stored in depleted geological reservoirs (fields, aquifers, salt formations) and it can be used to produce electricity when required.

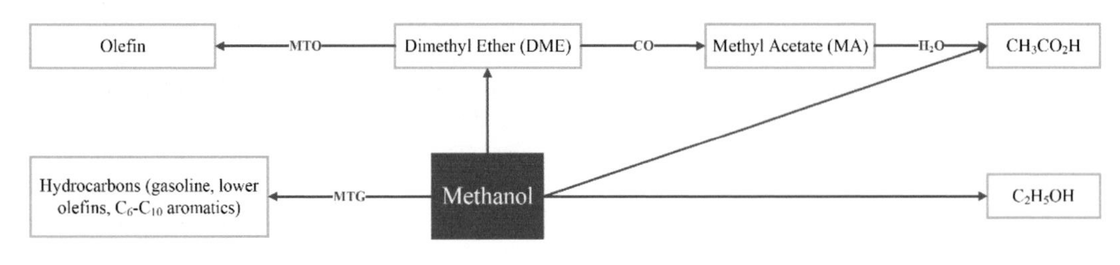

Figure 6. The possible transformation of methanol into value-added products (Andika et al., 2018).

Similar to the production of intermittent renewable energy sources, the electricity consumption is also characterized by variability. The fluctuation in the consumption can be balanced by the application of methane because the electricity production can be fitted to the needs of consumers by burning the required quantity of methane in turbines.

Renewable electricity can be stored in the form of liquid fuels, where the storage chain is known as Power-to-Liquid. The Power-to-Methanol (CH_3OH) chain is a favorable storage pathway as methanol is a versatile platform molecule and it can be upgraded to other chemicals (e.g., ethanol, acetic acid, hydrocarbons and ethers, as illustrated in Figure 6). Olah (2005) suggested a "methanol economy", which can be further supported with the combination of the storage of fluctuating energy. CH_3OH can be used in gas turbines, fuel cells and in combustion engines.

Methanol can be produced by various routes, such as coal-to-methanol, coke oven gas-to-methanol, natural gas-to-methanol and biomass-to-methanol (Zhuo Chen and Nannan Sun, 2019). The utilization of biomass is an important factor in decreasing the environmental effects of the fossil-based energy sector. Combining bioenergy with energetic carbon capture technologies offers negative CO_2 emission possibility, thus, it is considered as an efficient decarbonization aspect. Biomass can be transformed into gaseous energy carriers using thermochemical methods, e.g., atmospheric gasification (Lan et al., 2018; Fozer et al., 2017), hydrothermal gasification (Fozer et al., 2019) and anaerobic digestion. The produced biogas contains H_2, CH_4, CO_2, CO, and C_2-C_3 hydrocarbons. At this stage, a pure methane stream can be produced by upgrading the carbon dioxide and syngas (H_2 and CO) content of the biogas. During this methanation stage, several reactions, such as CO hydrogenation, water-gas shift and Sabatier reactions, can be considered. The components in the biogas can also be used to synthesize methanol through the hydrogenation of CO_2, Equation (5), and reverse water gas shift reactions, Equation (6) (Leonzio and Foscolo, 2019). A methanol-water mixture, which must be separated before long-term storage and utilization, is formed in these reactions.

$$CO_2 + 3H_2 \leftrightarrow CH_3OH + H_2O \quad \Delta H^0_{298} = -49.5 kJ\, mol^{-1}_{CO_2} \tag{5}$$

$$CO_2 + H_2 \leftrightarrow CO + H_2O \quad \Delta H^0_{298} = 41.2 kJ\, mol^{-1}_{CO_2} \tag{6}$$

Both PtG and PtL conversion routes have their own merits and limitations, as detailed in Table 5. The Power-to-Methane alternative is suitable to store energy using the already existing natural gas infrastructure. Carbon neutral operation can be achieved using the PtM pathway, but decarbonization opportunities must be further investigated since carbon dioxide is re-forming in the process as a result of the combustion of methane. On the other hand, Power-to-Liquid alternatives can be used for the production of chemical intermediates that provide suitable opportunities for decarbonization purposes. However, energy recovery cannot be assured by this way. The disadvantage of Power-to-Methanol is that it mostly requires biomass as feedstock. Methanol production can be achieved by synthetic way via Fischer-Tropsch synthesis, but the process requires carbon monoxide, therefore, the captured CO_2 has to be reduced to CO prior the production of synthetic methanol.

In conclusion, the Power-to-Methane storage chain offers large scale capacities and efficient variable renewable energy storage that can become a milestone in the transformation of the energy sector

Table 5. Summarizing the advantages and disadvantages of various Power-to-X alternatives.

Storage alternative	Advantages	Disadvantages
Power-to-Hydrogen	Supports hydrogen economy High efficiencies	High storage volume Lack of infrastructure
Power-to-Methane	Already existing infrastructure Closed carbon cycle can be applied High storage capacities	Pure carbon dioxide feedstock required
Power-to-Methanol	High energy density Platform molecule Decarbonization opportunity	Toxic storage compound Biomass feedstock availability

towards carbon neutrality. The Power-to-H_2 supports hydrogen economy, but hydrogen storage is still a challenging task, therefore, *in situ* applications are preferred. The Power-to-Methanol chain can be used for the production of synthetic, carbon-based materials, which highlights an up-and-coming opportunity for decarbonization.

6. Comparison and Conclusion

Humanity has always been trying to store energy, as well as finding ways to use the energy stored by nature. Photosynthesis is one of the oldest forms of energy storage, however, its efficiency is only about 1%. Besides biomass, people have also identified waste of nature, e.g., hard coal and crude oil, as fossil energy carriers.

The utilization of the natural reserves of the Earth jeopardizes the heritage of our children since our planet is a spaceship economy. Therefore we have to focus on using more and more renewable energy.

The problem of using renewable energy is that it fluctuates due to weather conditions. On the other hand, the energy consumption also fluctuates but with a different frequency. Energy storage can help, to solve this problem.

There are practically uncountable methods of energy storage and in Figure 2 the important methods that are capable of storing energy in a significant quantity are demonstrated.

In this chapter, we deal with the storage of thermal energy and electrical energy. The storage of electrical energy is the challenge of our age and, besides the well-known and applied mechanical and chemical ways, we introduce here the biological way, which is our novel development.

The biological way to carry out the Power-to-Gas and Power-to-X technologies proves to be the best solution, now having the highest efficiency. It has stirred great interest among the energy and refinery industries and the authors do believe that it is capable of efficiently solving the energy storage problem on large scale.

Acknowledgements

The authors are grateful for the financial support from the Hungarian National Scientific Research Foundation (OTKA) projects: nr.112699 and nr.128543. The research was supported by the European Union and the Hungarian State, co-financed by the European Regional Development Fund in the framework of the GINOP-2.3.4-15-2016-00004 project aimed to promote the cooperation between the higher education and industry. This work was supported by the NTP-NFTÖ-18-B-0154 National Talent Program of The Ministry of Human Capacities.

References

ADB. 2018. Handbook on Battery Energy Storage System. Asian Development Bank.

Alitalo, Anni, Marko Niskanen and Erkki Aura. 2015. Biocatalytic methanation of hydrogen and carbon dioxide in a fixed bed bioreactor. Bioresource Technology 196: 600–605. https://doi.org/10.1016/j.biortech.2015.08.021.

Andersson, Joakim and Stefan Grönkvist. 2019. Large-scale storage of hydrogen. International Journal of Hydrogen Energy no. xxxx. https://doi.org/10.1016/j.ijhydene.2019.03.063.

Andika, Riezqa, Asep Bayu, Dani Nandiyanto, Zulfan Adi Putra, Muhammad Roil Bilad, Young Kim, Choa Mun and Moonyong Lee. 2018. Co-electrolysis for power-to-methanol applications. Renewable and Sustainable Energy Reviews 95(December 2017): 227–41. https://doi.org/10.1016/j.rser.2018.07.030.

Banach, Anna, Sławomir Ciesielski, Tomasz Bacza, Marek Pieczykolan and Aleksandra Ziembińska-Buczyńska. 2018. Microbial community composition and methanogens' biodiversity during a temperature shift in a methane fermentation chamber. Environmental Technology (United Kingdom) 3330: 1–12. https://doi.org/10.1080/09593330.2018.1468490.

Benjaminsson, G., Benjaminsson, J. and Rudberg, R.B. 2013. Power-to-Gas—A Technical Review. SGC Rapport, 2013: 284, Malmö, Sweden.

Blake, Lynsay I., Alexander Tveit, Lise Øvreås, Ian M. Head and Neil D. Gray. 2015. Response of methanogens in arctic sediments to temperature and methanogenic substrate availability. PLoS ONE 10(6): 1–18. https://doi.org/10.1371/journal.pone.0129733.

Blanco, Herib and André Faaij. 2018. A review at the role of storage in energy systems with a focus on power to gas and long-term storage. Renewable and Sustainable Energy Reviews 81(July 2017): 1049–86. https://doi.org/10.1016/j.rser.2017.07.062.

BP. 2017. BP Statistical Review of World Energy June 2017. https://www.bp.com/content/dam/bp-country/de_ch/PDF/bp-statistical-review-of-world-energy-2017-full-report.pdf.

Burkhardt, M., T. Koschack and G. Busch. 2015. Biocatalytic methanation of hydrogen and carbon dioxide in an anaerobic three-phase system. Bioresource Technology 178: 330–33. https://doi.org/10.1016/j.biortech.2014.08.023.

Burkhardt, Marko, Isabel Jordan, Sabrina Heinrich, Johannes Behrens, André Ziesche and Günter Busch. 2019. Long term and demand-oriented biocatalytic synthesis of highly concentrated methane in a trickle bed reactor. Applied Energy 240(February): 818–26. https://doi.org/10.1016/j.apenergy.2019.02.076.

Chae, Kyu Jung, Mi Jin Choi, Kyoung Yeol Kim, F.F. Ajayi, Woosin Park, Chang Won Kim and In S. Kim. 2010. Methanogenesis control by employing various environmental stress conditions in two-chambered microbial fuel cells. Bioresource Technology 101(14): 5350–57. https://doi.org/10.1016/j.biortech.2010.02.035.

Collet, Pierre, Eglantine Flottes, Alain Favre, Ludovic Raynal, Hélène Pierre, Sandra Capela and Carlos Peregrina. 2017. Techno-economic and life cycle assessment of methane production via biogas upgrading and power to gas technology. Applied Energy 192: 282–95. https://doi.org/10.1016/j.apenergy.2016.08.181.

Deák, C., Á.B. Palotás, P. Mizsey and B. Viskolcz. 2018. Carbon-dioxide management network in hungary. Chemical Engineering Transactions 70: 2143–2148.

Electrochaea. 2018. Specifications of Electrochaea's BioCat Methanation Plants. 2018. http://www.electrochaea.com/wp-content/uploads/2018/03/201803_Data-Sheet_BioCat-Plant.pdf.

Fozer, Daniel, Nora Valentinyi, Laszlo Racz and Peter Mizsey. 2017. Evaluation of microalgae-based biorefinery alternatives. Clean Technologies and Environmental Policy 19(2): 501–15. https://doi.org/10.1007/s10098-016-1242-8.

Fozer, Daniel, Bernadett Kiss, Laszlo Lorincz, Edit Szekely, Peter Mizsey and Aron Nemeth. 2019. Improvement of microalgae biomass productivity and subsequent biogas yield of hydrothermal gasification via optimization of illumination. Renewable Energy 138: 1262–72. https://doi.org/10.1016/j.renene.2018.12.122.

Fuchs, G., Thauer, R., Ziegler, H. and Stichler, W. 1979. Carbon isotope fractionation by methanobacterium thermoautotrophicum. Archives of Microbiology 120(2): 135–39. https://doi.org/10.1007/BF00409099.

Ghaib, Karim and Fatima-Zahrae Ben-Fares. 2018. Power-to-methane: A state-of-the-art review. Renewable and Sustainable Energy Reviews 81(December 2016): 433–46. https://doi.org/10.1016/j.rser.2017.08.004.

Götz, Manuel, Jonathan Lefebvre, Friedemann Mörs, Amy McDaniel Koch, Frank Graf, Siegfried Bajohr, Rainer Reimert and Thomas Kolb. 2016. Renewable power-to-gas: A technological and economic review. Renewable Energy 85: 1371–90. https://doi.org/10.1016/j.renene.2015.07.066.

Guneratnam, Amita Jacob, Eoin Ahern, Jamie A. FitzGerald, Stephen A. Jackson, Ao Xia, Alan D.W. Dobson and Jerry D. Murphy. 2017. Study of the performance of a thermophilic biological methanation system. Bioresource Technology 225: 308–15. https://doi.org/10.1016/j.biortech.2016.11.066.

Jia, Chunmiao, Yihu Dai, Yanhui Yang and Jia Wei. 2019. ScienceDirect nickel-cobalt catalyst supported on TiO_2-coated SiO_2 spheres for CO_2 methanation in a fluidized bed no. xxxx. https://doi.org/10.1016/j.ijhydene.2019.04.009.

Kenning, Tom. 2017. EnergyAustralia Ponders World's Largest Seawater Pumped Hydro Energy Storage Plant. 2017. https://www.energy-storage.news/news/energyaustralia-ponders-worlds-largest-seawater-pumped-hydro-energy-storage.

Khemkhao, Maneerat, Somkiet Techkarnjanaruk and Chantaraporn Phalakornkule. 2016. Effect of chitosan on reactor performance and population of specific methanogens in a modified CSTR treating raw POME. Biomass and Bioenergy 86: 11–20. https://doi.org/10.1016/j.biombioe.2016.01.002.

Kim, In S., Moon H. Hwang, Nam J. Jang, Seong H. Hyun and Lee, S.T. 2004. Effect of low PH on the activity of hydrogen utilizing methanogen in bio-hydrogen process. International Journal of Hydrogen Energy 29(11): 1133–40. https://doi.org/10.1016/j.ijhydene.2003.08.017.

Klotz, D., Leonide, A. and Weber, A. 2014. ScienceDirect electrochemical model for SOFC and SOEC mode predicting performance and efficiency, 6–11. https://doi.org/10.1016/j.ijhydene.2014.08.139.

Kreitz, Bjarne, Gregor D. Wehinger and Thomas Turek. 2018. Dynamic simulation of the CO_2 methanation in a micro-structured fixed-bed reactor. Chemical Engineering Science. https://doi.org/10.1016/j.ces.2018.09.053.

Lan, Weijuan, Guanyi Chen, Xinli Zhu, Xin Wang, Xuetao Wang and Bin Xu. 2018. Research on the characteristics of Biomass GasiFiCation in a FlUidized Bed April: 1–8. https://doi.org/10.1016/j.joei.2018.03.011.

Lecker, Bernhard, Lukas Illi, Andreas Lemmer and Hans Oechsner. 2017. Biological hydrogen methanation—A review. Bioresource Technology 245(August): 1220–28. https://doi.org/10.1016/j.biortech.2017.08.176.

Leonzio, Grazia. 2019. ScienceDirect ANOVA analysis of an integrated membrane reactor for hydrogen production by methane steam reforming. International Journal of Hydrogen Energy, no. xxxx. https://doi.org/10.1016/j.ijhydene.2019.03.077.

Leonzio, Grazia, Edwin Zondervan and Pier Ugo Foscolo. 2019. Methanol production by CO_2 hydrogenation: Analysis and simulation of reactor performance. International Journal of Hydrogen Energy, no. xxxx. https://doi.org/10.1016/j.ijhydene.2019.02.056.

Li, Bin, Hui Min Wen, Wei Zhou, Jeff Q. Xu and Banglin Chen. 2016. Porous metal-organic frameworks: Promising materials for methane storage. Chem. 1(4): 557–80. https://doi.org/10.1016/j.chempr.2016.09.009.

Lunde, Peter J. and Frank L. Kester. 1974. Carbon dioxide methanation on a ruthenium catalyst. Industrial and Engineering Chemistry Process Design and Development 13(1): 27–33. https://doi.org/10.1021/i260049a005.

Luo, Gang and Irini Angelidaki. 2013. Co-digestion of manure and whey for *in situ* biogas upgrading by the addition of H_2: Process performance and microbial insights. Applied Microbiology and Biotechnology 97(3): 1373–81. https://doi.org/10.1007/s00253-012-4547-5.

Mickol, Rebecca, Sarah Laird and Timothy Kral. 2018. Non-psychrophilic methanogens capable of growth following long-term extreme temperature changes, with application to mars. Microorganisms 6(2): 34. https://doi.org/10.3390/microorganisms6020034.

Mizsey, P., Cuellar, A., Newson, E., Hottinger, P., Truong, T.B. and von Roth, F. 1999. Fixed bed reactor modelling and experimental data for catalytic dehydrogenation in seasonal energy storage applications. Computers and Chemical Engineering 23(SUPPL. 1): S379–82. https://doi.org/10.1016/S0098-1354(99)80093-3.

Olah, G.A., Goeppert, A. and Prakash, G.K.S. 2005. Beyond oil and gas: The methanol economy. Angewandte Chemie 44(18): 2636–39. https://doi.org/10.1002/anie.200462121.

Parra, David, Xiaojin Zhang, Christian Bauer and Martin K. Patel. 2017. An integrated techno-economic and life cycle environmental assessment of power-to-gas systems. Applied Energy 193: 440–54. https://doi.org/10.1016/j.apenergy.2017.02.063.

Rechargeable battery, https://en.wikipedia.org/wiki/Rechargeable_battery (last visited July, 2019).

Specht, M., Baumbart, F., Feigl, B., Frick, V., Stürmer, B. and Zuberbühler, U. 2009. Storing Bioenergy and Renewable Electricity in the Natural Gas Grid. 2009. http://www.fvee.de/fileadmin/publikationen/Themenhefte/th2009/th2009_05_06.pdf.

Strübing, Dietmar, Bettina Huber, Michael Lebuhn, Jörg E. Drewes and Konrad Koch. 2017. High performance biological methanation in a thermophilic anaerobic trickle bed reactor. Bioresource Technology 245: 1176–83. https://doi.org/10.1016/j.biortech.2017.08.088.

Teichmann, Daniel, Katharina Stark, Karsten Müller, Gregor Zöttl, Peter Wasserscheid and Wolfgang Arlt. 2012. Energy storage in residential and commercial buildings via liquid organic hydrogen carriers (LOHC). Energy and Environmental Science 5(10): 9044–54. https://doi.org/10.1039/c2ee22070a.

Thomas, Torsten and Ricardo Cavicchioli. 2000. Effect of temperature on stability and activity of elongation factor 2 proteins from antarctic and thermophilic methanogens. Journal of Bacteriology 182(5): 1328–32. https://doi.org/10.1128/JB.182.5.1328-1332.2000.

University, Battery. n.d. Understanding Lithium-Ion. https://batteryuniversity.com/learn/archive/understanding_lithium_ion.

Ursúa, Alfredo, Luis M. Gandía and Pablo Sanchis. 2012. Water Electrolysis: Current Status and Future Trends 100(2).

Wandera, Simon, M., Wei Qiao, Mengmeng Jiang, Dalal E. Gapani, Shaojie Bi and Renjie Dong. 2018. AnMBR as alternative to conventional CSTR to achieve efficient methane production from thermal hydrolyzed sludge at short HRTs. Energy 159: 588–98. https://doi.org/10.1016/j.energy.2018.06.201.

WINGAS. 2018. Greenhouse Gas Emission Figures for Fossil Fuels and Power Station Scenarios in Germany. 2018. https://www.wingas.com/en/media-library/studies/study-greenhouse-gas-emission-figures-for-fossil-fuels-and-power-station-scenarios-in-germany.html.

Wise, D.L., Cooney, C.L. and Augenstein, D.C. 1978. Biomethanation: Anaerobic fermentation of CO_2, H_2 and CO to methane. Biotechnology and Bioengineering 20(3366): 1153–72. https://doi.org/10.1002/bit.260200804.

World Energy Council. 2016. World Energy Resources 2016.

Zabranska, Jana and Dana Pokorna. 2018. Bioconversion of carbon dioxide to methane using hydrogen and hydrogenotrophic methanogens. Biotechnology Advances 36(3): 707–20. https://doi.org/10.1016/j.biotechadv.2017.12.003.

Zhang, Fang, Yan Zhang, Yun Chen, Kun Dai, Mark C.M. van Loosdrecht and Raymond J. Zeng. 2015. Simultaneous production of acetate and methane from glycerol by selective enrichment of hydrogenotrophic methanogens in extreme-thermophilic (70 °C) mixed culture fermentation. Applied Energy 148: 326–33. https://doi.org/10.1016/j.apenergy.2015.03.104.

Zhou, Jun, Rui Zhang, Fenwu Liu, Xiaoyu Yong, Xiayuan Wu, Tao Zheng, Min Jiang and Honghua Jia. 2016. Biogas production and microbial community shift through neutral PH control during the anaerobic digestion of pig manure. Bioresource Technology 217: 44–49. https://doi.org/10.1016/j.biortech.2016.02.077.

Zhuo Chen, Qun Shen, Nannan Sun and Wei Wei. 2019. Life cycle assessment of typical methanol production routes: The environmental impacts analysis and power optimization. Journal of Cleaner Production 408–16.

Zoss, Toms, Elina Dace and Dagnija Blumberga. 2016. Modeling a power-to-renewable methane system for an assessment of power grid balancing options in the baltic states' region. Applied Energy 170: 278–85. https://doi.org/10.1016/j.apenergy.2016.02.137.

CHAPTER 19

Lithium-ion Battery

Future Technology Development Driven by Environmental Impact

Mihaela Buga, Adnana Spinu-Zaulet* and *Alin Chitu*

1. Introduction

European and global energy policies are based on a competitive, sustainable and secure energy system in the coming years, promoting Renewable Energy Sources (RES) as a priority measure for long term sustainability of the energy sector (European Commission, 2011). Thus, the progress recorded up until now indicates that the increasing share of renewable energy translates not only into GHSs emissions reduction, but also a reduced dependency on energy related imports. However, a significant increase in the share of renewable energy in the mix will require an efficient solution for energy storage that is able to store and release a high amount of energy in a short time when needed in order to overcome the intermittent nature of RES.

Focusing on the world's largest carbon emitting countries in Figure 1 (CTC, 2018), it has to be highlighted that Europe has devoted relevant attention to climate and environmental policies.

Thus, in October 2014, the European Commission approved the 2030 Framework for Climate and Energy Policies document (European Council, 2014) which aims to achieve a reduction of the level of the EU greenhouse gases emissions (GHGs) by 20% by the year 2020, and by 40% compared to 1990 level values by the year 2030. According to EEA, in 2016, the EU road transport sector (including international aviation and international shipping) has contributed 72% of the EU-28 (EEA, 2018), being the only EU sector in which GHS emissions have risen since 1990.

In order to reduce the GHS emissions from the EU transport sector and reach the ambitious long-term decarbonisation targets, the EU focus has shifted on electrifying the transport sector, based on RES, through Battery Electric Vehicles (BEVs) and Fuel Cell Electric Vehicles (FCEVs) (EASE, 2019). In this way, the electrification of the road transport, based on RES, will play an essential role in decarbonising the transport sector, contributing to the improvement of air quality and reduction of noise pollution (ERTRAC, 2017). On the other hand, the massive infiltration of hybrid electric vehicles (HEVs), plug-in hybrid electric vehicles (PHEVs) and electric vehicles (EVs) calls for higher battery performances, safe electrode/electrolyte materials and less cost, with respect to current lithium-ion technology. Currently, the most commonly-used cathode chemistries for automotive applications are represented by lithium

National Research and Development Institute for Cryogenics and Isotopic Technologies – ICSI Rm. Valcea, 4 Uzinei Street, 240050, Rm. Valcea, Romania.
Emails: adnana.zaulet@icsi.ro; alin.chitu@icsi.ro
* Corresponding author: mihaela.buga@icsi.ro

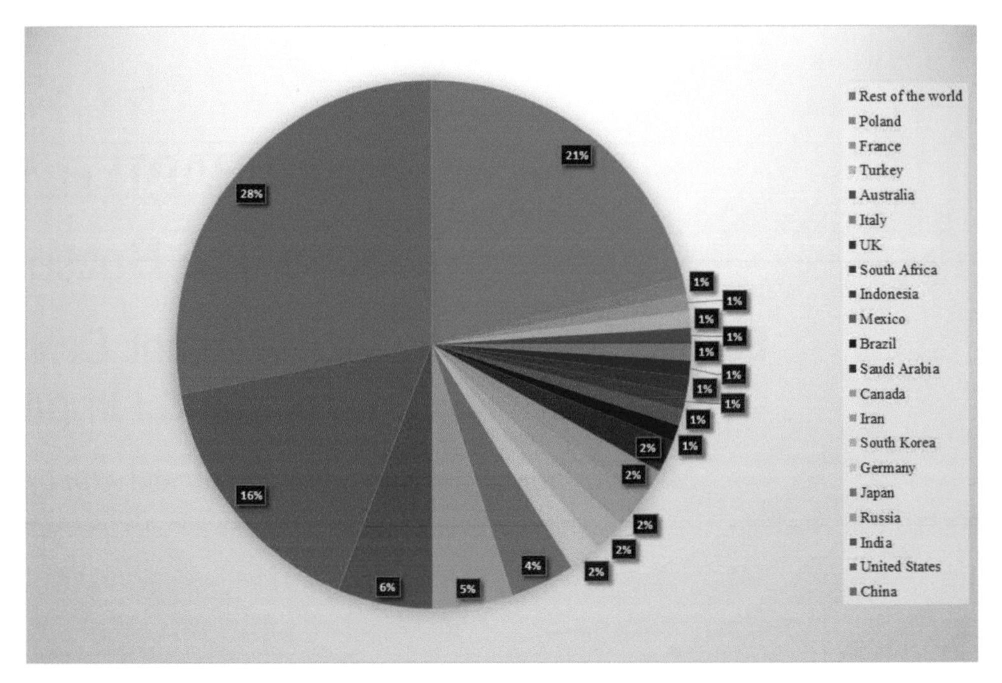

Figure 1. The world's 20 largest carbon-emitting countries, adapted from (CTC, 2018).

iron phosphate (LFP), lithium nickel cobalt oxide (NCM), lithium nickel cobalt aluminium oxide (NCA) and lithium manganese oxide (LMO) (Berman et al., 2018). Despite the fact that some cobalt-based cathode materials display a high energy density, from the EC point of view, cobalt is listed as "critical raw material (CRM)", among others materials, due to its issues of risk of disruption and economic impact (European Commission, 2017).

As Li-ion battery production is currently dominated by Asia, increasing the competitiveness of the European battery industry and research has become an important issue for the EU. Consequently, large initiatives and funding programs have been launched in this area in order to support research and development on a national and EU level, leveraging European activities to develop novel post-lithium-ion technologies that have not yet been established on the market and that have the potential to dramatically improve the overall environment profile of the battery manufacturing technology. Thus, the EU initiatives for the next generation of batteries developed and produced in Europe play a pivotal role in the implementation of electrified mobility. In this context, the chapter outlines the EU perspective on Li-ion battery value chain for automotive application in the transition towards a low-carbon economy.

2. EU Competitive, Sustainable and Innovative Strategic Battery Value Chain

The lithium-ion battery value chain for E-mobility comprises a wide spectrum, starting from raw materials production, cell component manufacturing, cell manufacturing, battery pack manufacturing, electric vehicle integration and recycling, see Figure 2.

The extraction and exploitation of the most common raw materials generally comes from resource-rich countries outside the EU. The process of extracting raw materials causes an environmental footprint through water pollution and release of GHG emissions, producing half of the GHGs from the battery production segment. Out of the most common elements integrated in Li-ion battery cells, like lithium (Li), cobalt (Co), manganese (Mn), nickel (Ni), natural graphite, silicon (Si), aluminium (Al) and copper (Cu), Co, natural graphite and Si display a high economic impact and a vulnerable risk of supply for the EU. In this regard, the European Commission included these materials on the list of the Critical Raw Materials for the EU (European Commission, 2017).

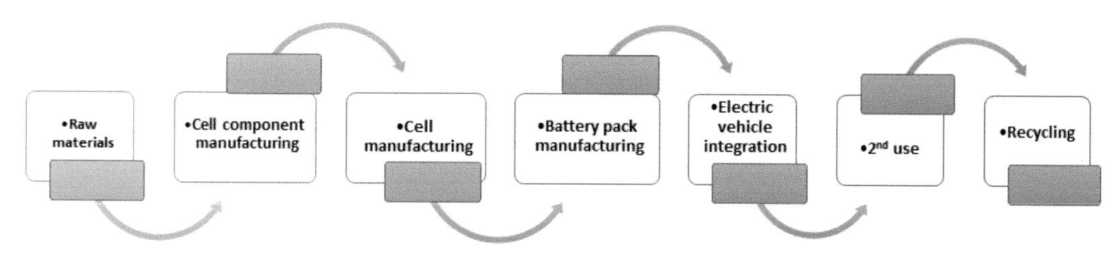

Figure 2. EV's Li-ion battery value chain, adapted from (Drabik et al., 2018).

Figure 3. Manufacturing steps of Li-ion batteries – ICSI ENERGY Li-ion battery laboratory.

Reviewing and updating the CMRs list once in three years is meant to support the implementation of the EU industrial policy not only in the battery sector. By 2020, all these measures should comply with the ECs targets and provide a growth up to 20% of industry contribution to GDP. Also, as the EU demand for electric vehicles is rapidly emerging, calling significant demand for Li-ion batteries and raw materials, the price of these materials will continue to increase. Thus, the EU should secure access to raw materials by boosting the battery raw materials production from European sources.

In terms of cell component manufacturing and cell manufacturing, China dominates the global market, while EU has a comparatively small share. The main process steps for the cell component manufacturing are represented in Figure 2. In the slurry preparation, active materials (e.g., LMO, LCO, LFP, NCM, NCA/graphite, silicon, LTO), binders (e.g., PvdF/CMC, SBR), conductive agents (e.g., carbon-based materials) and solvents (NMP/water) are mixed together, with different mixing procedures and mixing times, in specific mass ratios, resulting in a homogeneous slurry with the necessary characteristics to be used in the next process step. The electrode slurry is then coated onto the current collectors (e.g., aluminium or copper foils) using different coating technology with the achievable thickness. In order to eliminate the residual solvents, the electrodes are dried and further compressed to a desired porosity. Then, the electrodes are cut or punched into desired shapes for cylindrical or pouch cells. After the cutting/slitting/punching steps, the electrodes are dried for a second time in order to remove the water contamination. The assembly and electrolyte injection steps take place in special dry rooms.

With regards to the costly Li-ion batteries cell manufacturing chain, the Roll-to-roll (R2R) and slot–die coating technologies are the usual manufacturing processes for Li-ion batteries electrodes. The manufacturing process of the positive electrodes is still largely based on organic solvent (NMP) and fluorine-containing polymers (PvdF). When using organic solvents, one has to consider the environmental issues, the cost of the solvent and the cost of the necessary recovery system. Using organic solvents in the manufacturing process of the battery electrodes involves high energy consumption, due to the NMP recovery system, resulting in it being the main carbon footprint generating factor of Li-ion battery manufacturing. Also, cell formation (initial cycling) procedure is currently one of the longest steps in the cell manufacture and testing value chain. This adds to the costs, limits the battery production rate, and requires extensive facilities for testing and validation. The formation cycle and protocol are specific to the cell chemistry and performance metrics. Fast formation protocols can not only shorten the formation time but also improve the battery lifetime through controlled surface films formation and reduced irreversible electrolyte and Li-ions consumption. Taking into consideration that European and global energy policies are based on a secure, competitive and decarbonized energy system in the upcoming years (European

Commission's Energy Roadmap 2050), the required improvement of battery performance must be achieved through the development of green production processes for cathode, anode, binders, electrolyte materials and coating processes in order to reduce the overall environmental footprint. Therefore, future coating processes must involve less expensive and more environmentally friendly solvents, such as aqueous-based slurries. The use of water instead of organic solvents will be the required step in achieving low-cost production and environmental sustainability. Thus, the efficiency of the production processes will gain in terms of time, quality and energy consumption translated overall in cell costs, which represent approximately 70% of the total cost of the EV battery pack.

Regarding the next steps of the process, namely battery pack manufacturing and electric vehicle integration, the battery pack manufacturing, including the BMS, accounts for approximately 35% of the overall EV costs (Kupper et al., 2018).

In relation to the EV manufacturing market share, again, China is the leader, while the EU, USA and Japan share the same percentage with regards to the PHEVs and BEV producers.

Given the increasing number of the electric vehicles, recycling the automotive Li-ion batteries when they have reached their end-of-life will become imperative. The EU recycling industry is not yet prepared for the increasing volume of EoL Li-ion batteries, giving the opportunity for EU recycling companies to expand their processing capacity. Currently, the major players in the EU recycling market are represented by 10 leading companies, ERAMET – Valdi (France), Umicore (Belgium), ACCUREC GmbH (Germany), RECUPLY S.A. (France), SNAM (France), Euro Dieuze/SARP (France), AKKUSER Ltd. (Finland) and BATREC AG (Switzerland) (Alves et al., 2018).

Becoming a leading player in the battery manufacturing industry, at a global level, requires the EU to reduce their cost of Li-ion battery production and recycling. From raw material extracting to second life use or recycling, all sectors of the value chain must be strengthened through an action plan aimed at involving industry leaders as well as emerging actors. To support this challenge, the EC has identified the weak links in the value chain of the battery manufacturing process and reported them through JRC reports (Lebedeva et al., 2017). Furthermore, in 2017, EC launched the "EU Battery Alliance" cooperation platform, which plays a strategic role, from raw materials processing to recycling, bringing together more than 250 industrial, innovation and academia stakeholders, EU member states, EC, European Investment Bank, and is coordinated by InnoEnergy (European Institute of Innovation and Technology – EIT).

EIT has been working closely with key actors in the industry related field to reach the ambitious goal of EU Li-Ion production independency. Prioritizing their projects, educational and training programs will ensure cells manufacturing based on carbon footprint reduction, eco-design, 2nd use and an increased recycling efficiency. Also, funding opportunities for research and innovation were budgeted by the EU's Framework Programme for Research and Innovation. Projects for energy storage and low-carbon mobility were granted through the Horizon 2014–2020 program.

In order to guide Europe towards leadership in the battery ecosystem, in May 2018, EC approved a "Strategic Action Plan on Batteries" included in the "Europe on the Move" mobility package, which focuses on development in a circular economy, adding value to the environment and EU citizen's life standards (European Commission, 2019).

Sustainable battery cell production, including reduction of critical raw materials and the use of green and environmentally friendly manufacturing processes are of key importance to the upcoming Li-ion battery developments. For Europe, being far behind the technological battery developments in Asia, these points which require significant changes to state-of-the-art cell production, offer a golden opportunity of introducing new competitive approaches to get a more significant share of the Li-ion battery market.

Regarding the EU next energy batteries, a ten years large-scale and long-term European Research Coordination and Support Action grant under the Horizon 2020 programme, named Battery 2030+, has the grand ambition of developing novel research fields for future batteries in a sustainable framework. The Battery 2030+ consortium, coordinated by Prof. Kristina Edström, from Uppsala University, Sweden, includes five universities (Uppsala University, Politecnico di Torino, Technical University of Denmark, University of Münster, Vrije Universiteit Brussel), seven research centres (French Alternative Energies and Atomic Energy Commission, Karlsruhe Institute of Technology, French National Centre for Scientific Research, Forschungszentrum Jülich, Fraunhofer-Gesellschaft, Fundacion Cidetec, National Institute of

Chemistry – Slovenia, SINTEF AS), three industry-led associations (RECHARGE, EMIRI, EASE), and one company (Absiskey), and has full support of a number of European and national organisations, counting ALISTORE ERI, EERA, EIT InnoEnergy, EIT RawMaterials, EARPA, EUROBAT, EGVI, CLEPA, EUCAR, KLIB, RS2E, Swedish Electromobility Centre, PolStorEn, ENEA, CIC energigune, IMEC and Tyndall National Institute (Battery 2030+, 2019). The Battery 2030+ targets a series of challenges across the battery value chain, starting from battery materials mining toward sustainability, material resource efficiency in Europe, green chemistries and environmentally friendly manufacturing processes, and efficient battery recycling processes. An important issue in achieving the lowest carbon footprint through Battery 2030+ is represented by developing energy storage systems that enable e-mobility and grid flexibility, improving EU citizens' quality of life (medical devices, robotics, internet of things, etc.), advanced and safer battery-based control mechanisms and with considerably longer lifetimes, due to smart functionalities at battery cell level, and fast rate charging protocols considering EV applications and their impact on life cell.

Another important initiative belongs to some German research societies and institutes who are making ambitious efforts to build battery cell production in Germany. On this line, the German funding members ads-tec GmbH, BASF SE, Chemetall GmbH, Continental, German Accumotive GmbH & Co. KG, Evonik Litarion GmbH, Freudenberg Nonwovens KG, GAIA Akkumulatorenwerke GmbH, Leclanché Lithium GmbH, Li-Tec Battery GmbH, Merck KGaA, SB LiMotive Germany GmbH, SGL Carbon GmbH, Süd-Chemie AG, VARTA Micro Battery GmbH, Center for Solar Energy and Hydrogen Research Baden-WEUrttemberg (ZSW), set up a network of competencies for Li-ion batteries in Germany called "KLiB" (Kompetenznetzwerk Lithium-Ionen-Batterien). The goal of the national network KLiB, led by the Technical University of Braunschweig from Germany, is to strengthen the entire battery value chain, in order to build and exploit a common R&D, production and testing infrastructure by developing uniform standards (KLiB, 2018).

Northvolt Sweden started work on a demonstration production line for developing, testing and industrializing Li-ion battery cells before large-scale production through InnovFin Energy Demo Projects (EDP) facility and EIB. Northvolt Labs will allow process and product development to be fully validated, starting from active material to finished cell, enabling a faster and more efficient time to market for unique formulations, cell design and technology (Northvolt, 2017).

A 26 EU region (Andalusia (ES), Castile and Leon (ES), West Slovenia (SI), Aragon (ES), Austria (AT), Auvergne Rhone Alpes (FR), Baden-Wurttemberg (DE), Basque Country (ES), Bavaria (DE), Brussels-Capital Region (BE), Central Ostrobothnia (FI), East Netherlands (NL), East Slovenia (SI), Emilia-Romagna (IT), Flanders (BE), Kainuu (FI), Lapland (FI), Lombardy (IT), Metropol Region Eindhoven (NL), Navarra (ES), Northern Ostrobothnia (FI), Nouvelle Aquitaine (FR), Pohjois-Savo (Northern Savonia) (FI), Valencia (ES), Vestlandet (NO) and Viken (NO)) open partnership, focusing on developing R&D&I projects on advanced battery materials for e-mobility suitable for extreme working conditions, launched the Energy and the Smart Specialisation Platform on Energy (S3PEnergy) in October 2018. The Advanced Materials for Batteries Partnership (AMBP) identified six Priority Pilots, including sustainable raw material, extraction and processing in Castile and León, Generation 3b battery materials (cathode: HE-NCM, HVS anode: silicon/carbon) in Auvergne Rhône Alpes, generation 4 conversion materials (all solid state with lithium anode – Li/S) in Bavaria, liquid based-batteries (for stationary application) in Basque/Valencia, Li-ion recycling technology in Bavaria and research and testing infrastructure on West Slovenia (AMBP, 2018).

Important EU projects for research and innovation in batteries funded through HORIZON 2020. The Framework Programme for Research and Technological Development from European Commission in the battery field are listed below:

- ALION—"A low-cost aluminium ion-battery" (2015–2019). The main objective of the ALION project is to develop an aluminium-ion battery technology for energy storage application in decentralised electricity generation sources (ALION, 2015).
- eCAIMAN "Electrolyte, cathode and anode improvements for market-near next-generation lithium ion batteries" (2015–2018). The eCAIMAN Consortium, consisting of 14 partners (AIT,

the coordinator of the project, and Arkema, CEA, CERTH, CIDETEC, CRF, Eutema, Imerys, Lithops, Piaggio, POLITO, SP, ThinkStep, Volvo), had the grand ambition to bring European expertise together to develop an automotive Li-ion battery cell that can be produced in Europe (eCAIMAN, 2015).

- SPICY "Silicon and polyanionic chemistries and architectures of Li-ion cell for high energy battery" (2015–2018). SPICY project gathers the key scientific and technical researchers in Europe covering the complete Li-ion battery value chain, addressing the whole Li-ion battery value chain through the development of a new generation of Li-ion batteries, meeting the expectations of electrical vehicle end-users, including performances, safety, cost, recyclability and lifetime (SPICY, 2015).
- FiveVB "Five Volt Lithium Ion Batteries with Silicon Anodes produced for Next Generation Electric Vehicles" (2015–2018). The goal of the FiveVB project was to develop a new cell technology based on innovative materials, such as high capacity anodes, high voltage cathodes and stable, safe and environmentally friendly electrolytes (FiveVB, 2015).
- ZAS "Zinc Air Secondary innovative nanotech based batteries for efficient energy storage" (2015–2018) supported by the Horizon 2020 framework NMP program brings together the main EU players in the battery field and aims to develop a cost efficient zinc-air secondary battery as a promising option for stationary energy storage (ZAS, 2015).
- GREENLION "Advanced manufacturing processes for low cost greener Li-ion batteries" (2011–2015). The GREENLION project includes 10 industry members (8 Large, 2 SME), 3 research institutes and 3 universities, being a Large Scale Collaborative Project within the FP7 leading on the manufacturing of greener and cheaper Li-ion batteries for EVs (GREENLION, 2011).
- ELIBAMA "European Li-Ion Battery Advanced Manufacturing" (2011–2014). Based on multidisciplinary consortium representing the entire value chain from material developers to battery manufacturer and end user, the ELIBAMA project focused on the enhancement and the creation of a strong European automotive battery industry, structured around industrial companies already committed to mass production of Li-ion cells and batteries for EVs (ELIBAMA, 2011).

3. Next-generation Batteries

The state-of-the-art Li-ion battery cells (Gen.1-2b) can provide 200–630 Wh/L (90–235 Wh/kg), not yet close to the fundamental limits illustrated by their volumetric and gravimetric energy density. Li-ion battery cell Generation 1 and 2a is widely based on graphite anodes and LFP, NCA and NMC cathodes, while Generation 2b uses higher Ni content-based cathodes and natural graphite at the anode side. Generation 3 uses high-energy material NMC with higher Ni content and low Co content, focusing instead on Si/C anode materials, Si nano-structures directly on carbon particles, which reduce surface area available for side reactions, and improve cycle stability and lifetime (Table 1).

An increasing gap exists between the energy demands and what current rechargeable battery technology can supply, and the EU target is to improve cell-level energy densities to at least 750 Wh/L. The optimised Li-ion battery cells (Gen. 3b), which use Si/C composite anodes, are expected to reach 750Wh/L by 2025 at the latest (350 Wh/kg) (Steen et al., 2017). In order to bridge this gap, a change in the overall battery chemistry is required, including the electrodes (anode and cathode) and the electrolyte. Also, state-of-the-art electrolytes for Li-ion batteries are not stable at high operational voltages (> 4.5V). Higher working voltages lead to continuous electrolyte consumption unless stable surface films can be formed on the cathodes. Another step forward in the electrolyte composition is to be expected with the introduction of Generation 3, mixtures of carbonates and ionic liquids with tailoring electrolyte formulations specifically to the requirements of the future anodes and cathodes technologies. The European strategy aims at succeeding in the development of novel battery types which have not yet been established on the market, such as Li-S (Gen.4 – traction battery cells) or Li-air batteries (Gen. 5), well known as post-lithium ion technologies which have the potential to put the European battery industry and research community to the forefront of emerging technologies with great economic potential.

Table 1. Li-ion battery technology evolution (Steen et al., 2017).

	Cell generation	Cell chemistry
2030	Generation 5	Li/O_2
2025	Generation 4	All-solid-state with Li anode Conversion materials (mostly Li-S)
	Generation 3b	Cathode: HE-NMC, HVS Anode: S/C
2020	Generation 3a	Cathode: NMC622 to NMC811 Anode: C (graphite) + 5–10% Si
current	Generation 2b	Cathode: NMC523 to NMC622 Anode: 100% C
	Generation 2a	Cathode: NMC111 Anode: 100% C
	Generation 1	Cathode: LFP, NCA Anode: 100% C

Consequently, large initiatives and funding programs have been launched in this area to support research and development on the national and the EU level for the exploration of these post-lithium ion technologies which have the potential to exceed the performance of current Li-ion technology considerably. The latter requires entirely new approaches in battery development, targeted at systems which do not necessarily rely on Li chemistry at all. A break-through in the development of such batteries could have a disruptive effect on the battery market, providing an opportunity to push Europe to the forefront of battery development and production worldwide, based on such new technology. The EU could have a head start if strong and concerted research efforts are initiated now, both in academic research and pre-industrial development, so that know-how and intellectual property is generated early, serving as a basis for industrial production in the EU and market penetration worldwide.

In the area of next-generation batteries, at the end of January 2019, the EC launched a new call for proposals under the Horizon 2020 Programme, Research and Innovation Action (RIA). The H2020 call "Building a Low-Carbon, Climate Resilient Future: Next-Generation Batteries" brings together research and innovation efforts on the next generations batteries, under 7 different topics: LC-BAT-1-2019: Strongly improved, highly performant and safe all solid state batteries for electric vehicles (RIA); LC-BAT-2-2019: Strengthening EU materials technologies for non-automotive battery storage (RIA); LC-BAT-3-2019: Modelling and simulation for Redox Flow Battery development; LC-BAT-4-2019: Advanced Redox Flow Batteries for stationary energy storage; LC-BAT-5-2019: Research and innovation for advanced Li-ion cells (Gen. 3b); LC-BAT-6-2019: Li-ion Cell Materials & Transport Modelling and LC-BAT-7-2019: Network of Li-ion cell pilot lines LC-BAT-7-2019 (https://ec.europa.eu/programmes/horizon2020/en). Along the last call H2020, the consortium will enlarge the first circle of skills in order to reach the whole value chain of stakeholders concerned with the future generation batteries. The overall objective of this call is to elaborate and develop a disruptive scientific and technical approach of the new generation of viable high energy density and environmental friendly next-generation batteries.

References

ALION. 2015. http://alionproject.eu/. Last accessed on 20/04/2019.

Alves, D.P., Blagoeva, D., Pavel, C. and Arvanitidis, N. 2018. JRC Science for policy report. Cobalt: demand-supply balances in the transition to electric mobility 2018. Last accessed on 15/04/2019. http://publications.jrc.ec.europa.eu/repository/bitstream/JRC112285/jrc112285_cobalt.pdf.

AMBP–Advanced Materials for Batteries for Electro-mobility and Stationary Energy Storage. http://s3platform.jrc.ec.europa.eu/batteries. Last accessed on 20/04/2019.

Battery 2030+. https://battery2030.eu/. Last accessed on 20/04/2019.

Berman, K., Carlson, R., Dzuiba, J., Jackson, J., Hamilton, C. and Sklar, P. 2018. BMO Capital Markets Report, The Lithium Ion Battery and the EV Market: The Science Behind What You Can't See. Last accessed on 08/03/2019. http://www.fullertreacymoney.com/system/data/files/PDFs/2018/February/22nd/BMO_Lithium_Ion_Battery_EV_Mkt_(20_Feb_2018).pdf.

CTC. 2018. Carbon Tax Center https://www.carbontax.org/where-carbon-is-taxed/. Last accessed on 19/06/2019.

Drabik, E. and Rizos, V. 2018. Prospects for electric vehicle batteries in a circular economy, Research Report. Last accessed on 15/03/2019. https://www.researchgate.net/publication/326353145_Prospects_for_electric_vehicle_batteries_in_a_circular_economy.

EASE. 2019. European Association for Storage of Energy. Energy Storage: A Key Enabler for the Decarbonisation of the Transport Sector. Last accessed on 04/03/2019. http://ease-storage.eu/wp-content/uploads/2019/02/2019.02.EASE-Energy-Storage-and-Mobility-Paper.pdf.

eCAIMAN. 2015. http://www.ecaiman.eu/index.php?id=5. Last accessed on 20/04/2019.

EEA. 2018. Greenhouse gas emissions from transport—European Environment Agency. https://www.eea.europa.eu/data-and-maps/indicators/transport-emissions-of-greenhouse-gases/transport-emissions-of-greenhouse-gases-11. Last accessed on 04/03/2019.

ELIBAMA. 2011. https://elibama.wordpress.com/. Last accessed on 20/04/2019.

ERTRAC. 2017. European Roadmap Electrification of Road Transport, 2017. Last accessed on 04/03/2019. https://www.smart-systems-integration.org/system/files/document/2017%20ERTRAC%20Electrification%20Roadmap.pdf.

European Commission. 2011. Communication from the Commission to the European Parliament, the Council, the European economic and social committee and the committee of the regions. Energy roadmap 2050. Last accessed on 04/03/2019. https://eur-lex.europa.eu/LexUriServ/LexUriServ.do?uri=COM:2011:0885:FIN:EN:PDF.

European Commission. 2017. Communication from the Commission to the European Parliament, the Council, the European Economic and Social Committee and the Committee of the Regions.

European Commission. 2019. Report from the Commission to the European Parliament, the Council, the European Economic and Social Committee and the Committee of the Regions and the European Investment Bank. The implementation of the Strategic Action Plan on Batteries: Building a strategic Battery Value Chain in Europe. Last accessed on 20/04/2019. https://ec.europa.eu/commission/sites/beta-political/files/report-building-strategic-battery-value-chain-april2019_en.pdf.

European Council. 2014. Conclusions on 2030 Climate and Energy policy framework. https://www.consilium.europa.eu/uedocs/cms_data/docs/pressdata/en/ec/145356.pdf. Last accessed on 04/03/2019.

FIVEVB. 2015. http://www.fivevb.eu/. Last accessed on 20/04/2019.

GREENLiON. 2011. http://www.greenlionproject.eu/homepage. Last accessed on 20/04/2019.

Horizon 2020 Framework Programme (H2020). Last accessed on 23/04/2019. https://ec.europa.eu/info/funding-tenders/opportunities/portal/screen/opportunities/topic-search;freeTextSearchKeyword=;typeCodes=1;statusCodes=31094501,31094502,31094503;programCode=H2020;programDivisionCode=null;focusAreaCode=null;crossCuttingPriorityCode=null;callCode=H2020-LC-BAT-2019-2020;sortQuery=openingDate;orderBy=asc;onlyTenders=false;topicListKey=callTopicSearchTableState

KLiB–Kompetenznetzwerk Lithium-Ionen-Batterien. Last accessed on 20/04/2019. http://www.klib-org.de/veranstaltungen/.

Küpper, D., Kuhlmann, K., Wolf, S., Pieper, C., Xu, G. and Ahmad, J. 2018. The future of battery production for electric vehicles. Last accessed on 20/03/2019. https://eu-smartcities.eu/sites/default/files/2018-10/BCG-The-Future-of-Battery-Production-for-Electric-Vehicles-Sep-2018%20%281%29_tcm81-202396.pdf.

Lebedeva, N., Di Persio, F. and Broon-Brett, L. 2017. JRC Science for Policy Report. Lithium ion battery value chain and related opportunities for Europe. https://ec.europa.eu/jrc/sites/jrcsh/files/jrc105010_161214_li-ion_battery_value_chain_jrc105010.pdf.

Northvolt. Last accessed on 15/04/2019. https://chronicles.northvolt.com/.

SPICY. 2015. http://www.spicy-project.eu/. Last accessed on 20/04/2019.

Steen, M., Lebedeva, N., Di Persio, F. and Boon-Brett, L. 2017. JRC Science for Policy Report. EU Competitiveness in Advanced Li-ion Batteries for E-Mobility and Stationary Storage Applications—Opportunities and Actions. Last accessed on 20/04/2019. http://publications.jrc.ec.europa.eu/repository/bitstream/JRC108043/kjna28837enn.pdf

The 2017 list of Critical Raw Materials for the EU. Last accessed on 28/03/2019. https://ec.europa.eu/transparency/regdoc/rep/1/2017/EN/COM-2017-490-F1-EN-MAIN-PART-1.PDF.

ZAS. 2015. https://www.sintef.no/projectweb/zas/. Last accessed on 20/04/2019.

CHAPTER 20

Carbon Constrained Electricity Sector Planning with Multiple Objectives

Krishna Priya GS and *Santanu Bandyopadhyay**

1. Introduction

Over the years, the energy requirements of the human population have been rising consistently. A need for sustaining substantial economic growth has resulted in large-scale exploitation of energy resources, especially fossil fuels. The electric power sector plays a crucial role in enabling economic growth as it provides reliable and safe energy to industries, governments and residential consumers (Labandeira and Manzano, 2012). Adequate per capita electricity consumption has been linked to GDP levels (Bella et al., 2014) and human development index values (Martinez and Ebenhack, 2008) making a rise in electricity demand inevitable. The electric power sector, however, is also a major source of air pollution (IPCC, 2014). While producing electricity through the combustion of fossil fuels, such as coal, natural gas, and oil, pollutants, such as sulphur oxides (SO_x), nitrogen oxides (NO_x), particulate matter, mercury, etc., are generated (Swanson, 2008).

Research has shown that emissions from power plants have a detrimental impact on human well-being (Burt et al., 2013), ecological health (Maroto-Valer et al., 2012), and on cultural resources, such as statues and relics (Sinha and Agrawal, 2007). The emissions from power generation are harmful to human life and the environment. The effects of various pollutants, such as SO_x (Watras and Morrison, 2008), NO_x (Dise and Gundersen, 2004), mercury (Giang et al., 2015) and particulate matter (Pirani et al., 2015) have been the subject of many studies. Their impacts include reduced air quality, a decline in public health, low visibility, acidification of water bodies, harm to the sensitive forest and coastal ecosystems, and faster degradation of buildings and cultural artefacts. These effects are in addition to the repercussions that global warming is likely to have on the ecology (Bellard et al., 2012) and on human health (Moe et al., 2012).

A mix of energy sources that can effectively supply the electricity demand without exceeding the emission targets is, therefore, necessary. Power system planning problems which consider emission constraints try to identify an appropriate mix of energy sources. The objective of power system planning is no longer to just meet the energy demand, but also to ensure that the overall emission from power generation is within the target emission set by the corresponding country. In this chapter, a method for addressing the problem of emission constrained power system planning is presented and the Indian power sector is used to illustrate the method.

Department of Energy Science and Engineering, Indian Institute of Technology Bombay, Powai, Mumbai 400076, India.
* Corresponding author: santanub@iitb.ac.in

2. Multi-objective Optimization Problem

The aim of emission constrained power system planning is to meet the energy demand and emission target. In general, power system planning problems try to minimize the cost of electricity generation while meeting these requirements. However, power system planning is a complex problem that may involve objectives other than the cost minimization. This gives rise to emission constrained multi-objective power system planning problems.

A generic power system planning problem with emission constraint is as shown in Figure 1. There are N_s existing power plants and the jth power plant is characterised by an emission factor q_{sj} and energy flow F_{sj}. Additionally, there are N_r new power plants and the ith new power plant has an emission factor (q_{ri}), and the maximum available energy flow ($F_{ri,max}$). The existing and new power plants are used to satisfy N_d demands. Each demand k has a specified energy requirement (F_{dk}) and a specified emission factor (q_{dk}).

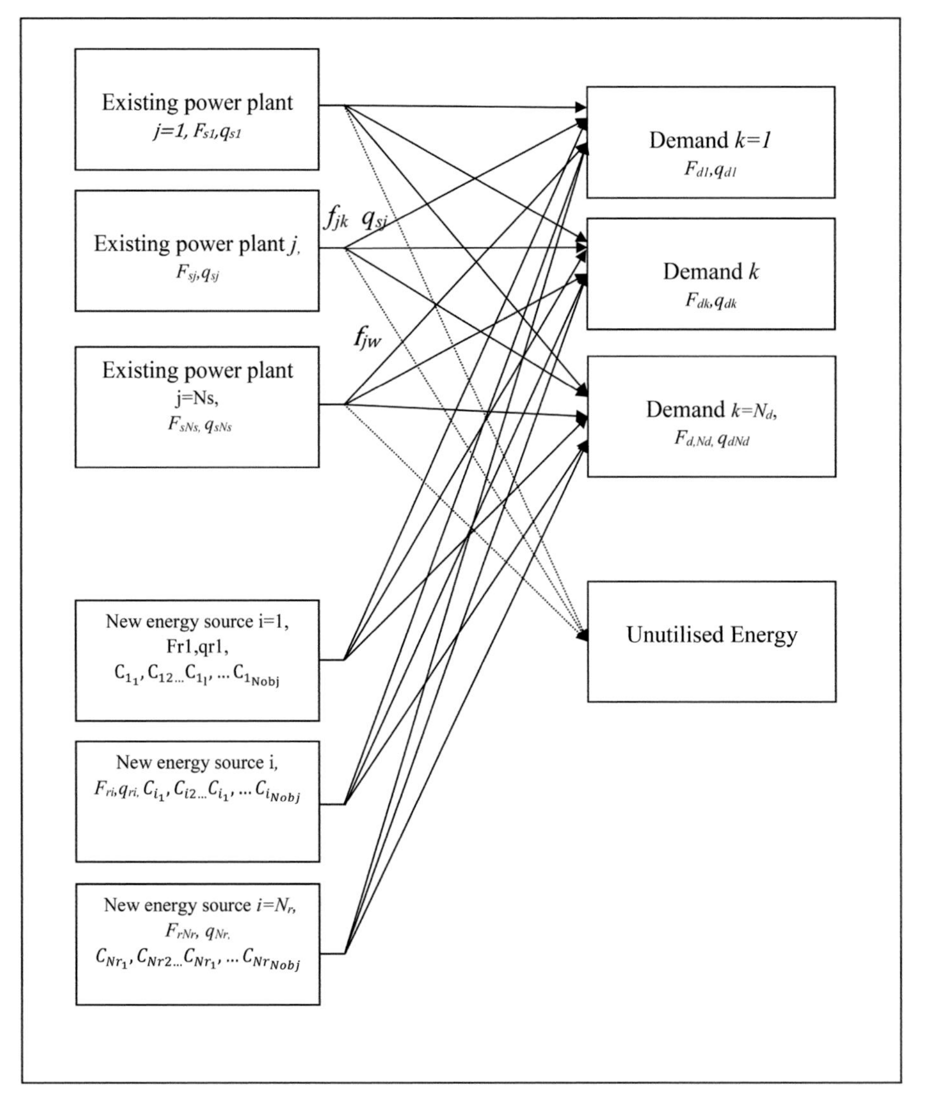

Figure 1. Representation of a generic emission constrained multi-objective power system planning problem (Krishna Priya, G.S. and Bandyopadhyay, S. 2017b. Multi-objective pinch analysis for power system planning. Applied Energy 202: 335–347).

For electricity sources, a higher value of emission factor indicates a lower quality source, i.e., quality is inversely related to emission factor (Bandyopadhyay, 2006). In essence, the purest new power plant (or the new power plant with least emission) has the lowest emission factor. This problem aims at minimizing the cost function (or objective function) while also satisfying the energy and emission constraints. A detailed mathematical model of the system is presented in (Krishna Priya and Bandyopadhyay, 2017a).

It is seen that the structure of the power system optimization problem is similar to that of a classical pinch analysis problem where sources and new resources are used to satisfy demands while adhering to certain quality constraints. While pinch analysis is commonly used to optimize water networks and heat exchanger networks, it has also been employed in power system optimization (Tan and Foo, 2007). In this case, it is necessary to modify the pinch analysis framework to accommodate multi-objective optimization, giving rise to Multi-Objective Pinch Analysis (MOPA).

The simplest optimization problem in power sector planning is to minimize the cost of generation. However, in general, power system optimization is a multi-objective problem. Multiple factors, such as cost, water availability, land use and other socio-economic parameters, need to be considered. Pinch analysis is traditionally a single-objective optimization method. Therefore, a weighting factor is used to modify the objective function in pinch analysis in order to accommodate the multi-objective requirements of emission constrained power system planning. In this modified framework, each energy source has N_{obj} objective functions associated with. Each objective function Φ_{il} is a linear function of resource flow, and the coefficient associated with it is C_{il}. The aim of the optimization is to study the variation in the optimal solution when the relative priority of each objective function is varied. This can be done by obtaining a Pareto optimal front of all solutions.

The *lth* objective functions associated with each energy source may be expressed as:

$$\Phi_{il} = F_{ri}C_{il} \; \forall i \in N_{ri} \; \forall l \in N_{obj} \tag{1}$$

The assumption for this formulation is that there is a linear relationship between each of the objective functions (such as capital cost), and the energy being generated. The *lth* objective function for the overall problem is, therefore, expressed as:

$$Min \textstyle\sum_{i=1}^{Nr} \Phi_{il} \; \forall l \in N_{obj} \tag{2}$$

The overall multi-objectives pinch analysis (MOPA) can be formulated as minimization of each individual objective functions given by Equation (2), subject to constraints associated with the capacity of power plants, capacity limits on new power plants, and emission constraints.

As mentioned earlier, in order to accommodate multiple objectives into the pinch analysis framework, weighted optimization is incorporated into pinch analysis to generate a composite objective function. The weighting factors ($\alpha_1, \alpha_2, \ldots, \alpha_l, \ldots, \alpha_{Nobj}$) are defined such that $0 \leq \alpha_l \leq 1$ and $\sum_{l=1}^{N_{obj}} \alpha_l = 1$. The values of each of these weighting factors indicate the relative importance of the corresponding objective function in the overall system. The weighted cost function can be written as:

$$C_i = \textstyle\sum_{l=1}^{Nobj} \alpha_l \Phi_{il} \tag{3}$$

Φ_{il} is a linear function of energy as per Equation (1), can, therefore, be modified as:

$$C_i = F_{ri} \textstyle\sum_{l=1}^{Nobj} \alpha_l C_{il} \tag{4}$$

The objective for the multi-objective optimization problem is therefore,

$$Min \textstyle\sum_{i=1}^{N_r} C_i \tag{5}$$

3. Prioritizing Power Plants

The method of limiting composite curve was introduced by Wang and Smith (1999). Later, an algebraic technique, called Composite Table Algorithm (CTA), was presented by Agrawal and Shenoy (2006) and Shenoy (2010). CTA extends the applicability of the limiting composite curve approach. In emission constrained power system planning, a limiting composite curve is a plot of net emission load against emission factor. The technique used to plot the limiting composite curve is explained in detail in Krishna Priya and Bandyopadhyay (2013). The limiting composite curve method can be employed to identify an order in which power plants can be added to an existing power system to meet the energy and emission requirements.

When multiple energy sources are available, there has to be a method for deciding which power plant is to be chosen and what quantity of energy is to be supplied from the new power plant. Adding the power plant that minimizes the cost function may violate the emission constraint, while adding the power plant with the least emission may increase the cost function. This trade-off between emission reduction and minimization of the cost is an integral part of MOPA. Additionally, the present state of the power system being considered also factors in. The concept called prioritized cost, introduced by Shenoy and Bandyopadhyay (2007), is applied in order to identify the order in which resources are to be prioritized in a pinch analysis problem. A similar prioritized cost can also be identified for power system planning using MOPA.

It can be shown that the quantity $\frac{C_{mwhi}}{(q_p - q_{ri})}$, which is the prioritised cost of the ith new power plant, can be used for prioritising power plants in a single-objective emission constrained power system optimization problem (Krishna Priya and Bandyopadhyay, 2013). When extending to multi-objective functions, the quantity e Multiple Objectives Prioritised Cost (MOPC) of energy source R_i is:

$$\rho_i = \frac{\sum_{l=1}^{N_{obj}} \alpha_l C_{il}}{(q_p - q_{ri})} \tag{6}$$

Here, is the pinch emission factor, which is a characteristic of the system. It can be concluded that substituting one energy source with the second makes sense only if:

$$\frac{\sum_{l=1}^{N_{obj}} \alpha_l C_{1l}}{(q_p - q_{r1})} \geq \frac{\sum_{l=1}^{N_{obj}} \alpha_l C_{2l}}{(q_p - q_{r2})} \tag{7}$$

This leads to the following lemma.

Lemma: *An energy source is part of the optimal energy mix if its multiple objectives prioritized cost is lower than that of all other energy sources with a lower emission factor.*

In essence, energy sources are to be added in increasing order of multi-objective prioritized cost. A detailed derivation of MOPC is provided in (Krishna Priya and Bandyopadhyay, 2017b).

3.1 Graphical representation of solution space

Visual representation allows for large quantities of data to be presented effectively, making it easier for the reader to decipher the necessary information (In and Lee, 2017). In prioritizing power plants, the numerical value of MOPC is of little significance. The values of weighting factors at which prioritizing sequences changes are more significant. Therefore, a graphical representation of solution space would provide a convenient way to represent the changes in solution space. A graphical solution space for two and three-objective optimization problems is, therefore, presented.

3.1.1 Two-objective optimization

The multiple objectives prioritized cost (MOPC) of resource R_i in a two-objective optimization problem is:

$$\rho_i = \frac{\alpha C_{\Phi_{i1}} + (1 \alpha) C_{\Psi_{i2}}}{(q_p - q_{ri})} \tag{8}$$

Here, α is the weighting factor for the first objective function.

In order to prioritize power plants for any weighting factor, it is enough to identify the order of prioritized cost at that weighting factor. However, Equation (8) shows that the MOPC has a linear relationship with the weighting factor α. Therefore, a plot of MOPC vs. weighting factor would be a straight line for all available energy sources. This also means that the prioritising sequence would vary with the value of the weighting factor. By plotting the MOPC vs. weighting factor lines for all possible new power plants, the changes in prioritising sequence with weighting factor can directly be identified. The fact that the relationship between MOPC and weighting factor is linear provides additional advantages and an opportunity to develop an entirely graphical solution space.

Since two straight lines can intersect only once, a system of N_r new energy sources can generate a maximum of $\binom{N_r}{2}$ intersection points. The prioritising sequence can only change at the weighting factors corresponding to these intersection points, given the linear nature of the graphs. In the interval between intersection points, the MOPC of various power plants follow a fixed order, ensuring that the prioritising sequence in each interval is a constant. This gives rise to a maximum of $\binom{N_r}{2} + 1$ possible prioritising sequences. Figure 2 illustrates how the prioritising sequence for three new energy sources can be identified using this method. The prioritising sequences are represented in square brackets.

In Figure 2, a system with three new power plants is considered. To identify the prioritizing sequence, consider that the energy sources are numbered in increasing order of emission factor (i.e., $q_1 < q_2 < q_3$). The maximum number of possible intersection points is $\binom{3}{2}$, indicated by weighting factors a, b and c in Figure 2. Consider the region where the weighting factor is less than a. Here, R_1, the energy source with the lowest emission factor, also has the lowest prioritised cost. This means that adding R_1 to the energy mix is optimal both for reducing emission and minimising the cost function. Therefore, in this region, R_1 alone can be part of the optimal energy mix. However, beyond the weighting factor 'a', the energy source R_3 has a lower MOPC than R_1. In this region, R_3 also has a lower MOPC that R_2. This means that its MOPC is now lower than that of all energy sources with a lower emission factor. Therefore, in this region, both R_1 and R_3 are part of the prioritising sequence.

This trend continues till weighting factor 'b', where the MOPC of energy source R_2 becomes lower than that of R_1 and, therefore, much like before, R_2 becomes part of the prioritizing sequence. Beyond 'c', the MOPC of R_3 is higher than R_2 and, therefore, no longer part of the prioritizing sequence. This example illustrates a case in which three new energy sources generate four or $\binom{N_r}{2} + 1$ possible prioritising sequences for the entire range of weighting factors: [R_1], [R_1-R_3], [R_1-R_2-R_3], and [R_1-R_2]. This may not always be the case. For example, if the MOPC of any given new energy source is consistently higher

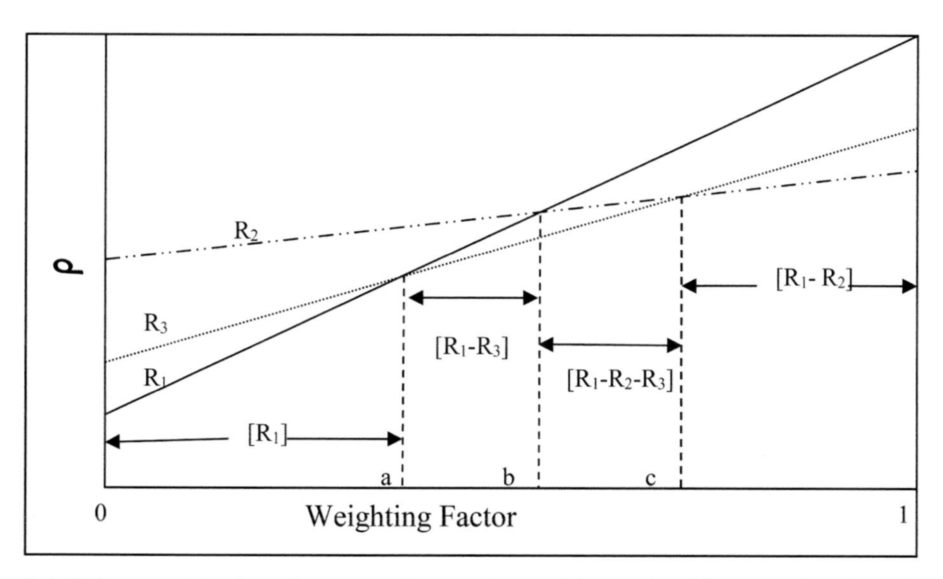

Figure 2. MOPC vs. weighting factor for two-objective optimization (Krishna Priya, G.S. and Bandyopadhyay, S. 2017a. Multiple objectives pinch analysis. Resources, Conservation and Recycling 119: 128–141).

than the MOPC of R_1 in this example, it would have never been part of the prioritising sequence. Such observations obtained from the MOPC vs. weighting factor plot can further reduce the number of possible solutions. These plots help make the two-objective optimization process efficient by eliminating unviable combinations and identifying the optimal prioritising sequences for various values of weighting factors

This method can effectively reduce the total number of combinations from 2^{N_r-1} to a maximum of $\binom{N_r}{2} + 1$. This reduction in the number of possible solutions is significant. In a system of just 6 possible new energy sources, this method reduces the number of possibilities from 32 (2^{N_r-1}) to 16 ($\binom{N_r}{2} + 1$). In reality, a national level power system planning problem is likely to have a significantly larger number of energy sources. Therefore, incorporating the MOPA solution method can significantly reduce the complexity of large-scale emission constrained power system planning problems. After identifying the pinch point, the MOPC plots can be obtained and a prioritising sequence can be identified for any given value of weighting factor. The new power plants are then added according to the prioritising sequence.

3.1.2 Three-objective optimisation

In a three-objective optimization problem, let the three-objective functions associated with the ith power plant be Φ_{i1}, Φ_{i2}, and Φ_{i3}. It is assumed that the relation between energy flow and each of these objective functions is linear. Let C_{i1}, C_{i2}, and C_{i3} be the cost coefficients associated with each of these objective functions. As described earlier, weighting factors are introduced to combine the various cost functions. Two weighting factors α and β are introduced corresponding to cost functions Φ_{i1}, and Φ_{i2}. Given that the sum of weighting factors is one, the overall objective function can be written as:

$$Min\sum_{i=1}^{N_r}(C_i = \alpha\Phi_{i1} + \beta\Phi_{i2} + (1 - \alpha - \beta)\Phi_{i3}) \tag{9}$$

As per Equation (6),

$$\rho_i = \frac{\sum_{l=1}^{N_{obj}} \alpha_l C_{il}}{(q_p - q_{ri})} = \frac{\alpha C_{i1} + \beta C_{i2} + (1-\alpha-\beta)C_{i3}}{(q_p - q_{ri})} \tag{10}$$

The equation of prioritized cost for a three-objective optimization problem is that of a plane. If and when the prioritizing sequence of any two new energy sources become equal, the two corresponding planes of MOPC will intersect to form a line, which will be termed as Prioritised Cost Intersection Lines (PCIL) in this study. Similar to the two-objective example, the prioritizing sequence can only change along these lines. These lines provide the prioritizing sequence for any weighting factor as the prioritizing sequence changes only along these lines.

In order to identify the prioritizing sequence, it is enough to understand the variations in PCILs for the entire solution space. A simple example consisting of just two energy sources, R_1 and R_2 (such that $q_{r1} < q_{r2}$), can be used to illustrate this point. The prioritizing cost of these new energy sources can be obtained using Equation (10). The prioritized costs of R_1 and R_2 will be equal at the PCIL where the prioritized cost planes of R_1 and R_2 intersect. Using this condition, it is possible to obtain the relationship between weighting factors α and β. This line of one weighting factor as a function of the other is the PCIL of R_1 and R_2 and a sample of such a PCIL is shown in Figure 3.

As the sum of weighting factors cannot be higher than one, half the region in the plot is infeasible. Of the remaining feasible region, the PCIL divides the solution space into two distinct parts, each with a different prioritizing sequence; one where R_1 alone is part of the solutions space as R_1 has a lower prioritized cost, and another were both R_1 and R_2 are part of the energy mix. In some cases, if the planes of R_1 and R_2 do not intersect for any value of weighting factors, then R_1 alone would constitute the optimal energy mix in this example.

In a system with three energy sources, the three planes of prioritized cost can, at best, generate three PCILs which will then determine the solution space. Figure 4 show a generic example of such a system. The shaded region represents the infeasible solution space and the arrows indicate the direction in which a given energy source is viable. It is interesting to note that the three PCILs will intersect at a common point. Since along line R_1-R_2, $\rho_{R1} = \rho_{R2}$, and along line R_1-R_3, $\rho_{R1} = \rho_{R3}$, at their intersection point

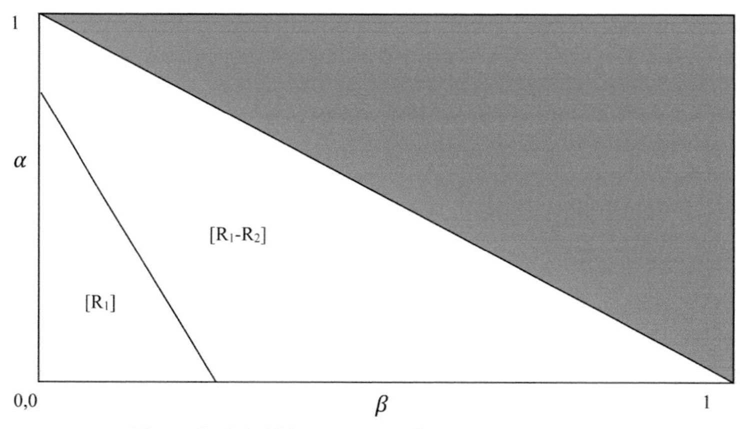

Figure 3. Prioritising sequences for two energy sources.

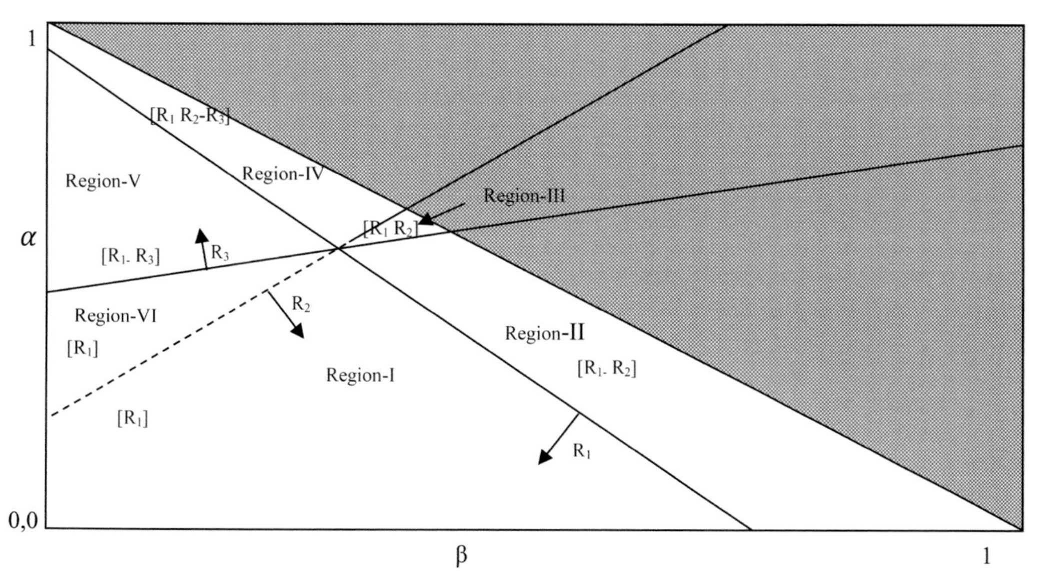

Figure 4. Prioritizing sequences for three energy sources (Krishna Priya, G.S. and Bandyopadhyay, S. 2017b. Multi-objective pinch analysis for power system planning. Applied Energy 202: 335–347).

$\rho_{R2} = \rho_{R3}$. This necessitates that the PCIL for R_2-R_3 passes through the same point as well. Therefore, it is safe to conclude that any three energy sources will generate three PCILs and one unique node in the solution space. Since these are straight lines intersecting, the solution space can be divided into a maximum of six regions.

In Figure 4, R_1 has the lowest MOPC, therefore, it alone will be part of the prioritised sequence. Hence, in this region, the prioritizing sequence is [R_1]. In Region II, $\rho_{R3} > \rho_{R1} > \rho_{R2}$. R_2 now has a lower MOPC than all other energy sources with a lower emission factor. Therefore, the prioritising sequence in this region is modified to [R_1-R_2]. Interestingly, in Region III, $\rho_{R1} > \rho_{R2} < \rho_{R3}$. Here, it should be noted that since $\rho_{R2} < \rho_{R3}$, R_3 cannot be part of the prioritising sequence though it has a lower MOPC than R_1. However, Region IV, $\rho_{R1} > \rho_{R2} > \rho_{R3}$ giving rise to the sequence [R_1-R_2-R_3]. In Region V, $\rho_{R1} < \rho_{R2} < \rho_{R3}$, changing the prioritising sequence to [R_1-R_3] and in Region VI, $\rho_{R1} < \rho_{R2} < \rho_{R3}$, making R_1 alone part of the optimal prioritising sequence. Therefore, in this example, the entire feasible solution space is divided between three possible prioritising sequences. Once the prioritising sequences are identified, the optimal energy mix can be identified.

The interaction of each of these energy sources with the energy source with the lowest emission factor (R_1) is of significance. In the given example, in Region I, R_1 alone needs to be considered, and all other interactions between energy sources can be ignored. Similarly, in Region II, R_1 and R_2 are alone part of the solution space, making it unnecessary to study the relationship between R_3 and various energy sources. However, in Region IV, both R_2 and R_3 are cheaper than R_1, making two prioritizing sequences likely. If the MOPC of R_2 is lower than that of R_3, then the solution would be [R_1-R_2], and if not, it would be [R_1-R_2-R_3]. A case study involving the Indian power sector is considered in order to demonstrate the applicability of this method.

4. Case Study of the Indian Power Sector

4.1 Present energy scenario

As the world's fourth-largest economy that is still growing significantly, the electricity demand in India has been constantly increasing. At present, India's power capacity is nearly 30% less than its demand (Engelmeier, 2015). India is largely a fossil fuel-based economy with coal and petroleum being the key energy sources. Indian Coal has a higher than average ash content and, combined with ineffective combustion, this results in increased air pollution, a higher concentration of particulate matter in the atmosphere, and global warming (Mishra, 2014). The dependence on fossil fuels is high and on the rise. Between 1990 and 2009, the share of electricity generation from coal increased by 10%, making fossil fuels responsible for 85% of electricity generation. Of this, coal alone is responsible for 71% generation in 2016. Among CO_2-free green energy, wind energy is significant in India, accounting for 14% of the total installed capacity (MNRE, 2016).

The initial state of the Indian power sector is as described in Table 1 in terms of installed capacity, capacity factor and operating emissions. In the Indian energy mix, the renewable includes Wind (25.18 GW), Solar (5.25 GW), Small Hydro (4.19 GW) and Biomass (2.68 GW) (MNRE, 2016).

Table 1. Details of existing power plants in India.

Source	Limit (GW)	Operating-capacity factor	Tonnes of CO_2/MWh[d]
Coal	175.24[a]	0.8[e]	1.08[g]
Gas	24.51[a]	0.8[e]	0.45[d]
Diesel	1.00[a]	0.9	0.65[d]
Nuclear	5.78[a]	0.82[f]	0
Wind	25.18[b]	0.14[c]	0
Solar	5.25[b]	0.18[c]	0
Hydro	42.66[b]	0.5[c]	0
SHP	4.19[b]	0.5[c]	0
Biomass	2.68[b]	0.7[c]	0.11

Compiled from data available on Ministry of Power, 2016 (a); MNRE, 2016 (b); MNRE, 2008 (c); Tan et al., 2009 (d); NTPC, 2015 (e); NPCIL, 2014 (f); CEA, 2008 (g).

4.2 Future of Indian power sector

India's electricity consumption has increased by 64% during the ten-year period between 1997 and 2007. This trend is likely to continue, and it is safe to assume that India's need for electrical energy will continue to rise at a rate of around 8 to 10%. This would make India one of the largest electricity consumers in the world (ICLEI South Asia, 2007). India's energy demand for 2030 from electricity is predicted at 2940 TWh (Khosla and Dubash, 2015). The need for increased electricity generation has to be balanced with the need to reduce greenhouse emissions. In this study, India's emission target is set to

1420 Mt of CO_2. This hypothetical target is a 25% reduction in emission factor from 2012 levels. India is committed to lowering its CO_2 emission per unit GDP by 33–35% of 2005 levels by 2030 (UNFCCC, 2015). As a signatory of the Paris Agreement, India is committed to lowering emissions by concentrating on renewable energy resources. Projections based on the existing National Energy Plan predict that India will reach its target of 40% electricity generation from non-fossil fuel sources by 2030. However, there is substantial uncertainty with regards to the translation of these policies into reality as the coal consumption continues to rise at 4.8% as of 2017 (Climate Action Tracker, 2019). Additionally, sector-specific targets, such as those for electricity generation, are not available. The actual CO_2 emission targets will be set by the Indian Government, considering various economic and societal parameters. A recent committee on energy highlighted this shortage, pointing out that the current allocation to Natural gas power plants is 70% short of allocation (Ministry of Power, 2019) and no new gas-based power plants are recommended for the next three decades (Pandey, 2018). Similarly, diesel-based power generation in India decreased 37% in 2018 due to fuel shortage (Rise in diesel, 2018). Under these conditions, this illustrative example assumes that diesel and natural gas will not be used for electricity generation in the future, given the present fuel crisis. The present case study is used primarily to illustrate the applicability of the proposed methodology; though realistic data has been considered.

Table 2 gives details related to resources available for future power plants. The lifecycle emissions are considered here for all renewable sources. For coal and biomass-based power generation, lifecycle emission and operating emission are treated as the same as the major contributing component is the fuel consumed. Also, all new coal and nuclear power plants are assumed to operate at a relatively high capacity factor of 0.9.

Table 2 lists the capital costs, in terms of its installed capacity, of various power plants. The option of carbon capture is also included with new coal power plants. Importance of CCS technology for the future power system cannot be neglected (Morgan, 2019). With regards to the cost of CCS technology, varying estimates are available in literature as no commercial plant is operational. In general, it is estimated that CCS enables power plants will cost 40–80% more than their traditional counterparts, though these percentages are expected to lower by 10–30% over the next decade (Heuberger et al., 2016). For the purpose of this example, it is estimated that CCS enabled power plants will have a 30% higher capital cost than conventional power plants, as suggested by World Energy Council report (Brendow, 2007).

It should be noted that the effective capital cost, per unit energy delivered (and not installed capacity), is considered in the analysis. This is because of varying capacity factors of renewable power plants. 1 MW coal power plant is expected to deliver 0.9 MWy of energy annually. On the other hand, a Solar PV plant of the same capacity is expected to deliver 0.2 MWy of energy every year. Therefore, in terms of energy delivery, the effective capital costs of renewable plants are significantly higher. The effective

Table 2. Future power sources for the Indian power sector.

Resource	Limit (GW)	Operating capacity factor	Tonnes of CO_2/MWh[e]	Water (m^3/MWh)[f]	Land use (m^2/MWh-y)[j]	Capital cost (10^6\$/MW)
Coal	NA	0.9[e]	1.08	2.56	0.433	0.67[g]
Nuclear	9.55[b]	0.82[b]	0.02	2.80	0.029	0.79[g]
Hydro	148.70[a]	0.50[a]	0.12	5.30	3.70	1.17[a]
Wind	47[c]	0.14[c]	0.07	0.015	0.006	1.13[l]
Biomass	19.50[c]	0.70[c]	0.11	1.80	0.126	0.50[h]
Small hydro	15[c]	0.50[c]	0.12	5.30	0.003	2.09[i]
Solar PV	NA[c]	0.20[c]	0.15	0.004	0.164	1.19[l]
Solar thermal	NA	0.20	0.185	1.08	0.366	1.92[m]
Coal with CCS	NA	0.9[e]	0.10	5.03	0.626[k]	0.87[g]

Compiled from data available on NHPC, 2012 (a); NPCIL, 2012 (b); MNRE, 2008 (c); NTPC, 2012 (d); Tan et al., 2009 (e); Wilson et al., 2012 (f); CEA, 2004 (g); Banerjee et al., 2006 (h); Nouni et al., 2008 (i); Fthenakis and Kim, 2009 (j); Carter, 2007 (k); CERC, 2015 (l); IRENA, 2012 (m) (1\$ = 60 Rs.).

capital cost of a power plant can be calculated by dividing the installed capital cost with corresponding capacity factor. Based on the capacity factors and installed capital costs, as provided in Table 2, the effective capital costs of these plants are calculated and used for further analysis.

It should be noted that the case study is used purely to illustrate the methodology, though realistic data has been used. The choice of one power plant over another is a complex one, involving socioeconomic factors, environmental considerations, and policy decisions in addition to constraints such as land and water availability, capital investment requirements, etc. This example considers three-objective functions related to land use, water requirement and capital investment. While the example is limited to three-objective functions, the methodology can support any number of objective functions, provided they can be quantified. Additionally, since the method can be employed to generate a Pareto optimal front of all possible solutions, policy makers can employ the same to identify discrete energy mixes based on complex requirements.

4.3 Two-objective optimization

As an example, the two-objective functions considered for the Indian power sector are the minimisation of capital investment (Φ) and minimisation of water footprint associated with new power plants (Ψ). This is considered purely as a hypothetical example as the deciding on one type of power plant over another involves multiple environmental, economic and societal parameters. The presently available power plants are already listed in Table 1 and the available energy sources that can be converted to new power plants are listed in Table 2. Pinch point of the existing system of power plants is identified as 1.08 tCO_2/MWh. Weighting factor corresponding to the capital cost objective function is introduced. When $\alpha = 1$, the problem is purely cost minimisation and when it is zero, the objective is simply to minimise water footprint. The relevant portion of the MOPC vs. weighting factor plot is shown in Figure 6.

The intersection points of the MOPC lines give the weighting factors at which the prioritizing sequence of the system may change. For low values of weighting factor ($\alpha \leq 0.24$), it is seen that solar energy is the resource with the lowest prioritised cost. Solar energy, which has the lowest emission factor, also has the lowest prioritised cost in this region. As solar energy is considered as an unlimited resource, in region O-A ($0 < \alpha \leq 0.24$), the entire energy requirement is supplied by solar energy. Beyond point A, Biomass becomes more cost effective than solar energy, and it too enters the solution mix and the prioritising sequence changes to [Biomass-Solar]. In region A-B ($0.24 < \alpha \leq 0.32$), biomass and solar energy supply the demand.

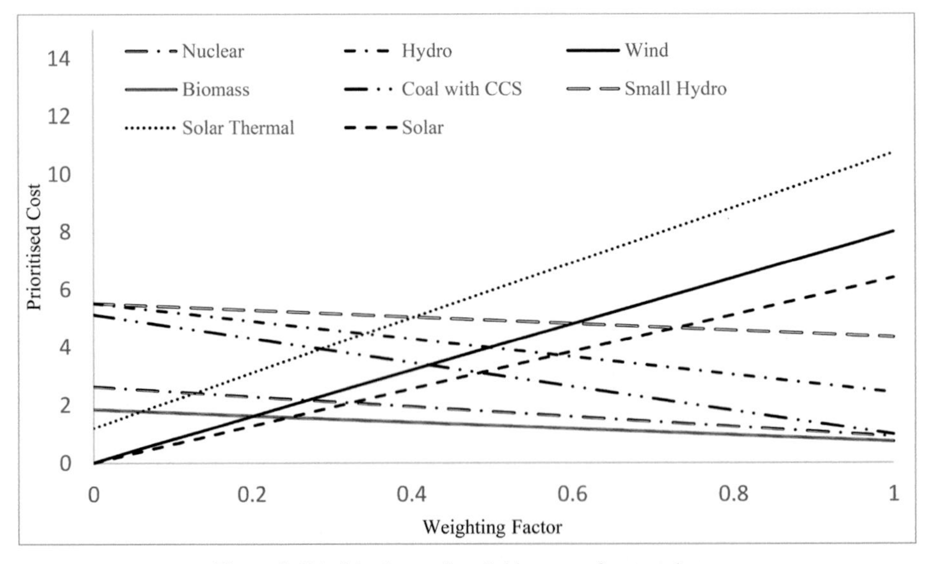

Figure 5. Prioritised cost of available types of power plants.

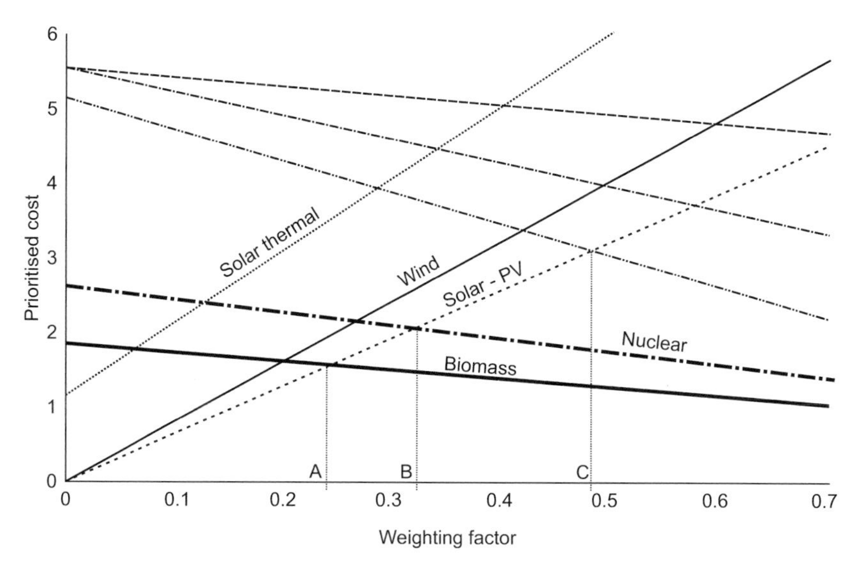

Figure 6. Regions with discrete solutions.

In this case, the maximum available capacity of biomass is limited when compared to the demand. Therefore, biomass is used to its maximum capacity, and solar energy supplies the rest of the demand. At point B, Nuclear energy becomes cost effective. Nuclear energy has a lower prioritised cost and emission factor than biomass. This changes the prioritizing sequence to [Nuclear-Biomass-Solar]. In region B-C ($0.32 \leq \alpha \leq 0.48$), the solution mix consists of nuclear energy, biomass and solar PV. The major change in the composition of energy mix occurs at a weighting factor of C ($\alpha = 0.48$), where CCS enabled coal power plants becomes more cost effective than solar PV. From this point onwards, the energy that was being supplied by solar PV is supplied by CCS enabled coal power plants, and the emission factor of CCS enabled coal is lower than that of biomass, meaning the prioritising sequence changes to [Nuclear-CCS enabled Coal-Biomass].

Energy generated from the existing coal power plants decreases as the value of w increases, i.e., the amount of unused energy (or waste) varies from 15.4% to 9.6% from point A to C. This is because, as the stress on water minimisation increases, the share of resources with low water footprint, such as solar energy, increases. In this example, solar energy has a slightly higher emission factor than CCS enabled coal, leaving less room for emissions from existing power plants. Note that not all points of intersection alter the constitution of the resource mix. For example, at a weighting factor of around 0.27, the MOPC lines of wind and nuclear energy intersect. As wind energy is not a part of the mix, this intersection point is of little significance. The Pareto optimal front in Figure 7 shows the four possible solutions for this system for a net electricity generation of 2940 TWh.

In essence, if there is stress on cost reduction, as opposed to water footprint minimization, the solution mix is dominated by fossil fuels (specifically new coal power plants with CCS). In fact, biomass is the only renewable that is part of the cost minimal energy mix. Nuclear energy also plays a significant part. However, for most weighting factors, the bulk of the energy requirement is met by solar PV and at lower weighting factors when water footprint minimization is more critical, Solar PV alone satisfies the entire energy requirement along with existing coal power plants or coal power plants with CCS. Improving the cost-effectiveness and water footprint of Solar PV and CCS enabled coal power plants can significantly affect the cost and water consumption associated with power generation in India.

4.4 Three objective optimizations of the Indian power sector

The three objectives considered in this example are the minimisation of capital investment (Φ_1) and minimisation of water footprint associated with new power plants (Φ_2) and minimisation of land footprint

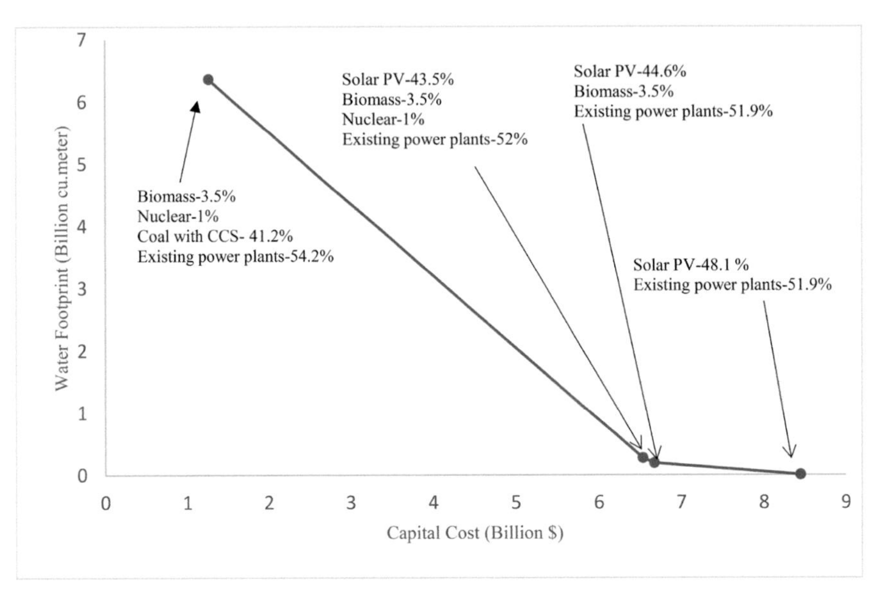

Figure 7. Pareto optimal front for the Indian power sector.

for the power plants (Φ_3). Two weighting factors α and β corresponding to Φ_1 and Φ_2 are also introduced. The power plants available and new energy sources are presented in Tables 1 and 2 along with their characteristics. The pinch point of the system is at 1.08 tCO_2/MWh. Knowing the pinch emission factor and the characteristics of the new energy sources, their MOPCs can be calculated for the entire range of weighting factors. It can be seen that solar thermal has a higher cost coefficient for all three objectives than solar PV (which is the energy source with least emission factor) and, therefore, will never be part of the optimal solution mix. The MOPCs of such energy sources can entirely be avoided, further simplifying the solution process. As explained earlier, it is possible to generate a graphical solution space for three-objective optimization problems. For this purpose, the lines along which the MOPC planes of any two energy sources are functions of α and β and they intersect when the prioritised cost of the two energy sources become equal. The lines along which they intersect are plotted as prioritised cost intersection lines or PCILs. Since the two planes intersect along a line, the prioritising sequence of the system is only likely to change along these lines. The PCILs for all new energy sources are provided in Figure 8. The sum of weighting factors cannot exceed one. Therefore, only the relevant and viable portion of the solution space needs to be considered as the sum of α and β cannot be greater than one. The solution space is divided into a number of regions by these lines. In each region, it is possible to eliminate certain PCILS as their MOPC is significantly higher than that of others.

Through proper understanding of the interactions between the PCILS, the prioritizing sequence for any combination of weighting factors can be identified. In order to better illustrate the process, consider an enlarged region, provided in Figure 9. The PCIL of wind and solar PV starts at a weighting factor of $\alpha = 0$ and $\beta = 0.94$ and meets the x axis at $\alpha = 0.1$ and $\beta = 0$. It means that wind energy is cheaper than solar energy for weighting factors nearer to the origin. However, for most of the solution space, solar PV is preferred to wind.

In the region surrounding $\alpha = 0$ and $\beta = 0.1$ in Figure 9, had wind energy been an unlimited resource, it alone would supply the entire demand requirement. As that is not the case, the prioritising sequence in this region is [Wind-Solar PV]. It can be seen that at weighting factors of $\alpha = 0$ and $\beta = 0.054$, the PCIL involving solar PV and nuclear energy enters the solution space. It divides the space into two regions, such that nuclear energy has a higher prioritised cost than solar PV in one region and a lower MOPC in the other. In this example, nuclear energy has a lower emission factor than wind energy. Therefore, depending on the PCIL between wind and nuclear energy, two prioritising sequences are possible here. If the prioritising cost of wind energy is lower than that of nuclear energy, then the prioritising sequence

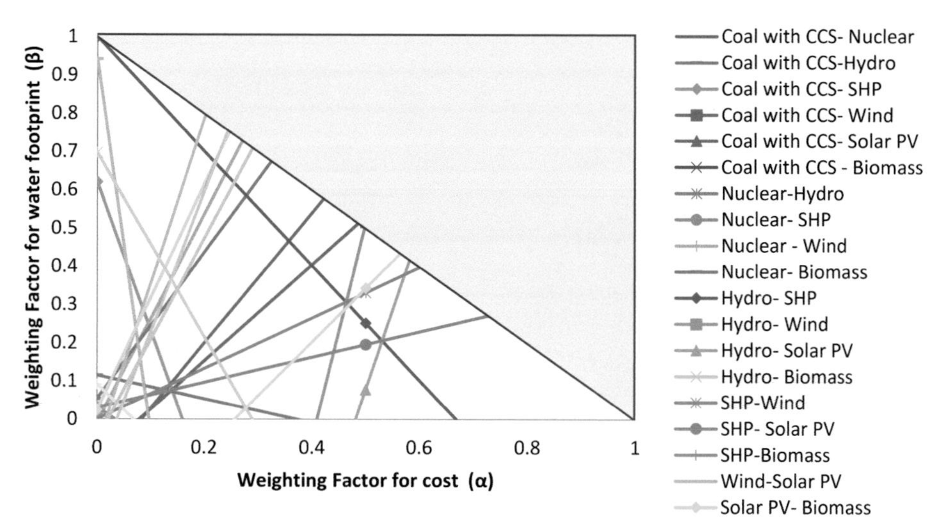

Figure 8. Prioritized cost intersection lines for Indian power sector (Krishna Priya, G. S., and Bandyopadhyay, S. 2017b. Multi-objective pinch analysis for power system planning. Applied energy, 202: 335–347).

is [Nuclear-Wind-Solar PV]. Conversely, if the prioritising cost of wind is higher than that of biomass, it will no longer be part of the solution mix, and the prioritising sequence is [Nuclear-Solar PV]. However, in this particular example, the demand to be supplied is much greater than the maximum capacity of nuclear energy. Therefore, it gets utilised to its maximum capacity and is no longer part of the available resources. The prioritising sequence then reverts back to include wind energy. In essence, in the second region, the prioritising sequence is [Nuclear-Wind-Solar PV]. This means that nuclear energy and wind energy are used to its maximum capacity first. The next low emission resource available is solar PV, which is capable of supplying the entire load. In essence, both wind and nuclear energy are used up and the rest of the demand is supplied by solar PV. The next major change occurs at weighting factors of $\alpha = 0$ and $\beta = 0.025$, where biomass becomes cheaper than solar PV. Beyond the PCIL involving Biomass and solar PV, the solution consists of [Nuclear-Wind-Solar PV-Biomass]. As biomass has a higher emission factor than nuclear, two prioritising sequences are possible here. The sequence can either be [Nuclear-Wind-Solar PV-Biomass] or [Nuclear-Wind-Solar PV] if biomass has a higher MOPC than nuclear energy. If nuclear energy or biomass had a maximum capacity comparable to that of the demand, it would have been necessary to study their interaction. However, as nuclear energy has a very limited capacity, it is used to its maximum capacity in either case, making its interaction with biomass unimportant.

The unique prioritizing sequence is unnecessary, as biomass, wind and nuclear energy will be exhausted in trying to supply the load. The solution in this region contains [Nuclear-Wind-Solar PV-Biomass]. Similarly, beyond $\alpha = 0$ and $\beta = 0.03$, the PCIL involving SHP and solar PV enters the solution space, and the solution mix changes to [Nuclear-Wind-Solar PV-SHP-Biomass]. It should be noted that in a small triangular region in the solution space with base $\beta = 0.025$ and $\beta = 0.03$, the solution contains SHP, but no biomass as biomass has a higher prioritised cost than solar PV in this region. At $\alpha = 0.096$ and $\beta = 0$, the PCIL relating wind and Solar PV intersect the x-axis. Beyond this line, wind energy is no longer a part of the solution mix. The next major change is at $\alpha = 0.16$ and $\beta = 0$, where the PCIL involving coal power plants with CCS and solar PV enter the solution space. Beyond this line, the energy being supplied by solar PV is replaced by coal power plants with CCS. Multiple low emissions energy sources are part of the prioritising sequence.

In this particular problem, the demand for electricity is significantly higher than the maximum capacity of most energy sources, such as wind, SHP and nuclear energy. The two energy sources of unlimited capacity are solar energy and coal power plants with CCS. Therefore, major changes in the composition of the optimal energy mix will be dependent on these two energy sources. At this stage, it

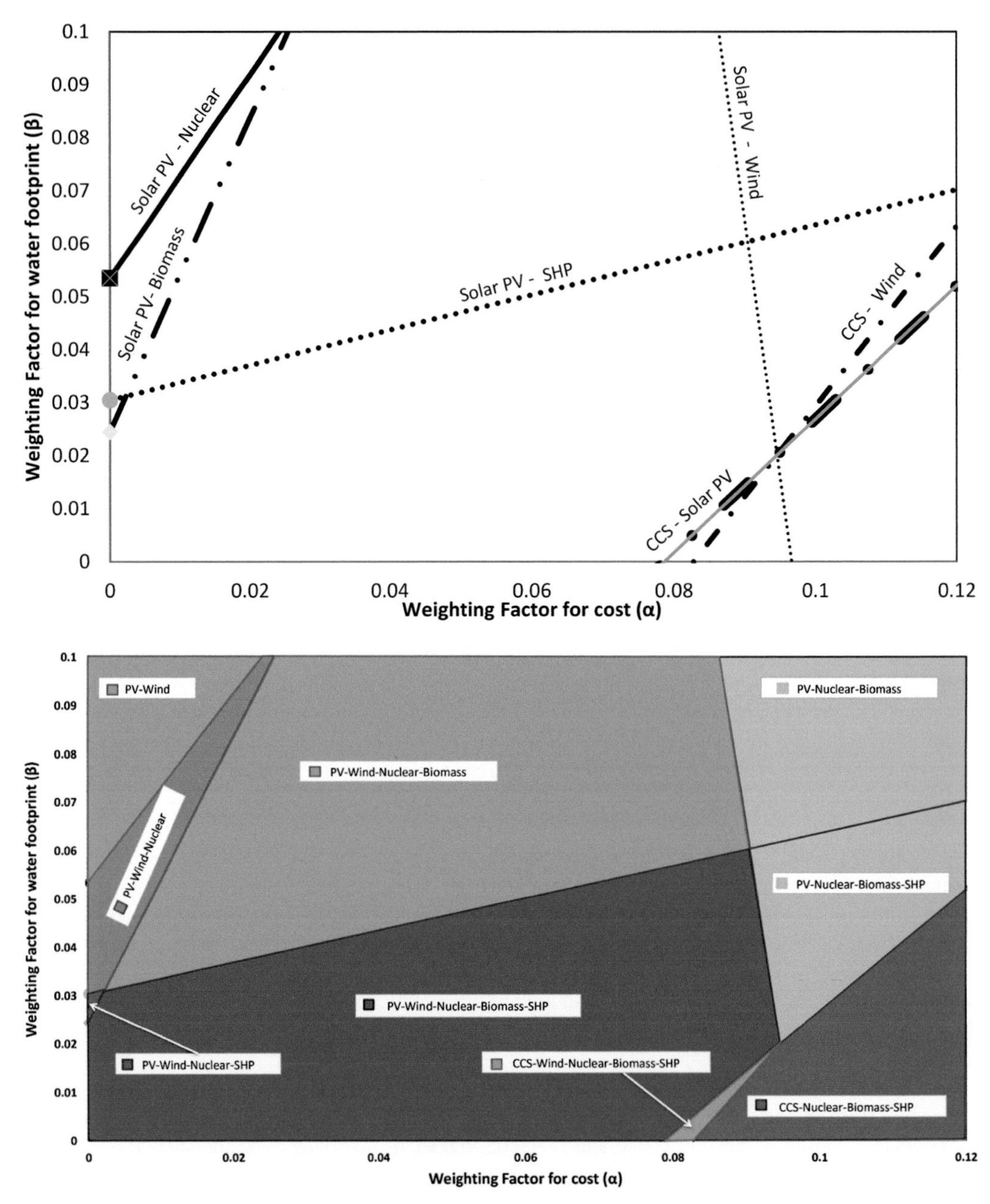

Figure 9. Identifying prioritising sequence using PCIL plot (Krishna Priya, G.S. and Bandyopadhyay, S. 2017b. Multi-objective pinch analysis for power system planning. Applied Energy 202: 335–347).

is necessary to study the MOPCs of wind, SHP, nuclear energy and biomass against that of coal power plants with CCS. In this case, by studying the complete plot of PCILs, it can be seen that they all have a lower MOPC than coal power plants with CCS for the specified weighting factors. The solution beyond this line contains [Nuclear-Wind-SHP-Coal with CCS-Biomass]. The PCILs along which the solution mix changes are highlighted in Figure 9.

Not all PCILs alter the solution mix. For example, the PCIL involving nuclear energy and wind is not as significant because both these energy sources have a small maximum capacity when compared to the demand, and will be used up entirely. For these resources, given their limited maximum capacity, they will be completely absorbed into the optimal energy mix, provided their MOPC is lower than that of solar PV. Using a similar analysis, it is possible to analyze the entire solution space and identify unique solution mixes. This is shown in Figure 10.

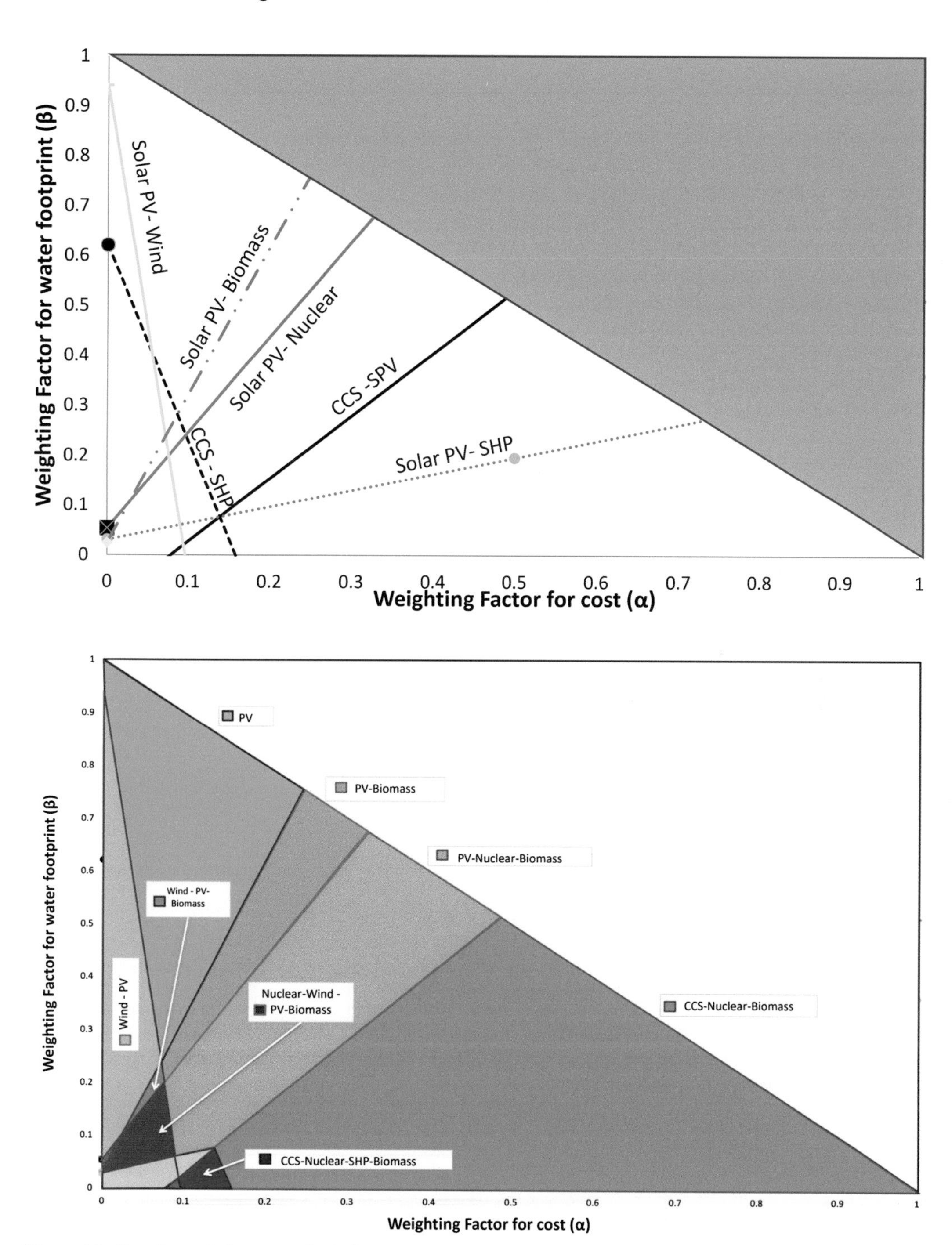

Figure 10. Complete solution space for Indian power sector (Krishna Priya, G.S. and Bandyopadhyay, S. 2017b. Multi-objective pinch analysis for power system planning. Applied Energy 202: 335–347).

A total of thirteen unique solutions are possible for this particular combination of energy sources. All these solutions have been identified using the method developed. Some of the regions are already explained in Figure 9 and have simply been marked in Figure 10. It can be seen that as solar PV and CCS enabled coal are treated as unlimited energy sources, they supply a significantly large portion of the overall demand requirement.

5. Conclusions

Pinch analysis is primarily a linear single-objective optimisation method. Power system planning, on the other hand, deals with multifaceted problems and a decision on power system planning may have to consider multiple conflicting objectives. However, the primary structure of a power system planning problem is similar to that of a classical pinch analysis problem. Therefore, this work explores the possibility of using the simple, robust framework of pinch analysis to address the problem of emission constrained power system planning. For this purpose, the weighted optimization technique was combined with pinch analysis to generate a composite cost function. The concept of the prioritized cost was extended to problems with multiple objectives, and a Multi-Objective Prioritised Cost (MOPC) is then used to identify a prioritizing sequence for any given value of weighting factor.

The method thus developed is capable of addressing any number of linear objective functions. However, having a graphical solution space can provide certain advantages. Therefore, this work presents graphical methods for identifying all possible prioritizing sequences for two and three objective optimization problems. The method is then illustrated using the Indian power sector as an example. The objective is to meet the energy and hypothetical emission targets for the year 2030. The solution spaces for two objective optimizations considers the variation of energy mix when cost minimization and water footprint minimization are considered. For three objective optimization, land use minimisation is considered as the additional objective function. Over all, it is seen that Solar PV and CCS-enabled coal power plants will be significant energy sources in the future. Similarly, renewables, such as wind, biomass and nuclear energy, also play a small, but consistent role in the overall solution.

This method provides a simple and effective means to identify all possible energy mixes for multi-objective optimization problems. Once the solution space is obtained, it will remain unchanged unless the pinch point of the system or one of the cost coefficients are altered. Therefore, it is possible to directly identify the optimal solution from these plots once the relative importance of each objective function is determined.

References

Agrawal, V. and Shenoy, U.V. 2006. Unified conceptual approach to targeting and design of water and hydrogen networks. AIChE Journal 52(3): 1071–1082.

Bandyopadhyay, S. 2006. Source composite curve for waste reduction. Chem. Eng. J. 125(2): 99–110.

Banerjee, R. 2006. Comparison of options for distributed generation in India. Energy Policy 34: 101–111.

Bella, G., Massidda, C. and Mattana, P. 2014. The relationship among CO_2 emissions, electricity power consumption and GDP in OECD countries. Journal of Policy Modeling 36(6): 970–985.

Bellard, C., Bertelsmeier, C., Leadley, P., Thuiller, W. and Courchamp, F. 2012. Impacts of climate change on the future of biodiversity. Ecology Letters 15(4): 365–377.

Brendow, K. 2007. Carbon Capture and Storage: A WEC interim balance. World Energy Council Cleaner Fossil Fuels Systems Committee, United Kingdom, London.

Burt, E., Peter, O. and Buchanan, S. 2013. Scientific Evidence of Health Effects from Coal Use in Energy Generation. Chicago, IL: University of Illinois at Chicago School of Public Health, and Health Care Without Harm. Available at <noharm.org/sites/default/files/lib/downloads/climate/Coal_Literature_Review_2.pdf>. Accessed on 20 February 2019.

Carter, L.D. 2007. Retrofitting Carbon Capture Systems on Existing Coal-fired Power Plants A White Paper for the American Public Power Association (APPA). Available at <uschamber.com/sites/default/files/legacy/CO2/files/DougCarterretrofitpaper2.pdf>. Accessed on April 4 2019.

CEA. 2004. Report of the expert committee on fuels for power generation.<cea.nic.in/thermal/Special reports/Report of the expert committee on fuels for power generation.pdf>. Accessed on January 21 2019.

CEA. 2008. CO_2 Baseline Database for the Indian Power Sector Available at <cea.nic.in/reports/others/thermal/tpece/cdm_co2/user_guide_ver4.pdf> Accessed on 2 Feb 2019.

CERC. 2014. Determination of Benchmark Capital Cost Norm for Solar PV power projects and Solar Thermal power projects applicable during FY 2014-15.<cercind.gov.in/2014/whatsnew/SO353.pdf>. Accessed on January 21 2019.

CERC. 2015. Benchmark Capital Cost Norm for Solar PV power projects and Solar Thermal power projects applicable during FY 2016-17. Available at <cercind.gov.in/2016/orders/SO17.pdf>. Accessed on 25th March 2019.

Climate Action Tracker. India: Country Summary. 2019. Available at <climateactiontracker.org/countries/india/>. Accessed on 10 July, 2019.

Dise, N.B. and Gundersen, P. 2004. Forest ecosystem responses to atmospheric pollution: Linking comparative with experimental studies. In Biogeochemical Investigations of Terrestrial, Freshwater, and Wetland Ecosystems across the Globe. Springer Netherlands: 207–220.

Engelmeier, T. 2015. How much power will India need in 2035? Bridge to India. Available at <bridgetoindia.com/blog/how-much-power-will-india-need-in-2035/>. Accessed on 10th April 2019.

Frunza, M.C. 2013. Fraud and carbon markets: The carbon connection. Routledge.

Fthenakis, V. and Kim, H.C. 2009. Land use and electricity generation: A life-cycle analysis. Renewable and Sustainable Energy Reviews 13(6-7): 1465–1474.

Giang, A., Stokes, L.C., Streets, D.G., Corbitt, E.S. and Selin, N.E. 2015. Impacts of the minamata convention on mercury emissions and global deposition from coal-fired power generation in Asia. Environmental Science & Technology 49(9): 5326–5335.

Heuberger, C.F., Staffell, I., Shah, N. and Mac Dowell, N. 2016. Quantifying the value of CCS for the future electricity system. Energy & Environmental Science 9(8): 2497–2510.

ICLEI South Asia. 2007. Renewable Energy and Energy Efficiency Status in India. Available at <local-renewables.iclei.org/fileadmin/template/Publications/RE_EE_report_India_final_sm.pdf>.Accessed on 2 Feb 2019.

In, J. and Lee, S. 2017. Statistical data presentation. Korean Journal of Anesthesiology 70(3): 267.

IPCC. 2014. Climate Change 2014: Mitigation of Climate Change Exit EPA Disclaimer Contribution of Working Group III to the Fifth Assessment Report of the Intergovernmental Panel on Climate Change. Cambridge University Press, Cambridge, United Kingdom and New York, NY, USA.

IRENA. 2015. Renewable Power Generation Cost in 2014. Available at <irena.org/documentdownloads/publications/irena_re_power_costs_2014_report.pdf>. Accessed on 23rd March 2019.

Khosla, R. and Dubash, N.K. 2015. What does India's INDC imply for the Future of Indian Electricity? Centre for Policy Research: The Climate Initiative Blog. Available at <cprclimateinitiative.wordpress.com/2015/10/15/what-does-indias-indc-imply-for-the-future-ofindian-electricity/#end1>. Accessed on 23rd March 2019.

Krishna Priya, G.S. and Bandyopadhyay, S. 2013. Emission constrained power system planning: A pinch analysis based study of Indian electricity sector. Clean Technologies and Environmental Policy 15(5): 771–782.

Krishna Priya, G.S. and Bandyopadhyay, S. 2017a. Multiple objectives pinch analysis. Resources, Conservation and Recycling 119: 128–141.

Krishna Priya, G.S. and Bandyopadhyay, S. 2017b. Multi-objective pinch analysis for power system planning. Applied Energy 202: 335–347.

Labandeira, X. and Manzano, B. 2012. Some economic aspects of Energy Security. European Research Studies 15(4): 47–64.

Maroto-Valer, M.M., Song, C. and Soong, Y. (eds.). 2012. Environmental Challenges and Greenhouse Gas Control for Fossil Fuel Utilization in the 21st Century. Springer Science & Business Media.

Martinez, D.M. and Ebenhack, B.W. 2008. Understanding the role of energy consumption in human development through the use of saturation phenomena. Energy Policy 36(4): 1430–1435.

Ministry of Power. 2016. Power Sector at a Glance (All India). Available at <powermin.nic.in/power-sector-glance-all-india> Accessed on 2 Feb 2019.

Ministry of Power. 2018. Stressed/Non-Performing Assets in Gas based Power Plants. Forty Second Report, Standing Committee on Energy. Lok Sabha Secretariat. Available at <indiaenvironmentportal.org.in/files/file/16_Energy_42.pdf> Accessed on 10 July, 2019.

Mishra, U.C. 2004. Environmental impact of coal industry and thermal power plants in India. Journal of Environmental Radioactivity 72(1): 35–40.

MNRE. 2008. Annual report 2007–2008. <mnre.gov.in/file-manager/annual-report/2007-2008/EN/index.htm>. Accessed on 29 Jan 2019.

MNRE. 2016. Physical Progress (Achievements). Available at <mnre.gov.in/mission-andvision-2/achievements/> Accessed on 2 Feb 2019.

Moe, S.J., De Schamphelaere, K., Clements, W.H., Sorensen, M.T., Van den Brink, P.J. and Liess, M. 2013. Combined and interactive effects of global climate change and toxicants on populations and communities. Environmental Toxicology and Chemistry 32(1): 49–61.

Morgan, S. 2019. Norway's carbon storage project boosted by European industry. <www.euractiv.com/section/energy-environment/news/norways-carbon-storage-project-boosted-by-european-industry/> Accessed on September 10, 2019

NHPC. 2012. <nhpcindia.com> Accessed on January 21, 2019.

Nouni, M.R., Mullick, S.C and Kandpal, T.C. 2008. Providing electricity access to remote areas in India: An approach towards identifying potential areas for decentralized electricity supply. Renewable and Sustainable Energy Reviews 12: 1187–1220.

NPCIL. 2012. Corporate Profile. < npcil.nic.in/pdf/Corporate_Profile_2012.pdf> Accessed on January 21 2019.

NPCIL. 2014. Plants under operation. Available at <www.npcil.nic.in/main/allprojectoperationdisplay.aspx> Accessed on 2 Feb 2019.

NTPC. 2015. Perfomance Statistics. Available at <www.ntpc.co.in/en/powergeneration/performance-statistics> Accessed on 2 Feb 2019.

Pandey, K. 18 Jan, 2019. Lack of gas, high cost 'stranded' more than half of India's gas-based power plants. Down to Earth. Available at <downtoearth.org.in/news/energy/lack-of-gas-high-cost-stranded-more-than-half-of-india-s-gas-based-power-plants-62854>. Accessed on 10 July, 2019.

Pirani, M., Best, N., Blangiardo, M., Liverani, S., Atkinson, R.W. and Fuller G.W. 2015. Analysing the health effects of simultaneous exposure to physical and chemical properties of airborne particles. Environment International 79: 56–64.

Rise in diesel prices shifts load to coal-based power plants, 2018. Financial Express, Available at <www.financialexpress.com/economy/rise-in-diesel-prices-shifts-load-to-coal-based-power-plants/1350568/>. Accessed on 10 July, 2019.

Shenoy, U.V. and Bandyopadhyay, S. 2007. Targeting for multiple resources. Ind. Eng. Chem. Res. 46: 3698–3708.

Shenoy, U.V. 2010. Targeting and design of energy allocation networks for carbon emission reduction. Chemical Engineering Science 65(23): 6155–6168.

Singh, A. and Agrawal, M. 2007. Acid rain and its ecological consequences. Journal of Environmental Biology 29(1): 15–24.

Swanson, H. 2008. Literature review on atmospheric emissions and associated environmental effects from conventional thermal electricity generation. Clean Air Strategic Alliance, Electricity Project Team. Available at <www.hme.ca/reports/Coal-fired_electricity_emissions_literature_review.pdf>. Accessed on 20 February 2019.

Tan, R.R. and Foo, D.C.Y. 2007. Pinch analysis approach to carbon-constrained energy sector planning. Energy 32(8): 1422–1429.

Tan, R.R., Ng, D.K.S. and Foo, D.C.Y. 2009. Pinch analysis approach to carbon constrained planning for sustainable power generation. Journal of Cleaner Production 17: 940–944.

UNFCCC. 2015. Climate Action Tracker. Available at <climateactiontracker.org/countries/india.html>. Accessed on 25th March 2019.

Wang, Y.P. and Smith, R. 1995. Waste water minimization with flow rate constraints. Chem. Eng. Res. Des.73: 889–1006.

Watras, C.J. and Morrison, K.A. 2008. The response of two remote, temperate lakes to changes in atmospheric mercury deposition, sulfate, and the water cycle. Canadian Journal of Fisheries and Aquatic Sciences 65(1): 100–116.

Wilson, W., Leipzig, T. and Griffiths-Sattenspiel, B. 2012. Burning our rivers: The water footprint of electricity. River network, Portland, Oregon. <climateandcapitalism.com/wpcontent/uploads/sites/2/2012/06/Burning-Our-Water.pdf>. Accessed on January 21 2019.

Index